THE SECRET OF THE STONES

The Gate to a Lost World

An extraordinary adventure of archaeological discovery which follows the footsteps of the Druids for 4000 years

Castle Rigg stone circle in the mist by Mike Shephard

GEOFFREY CROCKFORD
NIGEL HUGHES

First published in the United Kingdom in 2009 by
Biolocation Services

Typeset by
Graham Radband

Book design by
Graham Radband
Nigel Hughes

Cover production by
Graham Radband

Front cover image by
Nigel Hughes

Front Cover picture is of the prince and princess against the 6 + 3 design of the temple. The
stone is carrying the zigzag for land and the spirals and diamonds for water.

Rear Cover photograph of the Rollright Stones by
Mark Fearon
www.pitmatic.com

www.biolocation.com
www.thesecretofthestones.co.uk
www.thesecretofthestones.com

ISBN
978-0-9562921-0-0

Preface

In March 1994 I attended a dowsing course run by the British Society of Dowsers. My reasons for being on the course were first that I had seen a letter in 'The Times' where the correspondent said 90% or more of people could dowse. Secondly, being a health and safety professional I was aware of the health and discomfort problems being reported by users of visual display units, computing and electrical equipment. And thirdly it appeared that those people with dowsing skills could detect the deleterious energies generated by such equipment and prescribe control measures. The weekend course was a great success and I returned home able to locate water mains, foundations of buildings, drains and a host of other things hidden in the ground. However, my rods pointing, closing or opening as I walked across a buried object was one thing, knowing what the object was, was quite another. I was to learn from hard experience that it is the latter that dowsing is all about. Knowing what is there or how to find out what is there when the rods move defines a dowser and it was this that was going to set my research goals. Later I was to realise that most dowsers use what is called intuitive dowsing and so the use of the scientific method with dowsing would come to define what I called science based dowsing as a separate form of dowsing.

With an experimental research background I soon started on the tasks of finding out what it was that the dowser responded to and the properties of the sensory system that identified the signals received by the dowser. This work was based on the assumption that humans, like many animals and birds, have a magnetic sense. The result of the early work was the development of the 'Diamagnetic and Paramagnetic Theory of Biolocation'. These two magnetic fields are very weak fields and their use by the '6th sense' to identify unseen objects is referred to as biolocation. I was later to redefine biolocation to include the use of science based dowsing along with its underlying theory. The model or paradigm provided by the theory has been a very powerful research tool. It is able to handle dowsing phenomena derived from the atomic and molecular level, the health status of people, and animal navigation to lunar and solar magnetic fields. The reason for mentioning this is that my basic core research developed side branches, one of which was pre-Roman archaeology. This was brought about by what had been going on in my garden 2000 years and more in the past. Chemical traces left in the soil by past activities were interfering with my experimental investigations. I was not obtaining the results I was expecting and so had to find out why. Once my interest in the Iron Age was aroused I found that biolocation was able to reveal an incredible amount of detail about the pre-Roman people of the British Isles, their culture, houses and settlements. Biolocation and the wealth of data it produced brought the houses and settlements to life in a way that conventional archaeology could not. Through my magnetic sense I could also look at megalithic stone artefacts such as stone circles in a way that their builders may have possibly looked at them using their 6th sense.

When I started my research career many, many years ago my first boss said to me 'You do not need to know much about a subject to do research in it'. He was quite right but you do of course need to know what is important and how to do research. This means having an underlying theory or model, designing experiments, recognising and controlling variables and methods of measurement. Another aspect of research which many leaders of research teams know is that it is the outsiders, particularly if armed with new techniques, who can open doors and often make major contributions to the project. In the present case the nascent archaeological researchers came in from the outside, their basic and triggering archaeological knowledge and information often came from TV programmes and popular archaeological books. I use the term 'triggering' to describe information that raises questions and indicates what to look for, the provocation and challenge to develop the type of questions that lead to answers. For example, a set of ten irregular areas dowsed on my back lawn at first looked like the imprint of a stone circle. Then local archaeologists suggested wells. This did not fit so little progress was made until I learnt that Iron Age houses often had storage pits round them. This information immediately led to a number of answerable questions, not least of which were what was being stored and where was the house?

Early in 2004 I met another dowser who was interested in what I was doing. Nigel Hughes found himself drawn to archaeological biolocation and joined me. The advantages of working with a like minded biolocator very soon emerged. He quickly learnt to identify postholes, roads, paths, houses and henges. Importantly he developed witnesses, that is samples of the material being looked for which can be held in the hand to identify what has been found. When on site it was a great advantage to have somebody who could both check my own dowsing and challenge the deductions I was drawing and help with the development of theory and models.

During the project, major advances were made such as the recognition that objects which once stood on the ground leave a chemical stain which migrates down into the soil. These traces of material left in the soil act as powerful dowsable sources. Another was the power of witnesses to reveal detail and follow what people thousands of years ago were doing.

The presentation used is that of an adventure story as indeed it was, a most amazing adventure of discovery. The science and technology are only dealt with if required but some of the science has to be explained because it is relevant to the story. Some further detail is provided in boxes and appendices for those interested in the technical aspect of biolocation. However, the full science and technology of biolocation has to be the subject for another book where it can be dealt with in depth.

At the beginning of the project there was no reason to doubt the conventional academic view as to the age and origin of stone circles and woodhenges. This was that they were Neolithic and early Bronze Age and so the story starts at this point. As the study progressed incredible detail started to emerge. Using what I hope can be agreed as a reasonably sound scientific approach, sense and order has been made of the findings that started to emerge. A few of the pieces of the jigsaw have been identified and put in place. Enough for a great picture to start forming. Cultural traits have been followed back in time and the evidence accumulated is overwhelming in saying that the pre-Roman society of Britain was a highly developed society with a complex religion based on Druidism. The society was full of ritual and ceremony. It was intelligent and sophisticated but still locked in the Stone Age although using bronze and iron. The incredibly rich detail of ritual and engineering prowess that have emerged did not suddenly appear at a point in time. Much of it must have been there before the Bronze Age arrived in the British Isles. How far back in time still remains to be discovered but the scale and size, of the footprints left in the soil indicates many thousands of years.

The findings of archaeological biolocation will have to be taken onboard by conventional archaeology. Biolocation is a new technology for the future. It will considerably advance archaeological knowledge and understanding of the past and has the potential to become a science in its own right. It can be linked through laboratory analysis to main stream archaeology. This presentation is not a book about how to dowse or biolocate but how, by using a disciplined approach, it is possible to see into the past. The power of biolocation is truly impressive. It must be as it has opened the door to a 3000 year old civilization that still dictates where we walk, drive, often live and worship. The roots of our culture, Christianity and its religious icons can be traced back to the Henge Civilization which in turn can be traced back to the Neolithic.

G.W. Crockford   2009

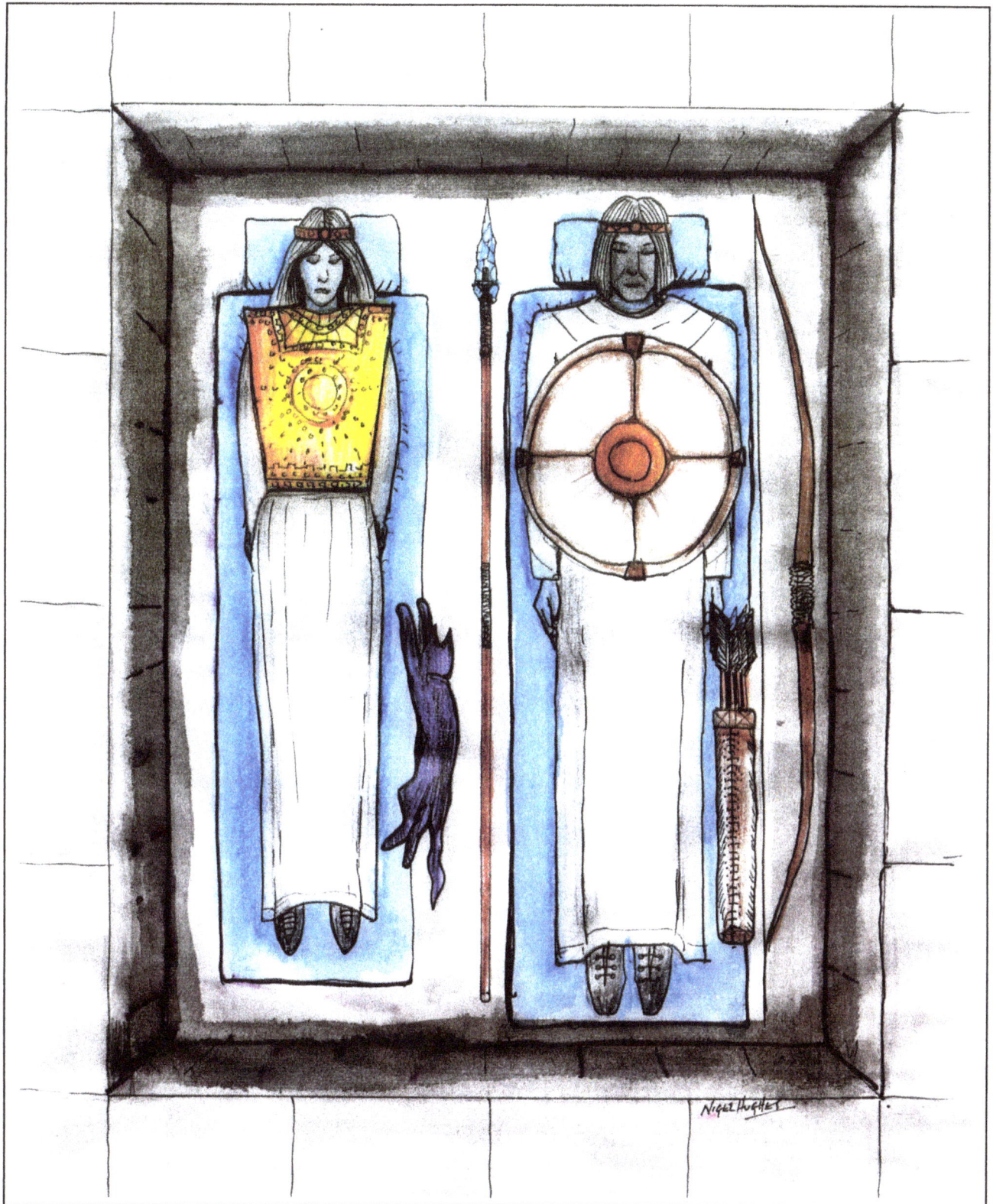

"The Prince and Princess"

Acknowledgements

The authors would like to thank the following for their help and support in so many ways.

David and Valerie Anderson, Marcus Bishop, Eleanor and David Burke, Maria Clark of Purgarvie Farm, John Christian, Susan Holmes and James, Bob and Anne Holden of Assington Mill, Mr Stoff of Magnet Applications Ltd, William Park Tonks, Rod and Ann Latham, Norman Kitchen, Alan Hayday and Peter Vincent, Horace and Wendy Dobbs

In particular I would like to thank Beulah and Philip Garcin who by teaching me to dowse changed my life. Thanks also to my mentors in the early years, the late Michael Poynder and Jeffrey King. Tim Johnson and his colleagues for their encouragement and support. Mrs Macleod Matthews for support and the use of gardens and fields for developing techniques.

Our special thanks to Rosemary and Graham Radband for providing computer support and more often than not moral support.

Our thanks to our wives Shirley for her hours of work at the word processor and to Debbie for understanding the time and commitment required during the project.

Nigel wishes to thank Karen and Ray, Vanessa and Steve of the family run pub, The Hollybush for their continual interest and support when listening to his dowsing exploits.

He also thanks The Hollybush regulars for their open and frank discussions of his findings over the years.

He thanks JJ Evendon for permission to use his aerial photographs..

Our thanks to the many members of the British Society of Dowsers for fruitful discussions and to all those who lent a hand on site or listened without complaint to the stories of dowsing achievements.

With a project spanning the country and lasting over a decade we have benefited from the help of a great many people. Our thanks to all those who rendered some assistance particularly to the farmers who allowed us to dowse their land, the organizations who have assisted us in some way and the National agencies who look after Britain's heritage.

We are grateful to Rob Williams for web site development.

Finally Nigel wishes to remember his late father Raymond Hughes who despite his senior years assisted on his field research in Pontypridd and contributed significantly to many discoveries there.

About the Authors

Geoffrey Crockford

Geoffrey Crockford is a former University researcher who became interested in the health effects of magnetic fields from electronic equipment. This led to work on the dowsing phenomena and the physics of paramagnetic and diamagnetic magnetic fields. He discovered human beings have a magnetic sense and developed an underlying theory to explain dowsing. This brought it into the rational world of science along with a powerful analytical methodology called Biolocation.

Nigel Hughes

Nigel Hughes is a science graduate and former airline pilot who became a keen dowser after meeting Geoffrey Crockford. His background added extra momentum to the research which led to the two authors developing new innovative ground breaking techniques in Archaeological investigation.

Sources of Quotations

Glastonbury Tor: A Guide to the History and Legends. Nicholas Mann 1993 Triskele Publications, Butleigh, Somerset www.britishmysteries.co.uk
New Scientist, London. *www.newscientist.com* The Caves and the Hell Fire Club by Sir Francis Dashwood, Bt.
West Wycombe Estate. *www.hellfirecaves.co.uk*
Exploring the world of the Druids by Miranda J Green © 1997 Thames Hudson Ltd, London. Reprinted by kind permission of Thames and Hudson. (Box 24.1)

Image credits

Images from several sources are used in the Figures in addition to and alongside the authors' own material. The images themselves have not been changed or modified in any way and are used to clarify and illustrate relevant points in the text. In some cases labelling and an overlay have been added. In using these images the authors acknowledge permission from the following sources:

Wikipedia: Figures 1.1, 3.1, 3.5, 3.6, 6.2, 10.12, 10.13, 10.2.1, 12.2, 16.1, Boxes 1.1, 1.2, 1.3.
DK Images Figure 3.4
www.penangbotanicalgardens.gov.my : Figure 10.1
Skyscan balloon photography: Figure 4.2.
National Monuments of Wales: Figure 18.3
Google Earth Satellite Images: Figures 12.1, 12.3, 15.1, 15.3, 17.1, 18.1, 18.6, 19.8, 20.4, 24.1
English Heritage: Figure 19.6
The West Wycombe Estate: Figure 21.2

The Secret of the Stones

Contents		Page

Preface		3
Acknowledgements		6
Quotation Sources		7
About the Authors		7
Image credits		7
Chapter 1	**Base Camp and a Mountain to Climb**	**16**
Chapter 2	**Some Secrets Revealed**	**25**
	The Circle of Killadangan	25
	Modelling the Circle	29
	The Blue Energy Line	32
	Splitting the Blue Energy Line	33
	Stones Outside the Circle	34
	Review of fFndings	34
Chapter 3	**Rings of Energy: by Chance or by Design**	**36**
	Stone Circle Sites	36
	The Rollright Stone Circle	38
	Drombeg Stone Circle	39
	Glebe Stone Circle	40
	Modern Stone Circles	40
	British Stone Circles	40
	Avebury Stone Circles	42
	Discussion	43
Chapter 4	**From Postholes to Woodhenges**	**45**
	A Roundhouse on the Lawn	45
	Woodhenges	48
	Dolmen Discovered	49
	From Woodhenges to Stone Circles	50
Chapter 5	**The Sleeping Prince and Princess**	**52**
	Snowdonia Reveals a Secret	52
	Confirmation of the Woodhenge Tomb	55
	Children's Graves?	56
	The Gold Cape	57
	A Visit to Woodhenge	59
	Review of 2003	61
Chapter 6	**An Extra Pair of Hands**	**63**
	A Slow Start to 2004	63
	An Abundance of Woodhenge Sites	64
	A Quick Visit to Ireland in April 2004	65

A Visit to Woodhenge and Stonehenge   65
A Saxon Church Points to a Druidic Temple   67
Problems with Magnetic Loops and Witnesses   69
Support for the Stain Theory of Dowsing   71
Temple Design: from Six to Nine Rings   72
Summary   72

Chapter 7   **The Pieces of the Puzzle Start to Fit**   **74**
A Stone Circle in Yorkshire   74
A Second Visit to the Yorkshire Stone Circle   75
Chenies Manor House Henge Temple   83
Summary   83

Chapter 8   **Pictures from an Ancient Past**   **85**
Bronze Age Boats   85
The Use of Witnesses   85
Ceremonial Paths   87
Mass Graves   90
Tree Auras   91
The Tracks Left by Wheels   94
Some Small Advances   94
Dolmen   94
Summary   95

Chapter 9   **Stone Circles Reveal More of their Secrets**   **96**
Avebury Stone Circles   96
Stonehenge   100
Stanton Drew   100
Biolocation Methods Move Forward   104
Blood Witness   104
Stonehenge Follow up   104
Execution Lines   106
Guards Through Time   106
Stanton Drew Follow up   108
The Rollright Stone Circle   110
Summary   111

Chapter 10   **The Henge Civilization Appears Through the Mists of Time**   **112**
A Winter Break in Penang   112
Return to Bovingdon   114
Berkhamsted Castle – a pre-Roman Past   114
Lodge Park and Chariot Tracks   117
Tunbridge Wells – a Druidic Centre   118
Indicators of Genocide   118
The Temples Tell More of their Story   120
Dyke Temples   120
A Stone Age Culture   123
The House of a Druid   125
Dating Tombs   126
Pictures on Walls?   127
Druidic Mortar   127
More Evidence of a Stone Age Society?   128

	Return to Yorkshire and More Discoveries	130
	Summary	134
Chapter 11	**Temples Provide a Time Line**	**138**
	A Temple Site in Stoney Lane	138
	A Stone Age Henge Temple	140
	A Review of Tomb Complexes	144
	A Druidic Temple Guard	145
	Stone Circle	146
	The Seven Altars	147
	The Clay Pool	147
	Temple Wall and Gates	147
Chapter 12	**Maiden Castle – Fort or a Druidic Temple?**	**150**
	Maiden Castle	150
	Return to Maiden Castle	153
	Stonehenge	156
Chapter 13	**Henge Temples : Matters of Life and Death**	**160**
	The Dolmen	160
	The Western Dolmen	160
	The Southern Dolmen	161
	The Eastern Dolmen	163
	The Northern Dolmen	163
	Review of the Dolmen	164
	Toilets	165
	An Iron Age Village	166
	The Temple Altars	166
	Temple Pools	168
Chapter 14	**The Stoney Lane Time Capsules**	**174**
	Many Temples on One Site	174
	Stone Age Tomb (SA1)	174
	A Review of Tombs	176
	Stone Age Tomb (SA3)	177
	Temple Walls	178
	The Chapel of Rest	179
	Dog Kennels	180
	A Temple Floor Plan	181
	Summary	181
Chapter 15	**Tunbridge Wells and Pontypridd – Unlikely Twins**	**183**
	Calverley Grounds Park	183
	Pontypridd	187
	Stonehenge	188
Chapter 16	**Giant Strides Start with Small Steps**	**192**
	Dykes Start to Take Shape	192
	A Stone Age Temple	192
	Subsidence into Druidic Tunnels	193
	Dells	193
	Hertfordshire Puddingstone	196

Henge Age Art: 197
    The origin of spirals, diamonds and zigzags
Summary   201

Chapter 17   **Following the Footsteps of the Archaeologist**   **202**
    The Destruction of Archaeological Information   202
    The Durrington Walls' Dyke   203
    The South Temple   206

Chapter 18   **Epitaph to a People**   **211**
    The Death of a Civilization   211
    Glastonbury Tor   218

Chapter 19   **The Power of an Ancient Landscape**   **224**
    Stonehenge : The trilithon powerhouse   224
    Cursi 1   228
    Cursi 2   234
    A Welsh Cursus   236
    A Ceremonial Tunnel   237
    Caerleon   238
    The Maypole   239

Chapter 20   **Things Get Spooky : Quantum Entanglement**   **240**
    Splitting the Blue Beam   240
    The Secret of the Heel Stone   245
    Woodhenge   247
    Druids and the Blue Beam   247
    Summary   249

Chapter 21   **Closing the Circle**   **250**
    The Reality of Dowsing   250
    Hell-Fire Caves   251
    The Stoney Lane Temple Site   256
    Summary   257

Chapter 22   **How Ancient are the Druids?**   **259**
    Avebury   259
    Beckhampton Avenue   261
    The Subterranean Stream   262
    The West Kennet Avenue   265
    Wansdyke   265
    The Sanctuary   266
    Silbury Hill   269
    Bell Chambers   272
    Marburgh Henge   272
    Central Scotland   274
    Brankam Hill   278

Chapter 23   **Some Surprises in Devon and Cornwall**   **282**
    The Henge Temples of Dartmoor   282
    The Merry Maidens   284
    The Michael Line   285

	The Round Pound and Kestor	286
	Scorhill Stone Circle	289
	Hell-Fire Caves	293
	Summary	293
Chapter 24	**The Dawn of Christianity**	**294**
	The Lost Well of St. Albans Cathedral	294
	The Stoney Lane Temple Site	295
	Return to the Lost Well	298
	The Access Shaft in the Horse Field	299
	A Crucifix at Three More Sites	301
	From Druidry to Christianity	301
	Crucifixes on Dartmoor	302
	Was the Crucifix a pre-Christ Logo?	303
	Review of Findings	305
	The Search for the pre-Christian Icons	309
Chapter 25	**The Cathedrals of the Druids**	**311**
	To South Devon	311
	The Hurlers, Bodmin Moor	311
	The Long Barrow on Bolt Tail	313
	The Henge Temple	316
	The HGV roads of Stonehenge	316
	Is Stonehenge a Christian Cathedral?	318
	The Processional Way Following the Split Beam	320
	The Bell Chamber	320
	Bolt Tail : A Stone Age Temple	322
	Review	327
Chapter 26	**Has History got it Wrong**	**330**
	The Hell-Fire Caves	330
	The Crucifix Presents a Problem	331
	The Rollright Stone Circle	332
	The White Horse at Uffington	336
Chapter 27	**A View Through the Stone Portal**	**341**
	Energy Engineering	341
	The Civil Engineering Legacy of the Druids	342
	Underground Temples	342
	The Sacred Landscape	342
	Woodhenges and Stone Circles	348
	The Development of Temples	351
	Druidic Roads	352
	The People of pre-Roman Britain	353
	The Artwork of the Britons – a written record of pre-history	358
	The Druidic Religion	360
	The Druidic and Secular Societies	364
	The Romans	365
	The Roots of Native Britain	366

Chapter 28	**A Never Ending Journey**	**367**
	The Knowledge of the Ancients	367
	Conjecture	368
	A Three Element Approach	371
Epilogue		**372**
Appendix 1	**Biolocation (Science Based Dowsing)**	**374**
Appendix 2	**The Concept of Earth Energies (EE) and their Possible Role in the Culture and Religion of the Henge Age**	**379**
Appendix 3	**A Synopsis of The Diamagnetic and Paramagnetic Theory of Biolocation**	**382**
Appendix 4	**Is the Blind Experiment to Prove that Dowsing is a Real Phenomena Possible**	**397**
References and Further Reading		**398**
Glossary		**402**
List of Figures and Colour Illustrations		**412**
Index		**415**

List of Boxes

Box 1.1	The Movement of Rods	19
Box 1.2	Stone Circles and Energy Rings	20
Box 1.3	Paths, Roads and Ley Lines	23
Box 1.4	The Mager Disc (Rosette)	24
Box 3.1	Dating Cultures	44
Box 8.1	Witnesses	86
Box 8.2	The Magnetic Fields of Trees	92
Box 9.1	Avebury Stone Circles	97
Box 9.2	Stanton Drew Stone Circles	101
Box 10.1	Studies of the Oak's Paramagnetic Field (Aura)	135
Box 10.2	Dykes	137
Box 14.1	Dating the Stoney Lane Tombs	182
Box 17.1	Tomb Evaluation at Durrington Walls	210
Box 20.1	Blue Energy Lines	243
Box 20.2	Polarized Fields and Quantum Entanglement	244
Box 24.1	Herne the Hunter and the Green Man	299
Box 24.2	The Grail Legend	307
Box 24.3	The Henge Age Society at the End of the Iron Age	308
Box 25.1	The Stone Age Tomb	328
Box 25.2	The Henge Age	329
Box 27.1	The Proto-Christian Crucifix Becomes a Birth Table	346
Box 28.1	The Interpretation of the Beltane Fire Pits	369

Photographs and Illustrations

The objectives of the visual material supporting The Secret of the Stones are to illustrate the principles and techniques of science based dowsing and show how the information it provides is interpreted and visualised.

They are a true record of the research and its findings as it happened during extensive travelling and fieldwork in all weathers throughout the year. Due to time and other constraints this imposed on the research schedule and the priority given to the dowsing analysis of study sites this record includes both low and high resolution photographic images.

Diagrams and sketches are simplified for clarity and are derived from surveys, measurements, mapping out and plotting form. The more detailed data underlying them are a matter for other publications.

Footprints in the Soil

We are a people, now long gone
who walked and raced the wind
across the land you now call yours.
For you who follow on four gifts
we leave.
Four gifts that will endure the years
till wisdom comes and leads to
understanding.
These gifts are round you now.

Symbols by which to define
yourselves.
Myths to make the dark nights
kinder
Music and art with which to share our
time.
Footprints in the soil, so you may
know our story,
as it was before the eagle came.

GWC

Stonehenge
Nigel hughes

Chapter 1

Base Camp and a Mountain to Climb

I was sitting on a low wall looking out to sea. Behind me a white washed cottage and its small stone walled back garden, then the green fields reaching up to grey dry stone walls on the lower slopes of a mountain. It was Ireland's sacred mountain, Croagh Patrick, (Figure 1.1) the character of which is a bit like my research projects involving dowsing. Sometimes it is hidden in cloud, sometimes distant and untouchable and occasionally its detail so clear it appears close enough to touch. As I looked out over Clew Bay back in June 2000 the mountain was grey, distant and partly hidden by cloud and I was hoping fervently that it was not an omen for my mission to Westport, Co. Mayo. At close on half a mile high it possibly represented the size of the task ahead of me so for good luck I had climbed it soon after arriving. After climbing nearly half a mile into the sky and then being chased back down by dark incoming clouds one tends to do a lot of sitting on whatever is to hand, hence the long contemplation of the sea and the work of the next two weeks.

Clew Bay looks as if it was a large plain which has been flooded (Figure 1.1). It is in fact easy to imagine it as a wide flat area with perhaps a valley going up to Westport and the myriad of islands once being the hills which now just poke above the water. It must have been an interesting countryside for the people who lived there. The water did not suddenly flow in and flood the countryside. A slow rise in water level over the years was responsible for creating what I was here for. Perhaps the last rise in sea level was about 1000BC or perhaps later and it may have been no more than three feet or there about. Just enough to create some salt marshes along the shoreline.

Many thousands of years ago the people who lived in the Westport area had built stone monuments and one was at Killadangan just to the east of where I was sitting and on the way into Westport. I had been introduced to the stone circles and ancient sacred sites of Ireland by one of my dowsing mentors in August 1999. For those learning to dowse there are plenty of introductory courses but few more advanced ones to help develop skills and knowledge. The best way forward is to find skilled and experienced dowsers and work with them, using them as mentors. I had several mentors with the top one living in Ireland at the foot of Croagh Patrick. A poet, painter, writer, a skilled dowser and mystic who could discern the energies of sacred sites and the countryside. He could identify earth stars and a balance or imbalance of earth energies. Only the best in the business was good enough for me. But hold on. I am a hard research scientist, earning a living as a consultant. Everything has to have an explanation based on science and mathematics. I have to convince others through reasoned argument that my advice is good. So if I am told a sacred site is on an energy line what is the energy? Where does it come from? Where is it going? How do you measure it? How do you prove that the dowser is responding to an energy field? How does this hard scientific approach fit in with the approach of mystics?

Energy is a measure of the ability to do work. In dowsing terms it is what affects people in someway. Makes them physically feel something. To me it was what made my rods move (Box 1.1) and as they did in fact move, the mystics' use of the term energy was good enough for me – at least for the moment.

Another reason for accepting the term energy is that if you want to research something it is as well to be able to 'see' or detect in some way the energies that the dowser responds to.

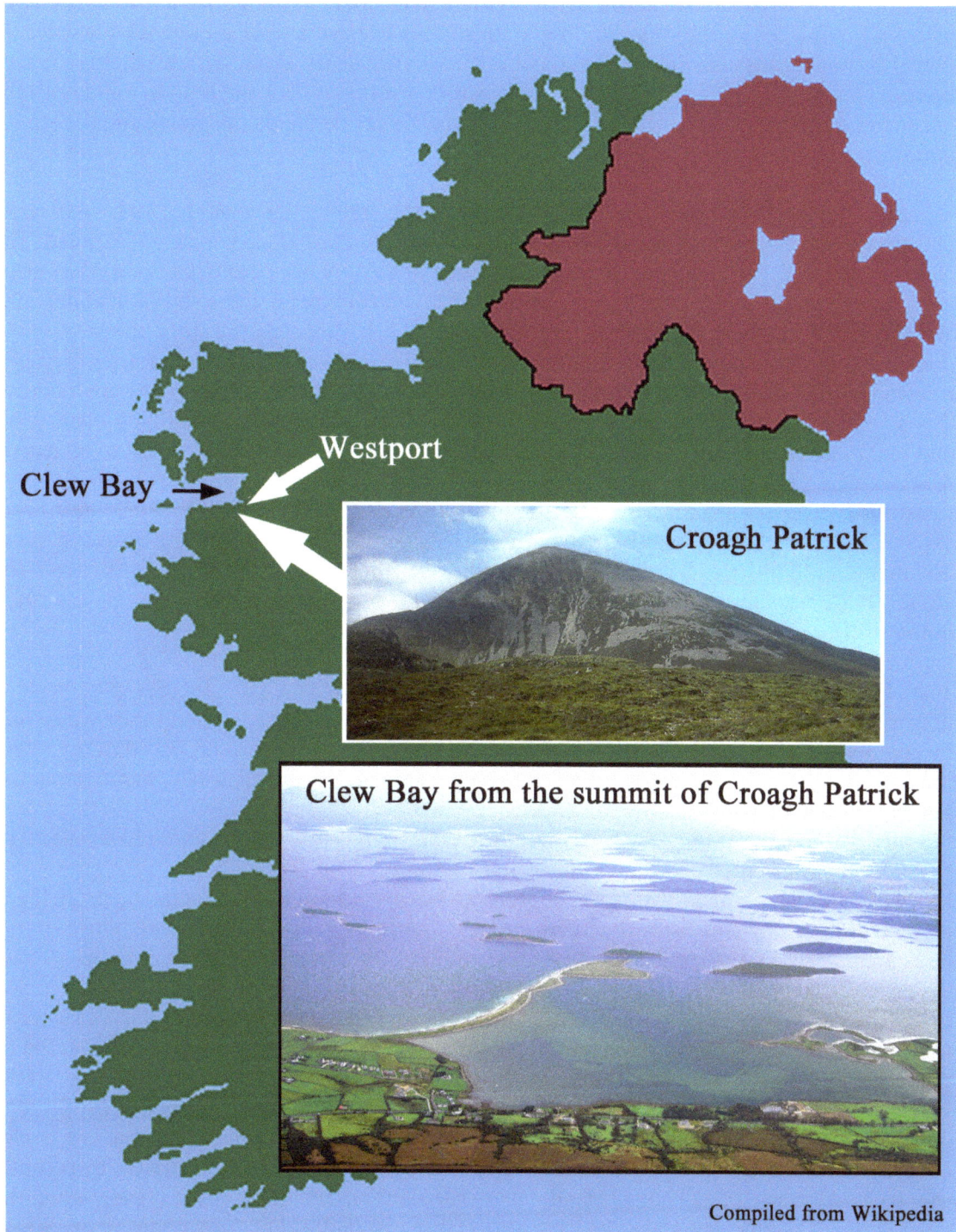

Westport

Clew Bay

Croagh Patrick

Clew Bay from the summit of Croagh Patrick

Compiled from Wikipedia

Figure 1.1 Clew Bay, Co. Mayo
Home to ancient Stone Circle builders.

My mentor was teaching me to 'see' what was there. This he had done with some success on my first visit. I had also worked with another dowser who was an expert on stone circles and what are called earth energies (Appendix 2). These are lines of energy running across the ground, sometimes for miles. After much tuition I was ready to research the magic of the stones. I say magic because large stones placed by our remote ancestors in configurations such as circles or even just placed upright as standing stones seem to have a hold on our imagination. To most this hold is a visual one but to some it is the 'energy fields' in and around the stones that hold the magic. The energy fields move, have subtle meanings, link the human mind with the earth and its energies. How did the ancients create the energy fields that surround the stones and the pattern of energies within them?

To help answer these questions and maintain some intellectual rigour, I had soon after learning to dowse in 1994 developed a model, or paradigm as we say in science, called 'The Diamagnetic and Paramagnetic Theory of Biolocation' (Appendix 3). This theory has great explanatory power and was to play a pivotal role in developing an understanding of dowsing and the sea of 'magnetic energies' in which dowsers work. However, the story about to unfold does not require a knowledge of the theory and I only mention it to show that the story as it relates to stone circles came about as the result of a mystic and a scientist working together. The scientist had to accept that what the mystic said he observed was real and observable and had to be understood, at least in part. The mystics' observational skills had to be learnt. An attempt also had to be made to go back in time and try to see the world as the builders of the stone circles had seen it. The models and theories that had been developed by mystics to explain their observations may be incomplete or over complex but they were the starting point. The models and theories developed by archaeologists about the peoples who constructed the stone circles may also have short comings but again they were the starting point for any study of the stone circles.

To most people stone circles are just a number of stones. Some have a lot of stones in them some only a few, some are small some are large. In 2000 the current opinions said they were built by late Stone Age or early Bronze Age people. All that most people are prepared to say about them is that the builders were not too hot on drawing circles and their arty arrangement of stones was not up to much. However, if you are 'into stone circles' and identify with them then the picture changes dramatically. The imperfect circles have meanings and when measured and figures produced, all sorts of ratios can be extracted. The arrangement of the stones can be related to astronomical events, the seasons and to prominent objects on the horizon. If you happen to have dowsing skills then a new world opens up as you identify and follow energy lines in and around the circle and across the countryside. Some of these energies come from the ground and are believed to have been there before the stone circles were built and are said to be the reason why they are built where they are. The ancients thought they were closer to the spirits or Earth Goddess at the selected points. Some of the energies have been created by the stones and some have been created by the builders of the circles. It was the energies of the stones that I had come to study. I had dowsed many stone circles both the ancient pre-Roman ones and those constructed over the past 200 years, the modern ones. The ancient circles always had a characteristic pattern of energy fields inside and outside the circle. The recently constructed stone circles I had studied never created the rings of energy not even when the most eminent of dowsers and stone circle specialist had built them. The modern circles that I had visited were always 'dead' easy to identify as modern and they never created the fields associated with the ancient circles. Where ancient circles had been repaired with new stones or old ones put back in place the additions and those stones incorrectly replaced could be identified. Provided a sufficient number of the original stones were still in place the magic and full power of the circle remained. But in my experience no modern builder had created the magic of an ancient stone circle. The ancient stone circles were based on the creation of

Box 1.1 Dowsing Responses and The Movement of the Rods

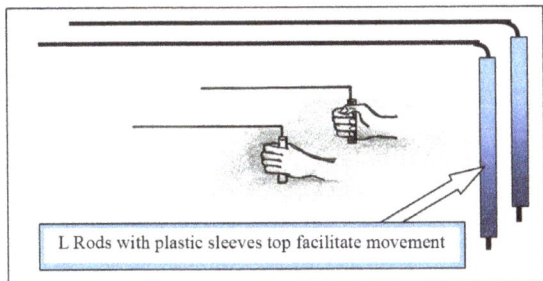

L Rods with plastic sleeves top facilitate movement

L Rods are easy dowsing devices to use. Response movements occur in a number of ways and are consistent between dowsers. The movements provide information about the magnetic fields being dowsed

Figure B1.1.1 Movements of the Rods

L Rods are useful because they give well defined responses

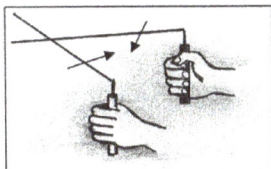

Some dowsers have reverse responses when dowsing North and South Poles.
Although generally stable polarities can suddenly reverse for short or long periods of time.

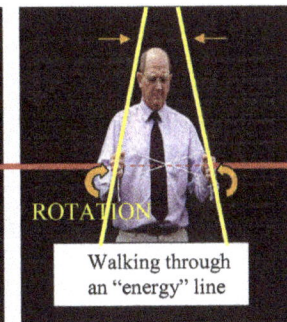

Rods Crossing
Rods swing inwards from the neutral position at the same rate and in doing so cross. This occurs typically when the field lines from the North Pole of a target are crossed by the dowser.

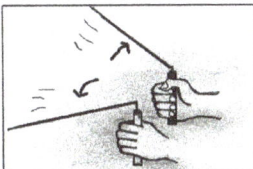

Rods Opening
Both rods swing outwards and in doing so both rods open out smoothly and at a constant rate. This occurs typically when the field lines of a South Pole are crossed.

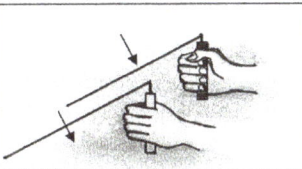

Rods Pointing
If a target with a strong magnetic field is approached to one side of it both rods point towards it. If when aside the target the dowser rotates through 360 degrees the rods will cross when the dowser is pointing at the target

Holding wood dowels show how the arms and hands rotate to cause rod movements.
In order to respond to an energy the dowser must walk through it at the correct angle.

Wood dowel

Walking toward an "energy" line

ROTATION

Walking through an "energy" line

Walking away from an "energy" line

Dowser responds to the field from the drain

North

Accurate dowsing locates two lines of responses corresponding to the shape or FORM of the drain.

Figure B1.1.2 Dowsing the FORM of a field.

A dowser traces out a magnetic field line from an underground object.

Underground objects may have other fields with them originating from other features associated with the object for example a pipe may have various material around it such as sand, clay or gravel as well as chemicals inside it. These can be useful in confirming the form of the object.

19

Box 1.2 Stone Circles and Energy Rings

Part of the fascination of stone circles, see figure B1.2.1, is that some people sense something, often referred to as 'energy', as they approach and walk into a circle. The circles create subjective sensations for these people. Many more can detect the 'energy' if they use dowsing instruments such as rods or a pendulum. If the 'energy' is investigated it appears to have the form of rings both round and inside the circle. When a colour wheel is used the rings come up on different colours such as black, red, yellow. They can therefore be referred to by colour. The actual colours will depend on the type of stone and the pigments used in the colour wheel. In addition to the rings, lines of energy can be detected. Some are associated with the stones themselves others with the ground.

The energies from the stones are derived from the material of which they are made and the crystal structure of the rock. These energies are therefore reasonably straight forward and can be measured and their origins identified. The earth energies coming up from the ground are not so easy to identify. Dowsers consider that underground water streams and blind springs are an important source of these energies. Michael Poynder probably gives the dowsers views on the 'energy' structure of stone circles in his book 'Lost Science of the Stone Age' (see page 48 & 62) However, it might be as well to keep an open mind on where the earth energy component of a stone circle's magic comes from. Any inconsistency in a magnetic field can be picked up by dowsing and such variation in fields can be caused by many people walking the same path, and creating a track, by wheeled transport (Box 1.3), cattle or stone walls now long gone. For example it is easy to see that any cultural or religious structure such as a stone circle can generate ceremonial paths and it may be that all the dowser is responding to are these paths. Fortunately earth energy lines can be analysed and their source identified. Unlike the stones however, a piece cannot be put on to a laboratory bench so they are not quite so easy to analyse.

Figure B1.2.1 Swinside Stone Circle, Cumbria

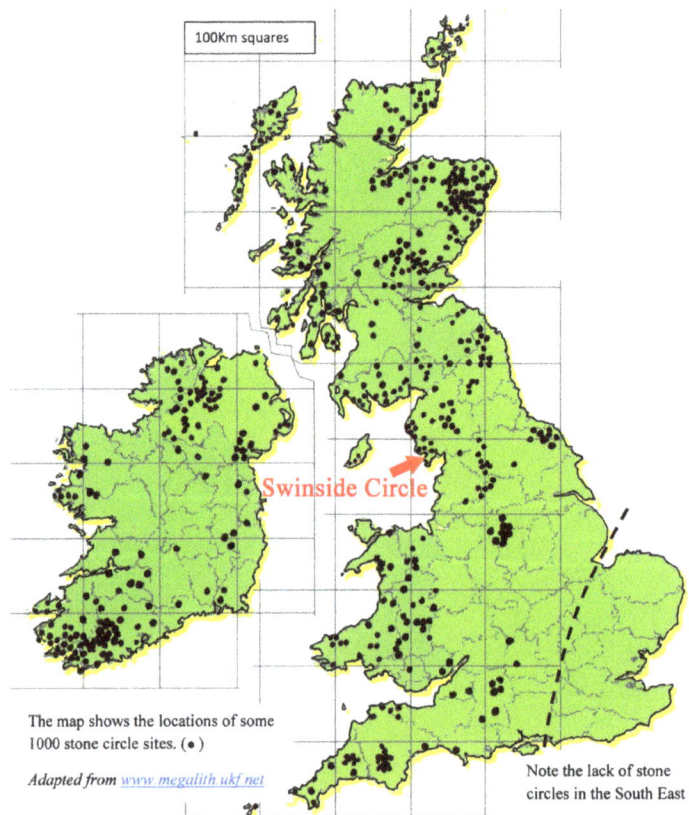

100Km squares

Swinside Circle

The map shows the locations of some 1000 stone circle sites. (•)

Adapted from www.megalith.ukf.net

Note the lack of stone circles in the South East

Figure B1.2.2 Distribution of ancient Stone Circles

Figure B1.2.2 shows the distribution of ancient stone circles in the UK. The presence of energies is often considered indicative of the authenticity of ancient circles. Generally it is true to say that of the stone circle sites investigated the ones erected in modern times lack the energies associated with an ancient circle. However to consider if a circle is ancient using such criteria it is as well to keep in mind that the stones in ancient circles can be moved , changed or interfered with over time. The need for a method for careful, rational observation and analysis of these energies is obvious.

a real mystic sensation. People could pick up and feel the energies of the circles and their construction was based on a real understanding of stones and their 'spiritual' energy.

My task was to find the answer to a number of questions such as a) were the stone circles built on pre existing energies? b) Were the stones set up to generate the energy system associated with them? That is was energy engineering involved and if it was c) how was the energy engineering achieved? The theory I had developed predicted that the stones would have been treated as magnets by the ancients. If this was so the magnetic axis of the stones would be set in a pattern that develops the energy rings and energy patterns. During my visit to Westport in 1999 I had found that the stones appeared to be laid in a particular and precise way. As I sat on the wall I read my log for the 26th August 1999 against the blue, well perhaps more grey than blue backdrop of Clew Bay "We (my mentor and I) made the discovery that the stones are laid out as if they are magnets". The brief note indicated that we had evidence that Bronze Age man was aware of the earth energies associated with stones. They were in fact sensitive to the diamagnetic and paramagnetic fields and recognised them in the rocks. (The magnetic sense is dealt with in Appendix 3) If those involved were Shaman priests they would probably consider the fields to be the 'spirit' of the stones and would be aware that there was a difference between stones being in harmony with each other and in disharmony. The ability of a trained group, say priests, to lay stones in magnetic harmony would be a cultural trait. If I could prove that the builders of stone circles had this trait it could be used to link geographically separated groups of people, say Ireland, UK, and France and follow a people through time. Were a people living in the shadow of Croagh Patrick 2000 years or more ago going to have this cultural trait and could it be used to link them with the rest of the British Isles and France. The presence of stone circles over this large area might suggest this was the case. However it is easy to lay stones in a circle but it is not so easy to produce the fields the ancients created (see Box 1.2). I knew that many stone circles would have to be visited and tested throughout the British Isles before the extent of circle building knowledge was known. But this is jumping the gun, I still had to unravel the secret of the ancients. At the time there was other evidence of what I term energy engineering knowledge being present in 'Celtic' times and that it was actually used by the early Christians in the construction of churches. Energy engineering is laying stones in such a way that the magnetic fields of one stone complements and reinforces the magnetic fields of another stone. In the 'Lost Science of the Stone Age' Pages 171-173 Michael Poynder describes the role of earth energies and their manipulation in the design and layout of early churches and cathedrals. Dowsers of a spiritual and mystic frame of mind believe that the earth energies were present before the original Church or building and were not created by the buildings or by human activity. For dowsers of a scientific bent, an open mind is the best approach and evidence has to be gathered which indicates the origin of the energies. That is, have they a geological origin such as faults in rock strata, local peculiarities of the earth's magnetic fields or are they simply created by human activity such as walking along a path or by wheeled transport travelling along an ancient road. Both the path and road are being generated because people need to travel to a built structure or a ceremonial meeting place. These energy lines would therefore be created after the meeting place had been agreed.

Ley lines connecting sacred sites are well known energy lines and are the subject of many books and learned discourses. John Timpson in Chapter 1 of his book 'Timpson's Ley lines' gives a good summary of the range of views on Ley lines. From my own personal experience of looking at what dowsers call Ley lines my view is that many of them are indeed track ways, prehistoric or otherwise. This was Alfred Watkins view. Once a track way has been established or a meeting area has become accepted it is difficult to change it so they could be used for hundreds if not thousands of years (see Box 1.3). The repeated foot or wheeled traffic leaves a stain of carbon, iron oxide and other materials in the ground. With time the stain moves deeper into the ground. This stain sets up the energy pattern the dowser responds to. Because the dowsable energy is set up by specific materials it is possible to use these materials to identify footpaths and roads for wheeled transport.

One of the interesting theories associated with stone structures is that they give off energy lines which travel miles across the countryside. Some of these lines, particularly the one that comes up as blue on the Mager colour disc (see Box 1.4), are believed to be associated with communications (telepathy) the passage of spirits and to enable travellers to find their way. They are a sort of magnetic beacon which is said to work both overland and across the sea. The Mager colour disc is very useful and a powerful technique for identifying specific energy lines. Once an energy line has been identified by colour or a signature of colours it can be followed across country to see where it comes from and is going to.

So much for musing on the task ahead. Across the Louisburgh road at the bottom of the front garden were fields, bungalows and then the sea shore. A length of sand and shingle stretching east from Leckanvy and Thornhill to Bartraw and White Strand. White Strand appeared to be a large dune projecting towards the islands which mark the outer limit of Westport Bay. White Strand is excellent for jogging, walking dogs and meditation. However, it is Thornhill Strand that is most interesting. Along the beach are a multitude of different rocks and stones and amongst them I expected to find the stones used by the stone circle builders. If I unlocked the builder's secrets I could gather the stones and construct my own Killadangan Stone Circle. There was only one problem, the weather. Wind and rain made careful work difficult. The dowsing response to a change in magnetic field is subtle and difficult to detect if buffeted by wind and rain. However, I was determined to rediscover the lost secret of the stone circles come what may. To help in this task I had brought the car across from the UK and was going to sit out any bad weather on site.

Box 1.3 Paths, Roads and Ley lines

Human activity and particularly movement along a road or path leads to a grinding of what is being trod on and abrasion of feet, shoes, wheels. Any animals using the road will leave droppings, hair, and chemicals from their hooves. The small particles produced and the chemicals released make their way down into the soil. They form a stain in the soil which the dowser can detect and analyse. The chemicals and particles of the stain are held firmly by the soil or rock and remain there for hundreds and even several thousand years. Particles, particularly the larger ones remain in place provided the soil is not disturbed. The chemicals and fine particles may make their way deeper into the soil over the years and gain some protection from disturbance. Ancient paths and roads can still be found today by looking for the chemical stains left in the soil. They can also be found by using the colour wheel. A typical road with one cart track will appear to be bands of black, red, white, red, black as you cross it. A blue and yellow band may sometimes be present. In dowsing the term 'Ley line' is not favoured and the preferred term is 'Energy Ley' to distinguish it from the Leys as originally defined by Alfred Watkins in 1925.

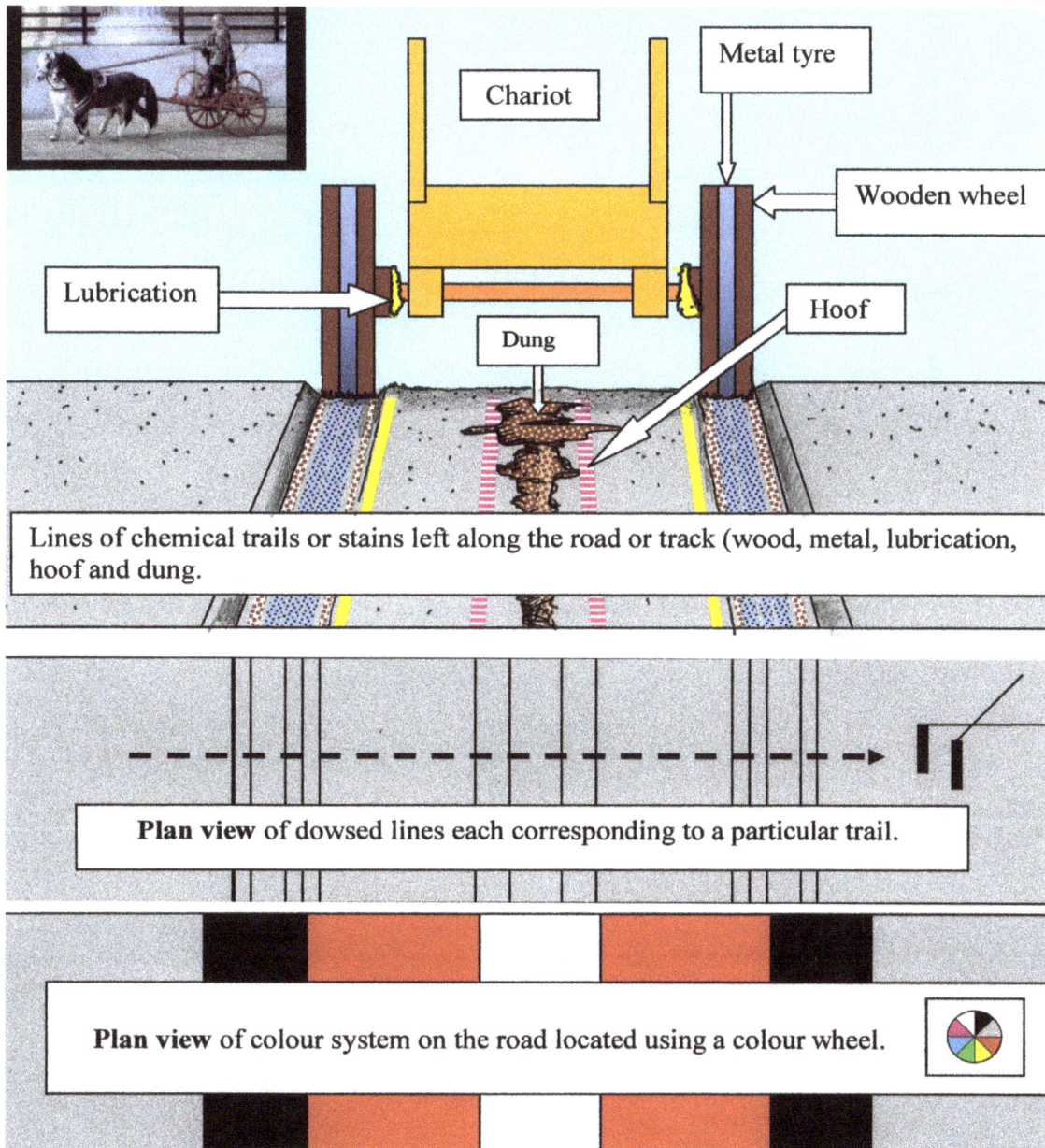

Metal tyre

Chariot

Wooden wheel

Lubrication

Dung

Hoof

Lines of chemical trails or stains left along the road or track (wood, metal, lubrication, hoof and dung.

Plan view of dowsed lines each corresponding to a particular trail.

Plan view of colour system on the road located using a colour wheel.

Figure B1.3.1 Chemical Stains

Box 1.4 The Mager Disc (Rosette)

The Mager disc or colour wheel can be adapted to provide serial numbers on each colour segment (Lost Science of the Stone Age, page 41) The colours are white, violet, blue, green, yellow, red, grey and black.

The method by which the colour disc works is not known but it does appear to be quite effective. The results obtained in terms of the colour signature of say a road depends on the pigments used for the colours. The eight colours are obtained from four pigments which may vary between printers. It is therefore important to standardise colour wheels for group work. If the wheels have been produced by a printing company wash them to remove any surface agent before use.

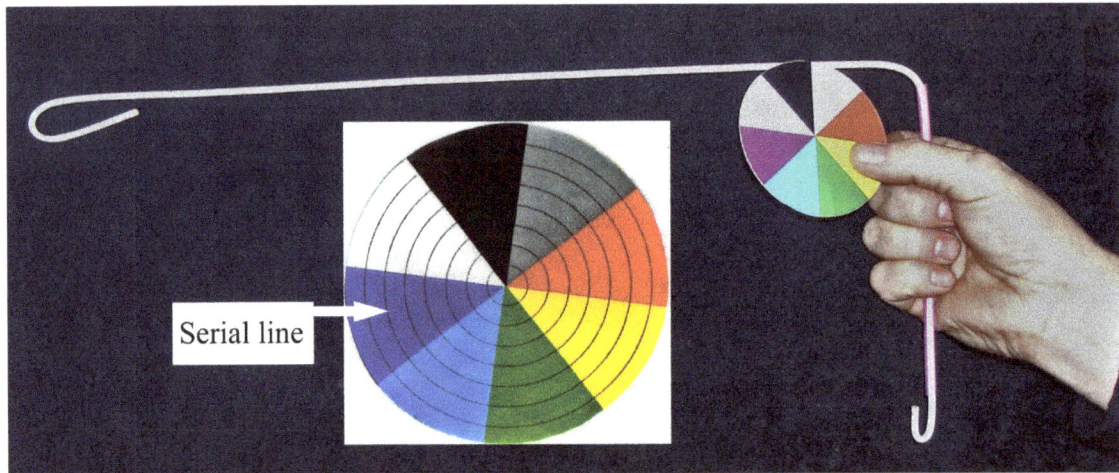

Serial line

Figure B1.4.1 Using the Colour Disc with Rods
The disc is held in one of the hands as shown on the appropriate serial line.

Chapter 2

Some Secrets Revealed

The Circle of Killadangan

Going east towards Westport there is a sign which points to the left down a narrow lane leading to Murrisk Friary. All that is left of the Friary are its grey stone walls and the grave yard but it is a fine example of an early Christian church which has been built on what is believed to be a pagan (Druidic) sacred site. That is if the many earth energy lines going through the building are anything to go by. That the area was home to a pagan community is indicated in the surrounding low lying land by what looked like paths and mounds. The land seems to say that an Iron Age people were here before the sea rose and displaced them. Sea levels were rising in the Neolithic and early Bronze Age so the settlement could perhaps have been earlier than the Iron Age. Returning to the road, walk of about a mile along Murrisk Sound brings the salt marsh of Killadangan into view. It is a wide expanse of grass and water, sometimes with cattle roaming about it and at high spring tide looking very wet indeed. Another half mile further along and round a bend in the road there is a place to park a car.

It is on the right hand side and opposite an 'access' point to the marsh and beach. From this point I could walk down onto the marsh which at first sight looks like a vast open expanse of course grass, reeds and muddy channels (Figure 2.1). It is bounded, one could say held in check, by the modern road on the left hand or southern side of the marsh. This side has probably always been the higher and drier part of the site and I was later to find that there are many Bronze or Iron Age roundhouses with their storage pits along this edge. Moving a few yards forward and to the west there is a raised area called a Cromleck with a bank round it. The bank seems to provide protection for a dense growth of reeds and other plants designed to make walking difficult. The Cromleck today appears to be a man made island within the marsh. However, when it was built there may not have been a marsh and protection against high tides not required. The Cromleck could therefore be defensive or religious. Whatever it was used for it is now overgrown with reeds and very difficult to dowse. Advancing into the Cromleck and towards the western side, a group of large stones rise from the reeds. They are a closely set row of four stones. The stone row, see figure 2.2, stands about three and a half feet high and is a major feature of the stone complex. At the southern end of the row there are the broken remains of a fifth stone. These stones are enigmatic in that they do not have an obvious use or function but they are amongst the largest stones on the site. Returning to where I entered the Cromleck the grass is grazed by cows and progress is easy. Walking further into the marsh, on the left hand side close to the stone wall and the road is a large lone standing stone. This is shown as Standing Stone 1 (SS1) on the plan of the site (Figure 2.1). It is also called the Blue Stone because of the beam of blue energy it emits. The larger stones are marked on the site plan as standing stones 1, 2 & 3. They are large upright stones between three and five feet tall. There are also smaller stones referred to as stone pairs. Stone pair 1 (SP1) looks as if it could have been the portal stones marking out a processional way towards the circle.

After walking about 110m from the gate with the road on my left, I arrived at the stone circle, which has probably about half of its original stones still in place. The stone circle is a complex of stones and the assumption is that stones lying outside the circle were deliberately placed and had a role to play in energy engineering. I therefore had to become familiar with all the stones. There are four stones lying on the eastern edge of the circle and possibly indicating what may have been the entrance to the circle, see figure 2.3.

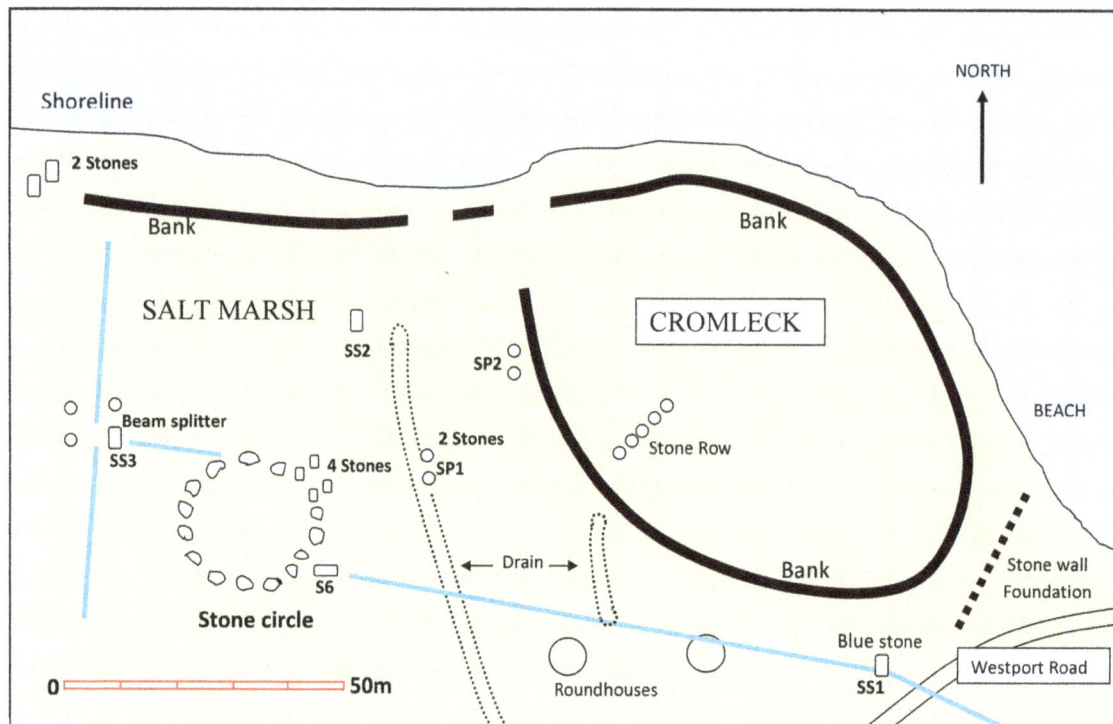

Figure 2.1 Salt Marsh and Killadangan Cromleck Stone Row and Stone Circle
The salt marsh, which extends to the west, floods at spring tides but still supports grass and cattle.

Figure 2.2 The Cromleck
Croagh Patrick is the conical mountain in the distance. In the foreground are the four remaining stones of the stone row in the Cromleck with Standing Stone 3 in the mid-distance to the right of the cattle.

Figure 2.3 Portal to the Circle
On the eastern edge of the circle are four stones that may have marked the entrance to the circle.

Figure 2.4 The Blue Line
The stone shown (S6) in the foreground lies just outside the stone circle. This stone attracts the blue energy field of the circle and then projects it to the Blue Stone (SS1) about 100m away.

There is a large partly buried stone (S6) on the south east side pointing towards SS1, the blue stone (Figure 2.4). On the Northwest of the circle is a large standing stone SS3 with a broken companion which may have matched SS3 in size and behind them at a distance of about 3m a pair of smaller stones (Figure 2.5). These stones, referred to as the beam splitter, are of considerable interest.

The Killadangan stone complex has attracted the attention of archaeologists as well as that of mystics and dowsers. A survey of the site by C. Corlett is reported in The Journal of the Galway Archaeological and Historical Society 49 (1997) pp 135-150. In the report the position of the site is given as being on the southern shores of Clew Bay, immediately NW of a rather bad bend in the Westport – Louisburgh road and just over 5km SW of Westport. Corlett says the stone row stands in a raised earthen enclosure marked as a Cromleck. The row is in line with the sun as it sets behind a niche in the mountain Croagh Patrick to the SSW at 1.40pm on the winter solstice.

Corlett gives a good description of the main stones of the complex in his paper to which reference should be made for an archaeological view of the site. However, he omits a number of stones including the complex of four stones based on SS3. The survey therefore looks as if it was a superficial one. It should be noted that in surveys of stone circles all stones are important and their position and orientation could be very important. Each one may be playing a role in the energy engineering designed into the site by the ancients.

At the centre of the complex is the stone circle. It is 13m in diameter and is best preserved to the east, south west and west. It is not too difficult to identify that some of the stones are misplaced and have fallen over or have slid out of position. There are two main gaps in the circle with no sign of any stones. However, there are numerous small and large pieces of stone scattered round the site. This may indicate weathering or vandalism over the past two thousand years. Corlett does not give a date for the construction of the complex but says it could have been in the period 1362BC to 794BC. Corlett suggests that the complex was designed as part of a pre-conceived plan and that the complete complex was more important than any individual element. He points out that the function of standing stone complexes is the subject of great speculation and that archaeological excavation sheds little light on what they were used for. That they served the community in some social, ceremonial, ritual way is generally accepted. Also that they formed part of a ritual landscape. It is easy to speculate, for example, the complex could have been laid out as the mirror image of the Orion constellation. It could have been constructed to deal with astronomical events and to also have a number of other functions. One such other function which is supported by dowsers is to create earth energy lines and to form a link with other centres and sources of earth energies. These energies are dia and para magnetic fields. (Appendix 3).

After familiarising myself with the site, not only its stones but also some of the flooded channels of the marsh, I started a survey based on dowsing. The weather for the first few days was wet and windy and limited what I could do. When the weather improved I decided to concentrate the study of the complex at the circle. It is a typical stone circle and I started by using a Mager colour disc (Box 1.4) to identify the different energy bands round the circle. I had done this many times with my mentor and as before the energy rings round it came up on black, yellow and blue. The circle also has what is referred to by dowsers as a blind spring at the centre. A blind spring is believed by dowsers to be an upward movement of water which is blocked before it reaches the surface. The blind spring gave a positive response on the black segment of the colour disc, the normal colour for the spring. Whether there are blind springs under the centres of the stone circles is open to debate as there is always a large area of organic matter at the centre. Organic matter also shows up as black on the colour wheel. The centre of the

Killadangan Circle is often flooded and most of the time muddy so there could be plenty of organic matter in it. Fortunately during the study it was possible to enter the circle and make measurements although not accurate ones. The blind spring was found to be circular and about 2 to 2.20m in diameter and the stone circle itself as already mentioned about 13m in diameter. Having identified the characteristic energy rings of a stone circle, so confirming that most of the stones were laid correctly, the next step was to try and identify the three magnetic axes of the circle stones.

There appeared to be one paramagnetic axis going round the stones and a second which was vertical. The diamagnetic axis ran along the radius of the circle. That is the field lines were going to or coming from the blind spring area to the stones. The fact that these fields could be identified indicated that the stones might be laid in a uniform way. This would be tested for later. The next step was to determine if the circular field had a particular component characterizing it. A Silica (sand, quartz) field was looked for to determine if it was responsible for the black field round the circle. Glass contains silica and so attracts and removes the field coming from sand or quartz in the stones. It was found that if glass was placed on top of the stones the black field round the circle disappeared. If the glass was placed between the stones the black field remained. It therefore looked as if the paramagnetic field produced by the silica was in the vertical magnetic axis. Checking most of the circle stones 8 of them did not have a silica side on top but 27 did. Doing the same with aluminium, if the sample of aluminium was on top of the stones the blue ring did not disappear. If the aluminium was between the stones the blue energy line disappeared. The paramagnetic field between the stones contained the aluminium field. The next step was to test stones for their diamagnetic axis. This axis was found on the radius of the circle with the South Pole towards the centre. If this was correct the magnetic field lines should flow from the spring to the stones.

In analysing the fields I only used a single chemical marker for each of the two paramagnetic axes, silica for one and aluminium for the other. It would of course have been possible to fully analyse both fields but this was not necessary. As there was only one diamagnetic axis there was no need to analyse it.

Modelling the Circle

Whilst working on the circle I picked up and labelled fragments of the stones and then used them as witnesses. These witnesses enabled me to go off to the beach and find matching stones on Thornhill Strand. I had to build my own working stone circle so after a few visits to the Strand I had the stones with which to build a mini circle on the cottage forecourt.

The stones "which dowsed as if they were circle stones" that is I could not say they were geologically the same but they dowsed as if they were, had their silica paramagnetic axis, aluminium paramagnetic axes and diamagnetic axis identified and then they were laid out to form a circle about 3m in diameter. Three metres is big enough to walk around and work with. Figure 2.7 shows how the magnetic axes of the stones were lined up and figure 2.8 shows the actual model of the Killadangan circle.

The mini stone circle produced the encircling black, yellow and blue energy rings and a circular area in the centre which dowsed as if it was a blind spring. By selecting stones to match those in the ancient circle, then identifying their magnetic axes and laying them to match the circle ones I had been able to achieve what the ancient engineers had done about two possibly three thousand years previously. For the first time as far as I knew a properly engineered stone circle had been created in modern times. To show that the mini circle behaved in the same way as the real one I did a few experiments. The mini blind spring would disappear if a piece

Figure 2.5 The Beam Splitter

The stone circle is to the right of the picture. The large stone is SS3 which with a companion stone, now broken and sunk in the mud, the two stones on the right and two more stones to the left (not visible) form a beam splitter. A blue beam comes in from the circle on the right. This beam is split to the left of SS3 at about the mid point between SS3 and the two outer stones on the left. One beam goes off to the far trees. The other beam heads past the photographer.

Figure 2.6 A Section of the Circle

The stones are not large and may have been set into the ground. There has been movement and loss of stones since the circle was abandoned.

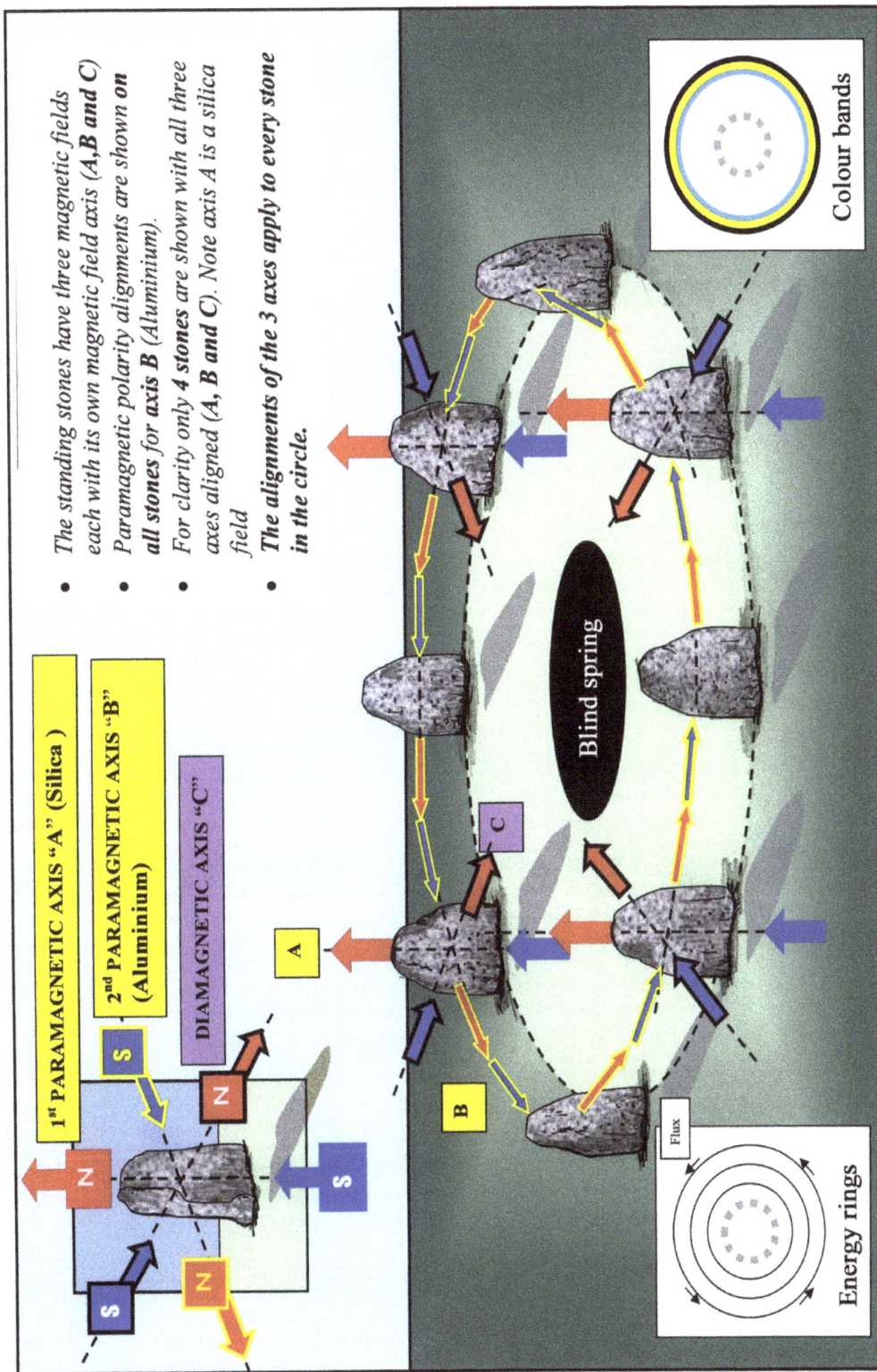

The standing stones have three magnetic fields each with its own magnetic field axis (A,B and C)

Paramagnetic polarity alignments are shown on all stones for axis B (Aluminium).

For clarity only 4 stones are shown with all three axes aligned (A, B and C). Note axis A is a silica field

The alignments of the 3 axes apply to every stone in the circle.

1st PARAMAGNETIC AXIS "A" (Silica)

2nd PARAMAGNETIC AXIS "B" (Aluminium)

DIAGMAGNETIC AXIS "C"

Colour bands

Blind spring

Flux

Energy rings

A

B

C

S

N

N

S

N

S

Figure 2.7 Killadangan Magnetic Alignments

The stones of the circle dowsed as if one magnetic axis, that had the field from aluminium as a component, was laid to complete a magnetic circle. This is done by laying a north pole to face a south pole. The other paramagnetic axis with fields derived from silica had its north pole on top of the stone. The third axis had the south pole facing in towards the centre of the circle. The dia-magnetic fields appeared to be based on iron, potassium, calcium and magnesium.

Figure 2.8 Energy Model of the Killadangan Stone Circle
The photograph shows the circle, the area of the blind spring, S6 for extracting a blue line from the circle and the four stones of the beam splitter which produced two blue lines. The circle was constructed by identifying the two paramagnetic axes and the diamagnetic axis of each stone.

of the stone used in the circle was placed in the middle and if oxides or salts of potassium (K), calcium (Ca), magnesium (Mg) and iron (Fe) were placed in the centre. It did as they are the materials in the stones attracting the fields of the spring. Their fields produce a circle because all the stones have the same magnetic pole facing in.

To summarise: It appears that the stones in the stone circle at Killadangan are laid so that one paramagnetic axis has its poles north to south round the circle and reinforce each other. The second paramagnetic axis is vertical with the north uppermost. The diamagnetic axis is along the radius and with the same pole pointing in, the field lines meet and appear to go up or down so creating the appearance of a circle. This area is often referred to as a blind spring by dowsers and it is said to be over crossing underground streams. The energy from the crossing streams is said to help create the mystic properties of the circle. The experiments on the patio showed that no blind spring is required. The stones could create the central area. It did not exclude something being at the centre but it was not necessary. Placing stones at the centre attracted the fields and prevented the circle being formed.

The Blue Energy Line

Once the ancients had created a circular paramagnetic magnet what did they do with its fields or energy? The reason why the ancients went to so much trouble to select stones and place them in precise ways to each other so that magnetic fields were created was going to dog me throughout the project. There was plenty of wild conjecture amongst dowsers but no well founded and testable ideas. All I could do for the moment was to study the stone circle complexes and work out the science behind their construction and the energy lines (magnetic fields) they produced.

Just by the circle on the SE side is a stone (S6) which plays a role in sending a blue energy line to Standing Stone 1 (SS1 the Blue Stone). Somehow (S6) attracts blue energy out of the circle's field. The presence of this blue energy line in the complex had been known to my mentor and to dowsers for many years. Blue energy lines are very common in structures built by the ancient world and are believed to have been engineered into buildings by the Romans and early Christians. Pulpits in the older churches and cathedrals are on blue lines. Mythology says that these lines are associated with communication, the movement of spirits and in Ireland the movement of the little people.

The next step was to determine if there was any evidence that the blue line had been deliberately engineered. Using the colour wheel, the blue paramagnetic field from the circle can be identified. When checking the fields of the circle it had been found that the blue field could in fact be blocked by using aluminium foil and removed from the circle's energy rings. That indicates that the blue of the circle is derived from the aluminium in the stones. The blue line could also be followed from the circle (S6) then to Standing Stone 1, the Blue Stone, and then on to some distant target, see figure 2.1. In theory it should be possible to block the blue line at a number of points. For example by removing the blue ring from the circle, by blocking the line at stone 6 and just before it reached the Blue Stone (SS1). The standing stone might have its own blue line so the bearing of the line emitted by SS1 was measured and found to be 120°. Next, the circle's blue ring was removed by placing a piece of circle stone at its centre. This caused the emitted blue line from SS1 to move to the south as the blue line from the circle disappeared. Next aluminium foil was placed between the circle and the Blue Stone to block the field from the circle and again prevent it reaching SS1. This caused the outgoing blue line from SS1 to move round to the south again. To summarise: Stone 6 by the circle appears to have the function of taking out from the circle a blue field which is then projected to SS1. To do this Stone 6 has to be laid in a particular position. This was confirmed on the forecourt model using a piece of stone that matched Stone 6. If the model's Stone 6 is set up a few degrees either way of where it should be the linear blue field is not produced. At the Killadangan circle, Stone 6 sends a blue line a hundred meters to a stone which is about 30 or more times its size. As magnetic fields are proportional to the mass of the magnet, the blue field from SS1 could possible be projected 100m x 30 that is travel for several kilometres to a distant target. The field from Stone 6 also determines the direction of the emitted line from SS1.

Splitting the Blue Energy Line

The circle has another feature. On the Northwest side of the circle there is a complex of stones in pools of water and mud. There is just one big stone (SS3) of this complex left standing. The complex is represented diagrammatically in figure 2.1 and the stones shown in figure 2.5. Part of the complex is by the circle and is possibly part of it. It consists of two stones (Figure 2.5) and then a few metres from them to the NW the large stone, SS3. A cluster of stones which were probably at one time a single stone matching SS3 is to the right of SS3. About another two to three metres out and away from the circle are two more stones which look as if they have fallen over. Although the stones were in some disarray a blue line could be picked up on the north and south sides of the rectangular complex. What appears to be happening is that the energy field from the circle goes into the rectangular area of what would have been four stones. These stones act as a beam splitter in that the incident blue energy beam from the circle is split into two and the energy beams are sent off in opposite directions and at 90° to the incident field. Using samples of the four stones on the mini circle I was able to construct a beam splitter for the mini circle I had made and send two blue beams off in opposite directions. From the measurements on site and the laboratory experiments it looks as if the builders of the circle deliberately laid out a pattern of stones on the NW side of the circle so arranged that field lines were sent off in opposite directions. From the size of the one remaining intact stone the field lines were intended to travel for kilometres. Using a compass the lines had rough

33

bearings of 40° and 220° which approximates to midsummer sunrise and midwinter sunset at Westport. The blue line from SS1 at 120° is someway off midwinter sunrise.

Stones Outside the Circle

Looking at the main stones outside the circle it was found that the diamagnetic axis of Standing Stone 3 was vertical with the North Pole at the top. The other stones in the group also had the diamagnetic north at what was probably the top when they were standing. SS1 and SS2 had north at the top, the stones of the row had north at the top, the stones of the stone pairs also had north at the top. The stones of the circle therefore differ from those outside the circle in that their diamagnetic axis is horizontal and along the radius of the circle and not vertical.

Archaeologists and students of stone circles are always trying to find alignments between stones and astronomical events involving the moon and sun. With this in mind I took a few magnetic bearings. The stone pair on the ditch are nearly due east of the circle and look as if they could be marking an entrance to the circle. I could not pick up a clear energy line between the stone pair and any other part of the complex. This would perhaps be the case if the stones were all that remain of an entrance. However, there is an energy line from the larger of the stones at 136° which is going east and could be bending towards the blue stone. There is also an energy field going west at 309° towards the circle. I was not convinced that they had any significance.

The stone row is a bit of a puzzle because it is in the embanked area and neither its magnetic bearing of 22° or the direction it faces 292° and 112° seem to tie in with important astronomical events, even allowing for measurement error and uncorrected magnetic bearings.

Review of Findings

To summarise, the weather conditions were not ideal for dowsing and the stone complex at Killadangan has been damaged. Nevertheless it was still possible to identify that the ancients were aware of the magnetic fields of stones and would set stones up so as to amplify the fields. They were also aware that field lines could be attracted out of stone circles and directed to other stones. The 'other' stone, if large, would amplify the signal and the orientation of the stone could be used to redirect the beam in a preferred direction. The ancients also knew that if an energy line was directed at stones, if the stones were laid in a particular way, the incident beam could be split into two beams. The use of this knowledge to achieve particular objectives is what I refer to as energy engineering.

In unravelling what the stone circle builders were doing the first and most important step was to know that stones have what I call magnetic axes with north and south poles. Most stones have at least one diamagnetic axis and two paramagnetic axes. Each axis consists of one or more fields derived from the materials the stone is made from. Provided the material of the individual stones is the same, the fields of the stones will interact, that is north and south will be attracted to each other and similar poles will repulse each other. If stones are laid in a circle or any other configuration with the same axis in line and with a north pole facing a south pole a giant magnet is formed with a giant field. A circle has advantages and appears to have been adopted by the ancients. It should be noted that the ancients could not have constructed stone circles if they only knew of one, say the strongest, magnetic axis. They must have been aware of three magnetic axes. If one of the axes is lined up the others could be at any angle to the linear one. However, the other axes have also been lined up. This means that the ancients were aware of them and indicates that they were very sensitive dowsers able to identify the different magnetic axes. This is indicated by how stones were set up in the complex. In the

Killadangan circle the diamagnetic axis is horizontal, with the South Pole facing in. The second paramagnetic axis is vertical, North Pole up. The effect of this is to produce a central zone where the field lines change direction going up or down and circular fields round the stones. The fields can be due to iron and quartz, giving black on the colour wheel aluminium-blue, phosphate-yellow, calcium and magnesium carbonates–white. Not every component of the stone will produce a field that can be picked up on the colour wheel. However, stone circles normally have at least three colour rings round them.

In 2000 I did not know if the ancients had colours to help them. They would have used their hands, pendulums or some other instrument to determine the magnetic axes and then probably trial and error to line the stones up and obtained the desired result. They were not novices when it came to dowsing. The use of stones to attract dowsable energy lines out of the circle and then to use other stones to direct the field lines in specific directions indicates a mastery of their art. What I call the beam splitter shows that they had a knowledge of the magnetic properties of stones, although they would not have called them such, which has not been equalled today. I can produce beam splitters but only because I found out how the ancients did it by subjecting their work to a systematic scientific approach.

The outlying standing stones of the complex all had a function. The only one that comes through loud and clear today is that of the Blue Stone (SS1). The Blue Stone somehow amplifies the signal coming from the circle so that it can be sent for miles and by rotating it about its diamagnetic axis, the beam was sent to a particular point on the horizon. The beam splitter would also have to be fine tuned to ensure its beams went in the right direction.

The Killadangan stone circle is in my opinion unique. Because the sea rose possibly three thousand years ago the circle and its outlying stones have been partially preserved. In mainland Britain and I suspect Ireland, the outlying stones of circles are no longer to be seen (at least I have not found them) and have been removed by farmers. The stone circles are all that is left of the energy engineering complex. Looking back over my stay in Westport the visit had been successful. It had shown the power of the rational scientific approach in dealing with the preserve of the mystic and the paranormal. What the mystics and dowsers claim in relation to stone circles is in fact there. The mystic's fields and energies do however have a rational explanation. They were created by a people with a knowledge that has been lost. A little of this knowledge has been regained and for possibly the first time in 2000 years a part of a stone circle complex was created. It is now clear that stone circles are energy engineered to produce the energy fields (magnetic fields) detected by dowsers. The complexity and scale of the stone complex must have had profound significance and relevance in the ancient culture. It is perhaps that significance and relevance that is the true secret of the stones. But Killadangan is only one stone circle amongst many. Were the people of Killadangan unique in having and using this technology? Many more circles would now have to be looked at in detail to determine how widespread was the knowledge.

Chapter 3

Rings of Energy : by Chance or by Design

Stone Circle Sites

The British Isles contain the remains of many stone circles (Box 1.2). The sites are spread all over the countryside with the exception of the south eastern part of England. The absence of stone circles in this part of the country is due to stone being in short supply in the area. The circle stone has therefore almost certainly been recycled and used for buildings or road construction. As a result the visible remains of stone circles disappeared from the south east long ago. There are records of the remains of one stone circle at Tunbridge Wells which unfortunately was removed to make way for modern development. Today, wherever they are, most stone circles consist of just a few of the original stones, some of which may still be standing, but most have often fallen or been deliberately pushed over. In some circles fallen stones have been re-erected by people who were not to know that the orientation of the magnetic axes is critical to the workings of the circle. It was indeed fortunate that I had the Killadangan complex to study as it had not been greatly disturbed over the millennia. The Avebury stone circles is a classic example of an ancient site.

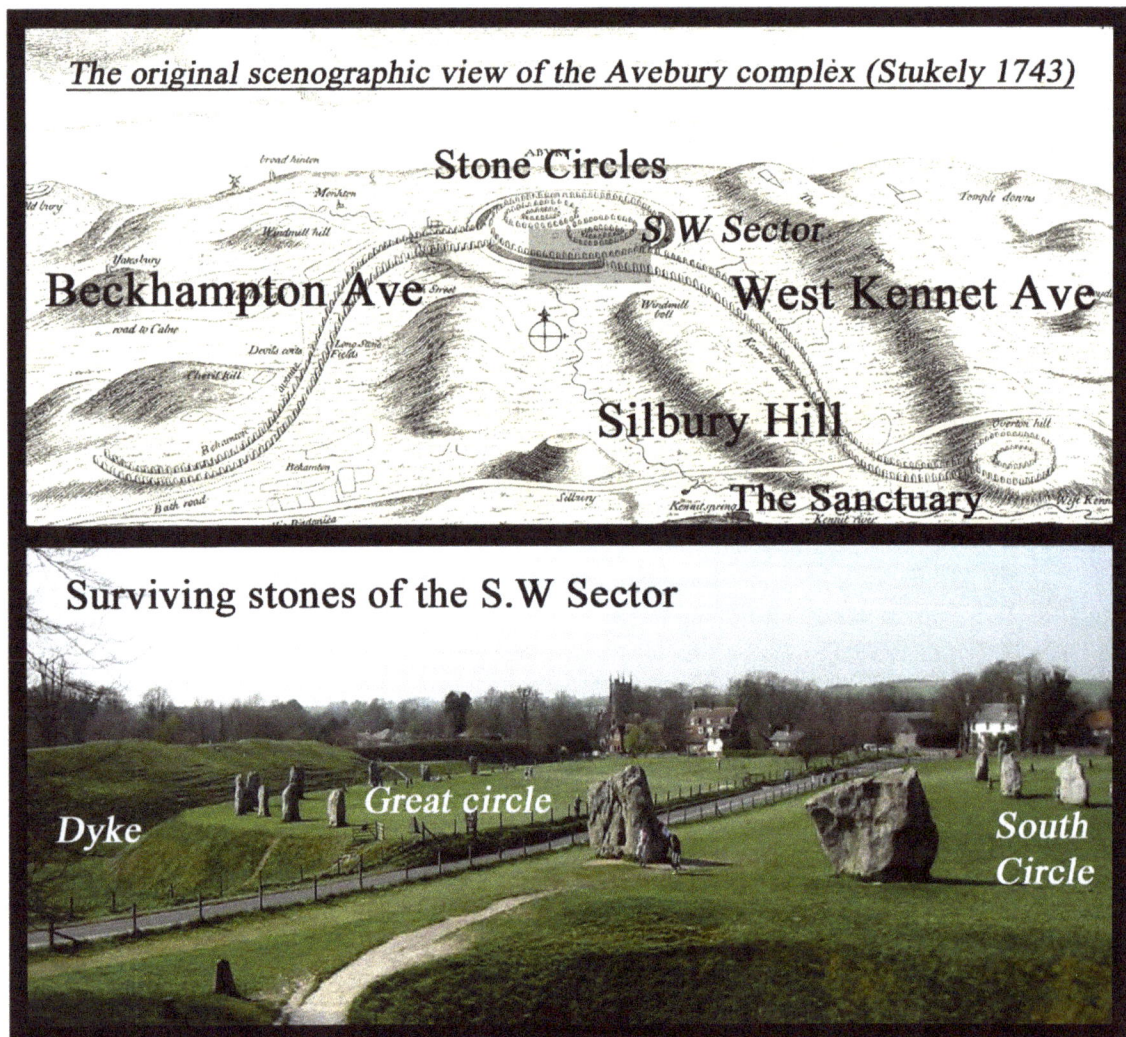

Figure 3.1 Avebury from an Aerial Perspective
The Avebury Stone Circles are a very large and complex site.

Some of the original stones (Figure 3.1) are no longer there having been broken up to use in construction work. Some have been re-erected and stand in their original place, and some have survived where they were first erected. In fact enough of the original stones have survived at Avebury for it to become a spiritual centre and for people to claim that it still has mystic properties and possibly power.

The most complete ancient circles are probably in Ireland where the people have treated them with some respect. Nevertheless after 2000 years or more most have been damaged to some extent or been renovated for the tourists. On a return trip to Ireland my mentor and I set out on a tour which provided me with an opportunity to see if other circles had their stones carefully positioned to ensure that the magnetic axes were working as a team. In other words were their builders energy engineers.

That ancient stones are set up in relation to magnetic axis has been observed by dowsers before. A German worker called Dr Bruno Fricke produced a report in the mid 1980's on 'The use of polarised stones (or crystals) as a method of clearing houses and communities of detrimental radiation'. The paper was translated by Nicolas Fricke and a copy was passed to me some years ago by a fellow dowser. In this report Fricke refers to stones and bricks having a positive side. He does not mention a negative side but they presumably have one. Fricke observed that in megalithic sites the stones are placed with the positive side turned inward. Positive is probably a north pole. Fricke also refers to ordinary building bricks being polarized. Bricks do indeed have clearly defined magnetic axes and the energies of a stone circle can be created by building a circle from bricks (Figure 3.2).

Figure 3.2 Magnetic Axes of Bricks
The photograph shows the magnetic axes of bricks which when lined up correctly in a circle .create the magnetic fields associated with an ancient stone circle. The bricks shown produced quite a strong circular field.

The bricks may acquire their magnetism as they cool down in the kiln and from their magnetic axes it should be possible to determine their position and possibly where they were in the kiln as they cooled. That is provided the kiln fields are known. I have successfully done this exercise with tiles from a small pottery kiln but not with brick kilns. However, the point to remember is that it is possible to lay ordinary house bricks to achieve a desired magnetic field. Again Frick refers this when suggesting that foundations, bricks and stones should be laid with the positive side up. The question I was interested in was 'did the ancients lay their stones with the positive i.e. North Pole of an axis uppermost as a standard technique?

The Rollright Stone Circle

My first stop was the Rollright Stones in Oxfordshire not far from the Cross Arms Public House on the A44. The Rollrights are a well known stone circle and used for a number of activities including weddings and celebrations. Most visitors to the stones would agree that as you approach the stones and walk inside the circle you can experience a change in the ambience. The surroundings, provided you are of the right frame of mind and tuned in, become more peaceful, you become at ease with yourself. Assuming that you were not in that state to begin with. However, when I visited the stones in July 2000 it was not for self therapy but to see if the 70 or more stones were set up with their magnetic axes following the same orientation. The sight of seventy stones made me realise that I had quite a task ahead of me unless I could prove my point with a sample. The main visual difference between the Rollrights and

Figure 3.3 The Rollright Stone Circle
The circle is an usual one in that the stones are positioned to create a 'curtain' wall or a protective wall round the outside of a building or sacred area. They do not appear to be structural.

Killadangan is that the Rollright Stones look as if they are standing on edge and form a much larger circle (Figure 3.3). The Killadangan circle stones were set into the ground, almost flush with the ground. It is said that some stones have been knocked and broken or have fallen down and then been put up again. This has not stopped the circle having the usual energy rings both around the outside and on the inside of the stones.

I found that most of the individual stones, in fact 56, have a diamagnetic north at the top and many have the correct north (say axis B) facing the correct south of axis B on its neighbouring stone as you go round the circle. Thirty one pairs of stones had a field between them. Most of the stones have a south pole facing out and a north facing in. The difference from Killadangan is that the diamagnetic axis of the stones is the vertical one. At Killadangan the diamagnetic axis was horizontal. Otherwise the Rollrights appeared to follow the principles of energy engineering found at Killadangan. This is interesting because the stone circle is quite a different type to the one at Killadangan. The stones were set up as a wall, possibly round a circular ceremonial site or structure. The stones were positioned as close to each other as possible with set access and egress points round the circle.

Drombeg Stone Circle

After visiting the Rollrights other aspects of dowsing took up my time and it was about a year before I returned to stone circles. Once again I was back in Ireland. This time down in the south of the country where West Cork has a number of magnificent stone circles. One of them is the Drombeg Recumbent Stone Circle. The term Recumbent is used to describe a circle where one of the stones appears to be lying on its side. The site is said to date from 1124 to 794BC and the circle of seventeen stones was restored in the 1950's. The circle averages 9.3m in diameter and the literature describing the circle indicates that some of the stones are lined up with astronomical events. On discovering that it was restored I just hoped that it meant no more than digging down and exposing the stones to view. The stones are said to be hewn from local sandstone and they have been worked to produce a smooth side that faces into the circle. Aubrey Burl notes this feature of circle stones in his book on the Rollright Stones. Referring to a book by T. H. Ravenhill he says that the stones of the Rollrights, Stonehenge, Avebury and Castlerigg had a smooth side facing in. The top of the recumbent stone is smooth as if it was designed as an altar. As a stone lying down would break the magnetic symmetry of a circle I was most interested in what I was going to find. On reaching Drombeg I found that the circle was in a fine state of preservation with easy access to the stones (Figure 3.4).

Figure 3.4 The Drombeg Stone Circle, Co. Cork
This circle has a recumbent stone with a dressed surface at a suitable height for ceremonies.

Once in the circle the first step was to determine the vertical axis of the stones and this I found to be a diamagnetic axis with a north pole at the top. The upper face of the recumbent stone was also a diamagnetic north pole. The one exception was stone 4 which had both a south and north diamagnetic axis on top. That is the stone appeared to be on its side so that the north and south poles were at either end of the stone. Normally the south pole is in the ground. The outside face of the stones followed the predicted pattern with a paramagnetic south pole facing out. Going round the circle the stones were laid north pole to south pole with the exception of stones 5, 6 and 7. Approaching the circle, black and red rings could be dowsed so a sufficient number of the stones were correctly lined up to create a ring magnet.

Near to the stone circle can be found stones that archaeologists believe to have formed the foundation of a roundhouse. These stones appeared to be laid in a haphazard way in relation to their magnetic fields. This would indicate, provided they have not been disturbed, that energy engineering to create magnetic harmony between stones was for religious or ceremonial structures and not for building huts or so it appeared in 2001. After studying the stones for a while I felt sure the seventeen stones of Drombeg Recumbent Stone Circle had originally been laid in a similar manner to those of the Killadangan Circle, that is with their magnetic axes in harmony to create the fields that modern dowsers associate with them.

Glebe Stone Circle

Further north on the West Coast can be found Glebe Stone Circle which is near Conge in County Mayo. It consists of about 24 main stones and is set in an open field protected by an iron fence. The circle has sustained some damage but it has not been repaired. All the stones appear to have diamagnetic north at the top and going round the circle there were 19 paramagnetic north's facing the corresponding paramagnetic south. There were two north to north and three south to south pairs. An energy ring is thus created round the circle as there are few misalignments. The stones had a south paramagnetic pole facing out of the circle and a north into the circle. The stone circle fitted the theory perfectly.

Modern Stone Circles

The next Irish stone circle to be analysed was a modern one. Towards the end of the twentieth century a stone circle building bug started sweeping the British Isles. Fortunately if you are a dowsing archaeologist, unfortunately if you are seeking to use them for contact with the spiritual world, the modern circle builders do not have the skill of their ancient ancestors. The only modern stone circle I have come across that has enough stones set up correctly to generate a passable circular field round it was built by a farmer using the geological features of the stones for alignment. This is interesting in that it can be argued that if the stone is quarried and such things as bedding planes are obvious it is possible to set up stones correctly aligned magnetically just by using visual features such as bedding planes to make the stones look good. It could be argued that this is what the ancients did and the fields are a by product of good artwork.

There were seventeen stones in the modern circle ten of them had a south pole at the top, five a north pole and two of them did not have a detectable pole at the top. They were probably side on. Only three pairs of stones appeared to be linked magnetically. Energy engineering is not easy and the builders had failed in that respect. Visually the circle looks great and I am sure a lot of people will get something from it.

British Stone Circles

Back in England, a visit to Devon enabled the Nine Stones Circle on Dartmoor to be visited. The circle has red, white and black magnetic field lines leading up to it indicating an ancient roadway. There were also red, yellow and black fields round the circle. Weather conditions prevented close study but on one side the stones appeared to be laid as predicted but on the other, the west side, they were not. Visually some of the stones did not look as if they were the original ones and it is possible that some repairs have been made to the circle. The circle still retains its fields due to a sufficient number of stones being magnetically aligned. In this respect it is supporting evidence for the theory.

In 2002 some more stone circles were visited and analysed. The first were Arbor Low and The Nine Ladies in Derbyshire. Arbor Low is a 45m diameter circle with unfortunately most of the stones of the circle displaced and pulled down. The circle is large, set out on high ground, in open country with splendid views of the Derbyshire countryside. In the centre of the circle there are two very large stones which lay either side of the centre point. It is said that they were once upright. However, the upper surface of both stones dowsed as a north diamagnetic pole. If this is correct then in the upright position two diamagnetic norths would be facing each other and two south diamagnetic poles would be facing out. This might take some reconciling with the north paramagnetic poles facing in from the circle stones. As far as I could judge from the now recumbent stones of the circle, when they were upright their magnetic axes may have been in harmony. Time and some careful work is required to establish how the stones might have been lined up in the past.

The Nine Ladies Stone Circle (Figure 3.5) is a small one only about eight metres in diameter. It is set in a wood and heather landscape some 20 minutes walk from the nearest road. The stones are so small that two or three people could have moved them. The stones are set into the ground and do not protrude a great deal. This has almost certainly protected them from damage over the years. Around the circle there are well defined magnetic fields. The stones have their diamagnetic north at the top and this is also the case with the recumbent stone. Going round the circle the stones are linked north to south. The Nine Ladies Stone Circle fits the theory perfectly.

Figure 3.5 The Nine Ladies Recumbent Stone Circle on Stanton Moor
This circle in the Peak District fits the magnetic theory well. The recumbent stone has the north diamagnetic pole at the top of the stone.

Further north in Aberdeenshire Loan Head Recumbent Stone Circle near Daviot was visited (Figure 3.6).The circle was 'dead'; there were no fields between the stones or round the circle although the individual stones had their fields. The circle had an air of unreality about it. The stones looked freshly hewn and the mass of small stones in the circle could easily be removing any fields present. The circle is on an ancient stone circle site but to me the present circle appears to be a modern one as I could not pick up any fields generated by the circle of stones. A word of caution here. Solar and terrestrial magnetic activity can abolish dowsable fields. When such activity is going on the dowser picks up nothing, that is no rods crossing or opening. Unless something alerts the dowser to what is happening the dowser can obtain a series of false negatives without realising it. The significance of solar activity had been brought home to me on one occasion when I was working in the laboratory. I had found that I could not obtain a dowsing response. It was not the odd few minutes of blackout which is quite common but it had been going on for about an hour. To check if it was me or something more general I phoned a friend and asked him if he could dowse and obtain responses. His reply was that the papers were full of news about a very large solar flare and that it might have something to do with my dowsing problems. Knowing that there was a solar flare with its magnetic fields and high energy particles immediately told me that if it was responsible then the most likely culprit would be a magnetic field from the protons of the flare. All I had to do was place a bottle of protons in the field from the sun and it would produce a shadow in which dowsing would revert to normal. Acids contain protons (hydrogen ions) so a bottle of vinegar from the larder should block the field streaming from the sun. It did and I was able to continue my laboratory dowsing. The episode illustrated one of the mechanisms by which the sun interferes with dowsing and the value of the diamagnetic and paramagnetic theory of biolocation in the scientific approach to dowsing problems. I noted that another visit to the Loan Head Circle was required.

Figure 3.6 The Loan Head Recumbent Stone Circle
The recumbent stone in a circle may be very large and not necessarily used as an altar.
The presence of rock fragments within the circle may destroy the circular magnetic fields.

The next site to be visited was The Easter Aquorthies Stone Circle near Inverurie. This circle has been renovated. It is on a small site not allowing much space for investigation. The circle is on a genuine stone circle site but I could not pick up fields round the circle or between the stones. The recumbent stone was at a height that would have made it impossible to use. My guess is that it is a recently renovated circle and perhaps, with the Loan Head Stone Circle, provides evidence that building stone circles is a very skilled operation involving knowledge of the 'spirits of the stones'. However, in case of solar activity I need to check the stones on another occasion. From North East Scotland it was then down onto the rolling countryside of Wiltshire in Southern England and the immense stone circle complex of Avebury.

Avebury Stone Circles

I visited Avebury to see what the state of play was with its stones. Many have been moved but most appear to have a diamagnetic north at the top and to have linking fields between them.

The stones are sandstone and I tried to determine on which magnetic axis the silica field appeared. I just could not find it although I knew that it should be there. I therefore decided to give up and took a piece of stone back to my bed and breakfast hotel. In the garden of my B&B there was a suitable table for studying pieces of rock and after a while I found the silica axis. It is on a 4th axis which is very close to being vertical on a standing stone. Consequently you have to be in the right position and close to the stone to pick it up.

One of the largest stones in the Avebury complex is one called the Swindon Stone. Visually this stone looks as if it has been put in upside down. That it is in fact upside down is confirmed by the diamagnetic poles. The North Pole is in the ground the South Pole is at the top of the stone. Again caution is required as the energy engineers may have deliberately placed the stone the wrong way up for engineering reasons. The stone masons may not have known what the stone was going to be used for and so did a normal north pole on top stone.

In 2004 a new stone circle was discovered by one of the authors in the North East of England having been put onto it by a farmer. See later for more details of this circle as it came to play a very important role in unravelling some of the secrets of the stones. The stones of the circle appeared to still be where they were 2000 years ago. The circle was of a type where the stones are set into the ground so in this case only the tops were visible. The tops of the stones were the diamagnetic north poles of the stones and each stone was laid so that the correct north and south poles were facing each other. It was a genuine ancient stone circle fitting the energy engineering theory.

Discussion

On the basis of the ancient stone circles visited and described here it is clear that their builders were sensitive to the magnetic fields of stones. They could feel or detect them in some way. The ancients would not have referred to them as magnetic fields but would have thought of them as perhaps the spirits of the stones. If not they may have given the fields some name to enable them to work with the stones, hold discussions with colleagues and to plan and execute energy engineering projects. What perhaps had not been appreciated before, although it may have been guessed at, is that the builders of megalithic monuments were accomplished dowsers. They could not have worked with stones small and large and placed them according to their magnetic fields without highly developed dowsing skills and in the case of larger stones engineering skills. They knew the different fields and magnetic axes associated with stones. They used their skills to achieve engineering objectives which were set by the society they were part of. Most probably the objectives related to spiritual, religious and other cultural aspects of community life. What was being sensed by the ancient dowsers may have been thought of as an extension of the spirit world and proof that the spirit world permeated everything. The use of such skills would not have been restricted to just building stone circle complexes. They would have been used in other aspects of the community's life. Those with the skills are likely to have had a special position in the community, they would have been trained, perhaps those with aptitude selected, knowledge would have to be passed down through the generations and this would have taken resources. Energy engineering was clearly a wide spread technology which required logistic support and in particular teaching and training services for the engineers.

Today dowsing is easy to teach at the introductory level but difficult and time consuming to teach at an advanced level. This is because there is no accepted underlying theory. This is in contrast to the ancient teachers of dowsing who may have used the spirituality of all things as their underlying theory and produced students who were able dowsers and performed to high standards in all the areas of the British Isles so far studied. At the present time, advancement in dowsing skills depends on self teaching and mentoring by more experienced dowsers. Without a theoretical framework it is very difficult to identify a standard method of detecting fields and agreeing what the fields are i.e. what is being detected. It is like teaching chemistry without the periodic tables. The diamagnetic and paramagnetic theory of biolocation dealt with in Appendix 3 should now provide a suitable physical and biological framework for training and research.

It is clear that if members of a society have well developed dowsing skills and the skills are being used by that society for community and social purposes it is not going to be possible for people 2000 years later to get close to the culture of that group without taking on board the human magnetic sense i.e. learning dowsing skills and trying to understand the role it played in the culture. The basic archaeological findings will not be understood if the human magnetic sense is ignored. Just as vision, hearing, smell, touch, taste play a role in interpreting archaeological findings so the magnetic sense must play a role and to do this it must be understood.

At the end of the first phase of the stone circle study late in 2002 it was clear that dowsing skills had been used over the British Isles for the construction of stone circles and, if the archaeologists were to be believed, by peoples of the late Neolithic and early Bronze Age. My own dowsing is on a broad front so while trying to work out how stone circles generated their mystic power I had been forced into studying Iron Age roundhouses. The wood postholes of such structures, their fire places, storage pits, paths and roads are fairly easy for a dowser to identify if they know how. It was the dowsing of the postholes left by wood structures that would lead to an even more amazing discovery about the culture and life of these ancient people.

Box 3.1 Dating Cultures

Cultures exist in both a geographical space and a time frame. Cultures are also multi stranded. There are technical, engineering, art strands for example and although some strands fade and disappear others continue and new ones appear. Archaeologists take advantage of this. They can take a particular method of making pottery or features of its design and trace the reach of a trait or cultural element in space and time. The same approach can be applied to dowsing skills. These skills are used for specific purposes whether it be for laying stones in a particular magnetic relationship to each other or for finding water. The result of using water divining is that thousands of years later the wells dug by a people can be found along underground water courses. The skills can also be used in the design of amphitheatres, shrines and churches. The result is that the use of the human magnetic sense can be traced through time and geographical zones.

Because cultural strands extend through time and because people tend to use what comes easily to hand and is easily made, there can be considerable overlap of artefacts in time. Being in the Iron Age does not mean that stone tools are not used. Consequently the presence of stone tools on a site does not mean that you are dealing with the Stone Age. The artefacts could be from the Bronze Age or even the early Iron Age. In the 21ˢᵗ century stone tools and artistic artefacts are still being made and used. The wood around people today in the form of buildings, artistic artefacts and furniture may date from several hundred if not thousand years ago. Great care therefore has to be exercised when using artefacts and cultural elements to date objects and place the activities of a particular group of people in time. For example the presence of iron or advanced artefacts and tools does not mean that the culture of a people has changed. The culture may remain a Stone Age one and have changed little over thousands of years. One late Iron Age house investigated had bronze and iron in its construction but work stations for flint knapping. Flint tools and weapons were still being widely used in the Iron Age.

Chapter 4

From Postholes to Woodhenges

A Roundhouse on the Lawn

One of my main areas of biolocation (dowsing) research is trying to solve the problems associated with demonstrating the reality of dowsing to people (Appendix 4). Some types of dowsing such as site dowsing in which the dowser is looking for water mains, drains, leaks etc is accepted as being reliable. Water divining is also very reliable. It is when dowsers start to do what are called laboratory type controlled experiments that dowsers fail and results appear to be random. However, what lies behind the failure of controlled experiments is another story. I mention it because in using the back garden for my studies my dowsing often failed for reasons that appeared to be associated with the garden itself. To a dowser the reality of dowsing is clear but to sceptics such as James Randi who through the James Randi Education Foundation had offered a million dollars (now withdrawn) to anybody who can prove the reality of dowsing, it is not. The challenge for the dowser is that the sceptics require a blind or double blind experimental approach and unfortunately such an approach is not possible for dowsing (Appendix 4). Amongst the problems I had met when doing blind tests was interference from what many dowsers call earth energies, that is magnetic fields coming up from the ground. Most of these energies in my experience are from ancient and not so ancient postholes, pits, wells, paths etc. The fields are due in the main to organic and carbon deposits in the ground.

If a target, that is something you are trying to find, is placed on top of an ancient storage pit it tends to disappear dowsing wise (Figure 4.1). Also when walking over a pit the dowser can obtain a response but it is a false positive as the target is not there. The false positive findings appeared at the time to indicate some large irregular areas. By using the black segment on the colour wheel, I started to plot them out on the lawn. It was not long before I had ten areas in a circle. My first response was that I might have evidence for the first stone circle to be found in Hertfordshire. However, when I consulted a few local archaeologists the consensus was that they might be wells. Further study soon indicated that they could be the storage pits associated with Iron Age roundhouses. Then one day, when sitting in the garden and pondering on what might be hidden in the lawn and flower beds the thought occurred to me that if the pits were dowsable the postholes of a roundhouse might also be dowsable. I had never attempted to find postholes as I thought that after 2000 years nothing would remain. In a matter of minutes I had found some postholes and soon had the outline of a possible Iron Age roundhouse and a small storage or toilet roundhouse pegged out on the lawn. The fine detail of a site can be obtained by using witnesses (Box 8.1). Incredible as it seems, by using witnesses such as animal fats, bronze, potassium, iron, hazel, straw and many others it is possible to 'see' the roundhouse as it was two thousand years or more ago. Carbon was used for the postholes, footpaths and pits, potassium for the fire place, bronze for the spear stand, black iron oxide for where iron has been used, for example wheeled vehicles, the iron spit by the fire, animal fats to determine what was cooked, straw for bedding, dog hair for where the dog slept, cattle dung for picking up the daub in the wattle and daub. Toilets and wash places were also eventually identified with the right witnesses.

An idea of the detail that can be found by dowsing is shown in the survey of Iron Age buildings in Berkhamsted Castle (Figure 4.2). There is a large version of a round-house which because of posts supporting the roof being arranged radially is referred to as a wheelhouse and a smaller roundhouse in the Castle grounds. The ease with which wood postholes can be identified

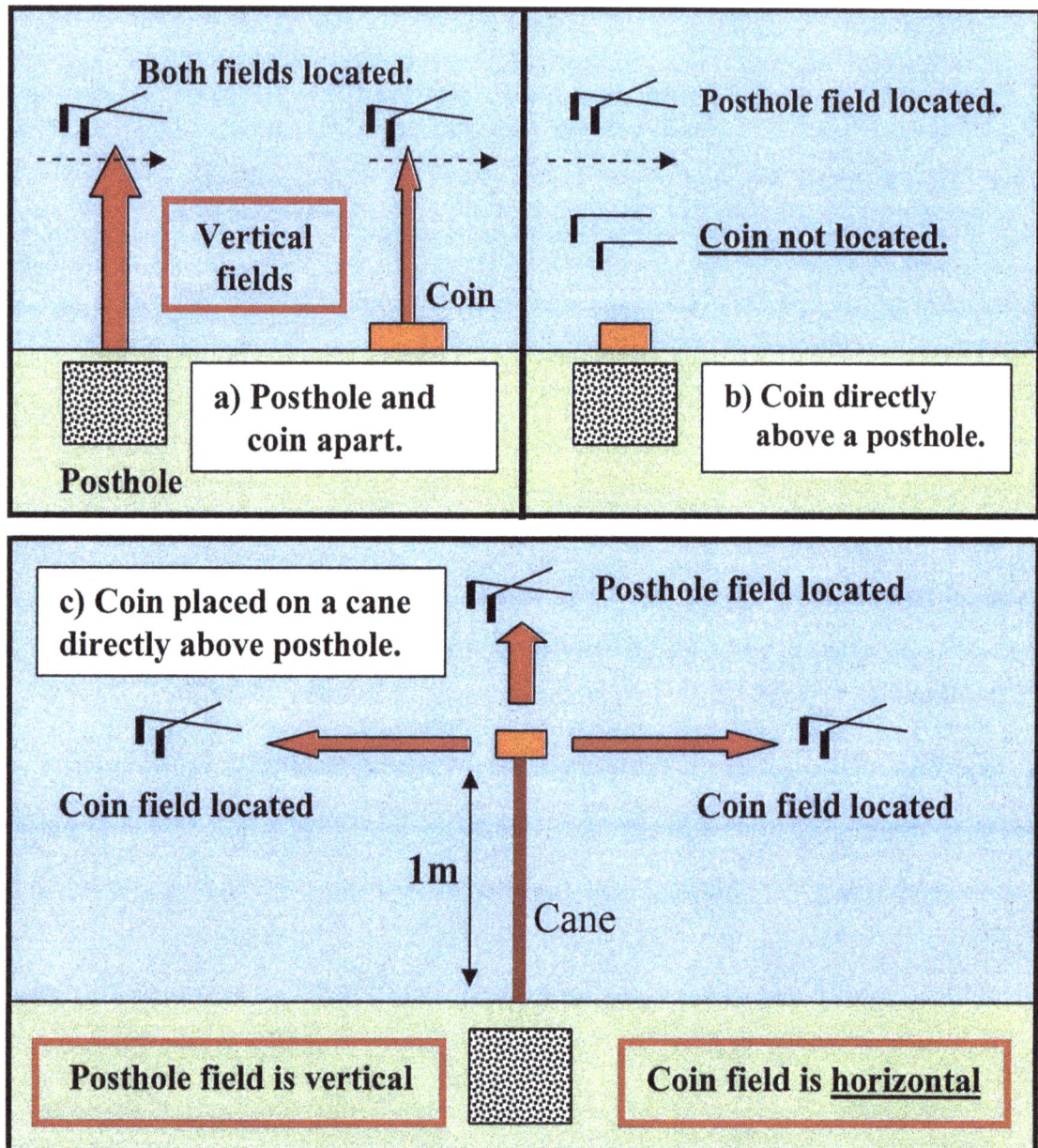

a) Posthole and coin apart.

Both fields located.

Vertical fields

Coin

Posthole

b) Coin directly above a posthole.

Posthole field located.

Coin not located.

c) Coin placed on a cane directly above posthole.

Posthole field located

Coin field located

Coin field located

1m

Cane

Posthole field is vertical

Coin field is horizontal

Figure 4.1 Field Interaction
If a target is placed on the ground
(a) a dowser will normally find it. If the target is placed on a posthole
(b) the field from the posthole forces the target's field to leave horizontally instead of vertically and the dowser fails to find it.
(c) Raise the target by about 1metre and the dowser can again identify it.

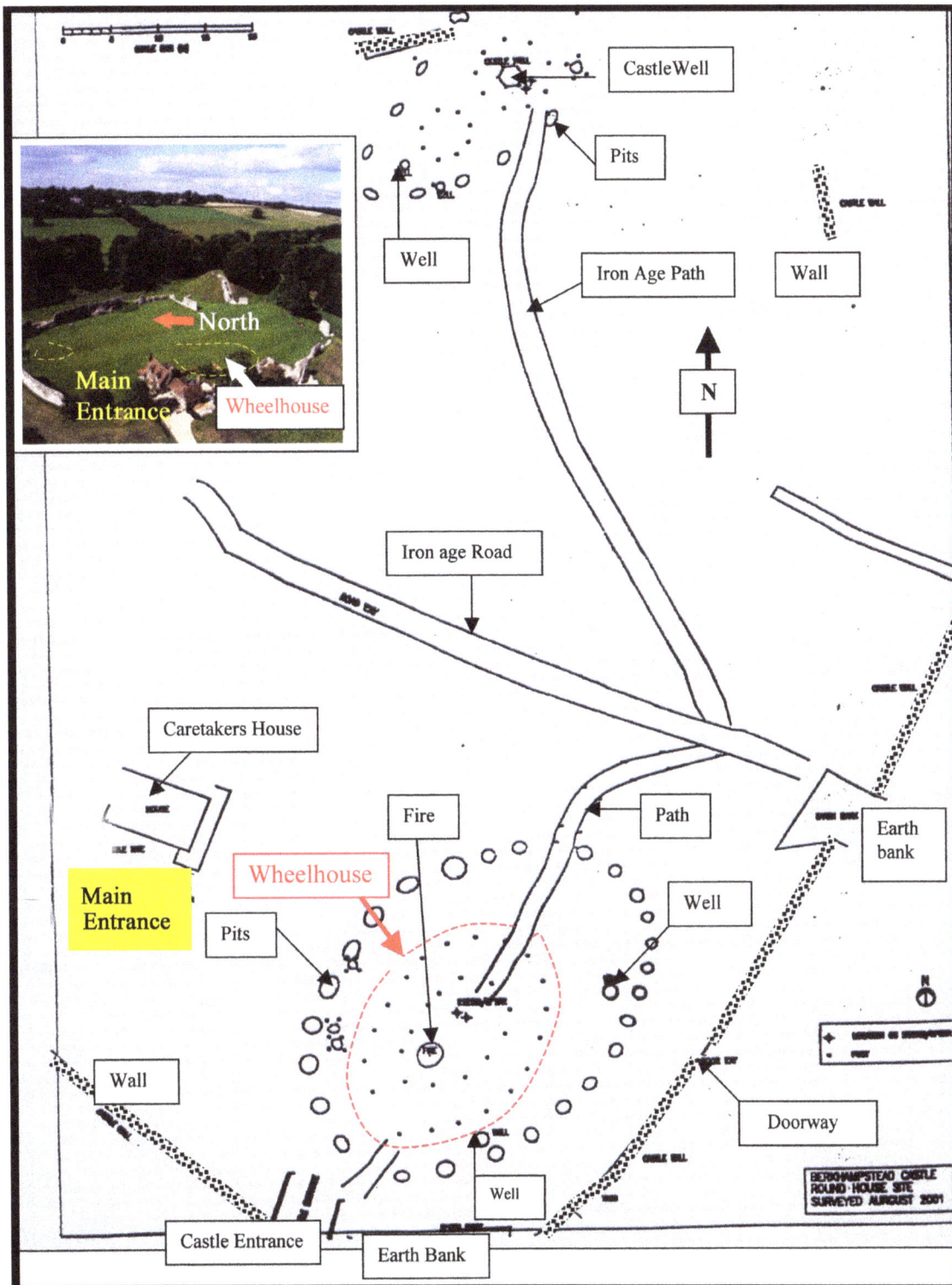

Figure 4.2 Berkhamsted Castle
Built on a Henge Age temple site the outlines of Iron Age Round and Wheelhouses can be found on the Inner and Outer Wards, now flat lawns. There are also washrooms, toilets, wells and roads.

came as a surprise. Soon roundhouses and Iron Age settlements were being found in the village and surrounding fields and wherever I looked for them. This fitted the Roman description of Britain as having settlements everywhere. Then in 2001 I started to find what at first appeared to be mega-wheelhouses. They had many rings of postholes and were of great size. It was not long before I realised that I had discovered something that looked very much like a Woodhenge. Confirmation that I had found henge sites had to wait but if I was right,ceremonial woodhenges were a common feature of the landscape.

Woodhenges

The book compiled by William R. Corliss on 'Ancient Infrastructures' describes 'woodhenges' as a roughly circular concentric arrangement of postholes. The woodhenges appear to be a feature of Britain and are frequently associated with stone circles. Corliss says that the henges, with one exception in the USA, do not have astronomy as their main objective. They are supposed to be the supports for large roofed buildings of some sort. Archaeologists are said to be unsure of the purpose of the henges. However, rituals and ceremonies are believed to be amongst their functions. In 1926 aerial photography identified the first woodhenge in Britain and in 1997 a magnetometer survey identified one at Stanton Drew, also in Britain. Corliss goes onto say that woodhenges may be older than stone circles and are probably much more common than suspected. He adds 'Magnetic Surveys may revise British archaeology in this respect'.

The first woodhenges I found were in the fields round the village of Bovingdon in Hertfordshire, England during 2001. The woodhenge as first dowsed consisted of a central area, which could be identified using black on the colour wheel or a carbon witness. The central area must have had something that left an organic and carbon residue. The area was circular and about 2m in diameter. At the time I thought it might be the remains of a tree trunk. Round this central area were six rings of postholes with the posts appearing to go out on radii. The structure was about 40m to 45m across. As a henge was considered to be a ceremonial site, according to dowsing 'law' or mythology, at the centre should be a blind spring, that is water rising or falling to different levels underground, and two underground streams should also be present, meeting and crossing at the centre. I was not too sure of this but I often found something dowsing as a stream at the centre of a woodhenge and on occasions I picked up two of them. The central black area is supposed to be the blind spring. However, at the time the best explanation for the central black area was that a large post or something made of wood had been at the centre. I never found any signs of a blind spring. The presence of woodhenges which are considered to date from about 2000BC in Iron Age in settlements which are 800BC or less presented a bit of a problem. How did ceremonial sites from the Neolithic or early Bronze Age come to be mixed up with Iron Age buildings? The henges should pre-date the Iron Age houses by a few hundred years if not a thousand years. Some of the henges did have Iron Age roads going across them and houses in them so although it was not possible to say which came first it was conceivable that the henges did pre-date the buildings. As this seemed to be in line with the wisdom of the day, there it rested for a while.

On 10th September 2001 I called in on the Rollrights (Chapter 3) to recheck the magnetic fields of the stones before going on to visit my brother at Evesham. After a few hours work I realised that dealing with 70 odd stones is quite a task and required more time than I had. I had confirmed my earlier 2001 findings and was not going to improve on them. I therefore decided to continue on my way. I had only gone a few miles down the road when I realised that I should have checked for postholes. If one theory about the Rollright Stones was correct, that is they were the walls of a building there should be the postholes characteristic of a wheelhouse. Later that day I broke the return trip home to check for postholes at the Rollrights. They were there. It was not the outline of a wheelhouse that I found but the central 'tree trunk' area and the rings of postholes characteristic of a woodhenge. The central black area and six rings of posts were present. The stones appeared to be between the fifth and sixth rings and I felt sure some stones may well have been placed on postholes i.e. came later than the rings of posts. I also felt certain that I was the first person to find out that the Rollrights were on an ancient woodhenge site as the literature I referred to on the Rollrights made no mention of a woodhenge. Some postholes like the large ones at the eastern entrance lined up with the stones. The 5th and 6th ring post did not appear to line up with the stones. Time did not allow for a detailed look at the relationship between the postholes and stones but they did not appear to be linked.

The following year, May 2002, when visiting Arbor Low Stone Circle in Derbyshire I found evidence that the site contained a woodhenge as well as the stone circle. I found a central black area and rings of postholes. During a visit to Aberdeen in November of that year I found that the Loan Head Stone Circle near Daviot and the Easter Aquorthies Stone Circle near Inverarie were both on woodhenges.

The evidence for stone circles being built on the sites of woodhenges was mounting. This was in keeping with what Corliss had said. It looked as if an early people possibly Bronze Age or Neolithic had identified a ceremonial site and its ceremonial or religious attractiveness had remained for hundreds if not thousands of years – that is if we are to believe the archaeologists datings for these structures. Although the four stone circles visited were all on what could be identical woodhenges there appeared to be significant differences in where the stones were placed on the woodhenge and the design of the stone circle as it appears today. For example Arbor Low is on an outer circumference, with stones now appearing to be well separated. The Rollrights have stones closely packed between circles 5 and 6. The Loan Head and East Aquorthies circles have large stones with gaps between them which stand closer to the centre of the woodhenge. Another point of interest is that there seems to be a marked contrast between the precision and elegance of the woodhenges and the circle of stones that have been placed on them. I think that part of the answer is the weathering and the deliberate destruction of stones over time. When looking at the protected base of some of the stones at Arbor Low and Stanton Drew for example there is evidence which indicates that the stones were dressed. Their magnetic alignment also indicates careful erection.

After my visit to the Rollright Stones and finding that the stones were not only on a woodhenge site but some of the stones were on postholes I knew that something was not adding up. If stone circles were Neolithic then when were the woodhenges built and by whom?

Corliss does say that stone circles may have come after woodhenges. If this were to be the case then stone circles could date from the later Bronze Age. It should be noted that if it had been common practice to put stone circles on woodhenge sites it would indicate a possible cultural or religious link between the people of the woodhenges and those of the stone circles.

Dolmen Discovered

Four stone circles had been visited in 2002 and found to be standing on top of woodhenges. From a dowsing point of view the presence of stones on a henge site made it easy to pick up footpaths going in to or out of the circle. The footpaths of interest are those made by bare feet and leather tanned with urine or tannin. The paths were traced using a carbon witness. Knowing where people walked made it possible to start exploring the ceremonial paths associated with the use of the woodhenges and stone circles. One of the important pathways associated with most woodhenges which had been found during 2002 was one that went out from the central area of the henge in a westerly direction. I had first picked up this path in the Henge Temple site in my garden and then again in temple sites in nearby fields. I had found that the path was straight and it continued outside of the temple to the west, still in a straight line, until it terminated in a structure of some sort. In due course I was to find that the structure was a Dolmen built for excarnation of the dead. The distance to the Dolmen from the temple is variable but normally between 50 and 300 metres. Most Dolmen were of wood construction and used to allow the dead bodies of people to be stripped by birds and or animals before burial of the bones. In some parts of the country the Dolmen were of stone construction and survive to this day. They were typically of two upright stones with a third stone lying across the top of the upright ones. Sometimes there may be three standing stones. There was always a phosphate halo round a Dolmen. This is most likely due to dispersal of parts of the body and possibly indicates that the bodies were eaten by animals or birds. When finding the Dolmen by dowsing the footpaths from the henge, you follow the carbon path out from the henge until it stops. The path actually goes round the wood Dolmen encircling it so the small postholes of the Dolmen and the phosphate halo are easy to find. Along the outward path and return path the points where the priest's staffs hit the ground can be identified as mini postholes. This indicates that they were in line and if right handed the direction they were going on each path. The wood Dolmen found so far appear to be quite small with the body supported at about three to four feet. This is in contrast to most surviving stone Dolmen.

Stone Dolmen can be very large, for example the Spinsters Rocks in Devon. The function of the rocks is easily demonstrated by the phosphate halo round them and the twin foot paths to the east which go straight into the centre of a woodhenge. The Cove at Stanton Drew is another stone Dolmen complete with phosphate halo but no phosphate is detected inside the stones. The stone the bodies were laid on would have protected the ground beneath it and cast a phosphate shadow.

The Dolmen and the path between it and the Henge Temple indicate priests moving, possibly in procession, between the Dolmen and temple. They would presumably carry a body out and bones back in to the temple. A presumption that would have to be proved.

From Woodhenges to Stone Circles

During 2003 a number of stone circles were visited, some for the second or third time but on this occasion I was looking for evidence of a woodhenge under the stone circle. The stones had generated another mystery which I had to solve.

Dealing with each circle in turn and starting with Stanton Drew, one of the largest stone circles in Britain. Stanton Drew is just south of Bristol, and has an abundance of postholes within the circle which can be identified using carbon. There is a central black carbon area about 2m across which is not in the geometric centre of the stones. The woodhenge is therefore not concentric with the stone circle. The visit in June 2003 was a short one so I could only confirm that the stones of the circle were apparently following the posts of a woodhenge. As the stones were not based on the same central point as the woodhenge it looked as if the construction dates of the henge and stone circle differed.

The next circle, Arbor Low in Derbyshire, is another large stone circle looking as ancient as the hills around it. The circle is unusual in that it has two large recumbent stones either side of what is probably the centre of the circle. In the centre between the two large recumbent stones is the circular black carbon area which is easily large enough to take a tree trunk. The postholes of 6 rings were identified with the outer ring lying outside the stone circle, which is between ring 5 and 6. A few miles from Arbor Low is the diminutive and neat Nine Ladies

Stone Circle set in woodland. It is about 9 metres in diameter so is well within the 6th ring of the postholes which are on a diameter of about 40m.

Going north to Castlerigg in the Lake District I found a stone circle that does not appear to be on a woodhenge. There is a woodhenge to the east of the stones so the question arises, have the stones been moved? There are paths going under some of the stones but I did not have time to determine if the paths were associated with the henge. Castlerigg is possibly a place for some detective work. At the moment it appears to be an unusual circle in that it is not on a woodhenge. I could not imagine the ancients ignoring a perfectly good woodhenge site unless there was a considerable difference in time between them and that all the infra structure associated with the woodhenge had gone. The alternative is that the woodhenge is later than the stone one. However, the possibility of it having been moved for farming reasons in relatively recent times also had to be considered.

Going south, the Rempstone Stone Circle in Dorset presents a foreboding atmosphere and is a difficult place in which to work. The trees, undergrowth and what are said to be World War 2 trenches help create the sites atmosphere. There are a few ironstone stones in a sort of circle near the road. Under the circle is a woodhenge. From the henge a double line of postholes go out across the road to the wood the other side where the Dolmen should be. In addition to the woodhenge, which is under the stone circle, there are three other henges close by. The place looks as if it was at one time an important ceremonial site.

By June 2003 I was satisfied that the case for stone circles being on woodhenges and following them in time was settled. In April 2004 I was to find that Killadangan, like Castlerigg did not appear to be on a woodhenge but Glebe Stone Circle was. The relation between woodhenges and stone circles therefore held in Ireland. Most of the stone circles looked at relate back to the one at Killadangan in having the magnetic axes of the stones lined up correctly. This magnetic alignment may create harmony which is appreciated by the visitor. The question arises as to where this cultural trait came from. Do the wood posts of the woodhenges set up magnetic fields and were stones used at a later date to simulate these fields. Another possibility is that stone circles with their magnetic harmony came in before woodhenges and became incorporated into them in some way. At this point in the study the answers to these questions were not clear. It was obvious, however, that from the range of sizes of stone circles and the way their stones were set out, for example compact or spread out, large diameters and small, that circles may cover a long time span. If this is the case it may be possible to trace them back to pre henge times. Another possibility is that different sized circles had different roles or functions.

During 2003 the study of stone circles and woodhenges had moved on at quite a pace. I had found they were elaborate designs and their dating was conflicting with conventional archaeological thinking. The footpaths indicated they were used in a ritualistic way and each Henge Temple was linked to a Dolmen. The temples were widespread, they must have been very important places for the religious practices, ceremony and ritual by the priests of the time.

The question of whether the circle succession was stone to wood or wood to stone or something completely different and whether all stages involved a magnetic sensory link had to wait. The visits to Avebury and Stonehenge had to wait.

The next part of the henge's structure to be discovered was right in the middle of the inner ring. Its exact position would show that the priests were doing considerably more than walking ceremonial ways.

Chapter 5

The Sleeping Prince and Princess

Snowdonia Reveals a Secret

Whilst I was trying to sort out the sequence of woodhenges and stone circles other developments were taking place. At this stage I was subscribing to the standard view that woodhenges and stone circles belonged to different ages and possibly different cultural groups. At the time there seemed to be no reason to challenge the standard view on such matters and suggest that they belonged to the same age and people. To do that would require some good sound evidence.

Early April 2003 found me at the Environmental Studies Centre, Snowdonia National Park. I had organised a water divining weekend for a group of dowsers keen to develop their skills at finding underground water. Water divining is what I call one of the 'hard' areas of dowsing. The diviner knows that somebody may spend a lot of money digging down or drilling to the identified water. The diviner therefore needs to get it right – the location, depth, quantity and quality of the water. Few wells are drilled without the assistance of a water diviner, unless the driller is going down onto a known aquifer, because of the cost of drilling. The accuracy of the dowsers in finding underground streams is, according to some research done in Germany, over 90%, way ahead of that achieved by geophysicists.

The Environmental Studies Centre is in a marvellous valley setting with its gardens overlooking and sloping down towards the river Dwyryd and its flood plane. Looking at the gardens they were clearly on land that had a long history of use. Iron Age roundhouses came to mind and ceremonial centres, possibly a woodhenge. The connection between water divining and ceremonial sites is that dowsing mythology says such sites are on two underground streams and a blind spring, a blind spring being where water rises but does not reach the surface. After my work at Killadangan and with laboratory models I did not go along with the mythology but being of a somewhat lazy disposition, if I could find a ceremonial site I might be wrong and find two underground streams. I could then trace them through the grounds and my homework for the course would be done.

The walk through the gardens soon resulted in the discovery of Iron Age or Bronze Age roundhouses. The double tracks of foot paths and possibly of wheeled transport were also present. A wide drover's way went through the site climbing up towards the high ground behind the Centre. It was not long before my ramble through the gardens had produced plenty of signs that the area had been inhabited during the Bronze or Iron Age or both.

Walking down the slope from the Centre to the flatter ground of the valley I was soon in a wood. When about twenty metres inside I picked up a posthole. The next step was to look for other postholes which would indicate the size and location of the building. It was not long before I realized that I had two rows of postholes going off towards the river. A causeway I thought as successive pairs of postholes were identified. However, I soon hit the two metre diameter 'posthole' at the centre of a woodhenge and knew that if my luck was in, two streams might be under it. The presence of a woodhenge indicated that in the country side round about there must have been a large Neolithic or Bronze Age settlement, traces of which I had already found. The centre of the henge was on the boundary between the wood and an area of lawn. This made it easy to plot out two lines of postholes across the lawn (Figure 5.1). The radius to the 6th ring was about 22m giving a basic structure which was over 40m across.

That is a large structure and as the henge was only one part of a larger complex of supporting structures Neolithic or Bronze Age man was into ceremonial centres in a big way.

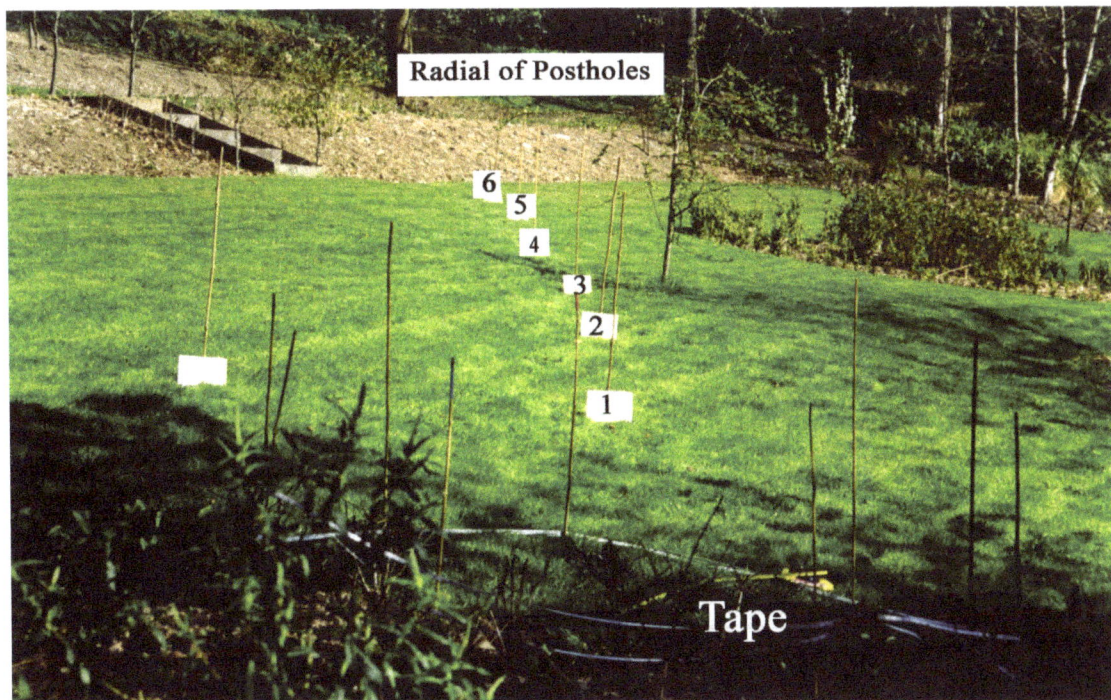

Figure 5.1 A Woodhenge with a Tomb
The central tomb of the Henge Temple found in the grounds of the Environmental Studies Centre in Snowdonia is indicated by the tape. One row of six posts is shown.

After searching the henge site for underground streams in vain I was standing at the centre looking at the lines of canes marking out the postholes to the outer ring. I did not know whether to be pleased because my view on ceremonial sites and streams appeared to be correct or whether to be disappointed as I now had to go and find some underground streams. Either way the central area of the woodhenge had always been an enigma. It was difficult to imagine that the builders always found such a large tree to chop down and stand at the centre of the henge. Another more feasible possibility was that the centre could have been a well or a pit that accumulated organic matter before being filled in. And yet another was that it was a grave. The grave hypothesis was easy to check on as bones contain calcium phosphate. Using a phosphate witness it did not take long to discover that there were in fact two bodies lying under the circular black area. The bodies did not explain the black area but it was a start. The term body is probably incorrect as all that is likely to remain are some bones and these may have broken down and decayed long ago. The dowser is so sensitive that even a chemical stain can be picked up. The 'bodies' may therefore be no more than a chemical stain of the original bodies, or so I thought. I took the magnetic compass bearing along which the bodies were lying and found it to be 340° so not too far off north/south.

Dowsing in addition to its sensitivity to chemicals is also very directionally sensitive. It is therefore possible to plot out on the ground the form of a skeleton, its length and position of head and legs. This was done to estimate the height of the two individuals.

In addition to all my home made witnesses, I have amongst my dowsing tools what is called 'The Element Collection' this is a set of small bottles containing specimens of all the non radioactive elements arranged in the form of the periodic tables (Figure 5.2).

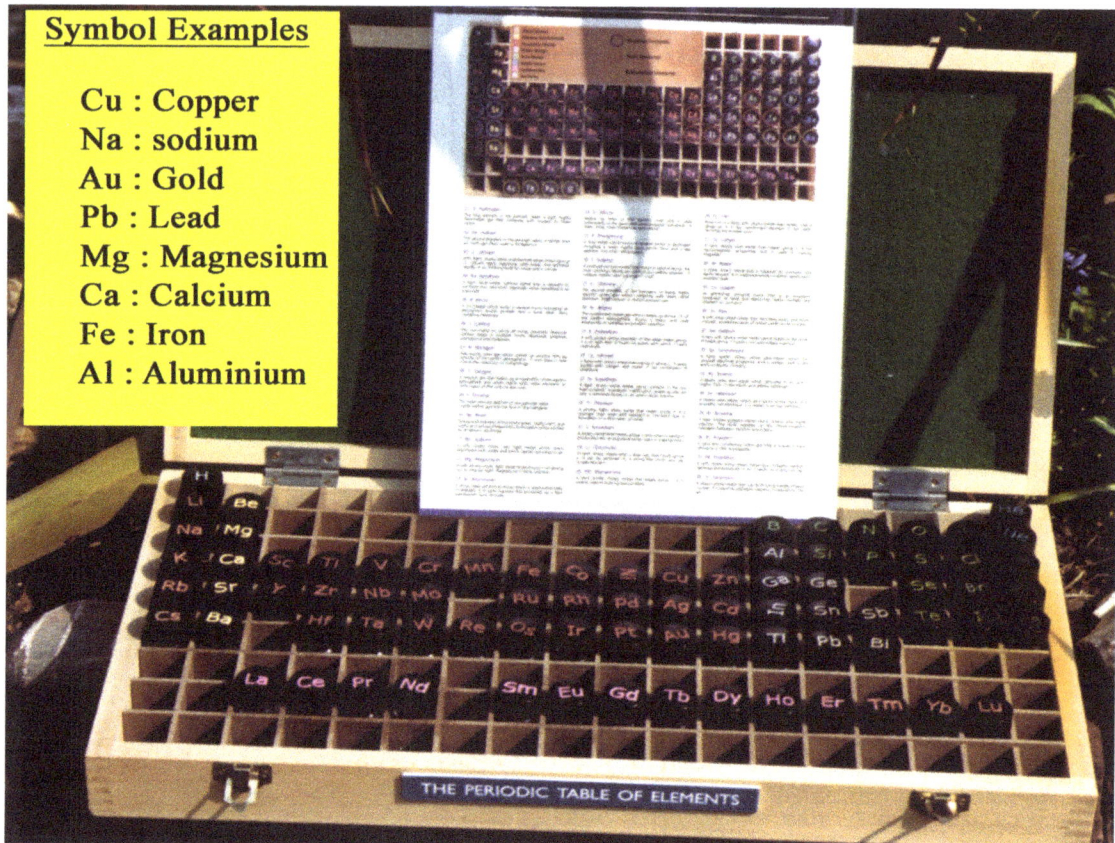

Figure 5.2 A Witness Set
The periodic table of elements is very useful if not essential in determining what is in a tomb.

For research into the scientific basis of dowsing it is a must. On this occasion I made use of its samples of copper, tin, zinc, silver, gold, arsenic and iron. Using the samples as witnesses I found that the shorter of the two bodies had bronze round the head, then gold from the shoulders to the waist. A witness of glass, that is silica, gave a positive response over the chest area indicating jewellery of some sort. Silver also gave a positive response over the chest. The other and longer of the two bodies had bronze at the head, then what looked like a large bronze shield covering the chest area and lower body. The bronze area also gave a positive response on arsenic, a contaminant of early bronzes. The use of a glass witness to indicate jewellery gave a positive response over the chest. Then came the shock. There was what appeared to be an iron sword along the inside of the larger body which had the shield. The grave could therefore be Iron Age and as it clearly appeared to be part of the henge, the henge could be Iron Age. This was going to take a lot of explaining and it was going to be some time before I could. At this stage I appeared to be dealing with a grave but a visit to the site before breakfast the following morning revealed a rectangular walled tomb sealed by what appeared to be three slabs of stone. The gap between the stones could be detected by dowsing across the stone slabs. The tomb was estimated to be 2.6m deep. The discovery of what I thought was an Iron Age tomb at the centre of a woodhenge put me on a high, fame at last I thought. However, life is not that simple. Hundreds if not the odd thousand years could separate the tomb and henge. There were also problems associated with dowsing metals that my main line research had revealed. Copper for example could appear to be gold or silver if the dowser hit the fields at a certain angle. The gold and silver on the Princess might therefore be coming from the bronze shield. On returning home after the course I wrote a report for the Environmental Studies Centre and their archaeologist. In the report I was careful to point out that the gold may not be there. To satisfy my own curiosity however I bought a sheet of copper and tested it at a distance of 2.5m, the dowsed depth of the shield. The gold and silver came up – but not where the Princess was. I therefore felt confident that gold had at least been in the tomb at one time if not now. Gold being a soft metal readily leaves a chemical remanence or memory of its presence. Only a trace of gold, removed by friction from the woman's dress, was required for it to show up thousands of years later.

Although I did not know it at the time, my iron witness would have picked up the iron oxide in red ochre. Red ochre was widely used by ancient peoples and if it was in the tomb it would have registered as iron. In this particular case even if the tomb was Bronze Age and not Iron Age it did not matter I had been made to question the age of woodhenges and that was the important step forward.

Another problem facing archaeological dowsers is that objects placed on the ground leave a chemical stains. Rain and weathering wash this stain into the soil. The Dolmen to the west of woodhenges are identified by the phosphate halo and the carbon in the postholes. The chemicals involved are now deep in the soil. A burial chamber built above ground would in time leave a photographic imprint of the tomb and what was in it. Work on other archaeological projects would in due course show that this does happen and that the stain appears to gradually move deeper and deeper into the ground. These problems had to be dealt with and the first step was to find more woodhenges and check them for the tombs of Princes and Princesses.

Confirmation of the Woodhenge Tomb

As a professional dowser one of my tasks is to find leaks in water mains for people. The dowser is called in after the plumber, water board etc have failed to find it or when there is half a mile of pipe work and nobody wants the job. On this occasion the high use of water on some allotments during a wet summer indicated that there might be a leak somewhere. I started by the boundary cock and meter and had only traced the main for about twenty metres when I came across what looked like an underground pool that is a somewhat circular area two metres in diameter. Ah, I thought, that did not take long and as there was a telegraph pole almost on top of the water main perhaps it was damaged when the pole was erected. There was one problem, pushing a metal probe down into the soil, a sort of dipstick, did not reveal wet mud. A chat with the man in charge of the site indicated that the telegraph pole had been there for far too long for it to be incriminated. I therefore continued the search. When I had finished checking the mains I returned to the suspect leak to try and find out what it was. The use of a carbon witness showed that the round area was carbon and round it were six rings of postholes. A search for the tomb using witnesses revealed a rectangular tomb covered by three slabs of stone and a Prince complete with shield and sword. The Princess had a bronze crown, a gold cape, dress and jewellery. As far as I could see the tomb and contents were identical to the one discovered is Snowdonia.

The next job that came along was to investigate the cause of the winter flooding of a cricket field. I was able to come up with the likely cause which was an ancient well piercing a water proof layer to an underground streams. In winter the pressure of water caused water to rise up the now permeable well onto the field. It was while sorting this out that I came across another woodhenge. Again there was a central tomb with a Prince and Princess in full regalia.

One of my tasks during 2003 was organising meetings for a water divining group. One of the meeting places was Ryton Organic Gardens near Coventry. The shop, lovely expansive gardens and restaurant make Ryton a very nice place to practice dowsing and look for underground streams. It was during one of the meetings that I found two woodhenges in the gardens. Both had central tombs complete with occupants correctly dressed.

By now I was convinced that the woodhenges and the tombs were part of the same ceremonial site. Closer study of the dowsing signals from the central tomb indicated that real gold and real iron may be present and that I was not dealing with a surface stain that had moved down into the soil. The depth of about twenty feet also indicated that if I was dealing with chemical stains they had originated from objects placed at depth in a tomb.

Children's Graves?

The next step could have been due to my guardian angel, (A dowsing group member has informed me that I do have one called Galano). It was during the summer of 2003 when I was discussing the discovery of the tombs with visitors to the house. We were sitting in the garden when my wife reminded me that visiting dowsers were always finding graves in the garden. In fact they were always finding the same ones. One was under the cherry tree and one in the lawn not so far from the cherry tree. Why did I not check them to see if they were tombs? The one in the lawn I had checked and it was a rectangle about two foot square. The phosphate and calcium were present but the rectangle also came up when I used my iron witness. It looked to me as if some pet had been buried in a metal box. The rectangular shape that the iron witness gave was well marked and angular. The grave under the cherry tree I had not looked at in detail so just in case I went down to check it out. It was not long before I had found two bodies lying north south, the shield, sword, gold, bronze headdress and walled tomb. I had my own henge tomb and temple in the back garden. A little while later I had found enough postholes to identify the six rings. Having at least part of a woodhenge in the garden was a gift. I could now look at it in detail, try out ideas, look at it when drinking my morning cup of coffee in the garden and ask Galano for a follow up. It was not long in coming. Suppose my interpretation of the tin box grave was wrong. I could not obtain a response when using a tin witness so perhaps a visit to the allotment and cricket field was indicated. The two henges, being in open space, are easy to access so it did not take me long to find that each tomb had a small grave at each of the cardinal points. Because of their size the idea that they could be animal sacrifices or children's graves came to mind and I referred to them as children's graves. There was still the problem of why they should come up on an iron witness. Who ever had built the woodhenges it was unlikely that they had iron boxes available to place children in. Visits to the henges that I was now finding under stone circles showed that graves at the four cardinal points were part of the design of henges. There must therefore be a reason for this. Sacrifices to the spirits of the four realms perhaps? Whatever the answer, it was another problem that had to be solved. But there for a while things rested. At the back of my mind was the problem of the bronze shield and the gold the Princess was wearing. I was finding more and more woodhenges. They were under churches and cathedrals, in the fields round about I was able to spot them visually by the depression in the ground. Whenever I checked the tomb I got the same picture. In some the gold response would be weak in others strong. I then realised that I knew of a gold cape that had been found in Wales but had no details. Even if it only weighed a few hundred grams there was going to be a lot of gold buried in the British country side if it was the sort of thing worn by the Princess. It was while musing on the problem of the gold buried in the tombs that I realised that if I could find the gold other dowsers could also find it and they had had plenty of time to find it. By this time I knew that the tombs were deep, twenty feet or more. Motor ways and airport runways could be constructed across them and nobody would know that they were there. If somebody were to dig down to the gold they might well leave a great big hole.

56

The Gold Cape

Sometime later I was watching a TV programme on archaeology and a gold 'cape' was shown and discussed (The Mold gold cape). It would have fitted over the shoulders of a small person and appeared to be designed to link to a fabric dress that would have hung down from it. The archaeological commentator explained that the cape would not have been easy to wear as the wearer could not move their arms in it. Dead people do not move their arms I thought. It fitted to a tee the gold garment the Princess might be wearing. A letter was quickly dispatched to The British Museum and the Curator of the Bronze Age collection sent a nice letter back saying that the original cape would have weighed 750g and could date from 1900 to 1600BC.

Figure 5.3 A Dell
The dell shown is a deep excavation with trees and bushes growing in and around it.

Now in the fields of Hertfordshire and many parts of England there are lots of big holes in the ground (Figure 5.3). They are referred to as 'Dells' or 'Borrow Pits' and are said to have been dug for clay, chalk or flints. Their names and the reasons for them having been dug probably vary over the country. It was while sitting in the garden, musing over a cup of coffee that I realised that there was a line of these dells marked by trees and scrub over a distance of about a mile in the fields near by. So if gold digging dowsers had been active over the past few thousand years the 'Dells' should be on woodhenges. Equipped with rods and witnesses I was soon on my way to the Dells. I started at one end and a few hours later had found two undug henges and six that had been dug down on. To determine if a Dell was on a temple site I looked for a circle of postholes. If I found them I assumed that the Dell was likely to be on a tomb. A few days later I had found another four that had been dug down on and another one that had not. Some of the excavations were enormous and they did not look as if the diggers had been aiming directly at the gold. However, it looked as if I may well have been beaten to the gold. The vision of 750g gold capes started to fade. The pits or Dells were in the main difficult to penetrate as they were overgrown with bushes and trees. The sides were often steep and they were used as a depositary for fallen tree trunks, branches and other rubbish. Although it was possible to accept that dowsers had identified the tombs and possibly dug down for the gold, the size of the excavations, their shape and the absence of a spoil heap did not fit. If I was to dig down I would have dug a shaft. Some of the pits were so big they had a sloped path or road into them. Something did not add up. And then there were the henges that had not been excavated. The gold was still there so why had people not dug down to it. In fact by the end of August 2003 I seemed to have as many unanswered questions as ever. Why such big excavations, why the range in size of pits, why not a shaft down to the tomb, why no spoil heap, why so many henges with a depression above them? One possible answer that could explain the shallow depressions and small pits was that the tomb was in a chamber that had collapsed. Swallow holes, a local name given to a subsidence when the ground suddenly collapses into an underground cavity, are well known. Perhaps the tomb was in a large cavity that collapsed and caused a depression in the ground above.

If there was an underground temple a 'burial chamber' it should be possible to find it by dowsing and the tomb in the garden was the place to start. I started by looking at the children's grave and trying to pick up a link, a tunnel, to the tomb. There was a tunnel or something heading for the tomb but then it went into what appeared to be a chamber. It took a while to work out that it was a chamber because the dowsable wall that the tunnel joined disappeared and it took a while to discover that the chamber was not rectangular in shape but was triangular. The top of the triangle, which was pointing at the tomb, joined another tunnel linking it to the tomb chamber. I then checked this finding out on other ceremonial sites with plenty of room round them. I found that there were four triangular chambers round the tomb which led from the tomb to the children's graves (Figure 5.4).

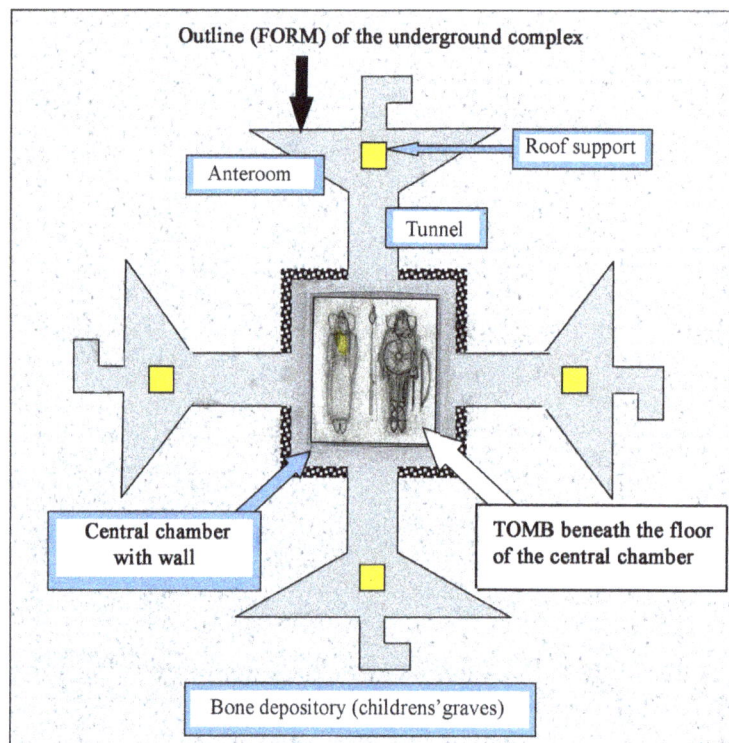

Figure 5.4 Children's Graves
The children's graves were found to be connected to a tomb chamber by tunnels and an anteroom.

If the dowsing was giving a true picture there was a large excavated space underground which if it collapsed could produce quite a depression in the ground. The builders would have been aware of the risk of collapse and would have taken steps to prevent it. A search soon found what looked like a square column of the type I had often come across when dowsing mine workings. It was small, it was set to one side but it could be there to support the roof of the triangular chamber. The tunnels were narrow and did not require support. A number of local woodhenges were dowsed and all had the underground 'Temple'. Visits to some stone circles confirmed that their woodhenges were complete with underground temples, Iron Age underground temples! Before jumping to the conclusion that this means the henges were dated 800BC or less it would be as well to consider the possibility that the Iron Age started much earlier in the British Isles. A visit to 'Scorhill', a stone circle on Dartmoor in November 2003 showed that it was on a woodhenge. Woodhenges were therefore being built before the climate change forced people off the moors if indeed climate change did force them off.

However, before dealing with the interface between the Bronze and Iron Ages there is still a problem. We have to account for all the gold and bronze that might be buried in the ground. We now have a temple that is quarried out of the ground in a particular shape common to Britain and as we shall see, to Ireland. There must be a reason for the shape, and its possible similarity, as shown in figure 5.5, to the Irish Shamrock and 'Celtic' Cross. Under the ground is a complex of rooms and passages which represents such a large investment that it is difficult to imagine that it was built and then sealed up. The temple looks like a complex that was used and if it was used the bronze, gold and jewels could have been recycled. The recycling could explain why all the henge sites and tombs have well dressed occupants but not why there are so many of them. There was also one other problem. Where was the entrance? If it was at the centre it would be difficult to 'see' because of the black area, the blind spring. But there could be a short cut to finding if there was one. The first woodhenge found in Britain is 'Woodhenge' which is only a few miles from Stonehenge. It was spotted from the air using aerial photography in 1925 and shortly after was subjected to an archaeological dig in 1926-7 by Maud Cunnington who is said to have dug to the bottom of every one of the 156 postholes at Woodhenge.

Some of the postholes have been replaced with concrete posts. Importantly, Maud dug out the central area possibly down to the bedrock. Down in Wiltshire there was therefore a woodhenge that had been dug over by archaeologists and its date was said to be about 2500BC. Were they right? If they were, there should be no Iron Age underground temple. The henge would still indicate a cultural link between the people of 2500BC and the Iron Age. If they were wrong, and there was an underground temple, then I could assume that the entrance was not at the centre of the henge. If it was, Maud Cunnington would have discovered it. All I had to do then was explore round the henge for a tunnel and then track it to its entrance. The presence of an underground temple might also indicate that Woodhenge was not 4500 years old.

A Visit to Woodhenge

On 3rd December 2003 I drove down to Woodhenge with my wife. It is quite an easy place to find and there is a small car park just off the road. The henge site is on elevated ground and even in early December the views were good and one could appreciate why the site had been selected for ceremonial purposes (Figure 5.5). The dowsing equipment required for the task was quite simple. A set of rods, a carbon witness for the central area and the postholes, a phosphate witness for the skeletons, bronze, iron, glass, gold and silver for the tomb contents, a piece of flint and limestone for the tomb wall.

The first task on reaching the henge was to have a walk round and get my bearings. Posthole markers whether cane or concrete always look confusing to start with. When the symmetries are appreciated the site takes on a form. To the west, about three hundred yards away, was the small Cuckoo Stone.

Figure 5.5 Woodhenge
This is the famous Woodhenge where the puzzle of access to the tomb temple was solved.

Next I had to confirm that the site had the typical wood postholes of a henge. When doing this I soon found some postholes unmarked by concrete posts. I then confirmed that the site had the typical six rings of postholes and looked like a woodhenge. The central black area, the 'blind spring', was missing. Maud had done a good job of cleaning out the pit or whatever was there. There was no trace of carbon left, unlike the postholes, unless the ones I had found were ones she had missed. The next step was to look for the tomb. It did not take me long to find the rectangular tomb and confirm that it contained two bodies complete with regalia. As often happens when dowsing passers by stop and enquire as to what you are doing. On this occasion the person responsible for the site saw me at work and called in to see what I was up to. I did not have the heart to tell him that the site was not Neolithic or early Bronze Age but an Iron Age ceremonial site. So instead we had a polite discussion about the ancients moving from woodhenges to stone ones when the wood ran out. Fortunately I had not pegged out what I had found. A tomb and Prince and Princes laid out on the ground may have taken some explaining. After the gentleman had departed I continued with exploring the rest of the temple. There it was, a complete 'Celtic' Cross with passages to the children's tombs as I still called them at this time. After establishing that Woodhenge was, as I thought at the time, a typical Iron Age ceremonial site I started the search for the tunnel. I now knew the entrance was not at the centre so I did a search by going round the site outside the area of the temple. Going round in an anticlockwise direction I had covered about three quarters of the circle before I hit the tunnel. It was coming in from the North Northwest straight to the centre of the temple complex. Following the tunnel out it finished at a square shaft outside the six rings of postholes. I now knew that the temple was almost certainly accessed and used for ceremonial purposes for a period of time. The valuable materials in it could be recycled and all that might be left is the chemical remanence of what had been there. No gold, no bronze shield or sword. I could put the JCB catalogue away.

The final check to confirm that the site was the same as all the others I had found was to find the Dolmen. The archaeological digging had not destroyed the paths running through the henge and out to the west. I picked them up and then followed them offsite and into the field. On and on they went until at last they arrived at and finished at the Cuckoo Stone. Round the Cuckoo Stone was the phosphate halo and next to it the imprint of the other stones that had been there. The Cuckoo Stone is the one remaining stone of a stone Dolmen. The missing stones could easily be near by performing some useful function in a wall, gate or house. Now that I knew that the temples could be accessed the role of the Dolmen started to make sense. The dead were left to decay or have their flesh removed by birds, until only bones were left, excarnation. These were collected using due ceremony by priests with staffs. The staffs were placed in the ground at precise points so that an imprint was formed which can still be detected more than 2000 years later. It looked as if the procession of priests went out on one path and returned on the other. Once back in the henge the bones could then be taken down to the underground temple. The children's graves were not children's graves after all. They were depositories for the bones of those who died. It was the bones that went through the funeral rights. The mystery of the missing Iron Age graves might have been solved.

The Prince and Princess were emissaries to the 'Celtic' Goddess who lived in the ground. The 'Celtic' heaven and afterlife was in the ground. The Druids were the priests of the goddess and it would have made sense to them to build their temples in the ground. The woodhenges, possibly referred to as 'groves' by the Romans, marked the entrances to the underground temples and were a focal point for ceremony. But why a central area which was not the physical entrance, why six rings, why a 'Celtic' Cross design for the underground temples.

Review of 2003

As 2003 came to an end one set of mysteries had been solved only to be replaced by another set. It looked as if the woodhenges may indeed predate many of the stone circles but do they predate the mega circles such as Stonehenge, Avebury, Stanton Drew. The relationship between the stone circles and the circles of postholes needed to be investigated further as the stone circles take on so many forms. The way the bones travelled from the Dolmen to the bone depositories had to be unravelled. The central black area is an enigma, what was it for and what would it have looked like. Another problem is why so many woodhenges have big pits and excavations on them. If somebody wanted the valuables they would have gone down the access shaft. In fact all the gold may have been removed from the temples by the Druids when the Romans fell out with them. They could then have taken it across to Anglesey. The presence of a Druidic hoard of gold would make sense of Paullinus's expedition to Anglesey. The Romans could easily have left the Druids sealed off in an island prison if it were not for the fact that they had something the Romans wanted, particularly Paullinus the Roman General. If there was a Druidic hoard in Anglesey did the Romans find it. The answer probably lies in an analysis of Roman gold coinage of the time. A sudden influx of gold from deposits in Wales and Ireland should be detectable by the trace metal analysis of Roman gold coins and jewellery. If there is no evidence of a sudden influx of 'Celtic' gold into Roman coins after their expedition to Anglesey does it mean it is still there? So much for wild conjecture. At the end of 2003 I appeared to have found a landscape covered with woodhenges each with a temple under it. The shear scale of the operation was pointing to a society that was rich and devoted a considerable proportion of its wealth to constructing ceremonial sites or they were masters of civil engineering or perhaps both. I had developed science based dowsing techniques to the stage where much detail could be determined. This detail applied to structures that were present, the chemical stains left by structures long gone and to the paths that people had trod and wheeled vehicles had used. The dowsing methods, if properly applied, were proving to be very powerful and it was clear that when a civilization and a culture did not leave a written history or a story in stone and brick the only way of finding out what went on is by using biolocation and chemical archaeology.

Using biolocation or science based dowsing I was able to 'see' the record in the ground, I was not dependent on shards of pottery or other artefacts.

Working on my own on both the theory and in the field was time consuming, nevertheless considerable advances had been made in the period 2000 to 2003. I had some amazing dowsing finds to substantiate and needed physical evidence both to prove dowsing theory and that henges were a physical reality. Furthermore I was the only person using the Diamagnetic and Paramagnetic Theory of Biolocation and its reality also had to be proved. The pace had been quite hectic. I was introducing my theory and findings by way of talks and lectures and this was to lead to a number of developments that would increase the pace of research tenfold.

Chapter 6

An Extra Pair of Hands

A Slow Start to 2004

It was an interesting end to the year 2003. I reflected on the many dowsing discoveries and the emerging patterns. I had substantial scientific evidence to demonstrate that stone circles were constructed on top of woodhenges, or possibly were part of them. I was aware that this connection had major archaeological and historical implications. Conventional archaeological thinking dates stone circles as Stone Age (Neolithic 3000 to 2000BC) and that they preceded the woodhenges. My evidence contradicted this. Stone circles stand on top and cover postholes in the henges so they must have been constructed later. Woodhenges are dated as early Bronze Age so I dated the stone circles as no earlier than the Bronze Age. More evidence and research would now be necessary. Furthermore it was becoming increasingly apparent that I was unravelling and seeing the physical basis of the Druidic religion. The Druids are well known in the historical written record but there is little detail, only vague references to their religious practices. My research was indicating that the Druid Temples were constructed consistently to the same ground plan with an elaborate design above and below ground. The indication was that by the Iron Age the design was being changed or added to with the construction of the energy engineered stone circles. To construct them obviously required considerable time, effort and planning so they must have been of immense importance in Iron Age life. So in the final week of 2003 the great stone circles of the UK started to beckon and in particular Stonehenge!

The usual Stonehenge tourist route is confined to paths through the henge. These paths are not suitable for dowsing due to the presence of all the visitors. However, there are occasions during the year when limited numbers of people are given access to the site and the freedom to roam round it. The next such occasion was going to be the winter solstice of 2003 so very early on the 22nd December I set out for Stonehenge. The sky was clear and promised a dramatic sunrise. The air was cold with a sufficient breeze to make the conditions very cold. The intention was to get onto the site long before sunrise and before anybody else would want to be there. In the event hundreds of people had the same idea so that it was impossible to do much dowsing and all I could do was a good imitation of a penguin on a cold Antarctic night and jostle myself to the centre of the flock. In the final stages of hypothermia and now on the outside of the flock I watched the sunrise just as the ancient scholars and priests must have done on many occasions. After seeing the sun clear the horizon and with only a few results I left the henge to the Druids, pagans, stone huggers and meditators, knowing that I had to return when the weather had warmed up. Frustratingly my return would have to wait as early in the New Year I had to fly to Penang, Malaysia, where I was to enjoy a sun that was already high in the sky. During this enforced absence from woodhenge research I started to try and find where burrowing shore crabs were hiding in their tunnels by dowsing. The results indicated that the burrowing crabs were better at dowsing where I was than I was at dowsing where they were and they made sure we never met up however many tons of sand I moved. The bane of the dowser is finding something which appears real whether it is a shore crab or an archaeological artefact and then on digging down finding nothing, not even a chemical stain. The dowsed target was in fact an image or phantom not the real thing. Other work involved studying the images of targets and trying to find ways of identifying them and how to use them to locate the true target. Penang therefore provided me with sun, warmth and an opportunity to do some basic research on images or phantoms as they are sometimes called.

An Abundance of Woodhenge Sites

During the last three months of 2003 I had set up a small informal group of dowsers in the St. Albans area. The approach of the Group to dowsing was analytical and scientific. The Group had held some successful meetings with demonstrations of the Diamagnetic and Paramagnetic Theory. The setting up of the Group followed on some dowsing courses I had given locally. Several people had asked if there was a dowsing group that would enable them to develop their skills and knowledge. There was no group in the St. Albans area so I found a few people who were interested in forming such a group and had it up and running by the end of 2003. During my absence in Penang the Group continued with meetings and it was not long before one member of the Group in particular took up the Dia and Paramagnetic Theory to develop his skills. This member, Nigel Hughes, emerged as a keen and very active dowser. He had four dogs to exercise so was out on the hills and fields round St. Albans twice a day. As the dogs did their sniffing Nigel Hughes set out to dowse the recreation grounds, parks and woodlands round St. Albans foot by foot. The dogs seeing a human in dowsing search mode, they would call it sniff mode, showed great enthusiasm for dowsing and kept a close eye on Nigel in case he found anything of interest to them and so did not wonder too far away. When I returned to the UK Nigel had taken on board the science based dowsing methods and had a wealth of discoveries and many questions for me. As I answered Nigel's questions and checked his discoveries the study of woodhenges and the Druids not only became a joint operation but a high speed one. The number of woodhenges discovered round St. Albans in the first few months of 2004 was large. Further exploration further west showed that there were at least three henges on the village green at Sarrat. Ponds were often found to be associated with the depression in the centre of a henge site and so ponds had to be checked. Most of the local ponds were on henges. As we became more and more familiar with the physical signs of henges we could spot them in fields and gardens. Some we confirmed by dowsing just to make sure we were right.

It had been recognised in 2003 that woodhenge building had been on a large scale. One concern I had on finding all these henges was did the scale of construction fit into the time scale of their existence.

On the face of it, if they were Iron Age as we thought at the time, it seemed they spanned possibly a thousand years so for a while I was satisfied with this. However, as more and more henges were found this became an issue. How could so many henges have been built in such a period of time? Our evidence was now presenting a considerable thinking challenge. It was beginning to look as if woodhenges and the underground temples were built for a reason and that reason came up frequently resulting in a massive building operation. What that reason was joined the long list of the Druids secrets waiting to be discovered.

In mid April 2004 on a visit to St. Albans I spotted the physical signs of a Henge Temple, a characteristic depression, in a very convenient position. The depression was in the middle of a recreation field very near to Nigel's house and his daily dog walk route. We had by good fortune a woodhenge almost in the middle of a field right on Nigel's doorstep. It was easy to get at, access was assured and it could be worked on even if Nigel only had a brief period available. Nigel marked out the central black area, the tomb under it and the underground temple. The paths, the Dolmen and the access tunnel were found and the six rings of postholes. With such a conveniently sited henge it was not long before more detail was discovered. This included the square shaft down to the tombs access tunnel. It had a number of postholes round it indicating that there had been a building or chapel protecting it. This little complex lay outside the sixth ring. We gathered a lot more detail to add to the design of the woodhenge. I was able to demonstrate to Nigel the use of form to enable him to research the henge himself.

Meanwhile the question arose as to whether Ireland was also home to the same design of woodhenge as Hertfordshire and other parts of Britain.

A Quick Visit to Ireland in April 2004

When I last visited Ireland the association between woodhenges and stone circles had not been discovered. The energy engineering link between the design of the stone circles in Britain and Ireland had been demonstrated but did the similarity between Irish and British stone circles go beyond that? Were Irish stone circles on woodhenges? What other similarities might there be? Conventional views considered Ireland to be Druidic so it was possible that there was a common culture over the British Isles. A quick visit was made to Ireland in April 2004. At Killadangan, although there were plenty of roundhouses and storage pits in the vicinity of the circle no sign of a woodhenge could be found round the stone circle itself. This does not mean that there was no woodhenge only that at the time of the survey one could not be found. The other site visited was Glebe Stone Circle. This circle is inland and on a slightly elevated site. It did not take long to find the rings of postholes, the central black area, the tomb with its two occupants, underground temple, access tunnel and shaft, Dolmen and paths. The second larger nearby circle was again on a woodhenge. The similarity of the Irish woodhenge to those found in England indicated a very strong religious and cultural link in the Iron Age if not before. At the moment it is not possible to say which way the cultural elements flowed but it could have been either way. If the Druidic culture is the same in Ireland as in Britain it would indicate early and good communication links between Ireland and Britain. Unfortunately time did not allow for a woodhenge search or visits to more stone circles. There was an important visit to Stonehenge coming up and I had to be back for it.

A Visit to Woodhenge and Stonehenge

English Heritage allow a limited number of people into the Stonehenge site for one or two hours before and after the public have admittance. I had managed to get permission for my local Dowsing Group to spend an hour amongst the stones after closing time on the 1st of May 2004. However, the rest of the day was not to be wasted and it started with a visit to the Woodhenge site which is about seven miles from Stonehenge. This would be my second visit but the first visit for most Group members. My previous visit had shown Woodhenge to be on an underground temple which had an access shaft and tunnel. At this stage my interpretation of Woodhenge was that it had been a single building standing on a single temple. The possible significance of having found postholes that the archaeologists missed had not registered. This time there were plenty of dowsers to confirm or deny previous findings and to explore other aspects of the site. Soon after arriving at the site it was confirmed that the entrance to the underground temple was in a building outside the sixth ring, the Cuckoo Stone across the field and to the west was found to mark only one of a number of Dolmen. This indicated the presence of more than one temple. The one set of footpaths to the Dolmen that was checked had staff marks on the outside of the paths. The staff marks are the stains identified by witnesses of carbon and ash wood. It is where the staff was placed on the ground when walking in procession. The fact that there are distinct marks and not a line from the repeated placing of staffs at random on the ground indicates that the procession was a precision affair. This fitted the findings on other sites where the staff marks always appeared along side the paths to and from the Dolmen and in some cases there had been indications that the staffs may have had an iron tip. When the staff marks had first been found there was the possibility that they were small postholes and indicated the presence of a fence. Searches for signs of fences or walls on other henges had failed to reveal any and the staff marks could always be followed inside the posts of the henge. It therefore looked as if we had good evidence that the priests used a staff as part of their ceremonial regalia.

After much walking and dowsing several woodhenges were found round the 'Woodhenge' indicating that the site had been used for a period of time. More underground access tunnels were found and the original one to Woodhenge itself appeared to have a side branch running from it. The impression of an underground temple complex with interconnecting tunnels at Woodhenge soon took shape. One side of the vertical shaft, nearest the henge, gave a response on carbon indicating that wood may have been along the edge at one time. The access tunnel dowsed as if it had been backfilled from the shaft to near where the tunnel branched. The backfill was identified by walking along the line of the tunnel with a chalk witness and picking up the line of the backfill in the tunnel. The backfill indicated that the access tunnel had been deliberately sealed. When checking the concrete postholes of the 'Woodhenge' many more postholes were found than were indicated by the concrete posts and they were on radii as would be expected. This was puzzling. There appeared to be a difference between the archaeologist's postholes and the dowser's postholes. The most obvious explanation is that there are at least two sets of postholes. It was going to be a while before we found the answer to this conundrum on another site. The visit to Woodhenge confirmed earlier findings and indicated that the archaeological 'Woodhenge' was only part of a much larger sacred or ceremonial site. The scale and complexity of the site raised the old doubts about what was being dowsed. Was it truly underground temples on a massive scale or was it the stains from some surface structure long gone. The dowsed depth of about 20 feet indicated that whatever it was, was well below the depth of normal archaeological digs. The number of the temples at Woodhenge and in other parts of the country was worrying. Why build more temples when you already have one, particularly when the new one is next door to the old one? Why dig another underground temple when there are already a number of underground temples possibly connected by tunnels? And again there was the question, where was all the spoil? The dowsing appeared to be convincing and it was being checked in a number of ways and by dowsers finding the patterns without knowing what was there.

By lunch time it was time to say farewell to Woodhenge and drive over to Stonehenge. We left Woodhenge with so much new information and so many questions. Now to Stonehenge for an hour with possibly the most famous stones in the world. After parking, the first stage of the study was to explore outside the fence whilst waiting for 6.30pm. The Processional Way was easy to dowse and it seemed to be straight forward. Going across the Processional Way from one side to the other there were first two footpaths which were identified with a carbon witness, then the tracks of a wheeled vehicle with a seven foot axle and drawn by two horses. A wheeled vehicle leaves a trace of its tyres in the ground, for example wood, bronze, iron and nowadays rubber. The trace of material enables the tyres and the axle length to be identified. The draft animals leave traces of their hooves and droppings so they can be identified. Then a second set of chariot tracks and then on the other side of the chariot tracks were two more footpaths. Alongside the footpaths were the carbon dabs of the staffs. About a hundred yards in a northerly direction away from the part of the Processional Way we had been looking at, a wide carriageway was found. The axle length of the vehicle that made the tracks was much greater than seven feet and it had at least four wheels if not six. It was the first indication we had of vehicles that were to become known as HGVs (Henge Giant Vehicles). They were clearly designed for very heavy loads but did they have wood, bronze or iron tyres? That would have to wait as 6.30pm was fast approaching.

Dead on 6.30pm a group of dowsers moved through the tunnel and into the henge enclosure for an hour with the great stones. My first task was to look at the magnetic fields of the upright stones. I found a south pole at the bottom of the stones but the tops were too high to measure but presumably would be north poles. The stones were also erected so as to have a side that was a north pole facing a side that was a south pole. The outer face of the stone was a south pole.

A limited number of measurements indicated that the stones of Stonehenge may have been laid to embrace energy engineering principles but being limited to one hour things were done in a rush and many more stones need to be studied to be sure on this point. In the meantime Nigel had been looking for signs of a woodhenge and had found the telltale postholes. Bone depositories were found and it looked as if some of the stones were on postholes indicating that they came after the woodhenge or at least one of them if there was more than one. The tunnel to the underground temple and the access shaft were found, also a tomb at the centre of the henge with its two bodies.

Based on what we thought we knew of tombs at this time it made it look as if Stonehenge was Iron Age. This would take some explaining but before jumping to the conclusion that Stonehenge is an Iron Age structure it might be as well to consider some alternatives. There is the possibility that the woodhenge is Neolithic and that the stone structure is early Bronze Age and was built on the site of a woodhenge. Then in the Iron Age underground temples were built under the stones. Because the stones were already there, no Iron Age woodhenge was built. Whatever the true explanation turns out to be on the 1st of May 2004 a group of dowsers found that Stonehenge is on what appears to be an Iron Age underground temple – or the stain of a surface temple. This note of caution is sounded because the results were obtained during a short and hurried one hour visit. The results did however indicate that another visit to Stonehenge was required. We now knew more clearly what to look for.

A Saxon Church Points to a Druidic Temple

Discoveries happen as often by chance as by design and chance played a role in the next discovery. It was during a dowsing course I was giving in St. Albans that a member of the class, who lived locally, mentioned that there were the remains of a Saxon church next to St. Albans Cathedral. As the dowsing practical was going to be round the Cathedral (Figure 6.1) I thought it would be a good idea to see if the church was on what dowsers call a sacred site. When trying to identify a sacred site the conventional dowser looks for a 'blind spring' and 'underground streams'. I look for a woodhenge and an underground temple complex. As the Saxon church was going to predate the Cathedral possibly by a few hundred years there was a good chance that it was on a pagan sacred site and that the site would be either a 'blind spring' and streams if the dowsers were right or a woodhenge if I was right. It did not take long to find the church which had been dug over by archaeologists about two years earlier. Fortunately the stains of ancient postholes and paths are about three to four feet deep and can survive modern ploughing and archaeologists who do not have JCBs. The movement of stains into the soil is useful and protects many sites. The temple is much deeper, possibly fifteen feet to the roof if not more and so will survive most digging activity. To my delight the Saxon church was easy to find and it was on a woodhenge. At one time it had puzzled me as to how hundreds of years after a sacred ceremonial site had gone out of use or had been destroyed by the Romans that people still knew where it was. Very often the centre of the henge is in the centre of the choir so the church or cathedral is placed on the sacred site with some accuracy. One reason maybe is that the central depressed area of the henge was considered sacred and was never touched by farmers. There are several of these very small dells in the fields near where I live in Hertfordshire. I now know that they are over two thousand years old and for that period of time the farmers have ploughed round them. Even today the farmers go round them.

I dowsed the henge and plotted out the form so that the group could see the structure and layout, on the way pointing out that they were probably the first people in 2000 years to see the outline of the henge and temple and to know what was at the centre. They looked suitably impressed.

The site was excellent for dowsing. The henge and temple were in the middle of acres of well mown grass. Ideal for marking out and looking at the structure of the henge and temple in detail.

A few days later I returned with Nigel and a lot of canes and white plastic chop sticks to mark out the details of the site. As the dowsing progressed the outline of a temple complex with the standard 'Celtic' Cross design emerged. There was an access tunnel coming in from the east with its rectangular shaft. The central tomb was easy to dowse in detail because of the short grass. After marking out the tomb and confirming that the Prince and Princess were there, complete with sword, shield, bronze headdress, gold dress from the shoulders to the waist, other artefacts were looked for. An object giving a positive response on a clay witness was found in each of the four corners of the tomb. This was a new find and strengthened our view that the tomb was in a chamber. A chamber appeared necessary if people were to move around and access the anterooms and the bone depositories. The tomb might therefore be set in the ground leaving a space above it, or a pathway would be left round it or perhaps both.

In the triangular chambers (anterooms) of previous temples we had always found what we believed to be a rectangular column. Being familiar with mining I had always interpreted this as a roof support. The St. Albans Temple was no exception and it had a rectangular support

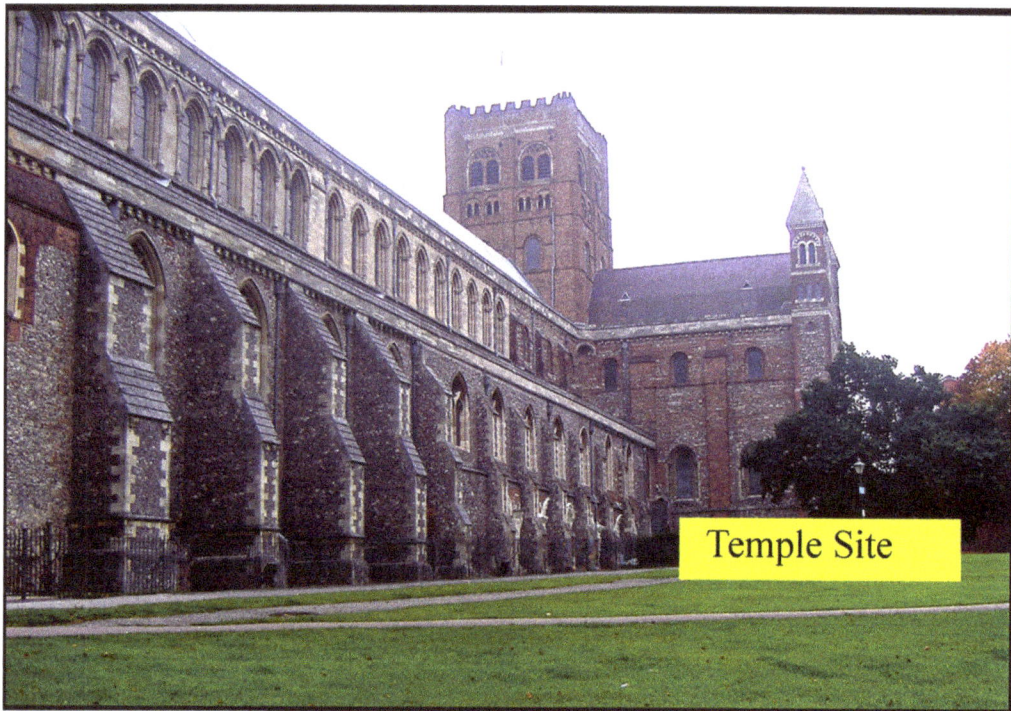

Figure 6.1 St. Albans Cathedral Temple Site
The Saxon Church on the south side of St. Albans Cathedral is on a Henge Age Temple. The position of the temple is indicated.

in each of the four chambers. The supports gave a positive response on a chalk witness. However, it was not possible to obtain a response using a chalk witness when crossing the walls of the tunnels. A positive response would have been obtained if there was a space in a layer of chalk. The temple did not therefore dowse as if it was hewn in a chalk layer. This indicated that the chalk objects were introduced into the chambers and were not left during excavation to support the roof. If the object was introduced, did it have a ceremonial role? To answer this question a range of witnesses were used and positive responses were obtained with potassium (carbonate), carbon and silica. This indicated that some burning may have been going on as the potassium comes from burnt wood (potash). The rectangular chalk objects may therefore have been small altars for offerings or supports for lamps. In the left-hand corner of each anteroom was found another small object which was identified using a clay witness. The presence of what might be called small altars in each of the triangular anterooms forced us to consider the practicalities of people using the temples for a period of time. The temples were enclosed spaces. People were not going to be very happy in what were quite small spaces for any length of time, particularly during the construction phase when hard physical work was involved. The space had to be ventilated and this meant that the design of the enclosed space – its shape, changes in ceiling height, had to be conducive to setting up ventilating currents of air. There also had to be a method of moving the air out and up the vertical shaft whilst another current of fresh air came in and along the floor. Checking on the design of the ceilings and ceiling heights was not possible but in the absence of mechanical ventilation air could be induced to move by using heat. The air is heated by lamps and body heat so that it becomes buoyant and floats to the top of the tunnel or space. The ceiling now has to act as an inverted river bed so that the air flows uphill along the ceiling of the tunnels to a vertical shaft. Only one vertical shaft or entrance had ever been found in a temple so it looked as if the shaft was both inlet and outlet.

The St. Albans Cathedral henge had shown us more detail of the temple and posed some very important questions about ventilation. The most obvious thing to use for supplying heat would be candles made from tallow or lamps burning animal fat. These substances produce smoke even when burning well and the smoke would be deposited on the ceiling and if left to itself on the inward wall of the shaft. Were these surfaces still contaminated by smoke? Within a few days Nigel had experimented with animal fat lamps and witnesses of animal fat smoke had been produced. The temples were then assessed using them. Positive responses were obtained from the underground spaces, along the access tunnel and from the inward wall of the shaft. In the anterooms the soot appeared to be missing from the two corners apposite the entrance. This made it look as if the ceiling curved down in the corners and may have missed the smoke. The smoke witness proved to be so good at picking up the tunnels and rooms it became the best method of identifying them. The class at St. Albans had introduced me to a Saxon Church which then introduced me to an underground temple from which we had learnt a possible use for the anteroom, and how underground temples were illuminated and ventilated.

Problems with Magnetic Loops and Witnesses

Although we now knew how the temple complexes may have been ventilated and the ventilation problem appeared to be solved, there was still another problem which had been bothering us for a while. While working on henges in the St. Albans area early in May we made a surprise discovery. Dowsing was showing us that in the recreation ground near to Nigel's house there was a Roman cemetery. The evidence for this was a considerable number of what appeared to be lead coffins lined up in rows. The presence of lead was taken to indicate Roman burial. There was however a problem – there were so many of them lined up in a regular pattern that hundreds of tons of lead would have been required and this was unlikely.

My ongoing research had revealed a phenomenon in which targets would produce images of themselves on the ground, both nearby and at a distance. Images are well known to dowsers and are one of the factors responsible for dowsers not finding what they have dowsed when they dig down. In other words the dowser identifies something in the ground. It dowses as a genuine target but on digging down nothing is found. I had done a lot of work, and digging, on this phenomena and had established that there was a magnetic loop between the actual target and the image (Figure 6.2).

Figure 6.2 Magnetic Loop

A magnetic loop forms an image of a target when the target's field lines are attracted to something in the ground. The dowser will see the image as a normal target unless steps are taken to identify it as an image. Magnetic loops are frequently observed on the sun.

The field from the target instead of going straight up is attracted to something in the ground and bends over to form a loop. Where it hits the ground the magnetic field lines form an image of the target. Another feature of the image is that it might either move with the sun or it could be locked onto something in the ground and remain stationary. However, there was a puzzling aspect of the images of the Roman graves and that was that a positive response was obtained with both a phosphate and a lead witness. The dowsing indicated that both lead and phosphate appeared to be present. Knowledge of the site and probes for coffins said there were no lead coffins on site. This observation also applied to a grave I thought I had in the garden. After augering down seven feet on the garden grave I came to the conclusion that there might be another explanation for the dowsed lead and phosphate. About forty feet away under the shed was a quantity of scrap lead from my diving days and this could be producing an image. I moved it up the garden and the Roman grave moved up the garden with it, both lead and phosphate. Now the scrap lead was scrap lead not scrap phosphate. It was just lead with its usual coating of oxide (PbO_2). Questions started to form and things began to speed up. A call was made to Nigel for some lead dioxide as he had access to some chemicals, then the acid cleaning of some scrap lead to remove its oxide coat and produce an oxide free sample of lead then some manganese phosphate. A new set of witnesses was produced and the problem was on the way to being solved. The normal lead witness used for identifying lead was in fact a mixed witness containing lead and lead dioxide. Lead quickly forms an oxide layer which is why flux is used when soldering to prevent the formation of this oxide layer. This had not been fully appreciated. Consequently when using a lead witness a response would be obtained with both a lead metal target and a lead dioxide stain left in the soil. The dowser would not know if the response was due to lead or the oxide. The latter would not however give a positive response with a clean lead witness. So it was possible to distinguish metallic lead from the oxide stain. On further investigation the lead dioxide witness was also found to give a positive response with the phosphate of bones and in turn the phosphate witness gave a positive response with lead dioxide. There was therefore scope for confusion and this is why when using witnesses one says 'it dowses as if it is iron etc.' just in case the target is mimicking the witness and is not the same material as the witness. Once the Roman lead coffins had been sorted out and we knew that we were dealing with images and that the images and real targets were phosphate and not lead it occurred to me that the bone depositories might be another case of mistaken identity. The presence of iron in the depositories had always been a puzzle. It seemed unlikely that people in the Iron Age or earlier would make up the equivalent of biscuit tins to bury bones in and yet a strong response was always found with the normal witness for iron which was plastic with iron oxide as the filler, or with an iron nail or other piece of iron, complete with its oxide coating. Following the pattern of the lead investigation the first step was to clean some nails with acid and then to obtain samples of iron oxides, iron (lll) oxide (Fe2O3) red-brown to black in colour and red ochre or red iron oxide. When the bone depository in the garden was tested with clean iron there was no response, that is there was no iron as metal present. The sword in the tomb did not produce a positive response with clean iron indicating that iron was not present. The iron oxides particularly the red oxide (red ochre) did produce a positive response indicating that the oxides of iron were present in the tomb and bone depository. This was not unexpected as it is well known that ancient people used red ochre for ceremonial purposes but if it was on the bones in the bone depository where was it being applied to them? I was confident that I knew and was on my way to the Dolmen in the nearby fields. It did not take long to show that the red ochre started its journey at the Dolmen. Unlike the phosphate halo which spreads some distance round the Dolmen the red ochre halo was under the Dolmen it was not scattered around and was restricted to a small area. It was easy to imagine the red ochre being applied to the bones and some of it falling through to the ground below. By chance we had now found a possible way of tracing the route taken by the bones on their way from the Dolmen to the depository. Every time the bones were moved or changed from one set of hands to another, some red ochre could have fallen off which it should be possible to detect. The discovery that the iron witness I had been using picked up red ochre invalidated my interpretation of tombs and associated woodhenges being Iron Age. They had to be checked out again and that included Stonehenge and Woodhenge.

Support for the Stain Theory of Dowsing

At this time, our confidence in the theory that we were able to dowse stains or trace amounts of material in the soil gained support from another archaeological project we were running. When digging down onto a target we would put the spoil into rubble bags. A vertical section of the soil became represented by a line of numbered rubble bags and amongst them, at a certain depth, a bag or bags would give a positive response with the witness being used for the target, for example phosphate, iron oxide, aluminium (Figure 6.3).

On some occasions there was physical evidence of the target, for example a stain or change in soil structure but on other occasions there would be no readily identifiable physical evidence. We now know that the stains held in the soil are as important as solid artefacts such as walls and pottery if not more so. Stains can give a detailed account of activity on a site. There could be almost a complete record in the ground of what had happened on it over the millennia.

71

Figure 6.3 Locating the Stains of an Archaeological Feature
*The dig is onto a Druidic water pipe to a pool. A point is selected for a
keyhole dig. Once near the anticipated depth, layers of soil are placed
in or on bags. The individual layers are then dowsed to determine which
layer the stain is in. This confirms the presence and depth of the stain.*

Temple Design : Six to Nine Rings

Towards the end of May a further discovery was made. A set of three rings of postholes were found outside the six rings of a woodhenge in St. Albans. The woodhenges were even bigger that we had thought. From about forty metres in diameter the henges had grown to about sixty metres in diameter and the shaft down to the temple access tunnel was now within the temple. The rings of postholes were divided into a set of six then a gap then a set of three. This information was available before a visit to the North of England to study water divining and meeting up with a farming couple who said they had a stone circle in one of their fields and I was invited to investigate it.

Before moving on, a summary of the first half of 2004 might help to keep track of developments. Fast progress had been made in a number of areas. This was due to being joined by another keen dowser who approached dowsing objectively. It both speeded up progress and made observations more reliable. Another step towards quality control in dowsing. With measurements being made by two people errors were more likely to be identified. It is also important to research errors and understand them. Again much easier with two people.

Summary

- The phenomena of magnetic loops and how they played a role in dowsing had been discovered. Magnetic loops are commonly observed on the sun but I had not heard of them being observed on the earth.

- Woodhenges and associated temples appeared to be so common that they were not only numbered in hundreds but possibly thousands. To be built on such a scale a henge must have both an important role and possibly a role limited by time. Once

72

its time slot came to an end another one had to be built.

- Woodhenges were present in Ireland with associated stone circles.

- The presence of underground temples at the 'Woodhenge' near Stonehenge was confirmed.

- The use of staffs in ceremony was indicated.

- At Stonehenge there are indications of iron rimmed wheels being used on ceremonial chariots and of very large transport vehicles having been used on the site.

- Energy engineering may have played a role in the design and building of Stonehenge.
- Evidence of an underground temple was found at Stonehenge.

- The tomb chamber and the triangular chambers (anterooms) were beginning to reveal some details indicating that they were accessed and used over a period of time.

- The method used to ventilate the underground temple was discovered and the nature of the fuel for the lamps.

- Lessons were learnt about the use of witnesses which enabled the use of red ochre to be identified.

- Identifying the presence of iron in tombs was refined.

- A posthole ring pattern of 6 + 3 was identified.

Although questions were being answered and discoveries made many more new questions were being raised. For example why does the henge have a 6 + 3 design for its rings? Why are the henges so far looked at very close in size? Why the central black area? What is it and what was it used for? Why the 'Celtic' Cross design for the underground temple? It seemed like one step forward and two back.

Chapter 7

The Pieces of the Puzzle Start to Fit

A Stone Circle in Yorkshire

Earlier in chapter 3 and again in Chapter 6 mention was made of a stone circle I had been invited to look at by a farming couple in Yorkshire. At the time (20 June 04) I was on a water divining course and had decided to book bed and breakfast on the farm in question. It is not everyday of the week that somebody says 'would you like to look at my stone circle'? mainly because not many people have stone circles in their garden or on their farms. Those who do, normally have a modern make believe one created by somebody who professed to be able to build them but does not understand energy engineering, or, and more commonly, the circle has been built by a garden designer who makes no pretence of being able to do energy engineering and just aims to create a visually pleasing circle of stones. It is often forgotten that the ancients would have used partially dressed if not dressed stone and are unlikely to have gone looking for stone weathered by millennia of frost and rain to create an 'atmospheric' effect. The circles they built would have looked reasonably neat and tidy when first constructed and possibly also geometrically pleasing to the eye. The ability to work stone and demonstrate the application of skills and material resources would have been important to those doing the building. Often forgotten by people 2000 years later, the stones would most likely have had a physical or practical function. For example load bearing or marking a boundary or portal.

Stone circles as we see them today differ in construction and size forming three distinct groups. What was becoming clear was that these differences made much more sense when the circles were considered as being part of the woodhenge temple instead of being separate structures. The position of the circle in the temple would determine its function and the size and shape of the stones used. Even after 2000 years of weathering and vandalism the position of the stones in the temple still determines how we see them today. The simplest and smallest of the circles are stones often set flush with the ground, close packed as if they are an edging round something. The individual stones can be of a fair size and set deep in the soil so they are not easily moved. Built this way the stones would have provided either a firm footpath round whatever was inside them or the foundation of a wall. The Yorkshire farmer's stone circle belonged to this type, (Figure 7.1).

The stones were close set with their diamagnetic north poles at the top. It was not possible to examine the fields between each of the stones but where it was they had a north pole facing a south pole and most of the stones must have been in harmony as they produced the characteristic circular magnetic fields that surround a correctly laid circle of stones (Figure 7.2). The fields both inside and outside the circle came up on different colours of the colour wheel. The stone circle therefore looked as if it had been constructed by someone skilled in identifying the magnetic axes of stones and who knew how to line them up to create the characteristic circular fields. The stones were on a small bank and there was a slight central depression as can be seen in figure 7.1. Having confirmed that the stones were laid correctly the next step was to identify the central black 'posthole' area. It was there, so providing additional evidence. Then the six rings of postholes, a gap and a further three rings of postholes. The postholes checked out and all nine rings were there. The Dolmen was also present at about 200m to the west of the stone circle. Finally the presence of an underground temple and tomb confirmed that the farmer's stone circle was a real one at least two thousand years old and also an integral part of a woodhenge!

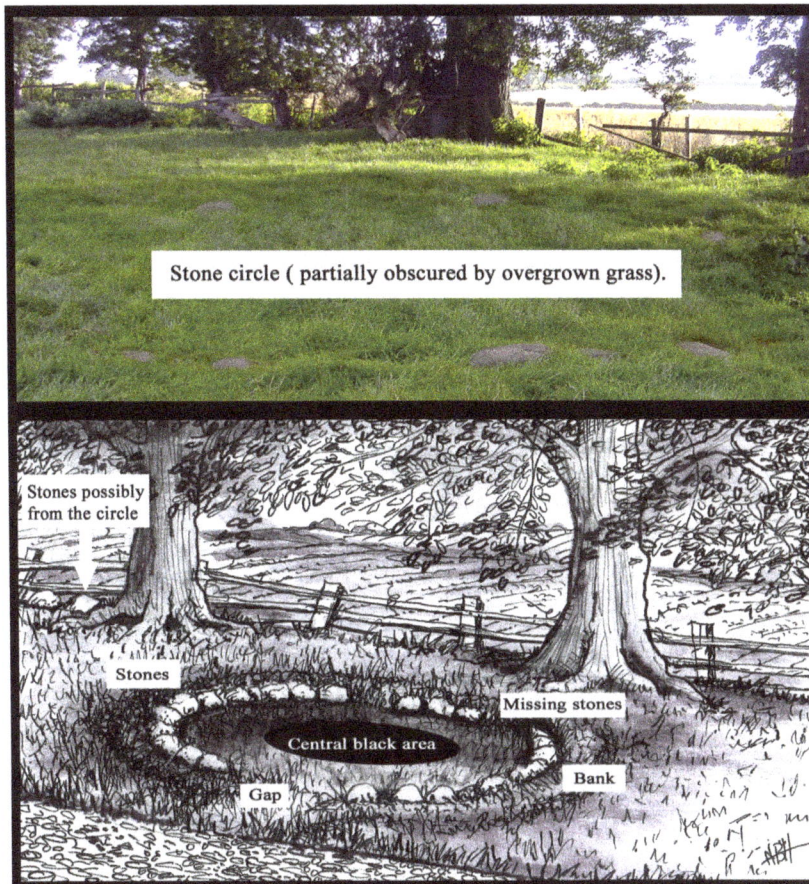

Figure 7.1 A Yorkshire Stone Circle

A diagram showing the main features of the Yorkshire Stone Circle.
The black area in the centre is the chemical stain from organic matter.
The stones were the foundations of a low wall round the clay pool. The
gap is possibly to give the priests access to the pool. The altars were
in a rough circle round the pool.

A Second Visit to the Yorkshire Stone Circle

The newly discovered stone circle was so important that a second visit was required to determine if the temple complex matched those in the south of the country. A few weeks later on the 9th August 2004 Nigel and myself drove through the worst conditions on the M1 in years to spend the best part of three wet days exploring the temple site, measuring and recording the underground tomb complex, locating the postholes that would help to tell us what the structure may have looked like and identifying some of the pathways that could indicate how the ceremonial site was used. The first new information and a clue as to how the henge may have been used came from a photograph. The photograph, which was hanging on the wall of our farmhouse B & B, was of a small archaeological dig round some of the stones. It was showing clearly what looked like reddish clay on the inside of the stones. The farmer's wife explained that on digging down she had found a thick layer of clay but nothing else. This indicated that the black central area was most likely the organic matter that had accumulated in a pool of water formed by the clay and it was not a large posthole along the lines of Seahenge. The central area of the henge was therefore clay that could either hold water that was deliberately placed in it or which may have accumulated water when the

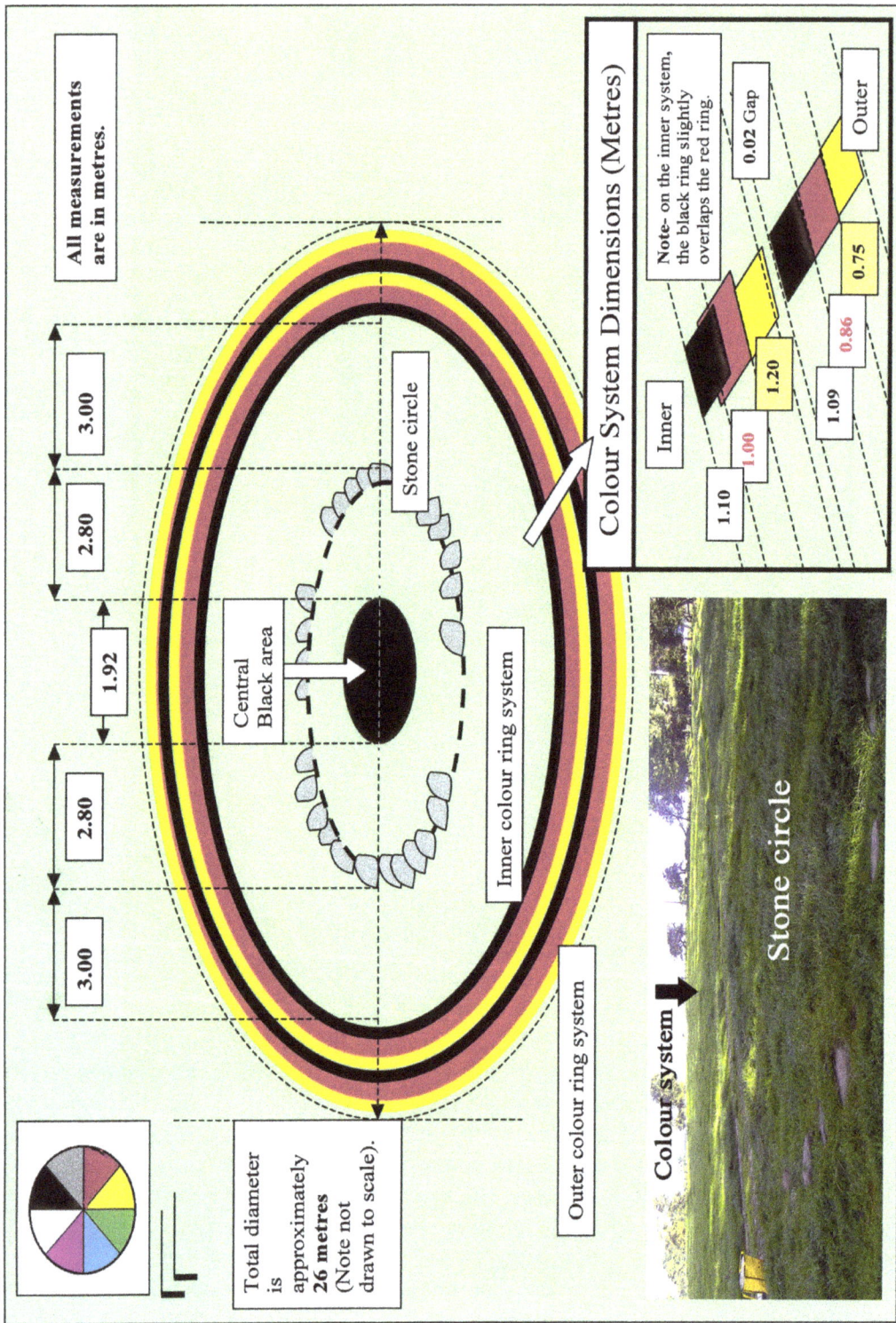

All measurements are in metres.

3.00

2.80

1.92

2.80

3.00

Stone circle

Central Black area

Inner colour ring system

Outer colour ring system

Total diameter is approximately **26 metres** (Note not drawn to scale).

Colour System Dimensions (Metres)

Note- on the inner system, the black ring slightly overlaps the red ring.

0.02 Gap

Inner

Outer

1.10

1.00

1.20

0.86

1.09

0.75

Colour system

Stone circle

Figure 7.2 The Magnetic Fields of the Yorkshire Stone Circle

The magnetic fields generated by the stones are shown as coloured circles. The colours are those found using the colour wheel which can be used to identify the different fields.

76

henge was no longer in use. The clay, the raised circular bank and edging of stones indicated that the centre piece of the henge was perhaps a pool of water, possibly used as a symbolic entrance to the underworld or in some ritual ceremony. Later in the survey when the altars had been identified, it was found that their drip trays also gave a positive response with a clay witness. The presence of the clay was confirmed when one drip tray was excavated. No trace of terracotta pottery drip trays has so far been found on any henge so the trays may always have been of hand moulded clay.

The discovery of clay at the centre of the henge was a big step forward. We have subsequently found that all the Henge Temples we had studied had a clay pool lining at their centre with a wall round it. The clay pool is now used as one of the diagnostic features of a Woodhenge Temple. The possibility that the clay pools may be the so called dew ponds has yet to be investigated. The circular clay area can still be identified in henges that have been ploughed for years. Traces of clay are taken down into the soil by weathering to a depth at which it is protected from most surface disturbance.

The original visit to the site and quick look at the henge and underground complex had shown that the triangular anterooms were separated from the tomb chamber by tunnels and the anteroom to the north and which was marked out at the time, was a large one. The central chamber had the two usual occupants, the Prince and Princess, with the gold signal well developed and iron oxide and bronze registering as being present. The temple followed the normal design but in some details it differed from the temples studied up to that time over 200 miles to the south. For example the tomb chamber appeared to be larger.The second visit to the site had a number of objectives the first of which was to confirm the original observations and make sure that it was a genuine stone circle. After the wet drive north the survey started by confirming that the stones were properly arranged to generate the circular magnetic fields.

Some dimensions were then measured. The outside diameter of the circle was 6.25m north to south and 5.75m east to west. The central black area was identified using a carbon witness. As already mentioned, a photograph had already provided a clue and the black area is probably the bottom of a pool where organic matter settled. It size was 2.0m north to south and 1.7m east to east. When measuring an object that is at least two thousand years old, too much must not be read into the figures as stones and other features are eroded and do move with time. The dowsed colour bands, black, red and yellow, showed up clearly round the circle, (Figure 7.2).

The central rectangular tomb was not set north south but was at 345°. The outside dimension of the walls were 2.72m and 2.82m top and bottom respectively and the east and west sides were 1.83m and 1.65m respectively. The area that was being called the tomb was large and so could be a chamber with the actual tomb set into the floor. The photographs of the dowsed chamber (Figure 7.3) shows the Prince and Princess on the eastern side, then a wall. There is possibly a lamp in the north east corner of the chamber and the tunnels from the four quarters can be seen coming into the chamber. The space to the west of the bodies appeared to contain bones. If the two bodies are set below floor level the tomb chamber provides quite a large space for people to move about in. Figure 7.4 shows the dowsed outline of the central chamber with the access tunnel from the shaft to the surface approaching from the south east. The tunnel going north from the chamber sets off at about 348° magnetic but then the tunnel, anteroom and bone depository seem to line up closer to magnetic north. The dowsed width of the tunnel entrances from the tomb chamber appear to be narrow at between 400 and 630mm. This is not the sort of space required for hauling spoil to the surface so it may not have come this way. The route to the northern bone depository at about 5.5m was a long one compared with most temples studied at this time. The anteroom has a small altar for offerings and what could be a lamp set in the left hand corner. The bones in the depository are in the right hand sector indicating that there may be a hole or some form of receptacle for them to the side of the tunnel floor. Dowsing the bone depositories with a witness of the slate found capping a nearby disused well indicated that there may be a large piece of slate in the depository, possibly acting as a top to a hole or providing a table top to place bones on. Whatever its position or function, the slate indicates that the depositories are not simple holes in the rock.

Figure 7.3 The Prince and Princess
The green rods show the position of the Prince and Princess in the tomb chamber.
The entrances to the 4 antechambers from the tomb chamber are shown.

Figure 7.5 is a straightened out plan of the underground temple complex. Each of the arms differ in detail although the basic design is followed. The four quarters of the compass were obviously important to the Druids and the four arms seem to home in on them. The underground temple construction shows variation from one quarter to the next which could relate to how the builders saw the gods or spirits of each quarter. They could also be due to more practical reasons such as geological factors.

Moving from the underground temple to the surface temple, the position of postholes over a segment of the henge was recorded starting with the inner ring and then working out to the west. Access to the eastern half of the henge was not possible as it was in a field under crop. Although the posts would have in the main been on radii it was clear that they were not arranged in a neat pattern. If the postholes are all of the same age their position should provide a clue to the structure of the building and what might have been going on in it. It is however possible that over the lifetime of the building posts were replaced in different places and this may be the reason why they are not in neat lines. An alternative explanation may be that the site started life as an open structure and then more posts were set up when a roof was added. In one sector of the henge, 300° to 346°, postholes extended out to about 50m. This area could belong to a different time or be the entrance to the Henge Temple or some other facility attached to it.

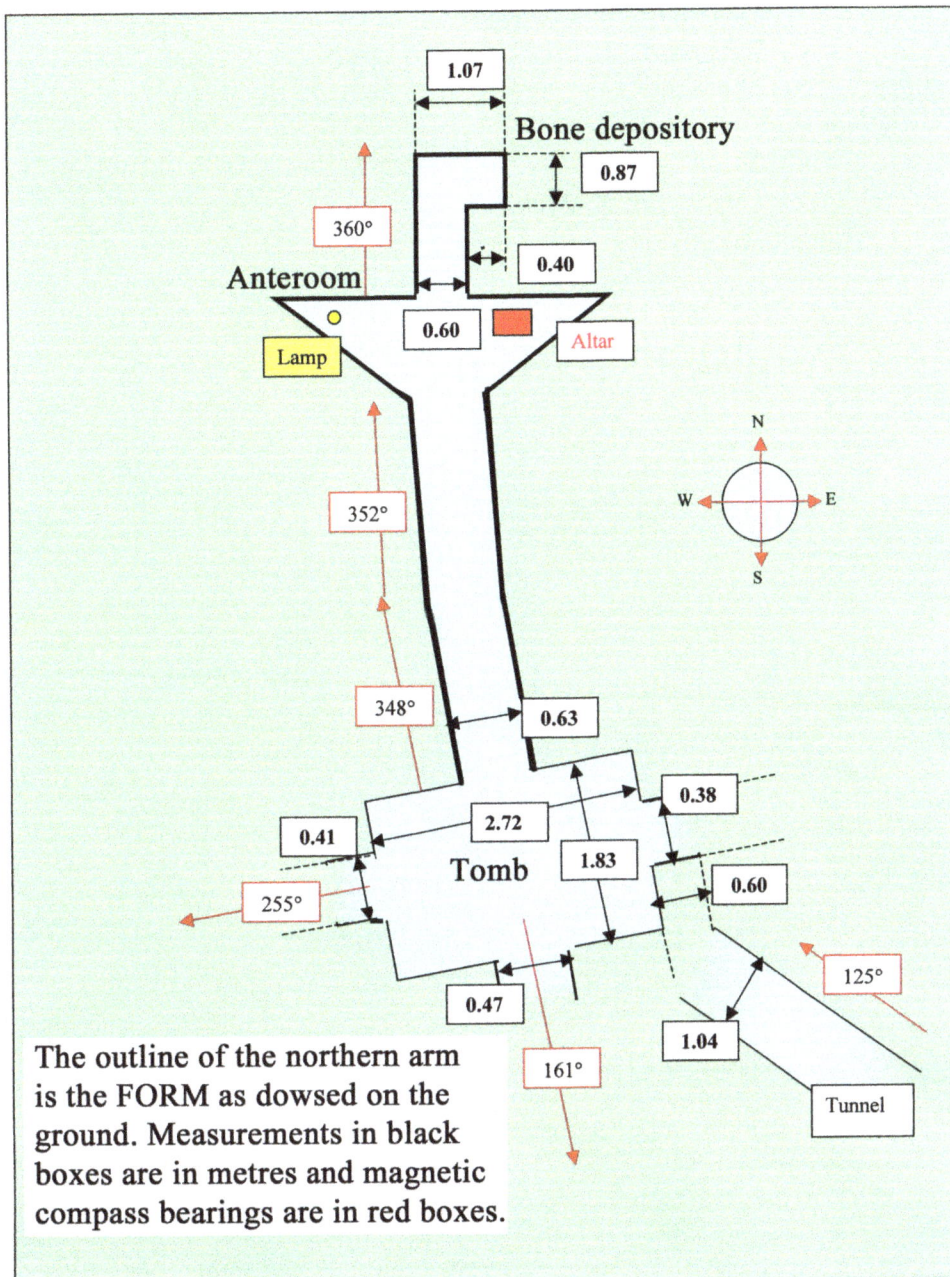

1.07

Bone depository

0.87

360°

Anteroom

0.40

0.60

Altar

Lamp

N

W ● E

S

352°

348°

0.63

0.38

0.41

2.72

Tomb

1.83

0.60

255°

0.47

161°

125°

1.04

Tunnel

The outline of the northern arm is the FORM as dowsed on the ground. Measurements in black boxes are in metres and magnetic compass bearings are in red boxes.

Figure 7.4 The Underground Temple
The tomb is shown with the access tunnel from the shaft and one of the anterooms and bone depositories.

The altars were identified, and as figure 7.6 shows, they do not fit a regular pattern. The bearings of the altars are magnetic so anybody wishing to read anything into them should correct them to what they might have been 2000 years ago at least. The largest altar is close to the bearing of the winter solstice sunset. None of the other six altars appear to line up with solar events. Five of the altars face in along the radius, the two largest ones are set at an angle. When dowsing an altar all that is left is the stain of the plinth, the clay of the drip tray, blood and other biological fluids and the stain of the altar stone. The stain the altar stone leaves is due to rain or dust falling to earth from its edge with some of the stone. The altar stones are big single pieces of stone and so of value to people doing building work in the area. It therefore made sense to look and see if any of the slate could be found on the farm. If so, it might have originally come from the altar tops. After searching the field a big piece of slate was found acting as a cover over an old well. More pieces were found in the structure of the farmhouse and its gardens. On one sample of slate blood could still be identified along an edge. The slate found on the site and the stain left by the altar stones was that of slate that had to be brought in from a quarry about twenty miles away.

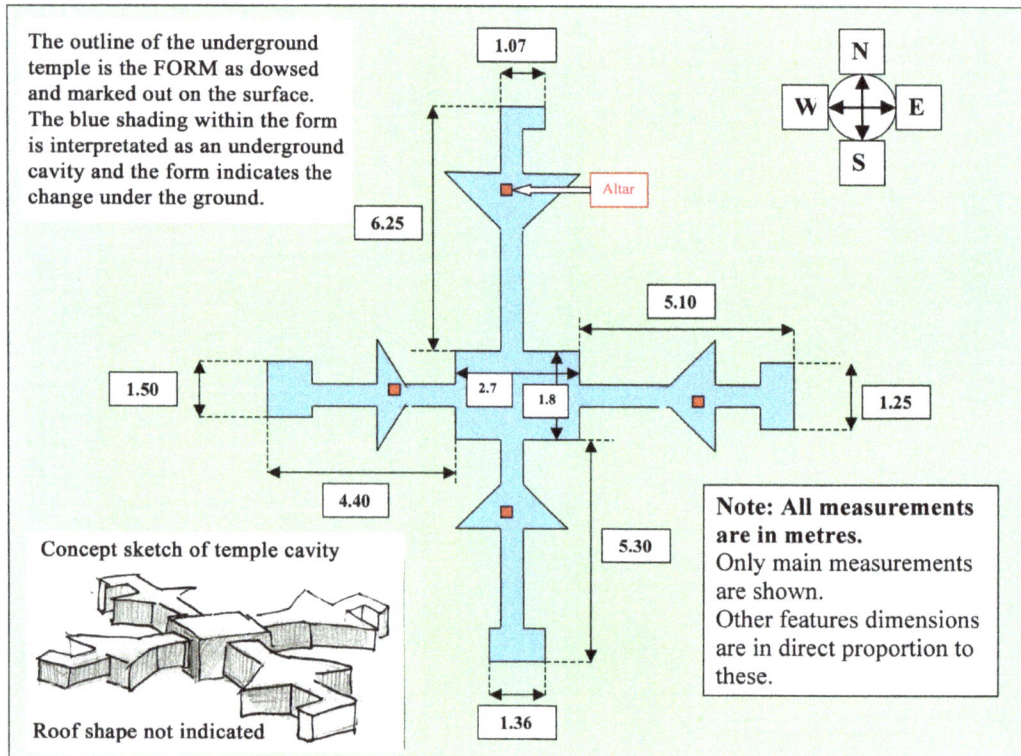

Figure 7.5 Schematic Design of the Underground Temple

The access shaft to the underground complex was found but unfortunately it was under a farm road and so not accessible. This was a big disappointment as we knew physical proof of access shafts was vital to the theory of the underground Druidic Temples. If one was found the idea was to remove the top half metre of soil and see if the outline of the shaft was visible. On one occasion when visiting a henge on private property we found that the clay of the central pool area had been excavated by the owner and removed in a skip. The resulting excavation indicated the size and depth of the pool. It was possible to follow the tunnel from the temple under the pool to the shaft by dowsing. At the end of the tunnel the outline of the shaft was still visible on the surface. However, the outline of a shaft alone is not good evidence of an underground temple and a real shaft is required. At this time we thought the underground system was about twenty feet deep. Later evidence started to indicate up to forty feet.

The Yorkshire henge was the first one we looked at in any detail. It was far from public roads and paths and canes could be left up overnight and work progressed from dawn to dusk. Each Henge Temple site so far investigated has provided more new data and with it the complexity of the Henge Temple, both above and below ground level, and the activities that went on in them has grown. The skill of the engineers involved was very high and the scale and cost of the operations that they were involved in seemed to grow every time we studied a new site or returned to an old one with new knowledge and to test new ideas. The Yorkshire underground temple was in hard rock, not chalk, the engineers dealt with it and knew where they were underground. They worked considerable distances from the access shaft but were able to ventilate the galleries. They were moving spoil long distances and had to raise it to the surface. Tens of cubic meters of spoil were involved. In chalk the flint was required for tools and buildings but the chalk and other spoil had to be moved and disposed of in some way. The most likely use was landscaping and road building.

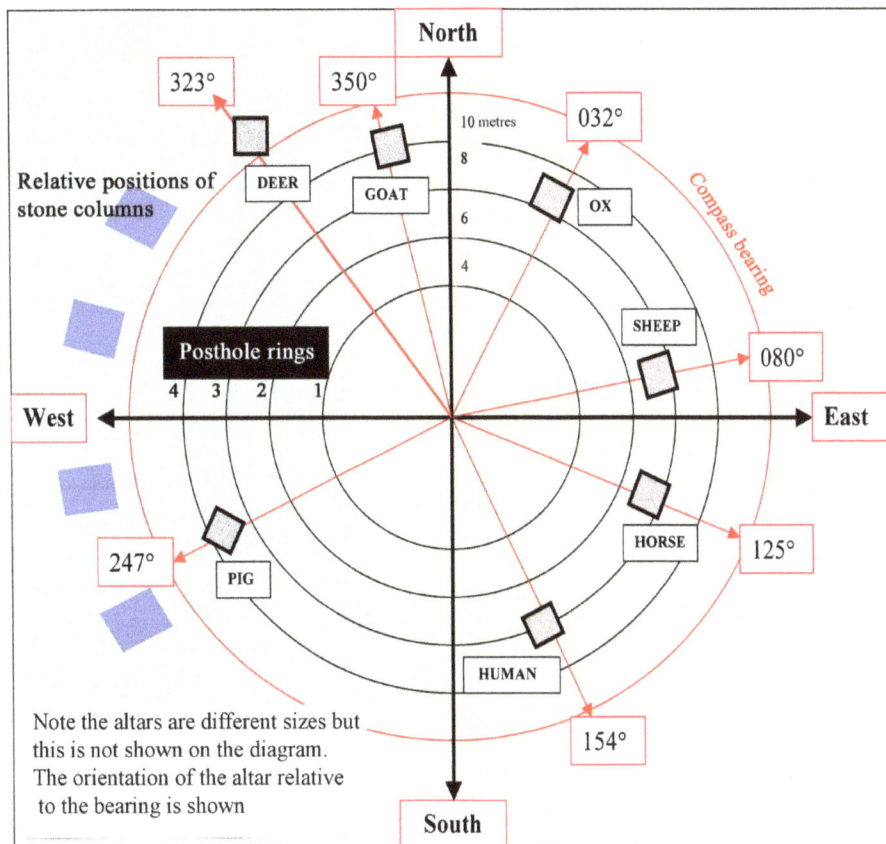

Figure 7.6 The Seven Altars
The position of some of the stone pillars used to support the roof are shown.

The Yorkshire Henge Temple may have had a side extension which would have turned it into a more massive structure than the previous temples studied. From one side to the other it could be about eighty metres across with the small diameter sixty metres. The Dolmen on the western side, which is a feature of every temple, appeared to have been made out of wood and under it there was a red ochre stain, spillage from dressing the bones. The ceremonial path between the Dolmen and the pool in the temple had staff marks along it. There was a lot of evidence that the Yorkshire Temple was almost identical in basic design to the henges over 200 miles further south and that similar ceremonies were practiced in it.

This was the first Henge Temple to have had a large number of postholes identified and a cane used to mark each post. It was the first henge to have its altars plotted out with any accuracy and the first one to have, still visible after 2000 years or more, the central pool area. Looking at the forest of canes it was clear that we were in the equivalent of a cathedral and knowing where the supporting columns and a few ceremonial features were was only the beginning. At this stage it was not known if it was walled, whether paint was used for decorative purposes, if there were internal structures and walls. Fortunately back south we had access to another henge site this time almost in the middle of a field so that it was accessible from all sides.

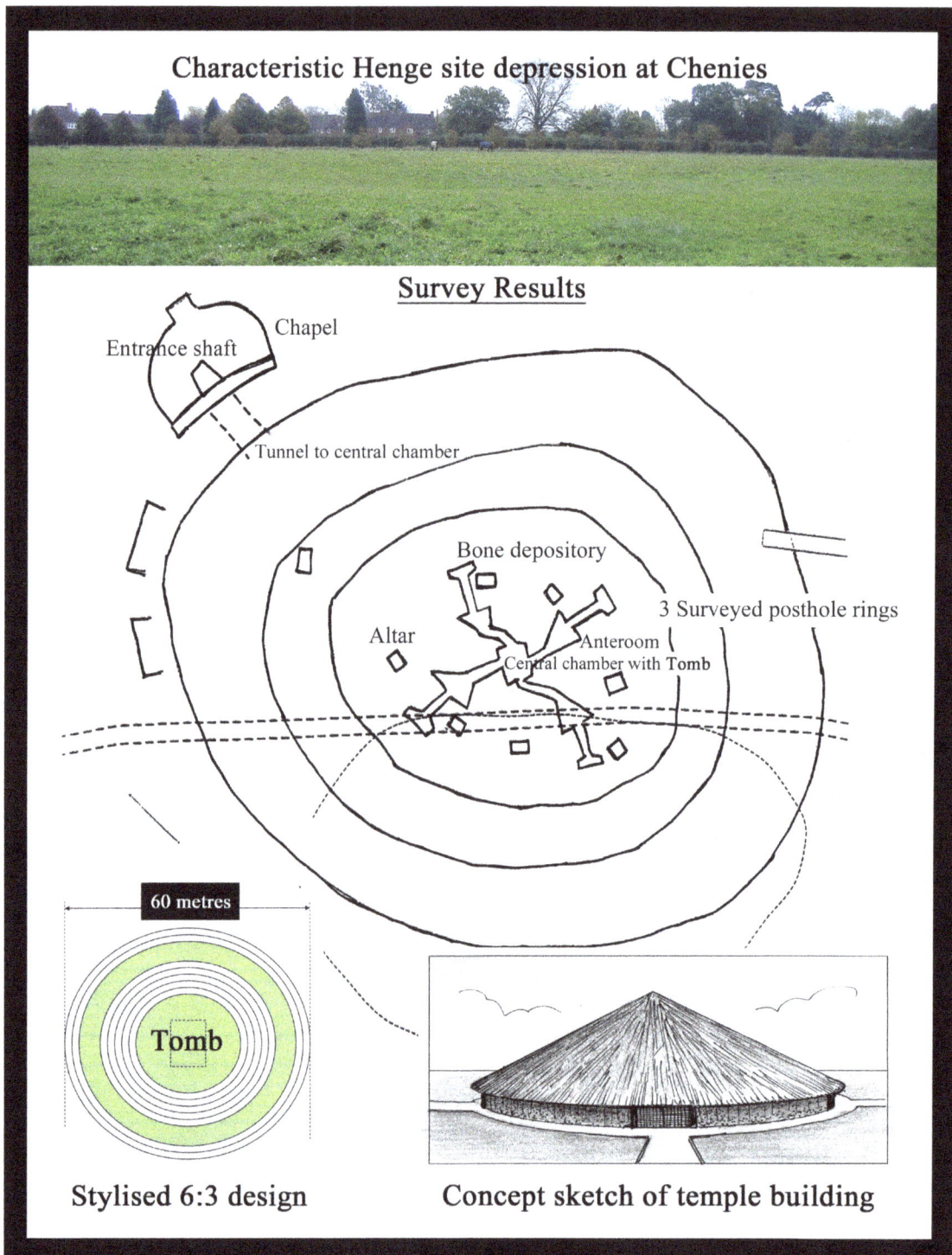

Characteristic Henge site depression at Chenies

Survey Results

Chapel
Entrance shaft
Tunnel to central chamber
Bone depository
3 Surveyed posthole rings
Altar
Anteroom
Central chamber with **Tomb**

60 metres
Tomb
Stylised 6:3 design

Concept sketch of temple building

Figure 7.7 Chenies Manor House Temple
The tomb temple at Chenies Manor House and the outline of the temple and chapel.
The Woodhenge is Iron Age and about 60m across with a 6 + 3 ring structure. The access shaft to the tomb temple is within the temple and within its own little chapel. The postholes of 3 rings are indicated. Above the tomb on the surface was a clay pool.

The henge was at Chenies Manor House in Hertfordshire. The depression it had left in the ground created the impression that a very impressive temple had stood there some 2000 years ago.

Chenies Manor House Henge Temple

The dowsed underground complex at Chenies Manor looked as if the engineers were not too sure where the bone depositories should be and after tunnelling out from the tomb chamber they had to change direction (Figure 7.7). The Idea that the engineers tunnelled out from the tomb is of course an assumption. If the access tunnel goes to the tomb chamber it may be a correct assumption but the access tunnel may also go to other parts of the complex. The odd shapes of some of the underground complexes we were finding presented a bit of a problem at this time. When marking out a tomb complex it was nice to find something like a 'Celtic' Cross appearing on the ground but too often this was not the case. A possible explanation for the 'twisted' crosses was to come later. At this time it could only be said that the presence of a tomb chamber, four anterooms and bone depositories were consistent elements of the Tomb Temple design but that a clean neat consistent plan was not. The seven altars of the temple followed the same general plan found in Yorkshire but some are closer in to the centre. The Henge Temple at Chenies Manor House had a clearly defined oak post and wattle and daub walled structure round the access shaft, creating a sort of chapel or 'confessional box'.

The shaft did not go straight down from the surface but there were a series of steps leading to it. These steps could be identified using a clay witness which at the time was taken to indicate clay soil. The top of the shaft was closed by planks of wood which left their imprint on either side of the shaft. The planks would have formed a ceiling over those entering the shaft and provided a floor for those receiving bones from the priest coming up to the front of the enclosure. Outside the front of the enclosure or 'chapel', a bone exchange point was approached by a path from the centre of the temple. At this point there was an area of red ochre. It looked very much as if the bones, on their ceremonial journey from the Dolmen, arrived at the 'chapel' where they were handed over to a set of priests who were inside. Some red ochre came off in the process to leave a stain in the ground. Within the 'chapel' the bones were taken down some steps to the shaft entrance and then once at the bottom it looked as if they would be taken along a tunnel to the tomb chamber then out through the anterooms to the bone depositories. This was a journey along a reasonably wide tunnel, easy to walk along to the tomb or a point near it. Then from this point to a designated depository or, if the bones were divided up, to a number of depositories perhaps all four or so we thought at the time.

Having identified the wattle and daub walls of the 'chapel', walls were looked for elsewhere in the henge. They were found running between the posts of the 9th circle. The temple was therefore enclosed by a wall. As our knowledge of the Tomb complex had developed considerably from the results of work in Yorkshire and at Chenies. It might be useful to review the main points.

Summary

After three days work in Yorkshire, much of it in the rain, many questions had been answered but there appeared to be many more unanswered questions including:

- Were the bones from a skeleton divided up or kept together in the depositories?

- There was a wall round the Henge Temple but were there any internal walls?

- Was the temple roofed?

- Was seating provided for an audience? Or perhaps the audience were also performers.

- Was paint used on the timbers and walls?

- Where were all the services for such a large ceremonial site?

- Where were the clues to solve the mystery of whether there was an underground temple or a surface complex?

However, some very important progress had been made.

- The association of clay bowls with the centre of the temple and clay with the drip trays of altars was the first physical evidence to be associated with the dowsed picture of Henge Temples.

- The circle of correctly laid stones round the clay bowl showed a clear physical link between woodhenges and one type of stone circle.

- The centre of the underground complex dowsed as if it was a reasonably sized chamber.

- The bone depositories could be more complex than just a receptacle for bones. The receptacle could be deep and be accessed from the side. It may have a stone lid.

- An enclosure or room for the top of the access shaft had been found at Chenies Manor House and walls had been identified on the 9th ring of postholes.

- The engineering and mining skills of the Druids or whoever were the Henge Temple builders was considerable.

- The basic pattern of the Henge Temple complex was the same in the north and south of the country but how much further north would the pattern hold?

Chapter 8

Pictures from an Ancient Past

Bronze Age Boats

On the way back from the Yorkshire stone circle in August 2004 we stopped to visit a site at North Ferriby, a village situated on the north shore of the Humber Estuary. During a previous visit to meet with some old friends I had been invited to look at a stretch of the foreshore and had found the outline of what might be two boats. A good test of dowsing skill is to see if you can find something a second time and if successful does it still have the same shape and size. Nigel was new to the site and would provide a further check on the original dowsing. Early boats have their planks sewn together. No bronze or copper is said to have been used in their construction. After a short walk along the foreshore the remains of the entrance to an old clay working was found. There were still remains visible on the foreshore to confirm the dowsing. The boats were found a little further on and in the same place as before. Both were lying under several feet of shingle and mud. The first boat was devoid of copper, bronze and iron but in outline it matched the boats that had previously been found in the Humber Estuary. The second boat was about 20m away and in a similar position on the shore i.e. drawn up bow first. It looked a bit like the first boat. However, there were signs of copper nails in the timbers and what might be a mast was lying along the length of the boat. The two boats as dowsed could be early Bronze Age, and three to four thousand years old. This is the same period that archaeologists believe woodhenges belong to. Buried boats on a tidal foreshore are not easy to dig out and have to wait for a storm to expose them. The land bordering the foreshore at this point is not readily accessible and it would not be an easy task to dowse for a local village and activities connected with the building of boats or of an export import trade. The interest of the site is that it is a reminder that 4000 years ago people in the British Isles were capable of building what might have been seagoing boats 40ft long. The boats were large and could be used for communication and trade over long distances. This does not fit in with the Roman description of the Britons who they say built boats from wicker and skins and did not build engineered boats from wood.

The Use of Witnesses (Box 8.1)

Along with the investigation of boats and a 16th century Manor House other aspects of dowsing were being developed during the summer of 2004. The most important development was the use of witnesses for identifying the magnetic fields the dowsing was picking up. Many dowsers do not use witnesses and it is possible to identify an artefact such as a posthole without one but a carbon witness makes the task much easier and ensures that the dowser is responding to a carbon stain and not to some other artefact. Witnesses help with quality control. The wood used for the post, say oak or pine, can be identified using witnesses. The use of stones as witnesses also enables the dowser to determine if the post was just 'heeled' into the ground without supporting stones, as might be the case if no load was to be placed on it, or alternatively, if the post was secured by supporting stones such as flint or limestone. This might be the case with a post that is going to support a load such as a roof or possibly just its own weight if it is a large and tall post. The possibility of a plinth upon which the post was placed to prevent rotting may have to be considered. The presence of preservative and or paints on the post can be checked for by using witnesses. The paint will have been taken down into the ground by rain and wood decay where it will leave a stain. However, people do not just paint pictures on walls and posts they also paint patterns on the ground. They mark out areas, they create pictures using stones and paints, the paths they tread leave pictures in the ground, their activities such as cooking, weaving, flint knapping, woodwork, brewing all leave marks in the soil.

Box 8.1 Witnesses

A dowsing witness is a sample of something, say a piece of bone, which is held in one hand when dowsing. The piece of bone appears to act by blinding the dowser to all fields other than that from bone. The result is that the dowser only responds to the fields from bone or the materials from which bone is made. This is of considerable help to the dowser who can now identify any bones buried in the ground using such a witness. If instead of bone the dowser is holding say iron or copper or flint they will only 'see' these materials. The use of witnesses requires some knowledge of how they work, practice and considerable care. Bone, for example, is calcium phosphate plus organic residues such as fat and proteins. A piece of bone from beef will therefore identify not only calcium phosphate but other phosphates and beef fat. When looking for a bone deposit the dowser can harden up the search technique by holding a phosphate witness such as manganese phosphate and look for the phosphate source by doing a circular sweep search (Figure B8.1.1).

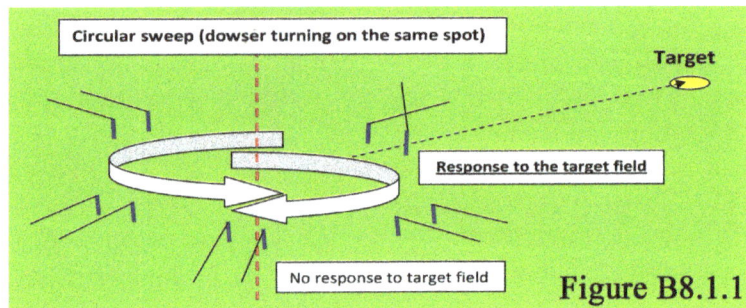

Figure B8.1.1

Bone is mainly calcium phosphate so is easily identified this way with the rods crossing when facing a source of phosphate such as a burial site or a septic tank. Most dowsers will respond to a large site, such as a church yard, at a distance of a hundred metres or more. If it is a small deposit of bones or just a chemical trace of bones long gone the dowser may have to walk over them in order to identify that bones may be there. The witness is not necessarily specific. For example a phosphate witness responds to lead oxide. For this reason all the dowser can claim when using a witness is that something dowses as if it is iron, flint, oak etc. They cannot say it is iron, flint, and oak without supporting evidence. The use of a witness will increase the dowser's sensitivity to the target material so that it is possible to identify incredibly small traces of material or pick up large deposits at great distances. As forensic scientists and environmental hygienists know, everything leaves a trace of itself on whatever it has been in contact with. It is virtually impossible to remove a hundred percent of a contaminant and in some cases what trace is left can survive for thousands of years. If the trace of material is left on the ground the stain, with time, moves down through the soil and in due course becomes deep enough to escape disturbance by surface activity. This is good news for the dowser who can now study a sort of soil colour photograph of what used to be there. The objects are long gone but the chemicals removed from them by rain and abrasion, are still in the soil.

The number of materials that can be identified using witnesses is very large but great care has to be taken when using them and it must be remembered that all that can be said is 'that it dowses as if it is x'. For this reason not only the discoveries made when using witnesses are given but also in some cases how we were led astray in the search for the secrets of the stones. To identify the witness to be used on a site it is necessary for the dowser to develop a model or theory upon which to base the selection. For example – where did the occupants go to toilet and to wash? What did they use, what happened to the waste? What was likely to be in the waste? Using this sort of reasoning witnesses can be identified, for example phosphate, blood from females, soaps, skin scales, hair, scents, drains made of clay or terracotta. Did the toilets and washrooms have a floor made from wood, tiles or stone? What did they dry themselves with after a wash, was it linen, wool or moss? Using this sort of approach it is possible to obtain a considerable amount of information about a site such as a roundhouse and the adjacent area by using witnesses.

Ceremonial Paths

The back garden of my house contains at least three main archaeological features. The 'Roundhouse' is one and its date of occupancy is known. The henge is one and could predate the house by a thousand years and finally the road. The date of the road is unknown. The only clue so far is that it was ceremonial during the Bronze Age and Iron Age and used by farm vehicles at sometime.

One day in July 2004 Nigel phoned me to say that he thought he had found patterns on the ground in a henge that was in the recreational ground near his house in St. Albans. This was exciting news as both art work and ceremonial paths could reveal a great deal about what the Druids got up to in and around their temples and about their culture. At this stage it was not known if the patterns were art work or ceremonial paths. Whichever they were, it would be possible to identify them using witnesses. Using the half of the Henge Temple that I have in the garden it did not take long to find segments of an inner ring going round the central pool and an outer ring lying between posthole rings three and four. Linking the two rings were loops. Later, when marked out on a henge in an open field, the picture appeared to be a bit like a shamrock leaf with one of the four leaflets in each of the four quarters. The portion of the picture that I had in the garden was so impressive that I invited the neighbours in to see it (Figure 8.1a). When I explained that they were the first people to see it for possibly two thousand years or more some were impressed and some asked which book I had got it from. Some people find it difficult to come to terms with the mystic power of the dowser and their ability to tune into the memories of the land. That evening as I sat looking at the pattern marked out in the ground I realised that it did take some believing and I still had the problem of deciding if it was artwork or a processional way. Although the 'paths' could be identified with a flint witness there was no aluminium associated with them as would be expected with flint. This indicated that a purer form of silica had been used than flint, possibly sand. Using sand the 'paths' could be identified but they did not respond to carbon which until this point we had always associated with footpaths. Carbon had been used to follow the pathway from the Dolmen into the henge so why no carbon on the internal henge 'paths'. Were they decoration only and nothing to do with a ceremonial route? Then the thought struck me, if the priests in the henge were barefoot they might not leave a carbon trail from shoe leather but as feet sweat, they may leave a salt trail. Using table salt as a witness the 'path' gave a positive response. Later it was to be discovered that human skin could be detected and later still bird feathers. For a long while the bird feathers were a puzzle but as other pieces of the jigsaw were found and fitted in they eventually came to make sense if they were part of the Druids ceremonial dress. The shamrock pattern that had been discovered was therefore a ceremonial pathway, the priests using it were barefoot and they had bird feathers as part of their dress. The salt that could be detected is no longer in the ground as salt is very soluble. It was the sodium sulphate in the table salt that was acting as the witness and giving the response. Sulphates tend to be insoluble and those from the sweat of the priest's feet have remained in the ground as the marker. After the euphoria of finding the Druidic ceremonial pathway in the temple and raising it from what we thought at the time was its 2000 year sleep (later it was discovered that the henge was more likely to be 3000 years old) I started looking for altars. I found one rectangular altar by using flint on the basis that the plinth may have been made from this local material and then a limestone witness on the assumption that it was the nearest available stone suitable for an altar top (Figure 8.1b). The limestone stain was round the outside of the flint plinth and indicated that the top extended about 10-cm outside the plinth base. Along one edge of the altar was an animal bloodstain. The construction of the altar from flints indicated that the builders used a mortar of some sort to hold the stones together. At this time it was not known how the mortar was made. The garden henge only revealed about a 120° segment so the next step was to explore the recreational ground henge at St. Albans. This was done two days later. First the inner circle then the outer circle followed by the curved paths linking the two. The paths formed a shamrock leaf pattern (Figure 8.2).

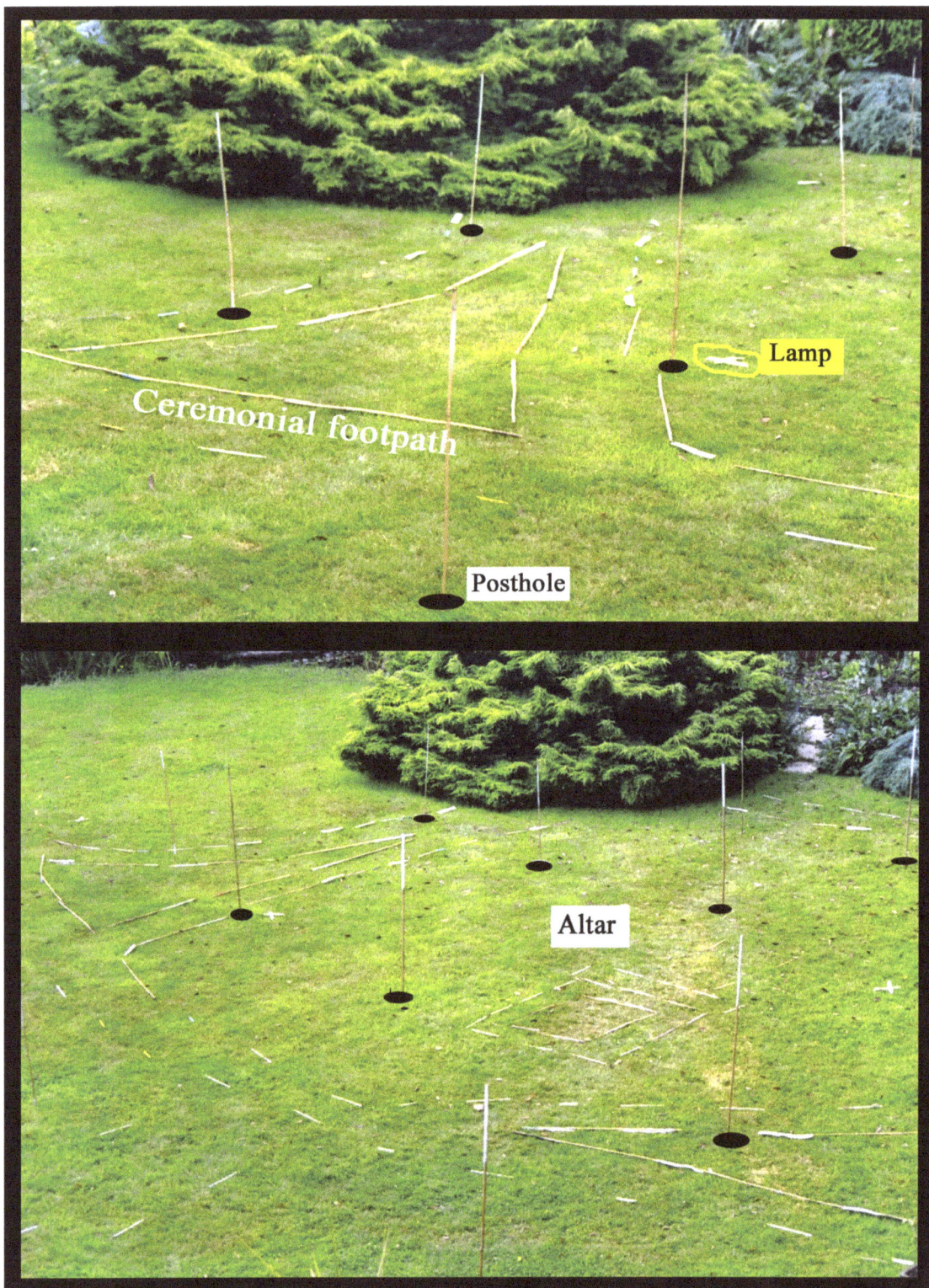

Figure 8.1a Ceremonial Footpaths
Two ceremonial footpaths are shown leading from the inner circular footpath to the outer circular footpath. The canes show the position of postholes and the crosses the position of lamps.

Ceremonial path

Altar FORM

Blood trough

Plinth

Altar top

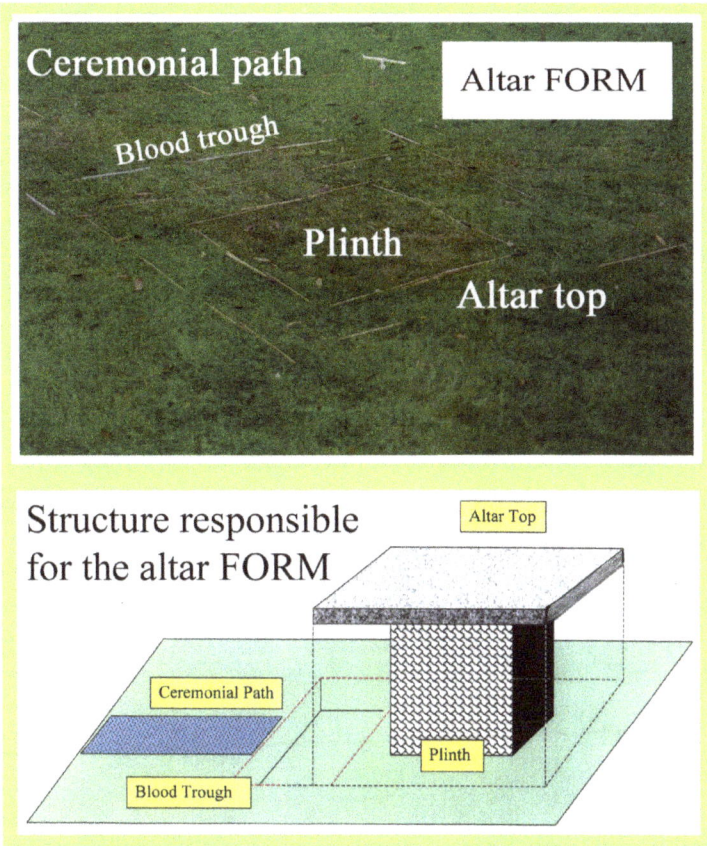

Figure 8.1b Altar
A ceremonial path comes up to the altar. The dowsing indicates that the altar top extends beyond the plinth. The trough for the blood runs along the front of the altar.

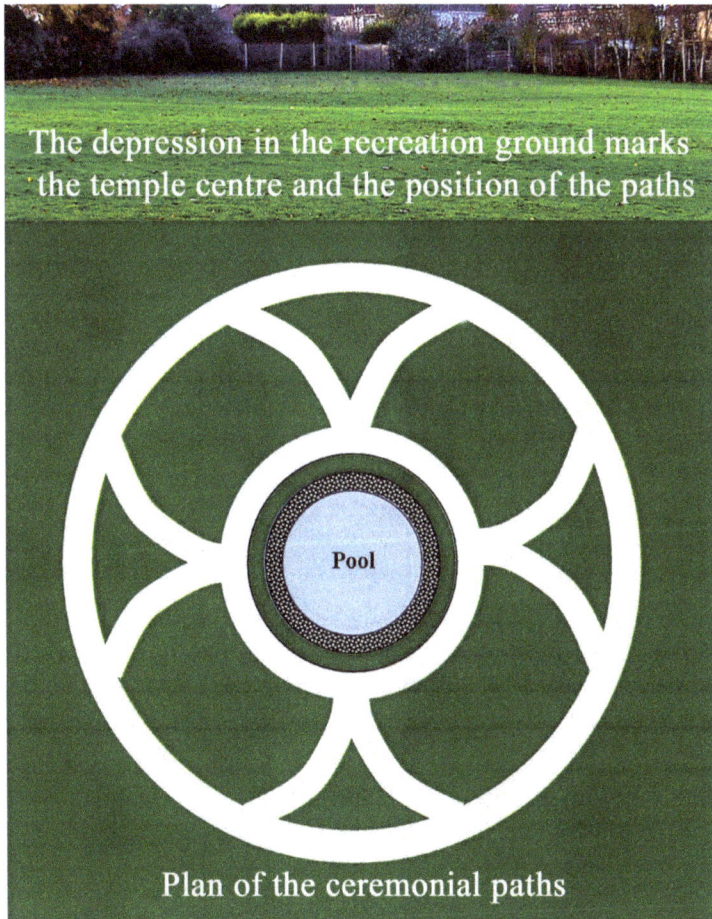

Structure responsible for the altar FORM

Altar Top

Ceremonial Path

Plinth

Blood Trough

The depression in the recreation ground marks the temple centre and the position of the paths

Pool

Plan of the ceremonial paths

Figure 8.2 The St. Albans Recreation Ground Temple
The ceremonial paths in the recreation ground temple, create a Celtic Cross.

Between the two circles there were seven altars of different sizes. Some of the altars were big and from the footpaths leading to them it looked as if the large altars, which could be two metres square, were served by two priests. Slabs of limestone were used as the top of the altars which were shaped so that the blood ran down towards the priests. This was puzzling as it indicated that the priests could be standing in blood. If this were indeed the case there would be a trail of blood from the altar, however, searches for this trail were in vain. There appeared to be no blood round the altar or in the paths leading to or from it. As far as blood spatter was concerned the area appeared to be clean. It therefore looked as if special methods were used to collect the blood. If the blood was collected, from what or who did it come from and what did the Druids do with it? The answers to these questions would have to wait but the discovery of altars enabled a start to be made on answering another question. In the garden henge it is just possible to pick up the two footpaths from the Dolmen coming in from the west and joining the inner circle. They then follow the circle for a little way before turning left towards the altar that had been discovered. The carbon track stops a few feet from the altar at which point the 'salt' paths take over. Where the carbon and salt paths meet there is a spillage of red ochre as might occur if bones are being passed from one set of priests to another. This observation was confirmed on the recreation ground henge. The bones can therefore be traced to a particular altar. The next place at which the bones can be identified is at the chapel housing the access shaft to the underground temple. A path goes up to the front of the 'chapel' from the central area and a spillage of red ochre occurs where the bones are passed into the 'chapel'. The route or routes between the altar and the chapel has not been worked out yet.

Mass Graves

Shortly after identifying the ceremonial pathways I was doing some water divining on a farm. Having found a nice underground stream, checked the purity of the water using witnesses, and estimated its depth and the flow of water the owner asked me to have a look at something else on the farm. At about a distance of three or four hundred metres from the farmhouse a 'sensitive', that is a person very sensitive to magnetic fields, had identified a large area of 'detrimental energy'. The sensitive had identified the detrimental energy from an even greater distance than the three hundred or so metres that I was standing from it. I knew that one of the magnetic fields that dowsers can detect at considerable distances is that due to phosphate and phosphate is closely associated with burials. So being very interested in what might be there I set off across the fields armed with witnesses. As I drew near to where the detrimental energy was said to be I did a sweep search using phosphate and found the area. The next step was to define its boundaries and size and then to see if animals were buried there or of it was an old manure heap, or perhaps a waste disposal area. The site was none of these, the area was a mass grave and it held about a hundred and twenty human bodies. Along side the first bodies to be looked at there appeared to be a length of iron oxide and the people had worn wool clothing and leather. It looked as if it was a grave for battle casualties. Surprisingly it was possible to dowse quite a lot of detail about the bodies and my bet after checking about ten of them was that they were Roman. This was our first contact with a mass grave. Nigel working in an area near St. Albans was later to discover that battle blood, that is the blood flowing from the wounds of a dying or injured person, leaves a dowsable stain that can be clearly identified two thousand years later. In Penang, Malaysia I had found blood spilt at the scene of a crime. The blood could be identified thirty years after the premises were cleaned. Not only could I find the spilt blood but the sex of the individual from which the blood came could be determined. Once we knew that it was possible to identify graves and battle blood our techniques for extracting information from bodies and blood rapidly developed. Some weeks later a return visit to the farm revealed who was fighting who, where the people had died and from drag lines, where the dead were buried.

Tree Auras

Returning to the Woodhenge Temples. Having at least partly solved the problem of what is at the centre of the temple, another of the many remaining puzzles of the temple design is its ring formation. Why six rings then a space then three more. It is of course well known that Druidism had its holy trees, plants and animals and we had at this time often wondered if the 'Celtic' Cross plan of the underground tomb complex, or an over ground structure if we were picking up a stain in the soil of what had been there, was modelled on the root system of the oak. We had searched in vain for evidence that the oaks roots followed a geometric pattern. However, early in September 2004 Nigel discovered that the aura of the oak tree consisted of a number of rings. From the trunk out there were six rings, then a gap, then three more with a radius of about thirty metres. This pattern was confirmed not only on oak trees but a number of other trees as well. The numbers of rings, their size and distribution in a Henge Temple fitted the rings of the oak's aura. This was more evidence that the Druids may have had dowsing skills and were using the human magnetic sense. Having discovered the nine rings it was back to the laboratory to find out what was producing them, see Box 8.2. Each ring appears to be the limit of the magnetic field of one or more materials.

Some of these materials are what the tree requires, such as minerals, in order to live and some are the products of photosynthetic activity which are being moved round the plant. The ancients would not have known this but they would have known that the rings were associated with the life of the tree and that as winter approached the aura disappeared followed by the leaves dying. The date the aura collapses and it is a sudden collapse, differs a little between deciduous trees but in 2004 it was early in October not long after the autumn equinox. The evergreen trees appear to keep their auras. The other sacred plant associated with the oak is the mistletoe. There are very few oaks infected with mistletoe today but on a nearby farm there are a number of lime trees heavily infected with mistletoe. The aura of these trees was checked almost weekly during the winter of 2004 and it was found that all the infected trees retained their auras. The trees were effectively being kept 'alive' during the winter, a period of death, by the mistletoe. If the Druids were aware of this it would be a source of wonder to them. A plant with power over death must have mystic properties.

Holy or religious trees were almost certainly selected as such for good reasons thousands of years ago. They may have been of great importance as a building material or for tools, they may have provided medicines or protected against some disease or pest. The tropical Nim tree, sacred to Hindus, is a source of aspirin, and a pesticidal resin and smoke if burnt. The root and bark are said to be anti-malarial and the seed produces oil still used as an antiseptic. The bean tree, so named because it produces long hanging bean pods that turn black, is a source of aspirin, quinine, a heart slowing drug, caffeine and tannin. Provided witnesses are available all these materials can be dowsed and identified in the aura of the tree or of its fruit and seeds. Somehow, the ancient peoples in different parts of the world became aware of what to them were magical properties of trees and so they became holy trees.

From the tropics back to Britain in the winter. Once the rings and summer aura of the oak had collapsed it was found that the tree still had an aura but an unusual one. There were eight rays with four of the resulting eight segments being closed across the top to form a triangle. When marked out on the ground a possible origin and source of the triangular anterooms of the underground complex and the 'Celtic' Cross became apparent. At this time it therefore looked as if two main aspects of the Henge Temple design might be based on the aura of the Druids holy tree, the oak. Later we were to find that our interpretation was not correct and had to be modified.

Box 8.2 *The Magnetic Fields of the Trees*

Trees have well developed magnetic fields or auras which contain a considerable amount of information about the tree as a living organism. Although reference has been made to the 6 + 3 rings in the oak's aura it should be remembered that a tree is a living dynamic system which may lead to changes in its aura. The tree is also subjected to the magnetic fields of the earth and sun which may again cause changes in the fields of the tree's aura. The result is that the aura is not a stable simple system of rings and rays. The three ring systems given in table 1 differ. If the aura is examined more closely a complex system of fields emerges. One component of the aura is found to be a set of triangular segments and rays, see figure B8.2.1. A segment may differ from its neighbours in terms of what fields are represented in it, that is the materials and metabolites contributing to the fields and their position in the ring system. A list of materials against ring number does not necessarily hold all the time or between different species of tree. Clearly the amount of glucose or other sugars circulating will depend on light levels and the time of day. The list of compounds given in the table shows that some rings are possibly stable, rings 1 and 2 for example, whilst the materials producing some of the other rings vary, see ring 5 and 6. The diagram, B8.2.1, shows some of the detail of the winter aura of a tree. Auras are complex because all the materials associated with the life of the tree produce a paramagnetic field and are consequently represented in it. The general shape or geometry of the aura is however reasonably standard so that in summer there are 6 + 3 rings and in winter eight rays and eight segments can normally be identified. The extent of some of the segments may be limited. This shows up very well in the oak. If the tannin rays are marked out, the closed segments identified and the outer boundary marked, the possible origin of the 'Celtic' Cross can be seen. The cross is due to tannin and is present during the winter when all the rings have gone. The Cross would have been very 'visible' to a Druidic dowser during the winter. At this point in the project we only knew of the rays, triangles and rings. Later in the study the full complexity of the oaks aura was discovered.

Table 1 The Aural Rings of a Eucalyptus Tree and a Flowering Cherry

Ring	Eucalyptus		Flowering Cherry
	Sept 04	*June 05*	*June 05*
1	Calcium sulphate	Calcium sulphate	Calcium sulphate
2	Magnesium carbonate	Magnesium carbonate	Magnesium carbonate
3	A plant fertilizer	-	Glucose, amino acids
4	Nitrogenous materials	-	Nitrate, blood fertilizer
5	Carbohydrates	Amino acids	Iron oxide
6	Tannin	Nitrate	Tannin
7	Sucrose	Tannin	-
8	Starch	Iron oxide	-
9	Iron oxide	Glucose	-

Table 2 The distance to the rings in metres from the trunks of three trees

Ring	Oak (1)	Oak (2)	Eucalyptus
1	2.50	2.30	2.40
2	6.00	5.00	5.00
3	9.30	7.70	8.00
4	12.50	10.15	10.70
5	15.70	12.95	13.30
6	21.00	17.40	15.30
7	24.40	20.25	18.80
8	27.80	22.90	20.70
9	30.50	-	23.00

Box 8.2 Continued

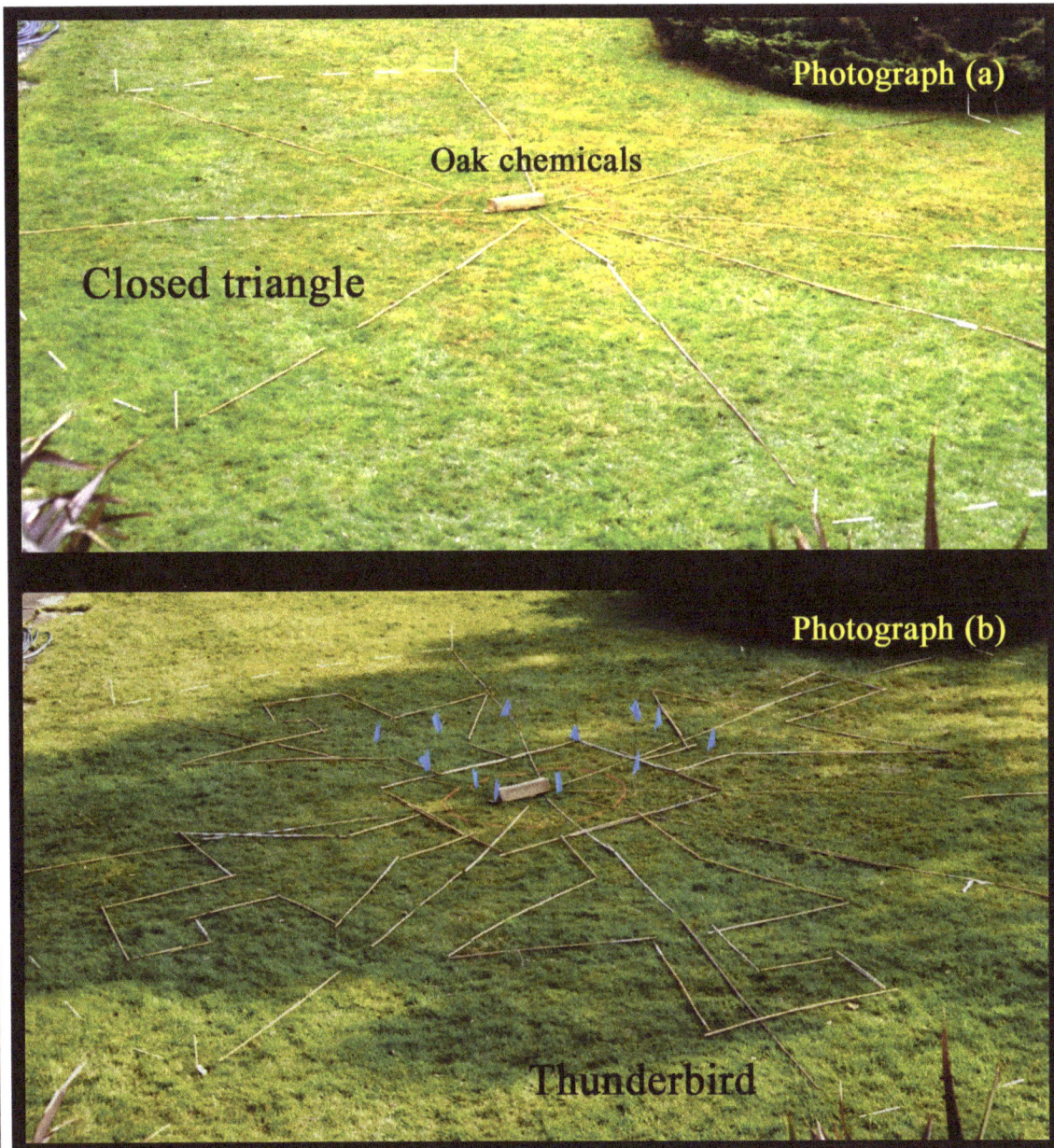

Photograph (a)

Oak chemicals

Closed triangle

Photograph (b)

Thunderbird

Figure B8.2.1 The Cross and Thunderbirds Photographs

a) shows the eight radial lines that form four closed triangles in the form of a cross.

b) superimposed on this cross are four Thunderbird heads which originate from a rectangle. The Thunderbirds have been used by the Druids as a symbol of death. The two auras have been created by using a model of an oak tree in winter seen as the cardboard box

The Tracks Left by Wheels

The ancient roadways, often referred to as Ley lines, show the wheel tracks of the carts and the tracks of the draft animals pulling them. Most ancient roads indicate that the wheels had iron rims and so when dowsing it was the iron that we normally looked for as an indicator of wheeled transport having used a road. However, wheeled vehicles were used long before iron was available and they evolved with time and the uses to which they were put. Whatever their use or design, wheels or the axles to which they are fixed require lubrication. In the ancient world the lubricant tended to be beef fat. During September 2004 the lubrication trails left by wheeled transport had been identified. We were therefore no longer dependent on finding an iron trail to identify vehicles. It was now possible to find the tracks of wood wheeled vehicles and from the position of the fat trail relative to the wheels, to say if the wheels were rotating on the axle or if the axle was rotating in its supports. The distance between wheels could also be measured. It was therefore now possible to identify some details of the vehicle. Next there was the question of who or what was pulling the cart. Were they pulled by people, oxen, deer or horses? It turned out to be quite easy to identify the motive power and also the point on the vehicles journey at which the cargo was loaded or unloaded. The load could also be identified by the stain created in the ground. The later part of 2004 saw some big steps forward in analytical techniques that could be applied to unravelling the mysteries of the henges, and Druids.

I mentioned that it was possible to identify horses and oxen pulling vehicles. The way to do this came about when it was decided to try and identify wattle and daub walls. The wattle is made from hazel which can be identified by using a piece of hazel as a witness. The daub contains cow dung. When it was tried as a witness the dung worked a treat. Horse manure gave no response on daub but it could pick up horse tracks. Horses were always associated with the large chariots that we had found up to that time. Later Nigel was to find that deer dung was also used in daub.

Some Small Advances

During the summer of 2004 a number of minor advances were made. The reason for iron appearing in the tomb and bone depositories had been traced to the misuse of witnesses. Ancient red ochre gave a response with black iron oxide and nails. Once acid cleaned iron was used as a witness what was thought to have been iron was no longer found to be there. This affected our dating of The Woodhenge and the one in the garden. There was no iron and they must be Bronze Age.

The triangular anteroom to the bone depository was becoming more complex with a possible offering table, the burning of incense and a lamp in one corner. One of the more important steps forward was the clear demonstration at Chenies Manor House that dowsers responded to stains (chemical remanence) in the soil. The underground tomb complex could be a stain but the stain would have to be deep as ploughing and archaeologists JCB's did not remove it. The pattern of pathways was identified with paths leading to the seven altars. The 6 + 3 ring system of the oak was identified and its size fitted the size of the temples. Some aspects of the temple design – size and number of rings, related to the oaks aura. Incense lamps had been found on temple sites, incense could be dowsed in particular parts of the temple complex. The position of lamps within the temple could be identified.

Dolmen

Early on in the study of the temples a double path had been identified running from the central area of the temple out to the west. This path could be followed if the temple was in a large field for sometimes a hundred metres and more. The path finished at a given point where the ground was rich in phosphate. At first it was thought that a body was laid out on a wood bier to allow animals to strip the flesh down to the bones. The bones would then be taken by the priests to be buried somewhere. At this stage the connection between the Dolmen as the bier came to be called and the central tomb and bone depositories was unknown. For a while it rested with just the existence of the path and the Dolmen. It then slowly began to dawn on us that animals run off with bones and the best strippers of flesh who can be trusted to leave the bones behind were birds. If they were birds, both the body and the birds would have to be protected from animals. The Dolmen could not therefore be a simple bier. On investigation the first thing to be discovered was an area of red ochre under the bier which was not scattered around to form a hallo as the phosphate was. The red ochre was limited to a small area under the centre of the bier. When the path to the temple was looked at in more detail it was found that there were regular staff marks on the outside of each path. These staff marks gave a response on carbon and ash wood, not oak or any other wood. Later and on other sites, iron would sometimes be found. Priests with leather shoes and staffs were in procession between the temple and the Dolmen. The next step was to find walls and other structures round the bier. This early work was done on one particular site in the village. Soon a wall was found round the bier and something inside running parallel with the bier. Again it rested there for a while until Nigel hit on the idea of using feathers and bird droppings as witnesses. These witnesses were so effective that the Dolmen complex came to life. The Druids were using birds to strip the bones. Not only that, the birds were of religious importance to the Druids because feathers were identified as part of their dress. It was also about this time that it was beginning to dawn on us that the nicely curved large depression or bowl at the centre of temples (not the small ones for pools) that were once temple floors were not nicely curved 2000 years ago but were almost certainly terraced. The use of terracing means edging and surfacing. A terrace is a road going round something and if the Druids could build them inside temples they could build them outside temples. Engineered roads now entered the scene. We were by mid 2004 aware of the prodigious use of wheeled transport by the Druids and roads for the vehicles made sense. The techniques for following paths and roads were developing all the time. The use of sulphate, leather, carbon, beef fat, hemp, iron, animal droppings, animal hair, materials being transported, staff marks, the materials from which the roads were constructed including mortar, what the vehicles were painted with. At the St. Albans site Nigel had even identified the garages used for the more important vehicles.

Summary

By the end of September 2004 we had learnt a lot about the use of witnesses. Their ability to dissect out the detail was truly impressive. Great care had to be used as few witnesses are pure, even the paper they were mounted on could act as a witness in its own right. The Henge Temple was no longer a simple system of circles. Its complexity was beginning to emerge. The scale of the engineering was becoming even more impressive. The ceremonial chariot tracks were often found leading to huge dyke and temple complexes and the first hint of a consecration ceremony prior to building the above ground temple was beginning to emerge.

Chapter 9

Stone Circles Reveal More of their Secrets

As work progressed on the Woodhenge Temples there was a need to check out new discoveries at stone circle sites. It was now very clear that there was a connection between stone circles and the Woodhenge Temples. There were good reasons for them being together and we had to find out what those reasons were. Earlier visits to the Avebury Stone Circles, the Rollright Stones and many others had shown that stone circles and woodhenge sites were connected in some way. It was not coincidence that placed stone circles on woodhenges. The imagination conjured up many possible connections but the aim was to find the real links between the two. Avebury (Box 9.1) was the first site on the list.

Avebury Stone Circles

The stone circles at Avebury are, weather permitting, reasonably easy to work on. Unfortunately on the day I had chosen for a visit, which was early in April 2003, the weather was not good and dowsing conditions were difficult. However, I was able to pick up footpaths and the tracks of wheeled transport, and to follow them into the main circle and round the outside of the big circle, (Figure 9.1). Along the West Kennet Avenue, double processional foot paths on either side of iron tyred vehicle tracks could be picked up. The Beckhampton Avenue along the village High Street also had double footpaths on either side of the tracks of an iron wheeled vehicle. On the West Kennet Avenue the iron wheeled vehicles could be traced up to the Sanctuary and round on its left hand side. After a difficult day I drove home from Avebury convinced that I could work out a lot of the ceremonial procedures used by the Druids and also knowing that the site was being used in the Iron Age. Regarding its beginning, there was only the standard archaeological view – Neolithic to early Bronze Age. I was, however, no further forward in terms of linking stone circles and woodhenges.

It was over a year before I was visiting Avebury again, this time the objective was to determine if the Woodhenge Temple sites fitted the normal pattern in having underground tomb complexes and were the small stone circles inside the Great Circle, built on tombs. At this time there was no reason to believe that the stones were not Neolithic or early Bronze Age. The site was in use during the Iron Age but that did not stop it being built in the late Stone Age or early Bronze Age. Both the stones and the size of the complex were too massive to be an afterthought on a Woodhenge Temple. However, when I started looking for the Woodhenge Temples in the Avebury complex they were there in abundance. After identifying a few opposite the village car park it did not take long to discover that the dyke cut into both the temples and their Dolmen paths and that some of the stones were also laid on postholes. The stones therefore came after the Woodhenge Temples which at this time I thought were Iron Age due to not realising that iron oxide in the ground did not mean that it came from iron and that it could be a contaminant of other materials.

Crossing the road to look at the small southern stone circle (Figure 9.1) I soon found the postholes of a woodhenge and then the underground tomb complex. There was also the access shaft and tunnel to the central tomb. Some of the stones appeared to be on the 4th ring of postholes. As far as the southern inner stone circle was concerned it fitted the pattern in that it was on a woodhenge complete with its tomb complex. There appeared to be no exceptions, stone circles were always on woodhenges and the question was 'what were they doing there'?

It was clear that Avebury was going to provide answers to many questions so three weeks later I was back with Nigel to try and answer the question 'was the Avebury Stone Circle Iron Age?' At this time we did not know how to date the tombs of the Prince and Princess. All we had to go on was the presence of iron or rather its oxide. The logic being that if iron was present it must be the Iron Age. The first task was for Nigel to check out what had been found on the last visit and to make sure that the stone circle had a lot to offer. Following this, the idea was to check on the transport of the stones. The tracks of iron tyred wheels are easy to identify and if iron tyred wheels were being used then the stones must have been moved in the Iron Age or the beginning of the Iron Age had to be redefined. It did not take long to pick up the tracks of iron wheeled carts leading up to the stones. The iron, axle lubricant, the cattle tracks are all easy to find. The stones were transported on carts which had four wheels to an axle, at least on the cart track identified at the time. Later, carts were identified with two wheels and six wheels on an axle. To date it has not been possible to identify the number of axles on a cart. There were four cattle tracks associated with the carts that had four wheels per axle. Finding the range of cart sizes and the use of large teams of oxen provided more evidence that the Druids knew quite a lot about transport engineering. Moving from the outer Great Circle to the inner South Circle complex (Figure 9.1) the cart tracks to the stones can again be identified. The next step was to look at the stones of the West Kennet Avenue (Figure 9.1). The stones on either side of the avenue would have been of a standard size and would have probably been brought along the same track from the hill slopes to the east. If there was a main track for the carts along the avenue it should be possible to follow the tracks as they branch off to individual stones. A main cart track was identified going along the eastern side of the West Kennet Avenue. Tracks could also be identified leaving the main 'road' then backing into the pit prepared for the stone.

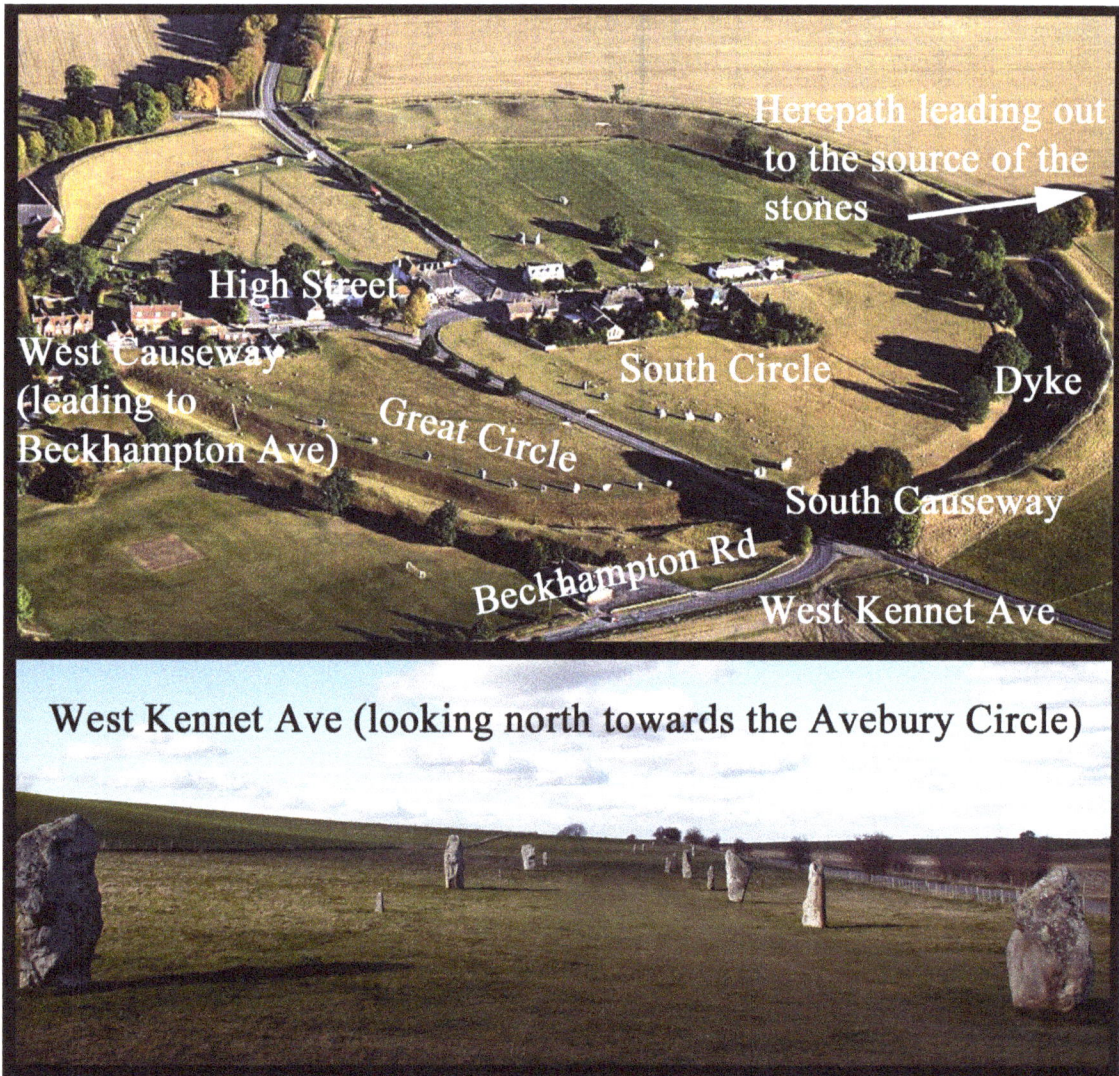

Figure 9.1 Avebury Stone Circles and Village

The tracks of wheeled vehicles were traced coming in, or out, at the South Causeway with iron wheeled vehicles turning and running between the Great Circle and the dyke. The West Causeway leaving the Great Circle leads to the Beckhampton Avenue. This avenue is on the left hand side of the village High Street. The West Kennet Avenue is in open fields and easily accessed.

Now it does not take an engineer to recognise that you cannot pick up a large undressed stone weighing many tons carry it on a wooden cart and push it off the end of the cart into a hole. If you try it you have got problems, not the least of which will be the loss of your cart as all of the weight moves to the rear axle. To move stones which had been conveniently left by the last Ice Age about a mile or so up a hill to a new location, certain things have to be in place. If the stone is to slide off a cart the stone requires a smooth dressed face. The ancient mason could not pick the easiest side of the stone to smooth down as the magnetic fields would dictate the face to be dressed. The stones also have to look good when set up. Other requirements may be to paint a surface, to reflect lamp light off a surface or reflect sound. Bearing in mind the importance of the magnetic axes of the stone, any stone dressing would have to be on the appropriate face so that when the stone is in place the required magnetic harmony is created. This means that there are at least two reasons for moving stones around in the fields where they were left by the ice sheet. The place to look for signs of heavy lifting gear is therefore going to be on the hills from which the stones came. Once the stones had been dressed they would have to be placed on the carts in the correct position for unloading so that there is minimal need for heavy lift gear at the stone circle site. It should also be possible to find the debris from working the stones on the hillside. After the stone was brought to the site it would be acceptable to slide the stone off the wagon provided it went onto a load bearing fulcrum upon which it could tip. The weight of the stone would not then go onto the rear axle. To stop the stone from tipping too far there might have to be a frame, sufficiently heavy and strong to hold a stone weighing tons and to prevent it from falling over. It would not be beyond the engineers to realise that jacking up the front of the wagon would help the stone to slide. Signs of this activity could also be looked for. On this visit to Avebury the exploration of the source of the stones on the hillsides would have to wait. The main aim was to try and identify how the stones arrived in the Avenue and how they were set up. Regarding the date of arrival of the stones on site, in August 2004 the only method available for dating the stones in the avenue was to analyse the wheel tracks and determine what the tyres were made of.

By following the cart tracks along the West Kennet Avenue it was easy to find tracks leaving the main track way and going up to the stones or places where stones once stood. The dowsing evidence indicated that the cart was probably backed up to the hole prepared for the stone and that there was some wood structure round the hole to stop the stone tipping. Backing a heavy load is cumbersome so this will have to be looked at again.

After a quick look at the Avenue the next step was to check the hill slopes to the east of the stone circles. Three tracks indicating four wheeled axles were found. The tracks were leading to where the stones came from. It looked as if the whole history of the construction of the stone complex at Avebury was laid out on the ground. I noted in my records that the place was a 'Godsend'. The first clear indication that the great stone circles were Iron Age and built by engineers who knew what they were doing and not by amateurs. The final proof that they were also Druids would come later. At this stage there was still no clear evidence to contradict the accepted view that stone circles were stone circles in their own right. They could have been built on the site of woodhenges during the Iron Age. However, the 'accepted' view was not unanimous and since at least the 1920's it had been suggested that stone circles were part of large wood temples. If they were right the stones should fit into the ground plan of Woodhenge Temples. Three weeks later at the end of September 2004 I was back at Avebury with Nigel. One aim was to follow cart tracks up the hill to where the stones were loaded onto carts or HGVs as they were now being called. Carts carrying the loads the Druids were dealing with would not be seen again in the British Isles for about 1800 years. Also, once the Romans had left, the Druids mastery of civil engineering would not be equalled for a thousand years or more. However, the first step was to confirm what had been found on earlier visits. The Woodhenge Temples, some being cut into by the dyke, were confirmed, wheel tracks up to the stones, some stones on postholes was also confirmed. The side of the stone laying on the cart was identified. It was as expected a dressed surface. Torches or fires on an area of clay or tile were on the outer side of the stones. The cart tracks to the stones were broken by the dyke indicating that the dyke came after the stones were moved or that bridges were constructed over the dyke. A bit of scrambling along the walls of the dyke and searches along its bottom revealed chariot tracks and footpaths with staff marks. On the sides of the dyke were postholes and signs of bone depositories and patches of red ochre. Then the inner southern stone circle was confirmed as being on a woodhenge. We then moved outside the circle complex. Wheel tracks going up the hill were identified and so on a nice sunny day Nigel and I started to follow them. About a mile further on we were into harvested fields and by good luck found the local farmer who was interested in what we were doing. He filled us in on the more recent history of local Sarsen stones. Briefly, they were in the way of farming activities so pits were dug and they were buried so although they are no longer visible the stones still lurk beneath the ground. With permission to roam over the fields it was not long before a set of wheel tracks had been traced to a loading point. The marks of the carts wheels went in between two large wood structures. The dowsed carbon marks indicated that a substantial structure had stood there and that it was involved in lifting the stone and lowering it onto the cart. There were signs that iron had been used in the lifting structure and that moving parts had been lubricated with beef fat.

So far what had been found fitted the theory that stone circles were associated with woodhenges and that in the case of circles with large stones the circles were constructed during the Iron Age.

Stonehenge

The next stop was Stonehenge. Not such an easy site to investigate and permission is only given for one or two hours access in which to dowse before or after the public have access. In early October 2004 a preliminary visit was made to the Stonehenge area. The actual stone circle was not visited but a friend living in the area of Stonehenge had pointed out that a Stonehenge bypass and tunnel was planned. The aim of the visit was therefore to look at the proposed route with our local guide and see what damage might be done by the proposed road works. Walking across the fields towards Stonehenge from the east and to the south of the A303, woodhenges were found on the planned bypass route. The route of the tunnel could possibly cut into the underground tomb complexes of the henges. Although soil stains have survived under old established roads and modern ploughing it is doubtful if they would survive today's methods of highway construction. In short if the proposed works were to go ahead they could do quite a bit of damage to the archaeology of the area leading up to the valley at Stonehenge Bottom. The information we had was that the road was going to cross this valley much as the present one does, that is by an embankment. Looking down into and along the valley, to the trained eye the valley looks as if it was the recipient of typical Druidic ceremonial route engineering. There were areas of the valley side where it had been cut away, terracing and the postholes of wooden structures one of which, because of the views it would have had along the valley, looked like a Royal Box. There were bronze chariot tracks running along the bottom of the valley. It was easy to see that the valley formed part of a large ceremonial system. On later visits to Stonehenge we would meet up with the ceremonial route again on the other side of the A303. The ceremonial routes round Stonehenge looked important but to understand them it was going to be necessary to understand Druidism and the scale and complexity of the landscape they engineered. The centre of the religion, the temples, were only just forming a mental picture in our minds and the picture was very hazy. The next visit to a stone circle was a return to the stone circle at Stanton Drew.

Stanton Drew

Three days after Stonehenge I was at one of the largest stone circles in Britain. (Box 9.2 and Figure 9.2) I plotted out the central tomb to try and identify the centre of the underground complex and then the north section of the complex. There was a short tunnel into the triangular anteroom. The tunnels to the other anterooms differed in length with the one to the south being about six feet. The whole tomb complex was there. The main difference from previous temples was that the bone depositories were much larger and consisted of a left and right section. I picked up the access tunnel which joined the tunnel to the northern bone depository. The access tunnel on its way to the entrance shaft turned through nearly a right angle (Figure 9.3). This raised the question of how did those digging the tunnel know where they were. At this time we had come across several instances of tunnels turning through angles for no apparent reason. Those digging the tunnels would have to know where they were and where their target was. How did they do it? Did they have a compass, did they dowse from the surface, or did they do it by dead reckoning? The scale of the underground workings is such that the engineers must have had a system by which to navigate underground at least 4000 years ago.

Having established that the main circle at Stanton Drew was on a woodhenge site (the following year we would find that there were a number of tombs each indicating a temple) the next step was to check the stones of the Great Circle. The stones had a lamp on their outside and they had been brought in by HGVs. As at Avebury there was some evidence of a wood structure to support stones as they were positioned. Away from the stone circle in a place called 'The Cove' nicely positioned in the beer garden of 'The Druid's Arms' (Figure 9.2) there is a stone Dolmen, now partially collapsed. The phosphate halo is easily identified and footpaths or cart tracks head off in the direction of the Stone Circle. The stones at 'The Cove' weigh many tons and with one of them having to be placed on top, some heavy lifting was required. Possible evidence for this equipment was found in the carbon stains round the stones.

Figure 9.2 Stanton Drew

The study was restricted to the Great Circle and the Cove (Dolmen) in the garden of the Druids Arms.

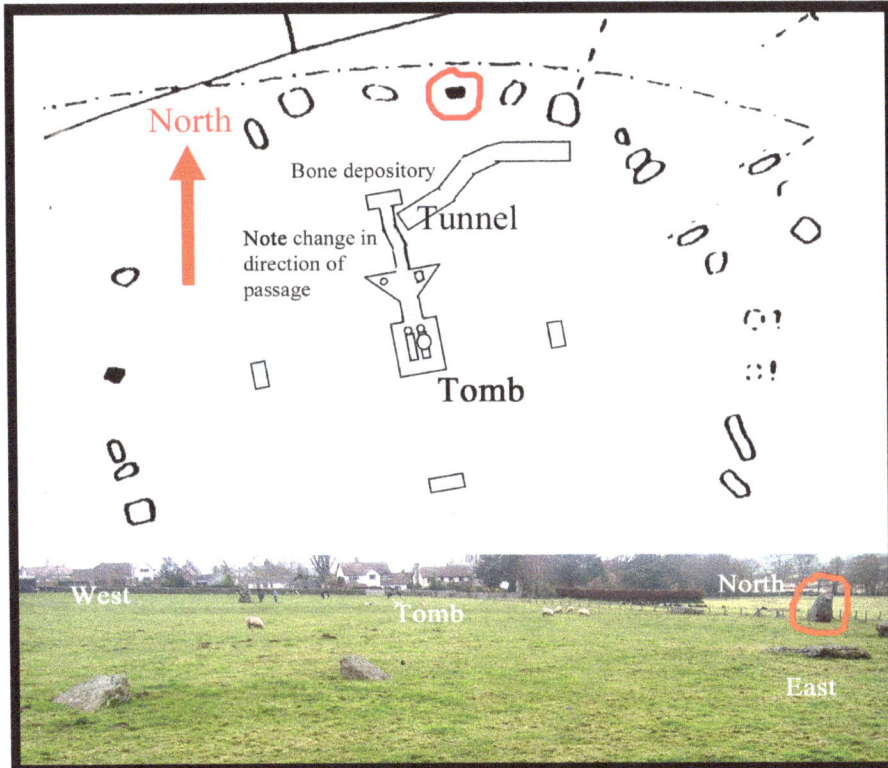

Figure 9.3 Temple Complex

A tomb was identified near the centre of the Great Circle with the Prince and Princess lying on a bearing of 350°. The bearings of the bone depositories from the tomb were North 353° East 103° South 178° West 252°. From the tomb there was a short tunnel into the anteroom which contained a small altar and a lamp. The passage to the bone depository was about 24ft long and changed direction twice. The access tunnel bent through nearly 90° before joining the bone depository.

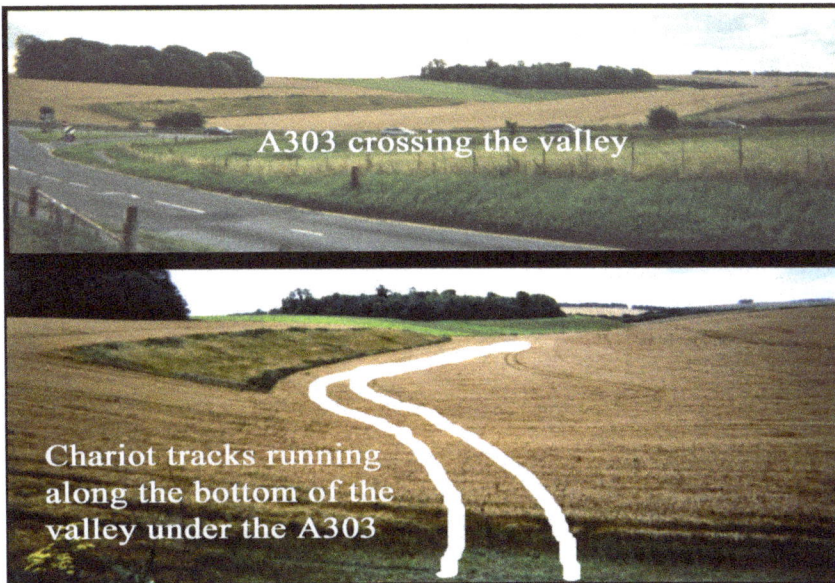

Figure 9.4 Stonehenge Bottom

The area shown is Stonehenge Bottom where the ceremonial chariot tracks go under the A303 and follow the valley towards The Avenue. To the south of the A303 there are signs of viewing areas and what could be a "Royal Box".

Biolocation Methods Move Forward

After some dowsing at Stanton Drew it was back to the development of methodologies. Nigel had found that it was possible to dowse for human hair. I tried the technique out on a tomb not far from my house and found that the Princess had hair down to her waist. The Prince had normal length hair in that it was not off the shoulders. This added one more item to the list of things to be checked in a tomb. Linen, wool, nettle, hemp, wood resin, feathers, dog hair, leather and method of tanning were already on the list along with weapons and metals.

The tomb was becoming more interesting but it always has to be remembered that dowsing is dowsing and all that is being picked up might be the stains left by objects long gone. If this were the case it is easy to see how bodies, clothing and wood objects could have left stains. But the shield had decorations on it such as amber and there were semi precious stones in the tomb. It is difficult to see how these objects could have left stains in the soil. It was becoming increasingly likely that the bodies are still there or at least had remained there for long enough for hard solid objects to leave a stain. The gold could be a gold leaf and so perhaps very little actual gold was ever in the tomb. Finally we had at last found the wall of the temple. There was a hazel wattle and daub wall round the temples on the 9th ring.

Blood as a Witness

At this time the bodies of the Druids were identified by using a phosphate witness but it was while work was going on at Chenies Manor House that Nigel made a very important and significant discovery. The discovery occurred while he was dowsing a site at which one of the battles of St. Albans had taken place. He found that by using witnesses he could identify where cannon had been placed and fired. The plume of sulphur and charcoal from the muzzle could be identified. He then tried human blood as a witness and found pools of human blood scattered around the site. The loss of a significant amount of blood normally indicates the place where somebody died. Horse casualties could also be identified using horse blood. The tracks of wagons, horses, men, cannon could all be followed, the direction of cannon fire could be determined. It was now possible through dowsing to bring ancient battle fields to life and say a cannon was fired from here and soldiers died there. The use of blood witnesses soon paid off and within days, when visiting a farm, I had identified three lines of blood patches with single patches in front. The line bent back in one sector. The blood patches acquired the name battle blood. Near by were mass graves for the casualties.

Stonehenge Follow Up

Later in October 2004 it was time for another visit to Stonehenge. Stonehenge is on the A303 and about 70 miles south west of London. The henge and the area round it is a part of an ancient pre-Roman landscape which has attracted the attention of archaeologists for generations. The most bizarre ideas as to the origin of the stone complex and the use to which it was put by the ancients have been produced over the years. By now we were very hopeful that we would find the origin and use of the henge or at least obtain some good pointers as to who built the henge. The day started with a visit to a long barrow which appeared to be in a reasonable condition. Tracks round the barrow were checked for and found. Bone depositories under the Barrow and altars on top of it were found and we picked up the first wooden wheeled vehicle for the transport of bones. The tracks of the vehicle came up to the entrance of the barrow and on the right side of the tracks there was a stain of red ochre indicating where the bones had been loaded/unloaded. The barrow looked as if it was early Bronze Age or even earlier but it was still in use in the Iron Age. At the time we did not pick up any specific indication that it was Stone Age apart from the use of wooden wheels for a ceremonial vehicle.

Figure 9.5a Dells

Dells are characterised by isolated areas of mature vegetation and trees in a field. The Dell itself is a sunken depression often quite deep and well below ground level. The vegetation grows undisturbed. At each of the four quarters and lying outside the boundary of the Dell shown are eight Druids in two rows of four. Their exact purpose is still unknown but are possibly gate guardians or honour guards. Behind this Dell is Bury Woods which has a number of magnificent Dyke Temples.

Figure 9.5b Dyke Temple Entrance

The dyke was in a building of impressive size. Execution lines were found outside the temple on the flat ground top left. Out of sight but leading up to the entrance of the temple are eight sacrificed Druids in two rows.

Moving onto Stonehenge we decided to look for the HGVs used to move the stones. We picked up one set of tracks near the fence and road and then traced it back away from Stonehenge to find that it took a loop round the depression so that it followed the contour of the land and wound its way past the car park before moving away from it and towards the Cursus, (Figure 9.4). Later when we were inside the Stonehenge site for our one hour of dowsing we found the HGV track. It went across a shaft entrance so the shaft had been filled in long before the stones came onto the site. The fact that the shaft infill had taken heavy vehicles indicated that it had been filled in so that it would take a load. There are many of these filled in access shafts round the country that are still settling and causing subsidence damage in property. It is possible that the Druidic engineers knew enough to realise that to take the weight of the stones the underground chambers and tunnels had to be back filled, but this needs to be checked for. We found that it was possible to follow iron cart tracks up to the stones, many of which were on postholes, and where the heavy lifting or tipping gear could be identified with its lubrication points clearly marked. Lamps were found both on the inside and outside of some stones. On this visit to Stonehenge one aim was to try and find signs of battle blood as the Romans were said to have killed the Druids. Not an easy task with big stones lying all over the place. In spite of this, in one area towardsthe centre traces of 8 bodies were found. The phosphate outlines indicated the bodies had been left at different angles. However, the question arises as to whether they were a sacrificed guard for the temple. Outside the stones a mass grave was found with possibly 100 bodies or more. There was battle blood scattered around the outside of the stones where the fighting may have taken place. Also there were three lines of bloodstain with a row of more widely spaced single stains in front or behind. At this time, October 2004, our interpretation was that we were looking at the Roman slaughter of Celts and Druids with the priests standing at their posts to be killed by the Romans. It was looking as if the Romans were responsible for the destruction of Stonehenge.

Execution Lines

Nigel on returning to St. Albans found that the temples he was working on had the same three row pattern of bloodstains round them as Stonehenge. I checked the Dyke Temples in Burry Woods. The entrance to two of the temples had the phosphate outlines of what we were now calling the Temple Guard. Later we were to find in the recesses of the large temple pools of blood as if they were slaughter areas and a few detached patches of blood. The second temple showed a similar picture. Outside the temples the treble lines of blood stains. Unlike the first sighting of this pattern most lines did not waver as a battle line might.

The battle blood picture was very similar to that found round the Druidic Temples in St. Albans and at Stonehenge. It was beginning to look as if the Romans killed everybody associated with the Druidic Temples. The question arose as whether the Romans destroyed the temples and those associated with them before of after the revolt by Boudicca, the Queen of the Iceni. The reason for the three lines of bloodstains outside the temples only slowly dawned on us and it was a while before we realised that they were execution lines. The eight Druids at the gate of the temple died at their station, or so we thought at the time.

Guards Through Time

A little while later I checked a small dell out in a field next to Bury Wood (Figure 9.5). I found the eight guards by the north gate and eight at each of the three other gates.

The second pair of Druids in each set of eight had a dog each lying beside them. Concentrating on the northern gate I found that the heads had been removed and lay a little way from the bodies and the pool of blood. By dowsing and using witnesses the long linen dresses could

Key

▌	Druid's body
◊	Dog
○	Head

Northern edge (Great circle)

Rays

Form plotted
and measure

N

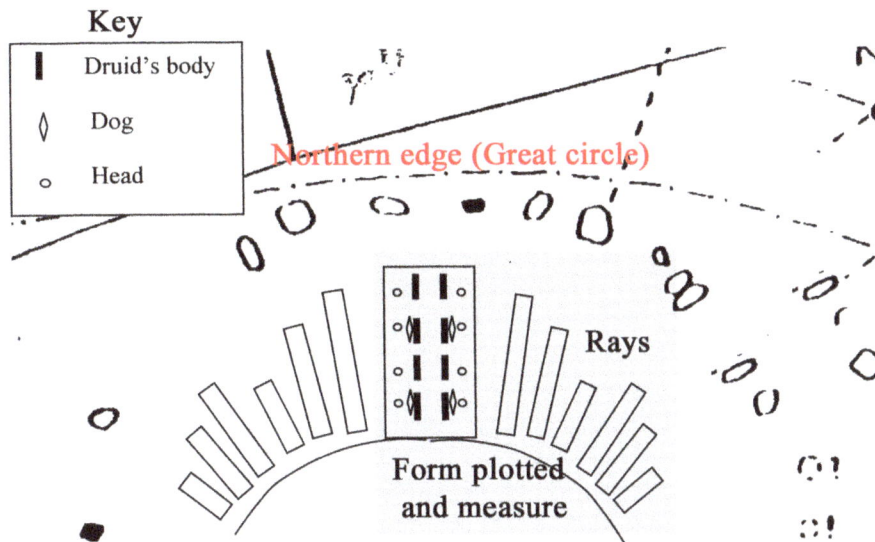

Figure 9.6 Stanton Drew North Gate Area

At the north gate area of the Great Circle were eight sacrificed Druids. Each of the second pair from the front had a dog beside them. The Druid's heads were on straw mats on the outside. To the eastern side a curved line was revealed by a manganese dioxide witness from which rays of different lengths projected.

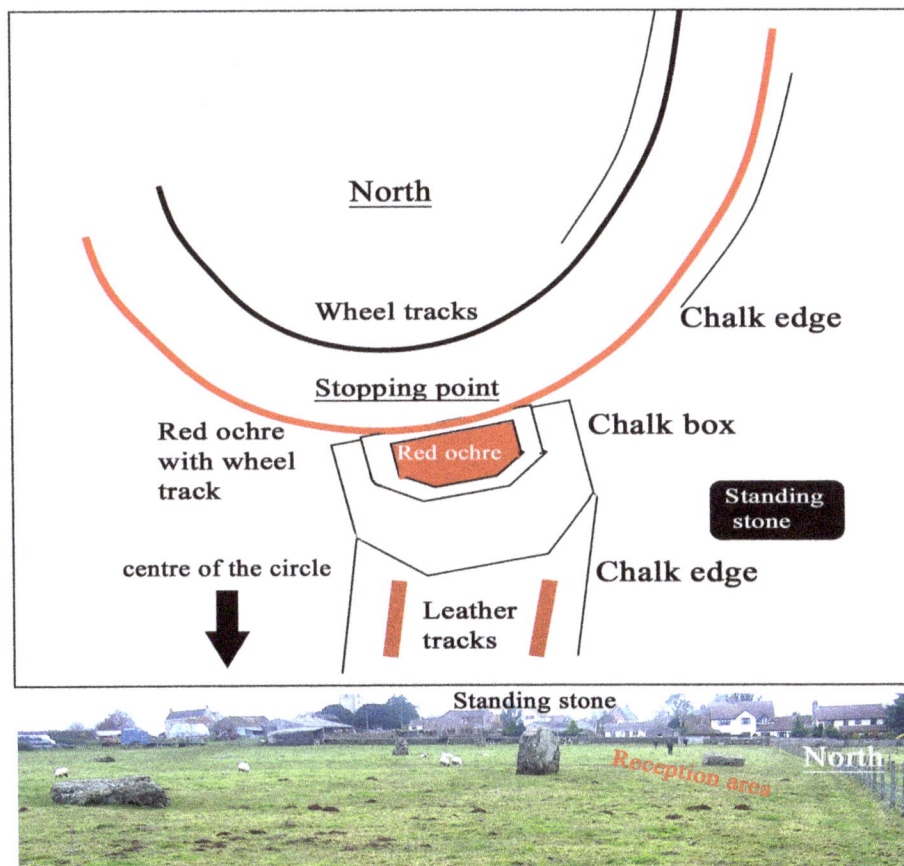

North

Wheel tracks

Chalk edge

Stopping point

Chalk box

Red ochre

Red ochre
with wheel
track

Standing
stone

centre of the circle

Chalk edge

Leather
tracks

Standing stone

Reception area

North

Figure 9.7 Bone Reception

The reception area for the cart bringing the bones to the temple and the path to the point where the bones are handed to another set of priests. The reception area is on the Great Circle between standing stones.

be identified, the dogs, the hair of the head, the ash staff. I dowsed one body in detail and pegged it out, the legs, body, arms, linen gown, leather belt, shoes, and staff. As the Druid took on the form he had when lying there over 2000 years ago, along with the other seven, it created a sight I will never forget and to this day I bow to them as I go past the dell. It was easy to see that people more sensitive to magnetic fields than myself would be able to see where the bodies had lain long after they had decayed and disappeared. To such people the spirit of the person was still there. Dowsing is a bit like relying on touch when diving in zero visibility. You convert information from touch or dowsing into a visual image which you then work with. The image created on this occasion, as is so often the case, was not totally correct but a more accurate image had to wait on more information becoming available. There were a number of things that did not quite fit the theory of temple guards standing there to face the Romans and being killed. I checked a total of ten temple sites and they all had eight Druids per gate. By this time Nigel had found out that one of the two rows at each gate were women. The precise positioning of the bodies, placing the heads on straw mats, the staffs always in the same position. It began to dawn on us that we may have done the Romans an injustice and that the Temple Guards were ritually killed. The Druids are said to have had a very strong belief in the after life. It may have been strong enough for 32 people in a community to volunteer for a particular task and be sacrificed. What that task was may never be known but there are at least two possibilities. The first is the consecration of a temple and the use of spirits to guard the entrances against the ingress of detrimental spirits. Much in the same way that Chinese people even today use guardians on the doors of houses. The guardian can be a gold talisman buried in the foundations under the door. The second could be the deconsecrating of a temple and guarding the underground complex. This is not so likely as the entrance to the tomb complex is the shaft not the gates to the temple. Yet a third possibility is that they are an honour guard to welcome something from the spirit world.

Stanton Drew Follow up

On the 19th November a final visit was made to Stanton Drew for 2004. Knowledge of woodhenges had been developing and new techniques were being applied. There was therefore a need to make sure that Stanton Drew still fitted into the general picture of woodhenges that had been developing. With permission from the farmer to put up canes in the main stone circle Nigel and I worked hard in almost continuous rain to mark out part of the temple. We made good progress and even had the occasional visitor who had braved the weather to visit the stones.

Four pairs of sacrificed Druids were found at the northern entrance to the temple (Figure 9.6).

We were now referring to them as sacrificed Druids as it appeared to be the only possible explanation if the Romans were not responsible. The lines of Druids did not line up with the centre of the ring of stones. We already knew that the tomb of the Prince and Princes did not appear to be central and the postholes that we had identified did not fit the circles of postholes the archaeologists said were there or the stone circle itself. This problem was partly solved by using a piece of local stone as a witness for finding postholes instead of a carbon witness. Using stone a different set of postholes showed up which looked much more like the archaeologists picture. A possible explanation is that the magnetometry survey made by the archaeologists picked up the stones used to support the posts of one particular temple and missed all the postholes without stones. As the postholes with stones in them appear to match the stone circle it may be that they indicate the temple at the time the Romans destroyed it.

Returning to the sacrificed Druids. At the St. Albans site Nigel had found artwork on the

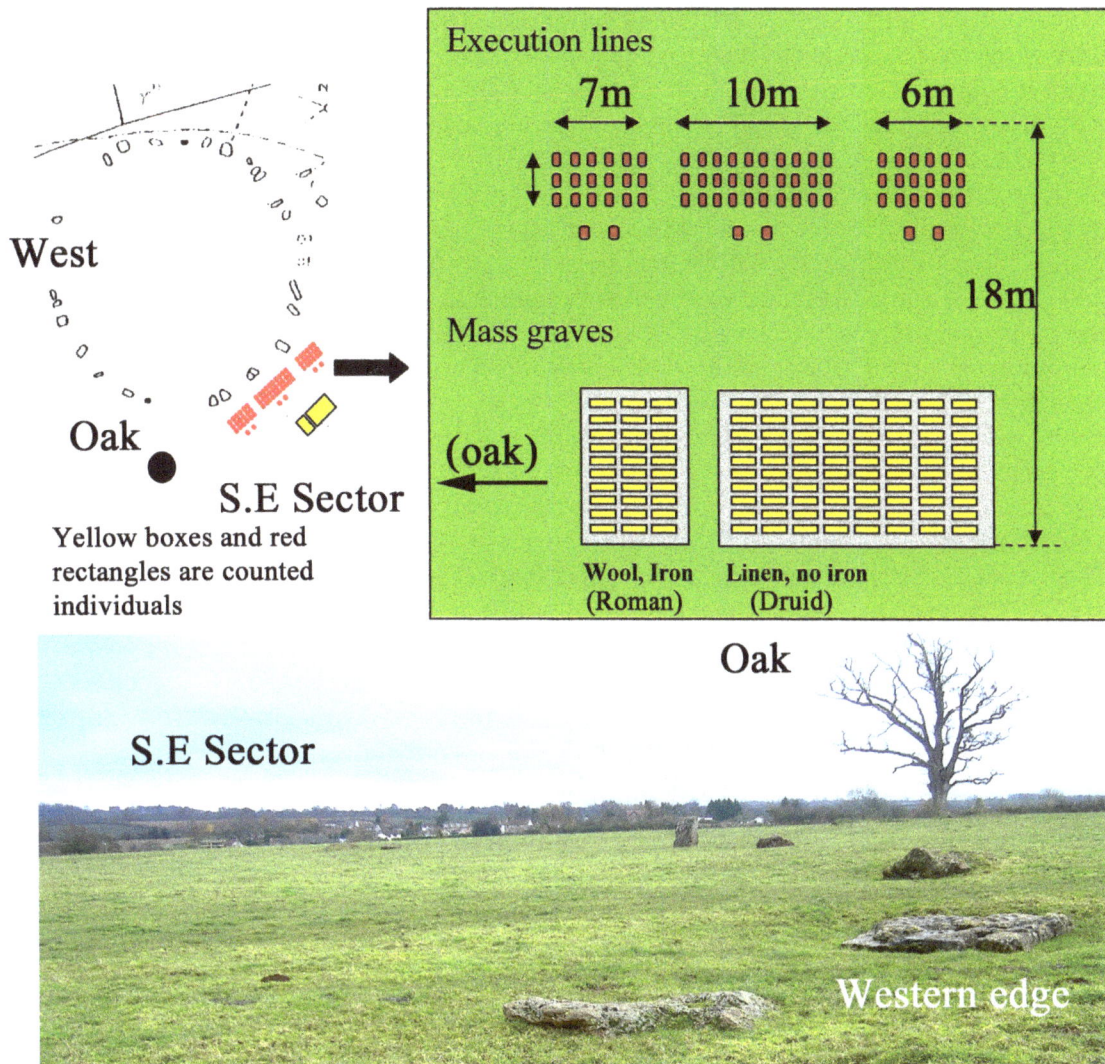

Figure 9.8 Execution Lines

On the south east sector of the Great Circle execution lines and mass graves were found. The smaller of the mass graves contains Roman soldiers as indicated by iron swords and wool clothing. The larger grave contains Druids indicated by linen clothing. Roman casualties indicate fierce fighting around the temple.

ground near the entrance to the temple that looked like a star burst. This was looked for at Stanton Drew and found at the northern entrance, see figure 9.6. A series of rays of different length radiated from a circle. The pattern of rays was too complex and extensive to do in detail so a small length was marked out on the ground and measured up. At the time, the pattern along with the sacrificed Druids was taken to be part of a consecration ceremony for the temple site. Later, as more details emerged, we were to find that the idea did not fit too well.

Using witnesses it was possible to find the paths and tracks of the priests and carts. One horse drawn cart came up to the north gate and unloaded bones covered in red ochre (Figure 9.7).

The priests then took the bones into the temple along a path that could be identified. Details of the building construction were identified. For example the roof structure of the building could be drawn out on the ground from the potassium lines produced by the burning timber joists.

109

However, some very careful work would be required to identify the roof belonging to the last temple which is represented today by the remains of the stone circle. Outside the stone circle there was battle blood and the three lines of blood pools stretching along the south eastern side with the high priests or officers in front. Mass graves were nearby (Figure 9.8).

Stanton Drew was therefore an active Druidic Temple up until the Romans closed it down in no uncertain manor between 43AD and 60AD. Later in the survey more tombs were discovered round the centre of the circle and so the site has a long history as a religious centre and has been home to many temples over the millennia. To find out just how old the site is the tombs contents would have to be analysed. Methods for doing this were developed in the following year. The temple represented by the stone circle is much bigger than standard temples if indeed the built structure goes out to the stones. This would have to be shown to be the case by identifying the wattle and daub walls. When dowsing postholes along radii in the main circle more than 9 were always identified and the 6 + 3 pattern could not be found. One possible reason is that when there have been temples on temples, as there have been at Stanton Drew, it is extremely difficult to identify the postholes belonging to a specific temple. The Great Circle is over a hundred metres in diameter and the large size of the circle presents a number of problems. For example if the temple was the normal size what was the space between the stones and walls of the temple used for. If the stones indicate the outer wall what was the extra space inside used for and what modifications were made to the roof to control its height. The only way to find out is by using biolocation. Any digging and the information held in the soil may be destroyed. (Chapter 17, Darlington Walls).

The Rollright Stone Circle

After Stanton Drew there was just time for one more site visit in 2004, so early in December 2004 the stone circle called 'The Rollrights' were visited. The Rollrights are to the Northwest of Oxford on the A44 (Chapter 3).

The stone circle is a well known one as it has an almost complete circle of stones. Although many of the circles stones have been chipped away at for hundreds of years enough is left to appreciate that at one time it was probably a complete circle of stone with entrances and exit. The larger stones are at least two metres or more high and look as if at one time they had been smooth on the inside. Using both carbon and stone witnesses there appear to be at least two sets of postholes which extend outside the circle. The outline of burnt roof timbers could be followed and there were signs of battle blood. The site is restricted by fences and hedges so it was not possible to look for execution lines and mass graves.

The clay bowl in the centre of the stones was about 4.8m in diameter and walking due west out of the henge and over a fence into a cornfield a Dolmen was found with its phosphate halo and the central area of red ochre. The staff marks of the priests came up on an iron witness indicating that iron tipped staffs were in use. If this was the case the site was used in the Iron Age.

The stones were checked again for their diamagnetic north pole (Chapter 3) and 56 of the standing ones had north at the top. Most of the stones were linked by an energy field, that is they had a north pole facing a south pole. The exceptions were the western and eastern pair of stones. On most of the stones a south pole faced out and there were signs at the eastern gate of the eight sacrificed Druids. The circle is, at about 33m in diameter, too small to be on the 9th ring of postholes and it appears to be on the 3rd or 4th. There have been a number of temples on the site so it is not easy to identify posthole rings and say which one the stones are on without some very careful work. The Rollright stone circle is an enigma as the stones being flat do not look suitable for load bearing within a temple.

Summary

As 2004 came to a close Nigel and I still had not found the link between the stone circles and woodhenges. We had ideas but as yet no proof. We were gathering evidence that the Romans may have been involved in the destruction of Druidic Temples. If this were the case woodhenges were standing at the time the stone circles were standing. We were finding execution lines and mass graves round Druidic Temples. The reason why bone depositories were associated with iron had been sorted out and we knew that we could no longer take the presence of iron oxides alone as indicating the Iron Age. The tombs and underground complex might predate the Iron Age. We knew that native red ochre with its mixture of oxides had been in use for a few thousand years before the arrival of the Iron Age. Stone supported posts or at least their postholes had to be looked for using stone witnesses. The postholes with stone in them could belong to the last temple on the site.

The winter aura of the oak tree had been identified as a set of four triangles much like the antechambers of the underground temple and they had possibly been the blue print for their design. The summer aura of the oak, 6 + 3 rings fitted the design and size of the temple. We were beginning to wonder if Druidism was descended from a Shamanistic culture in which nature and its spirits played an important role. Finally we were finding a system of dykes and dyke associated temples linked to the normal temples.

Chapter 10

The Henge Civilization appears through the Mists of Time

A Winter Break in Penang

In January 2005 I departed for the tropical climate of Penang, an island just off the West Coast of Malaysia, and left Nigel to continue dowsing in the cold grey weather of a UK winter. Officially I was visiting family, unofficially I was looking for some interesting dowsing with a bearing on "The Secret of the Stones" project. Once in Malaysia one of the first archaeological visits was to Lembok Bujang Museum which is about 50 miles up the coast of the Malaysian mainland from Penang and towards the Thai border. The museum is built next to a pre-Islamic era archaeological site. On the site are the ruins of an ancient temple which is accessed by way of a jungle path. Some of the temple has been revealed by excavation and is now a tourist attraction. Once on the site of the temple it did not take long to find drains, toilets, a well, graves and walls. Some artefacts such as granite carvings had been moved from the site to the museum. The most notable thing about the granite carvings was that they had been deliberately destroyed and broken into pieces. The attempt by the invader to destroy an old religious site does not appear to have been just limited to burning the place down and leaving it to the jungle. Even small artefacts had to be destroyed. Looking at the damage there seemed to be a direct parallel with what the Romans had done to Druidic buildings in Britain and the artefacts within them. Later in the holiday, a Neolithic cave site was visited. The site was called Tiger Cave and it was near Lerggong. The cave is in the jungle so once we had found the nearby village I hired some local children as guides and set off into the steaming heat of the Malaysian jungle. After twenty to thirty minutes of jungle tracks and scrambling up and down slopes a cave opening appeared in the side of a cliff. Fortunately the scree, tree roots and creepers enabled my Chinese assistant and me to climb up and join the children on a ledge just outside the cave. Looking into the gloom of the cave not much could be seen but on moving in to obtain a view of the floor it began to look as if a bomb had hit the place. My hopes of finding out what the cave had been used for started to evaporate. In a side 'room' I picked up human skin and there was possibly enough undisturbed surface left to get an idea of what might have gone on in the 'room'. The outside of the cave looked undisturbed and a phosphate area was identified not far from the cave entrance. The main lesson from the visit was that conventional digging does an awful lot of harm to a site. On the plus side a Neolithic jungle site could be assessed by science based dowsing – if you have the terrain ability of a mountain goat. From a site that may have been 10,000 years old to the twentieth century. In Penang there is a most magnificent botanical garden set in a small valley with steep jungle covered sides (Figure 10.1). In fact the gardens encompass some of the valley sides and the paths take you through dense jungle complete with snakes and monkeys. In terms of scenic setting it is one of the best if not the best botanical gardens I have seen. It is a fine place to jog round, which I often did as part of my heat acclimatization routine. However, it also has two dowsing attractions at least. One is that it is said to be the place the Japanese buried their gold and war booty at the end of the Second World War. The second is that it has a wide range of tropical trees many of which are considered to be sacred by one religion or another, possibly because they are the providers of pharmacologically active compounds. The trees and fruits studied in the Botanical Gardens had well defined auras showing rays, triangles (Box 10.1) and circles some of which could be related to such things as aspirin, tannin and caffeine. The staff of the Gardens provided help with identifying trees and explaining the religious connections. I did not find a 'Celtic' Cross or a 6 + 3 ring pattern. Some trees might show them but I did not find them. Having spent some time dowsing trees I moved onto treasure. The gold showed up and gave the strongest gold signal I have ever come across. Much stronger than the signals from any Princess in

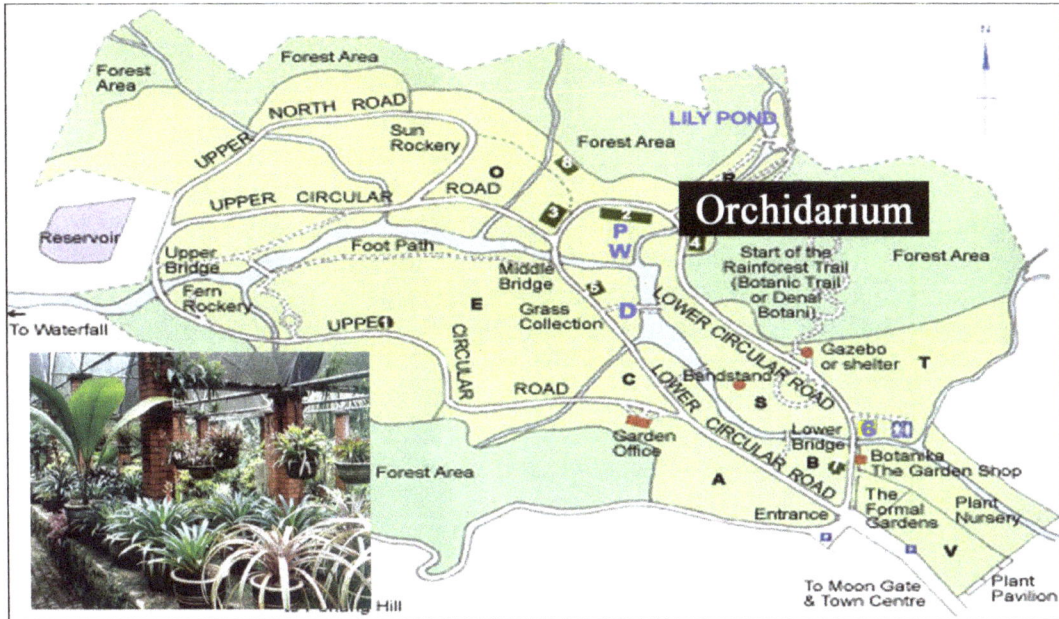

Figure 10.1 Penang Botanic Gardens

The Gardens are situated in a valley the slopes of which are covered by a rich jungle. There is a wealth of tropical trees and the lost gold is to the east of the Orchidarium.

Figure 10.2 Berkhamsted Castle - Historical View

The picture shows Berkhamsted Castle at about the time of King Edward IV. The ceremonial Druidic dykes became the moats and the round barrow provided the base for the Mott.

the tombs back in the UK. The compass bearings of the treasure are now on a map of the garden in a secret code waiting my return. One very interesting dowse was of a haunted house in George Town. The house had been derelict for thirty or more years. The reason for its abandonment was that a whole family had been brutally murdered and no one, even decades later, would live in the house. The house had been cleaned of all trace of blood. Not a trace was to be seen even under close scrutiny. However, when I dowsed the house using a blood witness the patches of blood showed up just as battle blood showed up in the Hertfordshire countryside. When the patches were tested with a nicotine witness some blood patches showed its presence some did not. Nicotine was used as a not very accurate method of identifying the men. I had been trying to obtain some birth control pills which should enable me to identify the graves of women from their hormones but all my young female friends, and Nigel's, refused to own up to their use. The problem was later solved when I realised that all we needed was male and female urine plus urine from children for good measure. Once Nigel's female family members had donated samples the sexing of bodies or their remains in graves and battle blood became routine.

Return to Bovingdon

By mid March I was back in the UK and on 22nd March 2005 the summer aura of the oak returned, at least in Hertfordshire. One day the aura was not there the next day it was. Exactly the same was to happen in 2006 and 2007. We considered that the Druids would be able to pick up this event just as easily as a 21st century dowser. The significance of the event is that the Druids may not have needed sight of the sun to know that spring had arrived and the oak coming to life would have had far more significance than a sunrise to the ancient people.

At the Bury Wood site three lines of bloodstains were found with the partial fourth line in front. They were running along the outside of the main temple. Inside the temple it looked as if people, possibly children, had been driven back into two alcoves to be killed. By standing just outside the temple I was able to pick up the direction of the mass grave which was 50m away. It contained Romans as well as Druids so some fighting had taken place and the Romans had taken casualties. The total number of bodies in the grave exceeds a hundred.

Berkhamsted Castle – a pre-Roman Past

About five miles from Bury Wood is the Country town of Berkhamsted with its well known Norman Castle. Well known because the castle is described as having three moats and a Mott which is a big mound upon which the final defensive position of the castle rests. At the top of the Mott is the entrance to what is said to be a well, sunk deep into the mound of earth. As far as I know nobody has questioned the role of the building and the use of three moats. The literature just says that three moats is an unusual feature of castle design. The Mott is perhaps not such an unusual feature.

However it does not take a military expert to realise that if a defensive position is on a pile of earth and somebody digs away the earth from a steep earthen face the top is soon going to follow. Perhaps a more likely explanation is that the Mott was built for the view of the surrounding hills, valley and marshlands which is well worth the climb to the top. Rather than a defensive castle, Berkhamsted Castle was perhaps a country retreat for somebody partial to water fowl and with an army nearby to rescue them if need be. This idea is supported by the fact that one of the fires in what is left of the castle's kitchen was set aside for fowl such as ducks and geese. The Castle was at one time a large and palatial place which now provides a very extensive sheltered lawn area as well as the Mott set at one end (Figure 10.2).

Figure 10.3 The Castle Ruins
The present day remains of the castle are shown. The central area is home to wheelhouses, roundhouses and temples. Two remaining moats are part of the Druidic dyke system with chariot tracks easily identified. The wheel tracks going to the Mott are of a small vehicle with wood wheels.

Because the standard view of the origins of the site do not sit too well with what is there, that is three moats that do not look like moats and a Mott that does not look like a purpose built Mott, other ideas about the Castles origins have developed over the years. One is that the Mott is on an ancient barrow. The moats also look much more like an ancient ceremonial dyke system for chariots. I had dowsed the lawns at Berkhamsted Castle for signs of houses in 2001 and had found some. In 2005 with a small group of dowsers we decided to look more closely at the 'moats' and Mott. The aim was first to see if there was any evidence for the moats being chariot routes and if the 'moat' walls had any artwork on them? Artwork had been found on dyke walls at St. Albans. Secondly to see if there was any evidence that the Mott had started life as a round barrow (Figure 10.3).

The moats were dry and it was too early in the year for the vegetation to have got going so it was easy to pick up the seven foot gauge of the iron tyred wheels of chariots. They could be followed round the moats. Iron tyred wheels were used on some chariots with the lubrication lines indicating that the axles were rotating. This seemed odd but it will have to wait for a closer look to resolve it. By the Iron Age local people were quite capable of making rotating wheels on stationary axles. The lubrication lines may therefore be indicating an axle or wheel design feature. On each side of the chariot tracks were the tracks of people walking with staffs. A zigzag pictogram was found on the wall of the dyke. Then one member of the group picked up blue lines using a colour wheel. This was the first time a blue energy line had been detected going to the entrance of any Druidic structure. Mainly because we had no reason to look for them. The significance of a blue energy line is that it is the line of communication, the line along which spirits are said to move. The Druids, Romans and early Christians are said to have known how to construct and generate blue lines as they appear in their Temples, Shrines and Churches. Blue lines normally mean the presence of a Roman shrine but in this case they lead to the entrance of the barrow. The Dowsers then found where incense had been burnt, where the bronze lamps had been placed and where the limestone blocks of the entrance door to the barrow were or had been. Somebody found the tracks of a small cart going to and from the barrow entrance and the telltale red ochre stain of bones being loaded or unloaded. A possible garage for chariots was also found. At the end of the morning we knew that we had found a Round Barrow with a ceremonial dyke system. It was used during the Iron Age but it might go back to a much earlier date.   Berkhamsted Castle now became a much more interesting place. The anomaly of its three moats and earthen Mott had been solved. The site had roundhouses and at least one wheelhouse on it. There were also signs of a Druidic Temple with its 6 + 3 rings of postholes and an underground tomb complex. The possibility came to mind that the Berkhamsted site might answer another question. By mid April 2005 we were aware that if the middle sized stone circles in terms of diameter were part of the mechanical structure of Druidic Temples then there should be signs of stone columns in all later temples at least. In fact the stone columns might go back to Stone Age Temples. The idea was that there was a ring of stone columns with a circle of either stone or wood lintels on top of them. The lintels would be linked as they were at Stonehenge so providing strength and rigidity. The lintels could be of wood or stone. They should also be at about the pivot point of any wood beams reaching from the wall at the 9th ring to the centre. In other words about the 3rd or 4th ring. The reason for using stone could be to take most of the weight of the roof as stone is very good in compression. Wood is not as good as stone in compression and there would be a tendency for tall posts carrying the weight of a massive roof to bend and become bowed with time. Perhaps another reason is that the temples were maintained and repaired over the hundreds of years they were in use and it would help if the main structural elements were in a material that did not rot.

On the 13th April 2005 the circles of postholes of a Druidic Temple at Berkhamsted Castle were marked out. Then using a stone witness the remains of stone columns were looked for. A complete ring was found. On a later visit a second circle of stone columns was found belonging to another temple. A number of other henge sites were also checked and all the temples had a circle of stone columns round about the 3rd or 4th ring of postholes. The evidence was mounting that all Druidic Temples of the 6 + 3 rings type had a stone circle at about ring 4. In some temples the stones have survived for 2000 years and are still to be seen to this day. However, most temples lost their stones which were almost certainly recycled either by the Druids when they built the next temple or by the Romans and Christians when they built their roads and churches. If present day stone circles are derived from Druidic Temples their sizes would be expected to cluster round the sizes of pools, the 4th ring and the 9th ring. The sizes of stone circles are mentioned in Aubrey Burl's book "The Stone Circles of Britain, Ireland and Brittany" the book has a section on the size of Stone Circles, page 44-46, in which Burl gives the diameters of circles averaging about 7-7m, 22.4m and 64m for some circles in Brittany. There are circles such as Stanton Drew in Somerset which are about 100 to 112m in diameter. Most of the larger circles cluster around 50-60m in diameter. From Burl's diameter data it looks as if stone circles do fit into the structure of temples. However, in addition to circle diameter the shape and size of the stone would also have to fit. For example, the use of small stones in the small circles which are the remains of the pool wall. There may be larger portal stones outside this ring as at Pontypridd (Chapter 15) but the stones will be small and can be seen to fit in as part of a pool wall. The structural stone columns on or about ring 4 could have been built from a lot of small stones cemented together or from large megaliths as at Stonehenge. The columns will have wood or stone lintels. The stones round the 9th circle and outer wall of the temple are likely to be flat unless they are portal stones or have some other function that flat stones are not suitable for or perhaps they were not available locally. Later in April and following a visit to Lodge Park (see later) the castle gave up another secret. There were bronze chariot wheel tracks in the moat/dyke system and the cart going to the door of the barrow had not only wooden wheels but the wheels were fixed on the axle so it was the axle that was rotating. Does the barrow go back to the Stone Age? It is not possible to say, supporting evidence will be required but round barrows are said to go back to the Neolithic that is the late Stone Age.

Whilst I was looking for evidence of the use of stone in Druidic Temples Nigel was checking the underground tomb complex. He found signs of rooms coming off the tunnel that runs from the access shaft to the central tomb. The tunnel of another temple was checked and again what appeared to be side rooms were found. A long tunnel from the main tunnel, going to some distant point was also found. There is therefore some dowsing evidence that the underground complex may contain more than the tomb and its associated rooms and also that a wider range of activities may have gone on underground. However, the same question remains, are we looking at stains which started life on the surface and which moved deeper over the hundreds of years. One set of stains that we investigated which looked as if it could belong to the tunnel system later turned out to be dog kennels.

Lodge Park and Chariot Tracks

On the 17th April 2005 I was at an English Heritage site called Lodge Park which is in Gloucestershire and about 3 miles east of Northleach. The Lodge is set in extensive grounds on the eastern side of the River Leach Valley. The main task with other dowsers was to find the site of a sunken fence which is called a Ha-Ha and had been dug round the front of the Lodge. The outside of a Ha-Ha is a slope from ground level down to four or five feet then a vertical face up to ground level. The ditch part would have been filled in when the Ha-Ha was no longer required but the vertical face, which was possibly faced with stone, brick or wood should remain dowsable and identifiable. However, the reason for mentioning the visit to Lodge Park is not to discuss Ha-Has. Lodge Park is in a fantastic Druidic landscape. I say Druidic and not Iron Age because the site can be traced back to the Bronze Age and I suspect that in due course it will be traced back to the Neolithic. The site ceased to be used sometime between 43AD and 60AD when the Romans paid the Druids a visit. Some of the execution lines and mass graves are down by the river. It was later in the day when the attention of the dowsers had wandered from finding the Ha-Ha that one said to me that he had found something that was linear but did not know what it was. I had a look and (without witnesses) found a straight line running off into the distance either side of me. It looked a bit like a track of some sort so I walked over it looking for the other wheel and seven feet away found another track running parallel with it. To me it looked like a chariot track but without an iron witness to confirm tyred wheels I could not be certain. However, I borrowed a 2p coin from the dowser, most of which are now made of steel, and used that. Up came a response so I announced that the dowser had found a chariot track. However, I kept the coin with the intention of checking it just in case I had got it wrong. If it was a bronze coin it should not have given a response with the iron oxide stain left by iron wheels. That evening when I had returned home I found that the coin was made of bronze. I checked for iron, no iron. I then checked the road in the garden where the tracks always come up on iron. This time I found 7ft wide chariot track on bronze not iron. The iron tracks belonged to different vehicles. I had never checked the tracks of wheeled vehicles with bronze only wood, leather, fibres, iron. After all, who could afford to put bronze tyres on their chariots? It appears that somebody could and did, at least on their ceremonial vehicles. The new discovery meant that once bronze has been identified it is then possible to determine if it is copper and arsenic, that is early Bronze Age, or copper and tin, the later Bronze Age but which in fact covers most of the Bronze Age. Because the chariots are ceremonial it enables ceremonial roads to be identified.

So, by making a mistake I had found a method of identifying ceremonial chariot routes. This was quite important because once iron tyres came into use it is not easy
to distinguish ceremonial chariots from farm and commercial carts. Chariots are drawn by horses, other vehicles normally by oxen. There are some smaller ceremonial vehicles drawn by deer but at this time we had not found them on what might be called the open ceremonial roads running across the landscape, only in the dyke system of roads and short lengths of ceremonial road.

Tunbridge Wells – a Druidic Centre

On the 24th April 2005 I was invited to visit Tunbridge Wells by two dowsers who had been doing a lot of archaeological work in and around the town. Much of their work focused on Calverley Grounds Park. The park is in a Druidic sacred valley full of pre-Roman archaeology. Woodhenges, ceremonial roads, burial grounds and there are even records of the remains of a mini Stonehenge with linteled stone circle which was removed to make way for a car park. There was so much within the park and outside it that I decided to return at some future date with Nigel and see if we could unravel some of the pre-Roman archaeology of the site. The report of a stone lintel on stone uprights was tantalising. Was it a garden ornament from an opulent time or was it 2000 years old. Round the park and incorporated into garden walls and the stone borders of flower beds were masses of stone which my guides assured me was not local and had come from the stone circles that had at one time graced the grounds of the Park. I knew how to identify stone derived from Druidic structures so another good reason to return. Tunbridge Wells, at least the open areas that are accessible, is rich in archaeology and Calverley Grounds Park looked as if it might be home to 3000 years at least of pre-Roman archaeology.

Indicators of Genocide

Back in Bovingdon, April finished with a few more discoveries. The temple, in what is referred to as the horse field, was checked for battle blood and outside the temple walls were three blocks of blood stains. Each had three ranks with a row of officers or senior Druids in front. In terms of pools of blood the block in front of the temple was 25 long and 3 deep. The left side 20 long by 3 deep and the right hand side 10 long by 3 deep. In front at about 6m intervals were the senior people. About another 20 making a possible total of 185 to which must be added the scattered patches of battle blood from warriors. Two mass graves were found a little way from the temple (Figure 10.4).

At first we had in our minds eye a sort of last stand by the Druids with the warriors fighting the Romans as they approached the temple and the older and young ones standing there waiting to be killed. To them death was no big deal as they would when asked, sacrifice themselves to their Gods. There was for them a life the other side of the grave which they would from time to time visit using meditation and hallucinogenic drugs. Their dowsing abilities, that is their magnetic sense, revealed the spirits in another world whether it was the spirits of trees, animals, stones or what was in the ground. They would be able to see the spirits of the temple guardians at each of the four gates. This was the image we had in April 2005 but more information was coming in and the model would soon change. The scale of the killing was beginning to become apparent and it was presenting problems. If Boudicca was going to lead a revolt she was going to need large numbers of men, and as according to the historians she managed to obtain men in their tens of thousands the Romans could not have killed all of the available men round the temples.

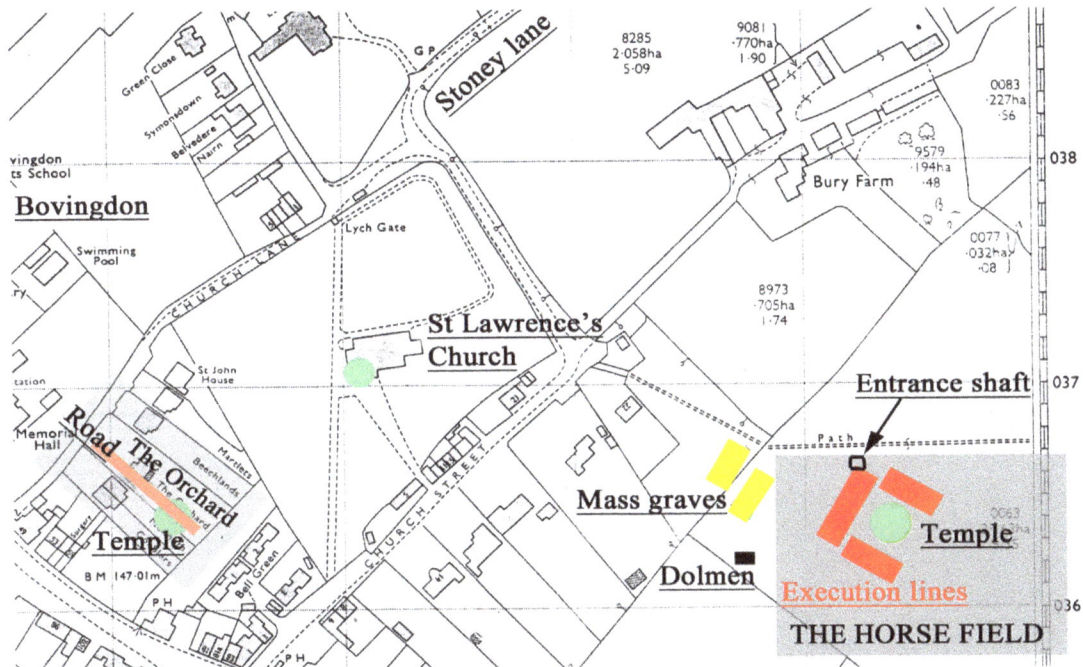

Figure 10.4 Bovingdon

The map of Bovingdon shows the position of The Orchard with its early Bronze Age temple, ceremonial road and a Druid's house. Church Lane running past the house is Druidic with very clear diamonds running along it. Stoney Lane is also Druidic and at least 4000 years old. St. Lawrence's Church stands on a Druidic Henge Temple site. In the Horse Field there is a depression in the ground marking the site of a temple. When the Romans arrived the large temple building had gone and had been replaced by a smaller one. The execution lines were on either side of the new building and across its front. One of the execution lines goes across the access shaft which had been sealed sometime prior to the arrival of the Romans. The mass graves are to the right of the entrance to the field.

Yet Suetonius Paulinus and the other generals were well on their way to clearing the country of Druids. In fact Suetonius Paulinus was in 60AD clearing Anglesey of Druids (Chapter 18). As Anglesey is the most westerly point it is reasonable to suppose that the Druids had been dealt with on the way to Anglesey which was the last area to be cleared. If this supposition is correct it looks as if the Druids were sufficiently separated from the secular society for the Romans to line the Druids up for a ceremonial execution in front of an audience, presumably drawn from the secular society who were not too worried by what was happening. The size of the execution lines, as we were shortly to call them, and the size of the mass graves indicates the size of the Druidic population at the time. At the moment we do not have a reliable method of dating the killing of the Druids round their temples. The executions could either have taken place between 43AD and 60AD or after the Boudiccan revolt was put down in 61AD. As the killings look ordered, neat and tidy and underground parts of temples have been deliberately sealed either after of before the event it looks as if the Druids and their temples were destroyed prior to 60AD. This would fit the picture of a methodical clearance of Druids from Britain starting in the south east and then working north and west. Also after the revolt the Romans would not have been in the mood to do everything by the book so as to speak. After the revolt there were not many people left in the Hertfordshire area as the Romans killed off the secular and tribal section of society. The Emperor Nero is said to have intervened and put a stop to the killing as many Romans were worried by the shortage of labour that was developing.

The Temples Tell More of their Story

April finished with one final discovery. In Bury Wood where there is a Dyke Temple complex, I found by dowsing two tunnels leading out of the rear of one of the temples. In one of the tunnels I dowsed the phosphate outline of six bodies. They were children from about 2 – 3 years old. Mortally wounded they had died on their way to the rear chamber which three of them reached. There was blood all along the tunnel (Figure 10.5). It looked as if the Romans knew the layout of the temples and made sure every child died.

Dyke Temples

May 2005 started with some fine weather and a speaker on the radio talking about the celebration of a festival called Beltane when one of the rituals was to pass cattle between fires. Even listening to the radio first thing in the morning there is a risk of being reminded of the Druids. How old is Beltane 2000, 3000, 4000, may be 6000 years old? If Beltane was one of the festivals celebrated by the priests of the Henge Temples it may go back to the date of the earliest of these temples. Perhaps a reason if not a good one for trying to sort out the dating of temples.

Whilst I was pondering on dating temples Nigel was pondering on tunnels. Our experience with tunnels was that they were very difficult to identify and generally when a hole was dug or an auger sample taken they turned out to be ditches or drainage channels. However, I had mentioned to Nigel that the concentration of carbon dioxide in the soil was higher than that in the air. Nigel spotted that if that was the case the carbon dioxide level in tunnels might be elevated. He therefore made up a carbon dioxide witness and soon found that by using it he could pick up tunnels. I found that by using a plastic bottle filled with exhaled air in one hand and a bottle with atmospheric air in the other I could also pick up what I thought were tunnels dug by the Druids. The first main study in May was of the Dyke Temples in Bury Wood known locally as flint mines. According to one passer by who stopped to chat, lorries had taken flint for local houses in recent times. With two of us dowsing we were able to discover enough detail to give the site some structure. Looking at the first and biggest temple (Figure 10.6) it was possible to see the remains of flint walls, pieces of dressed flint were on the ground, (Figure 10.7) signs of where the terraces had been could still be picked up and an altar with human blood on it was in a prominent position.

Stairs, or where they had once been, were found, entrances to tunnels and the postholes of the posts supporting the beams and rafters of the roof. These postholes were inside the dug out interior of the temple and along the edges at ground level. The building was enormous and clearly the forerunner of the Christian Cathedral in size and design. There were many similarities between the temple and a Cathedral. For example the use of an enclosed large building for worship, a pulpit area on the left, an 'altar' area at the far end, a side chapel on the left, an underground crypt and with places for the bones, a sort of charnel-house.

120

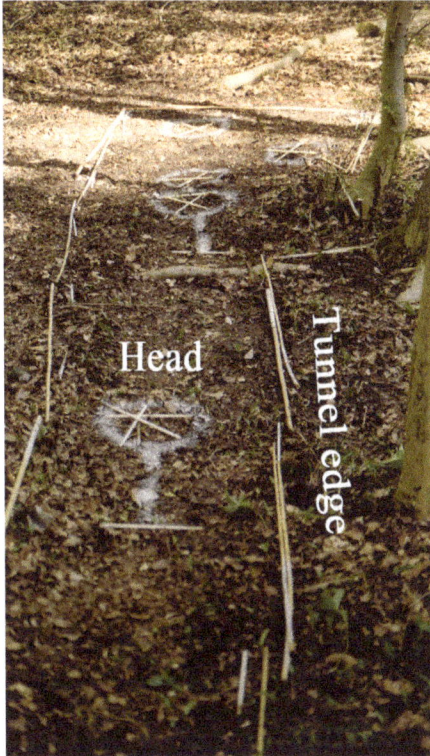

Figure 10.5 Tunnels
The outline of a tunnel from a Dyke Temple in Bury Wood showing where the bodies of children lay.

Figure 10.7 Dressed Stone
Dressed flint can still be found in the Dyke Temples and is still removed by local people requiring flint.

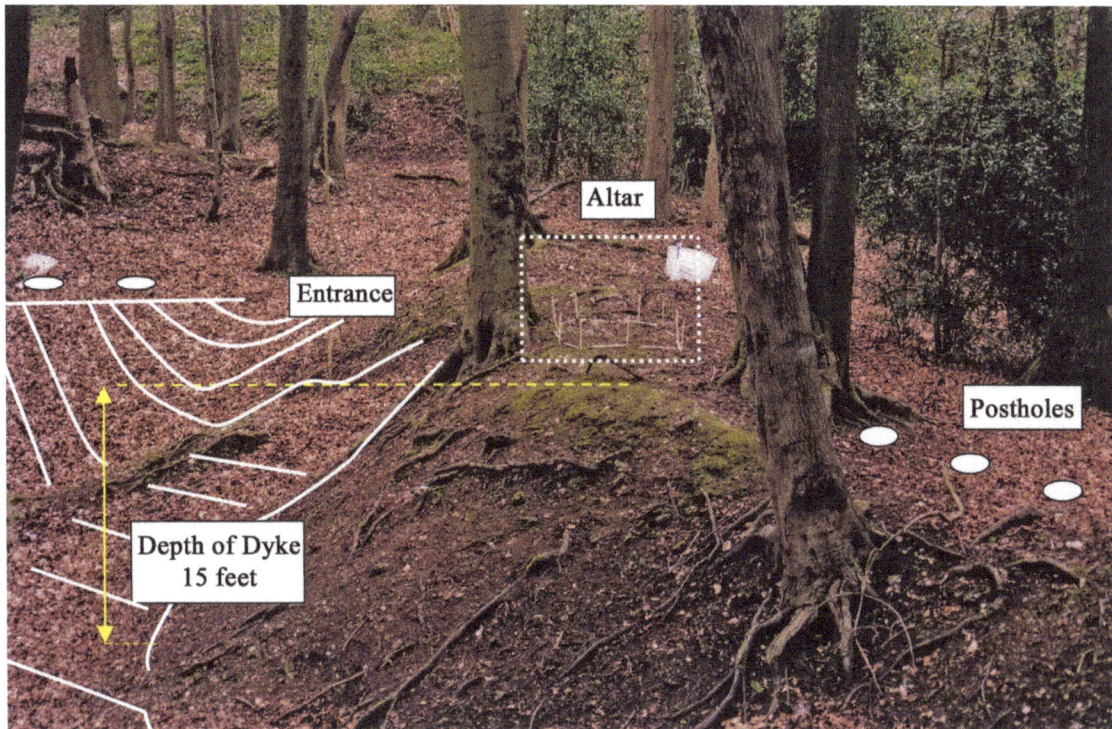

Figure 10.6 Dyke Temple
The remains of Dyke Temples are now often called flint mines. The structure of the temple is still discernable in this photograph with the position of an altar shown.

121

The main difference between the Dyke Temple and Cathedral was the absence of the pool, in the latter represented today by a font, and the entrance to the building. The temple had steps going down into the ground whilst a Cathedral has steps going up. Also the lack of daylight. The temple would have relied on lamps. Using potassium as a witness or the ash from burnt timbers it was possible to pick up some of the roofs structure. Stone, brick and tile were used extensively so the temples were not simple wood buildings. The idea that the Druids did not build in stone or brick was clearly wrong. Even in the 20th Century lorries were calling in to pick up building materials left by the Druids. Their stone work was extensive and still being plundered and recycled 2000 years after they died in the execution lines. The Romans and later the Christians would have used the materials left by the Druids in much the same way as the materials of the monasteries were used after the dissolution of the monasteries by Henry VIII. The Dyke Temple in Bury Wood was the first Dyke Temple that we looked at in at least a little detail. The final demise of the temple was in about 50AD when the Romans destroyed the temples as they worked their way across the country to Anglesey. Although it looked as if the Romans killed everybody associated with the temple somebody still appears to have buried the bodies with some degree of reverence. The graves for the Druids were also good enough for the Roman dead who were buried with them. At this time our image of events was of Druids and their families waiting to be killed. We could see no reason why the Romans would have first captured them before killing them. The scattered battle blood of warriors and Romans made it look as if warriors died in combat and the rest died because they refused to move or flee. They stood there and watched the warriors die then their chief priests then the row in front of them. Only the children fled into the temple. Their belief in their gods must have been rock hard. Two thousand years after the event I walk or jog through their ranks. I know where they stood or kneeled and died. I see the patches of blood as if it happened yesterday, such is the power of the human magnetic sense. On the basis of what happened in the horse field, the Dyke Temple and other sites round the village and St. Albans and also in distant parts of the country, people died in their tens of thousands and hundreds of thousands. It makes a person wonder what propaganda the Roman soldiers were given to make them kill everybody – a 2 year old child, old men and women, men and women suitable for selling as slaves. I doubt if there has been anything like it in European history until the 20th century. These were the notes I wrote early in May 2005. A full appreciation of the scale of what the Romans did in Britain was yet to come.

Over at St. Albans, aided by some new witnesses, Nigel was working out how the Romans attacked temple complexes. A dowsing friend had mentioned that Roman soldiers were partly paid in garlic. This sounds very odd but garlic and onion have the same effect on magnetic fields as salt, they deflect or attract them. The eating of garlic could therefore protect the soldiers from spells and spirits as well as providing chemical protection against infection. It was not long before Nigel had discovered that Roman blood could be identified using a garlic witness. For Druidic blood we had been using beer yeast and toadstool and for dogs, dog blood and dog hair. The Druids used hunting dogs or war dogs when fighting the Romans and pools of dog blood are found with the warrior's blood. Once horse tracks, Roman blood, 'Celtic' blood and dog blood could be identified the pattern of a skirmish could be worked out and also who came off worse. Patches of Roman blood in a cluster showed that their shield wall could fail and the crude weapons of the 'Celts' get the better of them. At the time all this was going on along the western edge of the Roman Empire about thirty years earlier on the eastern edge there had been a man preaching love and forgiveness and a new religion professing tolerance was taking shape. Such was the range of culture and social conditions in the Empire.

A Stone Age Culture

By the middle of May 2005 we were picking up evidence of paintings or pictures on walls. A line of egg white and pigments such as manganese dioxide and gypsum along a section of wall, either in a tunnel or a wattle and daub wall, indicated that there may have been a painting there at one time. Also at about this time we started analysing flint tools. We were finding a lot of flint artefacts such as knives, axes, saws, incense burners and stones selected for their artistic merit. Flint, like other materials, is contaminated by what it comes into contact with. Although the surface of knapped flint looks smooth, at the microscopic level the surface is covered with small holes and pores into which molecules can diffuse. Once inside the surface of the material the contaminant is protected and two thousand years later can still be detected and analysed by a dowser. What we found was that stone tools tended to be used on one material only, for example a stone axe or knife was only used with oak, another one only with ash, another only with yew. If a knife was used on an animal it was one type of animal. It was during the analysis of axes that some were found to be contaminated with Roman blood. This indicated that they were possibly war axes and had been used in combat. The impression was growing that the Druidic culture was a Stone Age one. They used bronze and iron widely but they were dependent on stone tools and weapons. It looked as if their approach was 'if a piece of stone does the job, use it'. This applied to lamps and items used in the temple. Another stone item we were coming across was what we called prayer stones (Figure 10.8)

They were in the main rounded pebbles which had been painted at one time. Some of them still had visible traces of paint on them and when dowsed egg white and pigments could be detected.

Figure 10.8 Prayer Stones
Pebbles marked with paint are found around the sites of temples. They are called prayer stones although a clear use for them has not been identified.

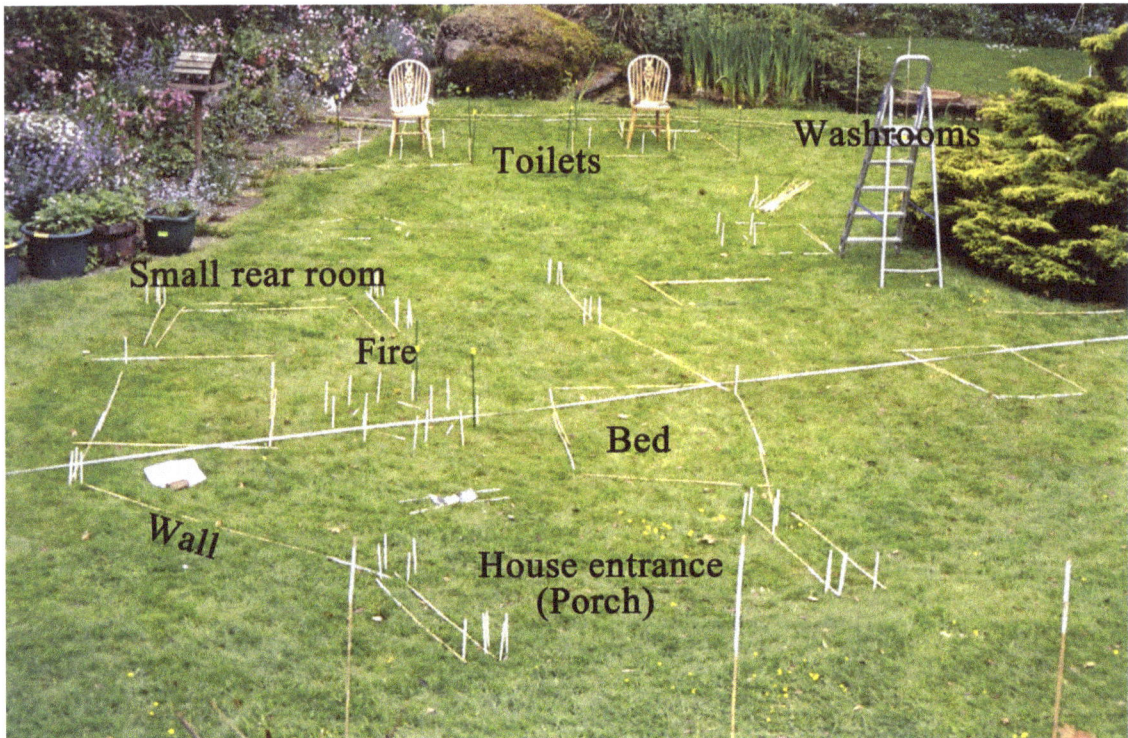

Figure 10.9 Druid's House
The Druid's house found in the back garden has been marked out. The canes are laid to show the entrance porch to the living area with its beds and fireplace. Beyond them on the left is an area for brewing and clothes storage. On the right is the back door to the workstations, toilets and washrooms.

Figure 10.10 Workstations
The photograph shows the front entrance to the house and to the right are the work stations.

The House of a Druid

On 23rd May 2005 I got up early to look at the recreation ground across the lane from my house before the dog walkers arrived. The ground is about a hundred meters from the house. Over the years I had looked at an Iron Age road and houses there. On this occasion I found Druidic houses with their plumbing systems but also three lines of blood stains, execution lines. This was the first time execution lines had been found that were not associated with a temple. It looked as if the Romans may have cleared people out of the houses and lined them up on the road outside before killing them. This may be what was going on in front of the temples, that is the people were brought out and lined up. Because the recreation ground is a public place it was not possible to mark out the Druidic houses so I decided to look at the one in my back garden in detail. I had always thought of the house as a roundhouse but it was not long before a much more rectangular shape emerged. Figures10.9 and 10.10 show the house I have in the garden and the area round it. It was a small house and contained two double bed areas revealed by skin and a number of other witnesses. It is possible to identify that the adults had the right hand bed as you enter from the front door. The dog slept at one end of the children's bed. The fire had an iron spit which was used to cook beef, lamb, pig, deer, and pheasant. The fire place was constructed from stone with a tile or brick oven on one side in which a type of wheat bread was baked. In front of the fire as you enter the house is a small area which was used for flint tool or weapon storage. This small area is of considerable interest both for what materials are present and for those that are not. For several years I thought that this area was a sort of umbrella stand for bronze headed spears. It was a feature of every roundhouse that I looked at. Then in May 2005 I discovered that the bronze field was a phantom or image coming from a nearby henge tomb. If the magnetic loop between the tomb and tool stand was blocked then the bronze disappeared and the umbrella stand only showed traces of stone implements. There was no bronze and no iron or iron oxides only flint. The absence of bronze and iron is interesting as it indicates that the occupants did not use tools made from them or if they did they were kept at a separate location. Bronze and iron were readily available to and used by the occupants of the house. To the left of the backdoor is a sort of alcove with the rear wall used for storing garments or fabrics made of wool and the two side walls for storing linen fabrics or garments. In the centre of the alcove was a clay bowl and round it traces of barley, honey and yeast. Traces of toadstool or mushroom was also found. From the dowsing results it looks as if the occupants brewed their own beer or mead. Both the front door and backdoor had bronze hinges (copper and tin) and the doors were made from ash. The house is small but it provided adequate room for two adults and two children, who were girls. Going out of the backdoor there is a work station for grinding grain, wheat, barley and oats. Toadstool or mushroom was dowsed and the sandstone of the grinding equipment could be identified. The scale of investment shows that the home and village were permanent. The workstation had some protection provided by hazel wattle. A few yards further on are two toilets. The left hand one was for women the right one for men. The floors were made from clay tiles and a clay drain went out of the rear of each toilet to a gully. To the right of the toilets is a wash place. There are two washrooms but it was not possible to get at the second one or determine which was for the ladies and which for the gents. The washroom had a clay tile floor. The gully was linked to the washrooms by clay drains to take the water away. An open clay drain would not smell very sweet and just before it reached the first toilet it was in fact covered with stone slabs. What goes down a drain leaves a trace that the dowser can pick up so drains can tell you a lot about the people they served. When investigating a deserted house in which violent murders had taken place thirty years prior I was able to trace the blood in the drain from the house for about 50 metres until it joined the main sewer in the road. Two thousand years after the gully from the 'Celtic' toilets and washrooms were last used it was still possible to find traces of urine, phosphate, camomile and blood from the female toilet. There was no trace of faeces which must have been removed as solid waste. Moss gave a positive response indicating that it may have been used as toilet paper. It is easy to jump to the conclusion that the solid waste was collected to use on the land. This may have been the case. However, on the chalk of the Chilterns water is in short supply and liquid waste could be useful for a number of processes. At about the time the house was being studied it was found that the Druids or 'Celts' used a chalk mortar when building flint walls. The liquid waste was used for making this mortar. The result of urine having been used for making mortar is that 2000 years later the easiest way of identifying a wall or any stone construction of this period or earlier is by using a urine witness. Other materials that could be identified in the gulley included mushrooms or toadstool from the toilet branch and from the washroom it was possible to detect wool (towels) olive oil, camomile, cedar wood oil, iron oxide. In the washroom there were signs of an ash bench. There was no sign of door hinges. One important thing to remember when using witnesses is that it is only possible to say that something dowses as if it is say olive oil. At this stage it is not possible to say the family were using olive oil although it is likely, the olive oil will have to be checked against other vegetable oils and lanoline. An olive oil witness also contains iron which has to be cancelled out. Turning back towards the house from the toilets there is a small work station for working flint. It is about two feet by four and has a stool for sitting on. The materials dowsed as being present at this work station included: skin, ash, oak, flint, deer antler, leather, nettle, tree resins, and bluebell glue. All the materials associated with flint knapping and tool making are present. Other materials could be present but were not tested for. The next item was a small dog kennel approximately two foot square with a feeding area in front of it. The dog ate what the occupants were eating. Following the workstations round the house there are three storage pits. Each is lined with clay and they stored wheat, oats and barley. Note the presence of oats. It is said by some archaeologists that the Romans brought oats to Britain but oats were here long before the Romans arrived. Just outside the pits is another work station for flint tools, four feet by three and a half, and next to it a weaving loom protected by wattle walls and thatch roof. There is a well which is much more difficult to link with the house. Because of the present day garden it is not possible to dowse the eastern side of the house to identify other work activities.

This study of a small house that was in use when the Romans arrived shows how powerful the use of witnesses can be in revealing what people were doing nearly two thousand years ago. It is also possible to see that the house is linked to a drainage system and that if other houses are also going to be linked to it planning is required. The house is not a personal house developing as the owner acquired the necessary money. It is much more like a house provided by a commune or cooperative or the Council of the day. The house and its outbuildings appear to have a stamp of quality – bronze hinges, tiled floors and work stations with weather protection. From the signs of pigments and gypsum (plaster) on the walls it may even have been decorated. The postholes on either side of the front door indicate large substantial posts when compared with the rest of the house.

Dating Tombs

Back at Chiswell Green Nigel had found a tomb with no sign of iron in it only bronze. I went round to check and also could not find any iron. We began to realise that it may be possible to date the tombs of the Prince and Princess by their contents. That evening Nigel phoned to say that he had found a tomb with no bronze or iron. We immediately thought of the late Stone Age or Neolithic. However, religions tend to be conservative and just because a tomb has no bronze in it does not mean that bronze was not available. There was however the distinct possibility that the Chiswell Green site contained Druidic tomb complexes dating from the Iron Age 43AD back to the Neolithic possibly 3000BC. It was beginning to look as if Druidism could be traced back for 2000 to 3000 years. And if it did go back to the Stone Age, what did Druidism develop from?

June 2005 started with the analysis of a few more stone tools to determine what they were used for. As previously found each tool was specific to a material. A tomb was identified with copper but no arsenic or tin and no iron. I had another go at the Henge Temple I have in the garden. The bronze in the tomb is a copper arsenic bronze. The stains of bronze lamps and wheel tracks are also copper arsenic, that is the early Bronze Age.

During a visit to Berkhamsted Castle to investigate a temple I decided to mark out the outer 9th ring of the Woodhenge Temple and started off in one direction. Nigel helped by doing the same in the opposite direction. As we made our way round to the far side and began to close on each other it became apparent that we were not going to meet up. We then realised that there were at least two Woodhenge Temples in the Castle grounds instead of the one found previously. This meant that there should be two tombs and they would not be far apart. When you know more or less where a tomb is it is easy to home in on it using a phosphate witness or by using the north south bearing to guide your angle of approach i.e. normal to the target and look for the walls of the tomb. The two tombs were soon found. One of them contained copper/arsenic bronze and the temple gate hinges were also a copper/arsenic bronze. The second tomb contained copper/tin bronze. This meant that the site was in use early in the Bronze Age until the later Bronze Age at least.

Pictures on Walls?

We were able to identify what we thought were pictures on walls by looking for a line of egg white and paint pigments. These lines could be picked up when dowsing wattle and daub walls and along tunnel walls. In one of the anterooms of the tomb complex in my garden I had identified what appeared to be pictures on two of the walls. I decided to investigate further. If they were painted on the chalk walls they could still be there. If they were painted on something and then hung up they would have rotted long ago and now be no more than a stain on the floor. I checked using wattle and daub and obtained a positive response with the wattle, that is hazel wood. This made the possibility of finding anything on the walls remote as hazel would soon rot in the 100% humidity of the anteroom. As wattle and daub is not very dimensionally stable I checked for a frame and obtained a positive response with an ash witness. Then I checked for preservatives and obtained responses on linseed oil and tree resin. Finally I checked for nails and found signs of copper/arsenic bronze. I obtained similar results in the access tunnel to the tomb. It is necessary to remember that dowsing is dowsing and there could be other interpretations of the results. All that can be said is that the walls were marked with something that has left a dowsable trace, or alternatively something vertically above them on the ground has left such a trace. At this stage we were well aware that the Druids painted things including the ground, stone columns, posts and chariots. It is not inconceivable that they had pictures that they would hang on walls and move around. By dowsing the pigments it was possible to show that the colour pallet of the pictures differed.

Druidic Mortar

Another advance that occurred about this time was the discovery that the Druids used a chalk mortar when building with stone. The mortar was made using urine which can be dowsed for by using a urine witness or a material that is in the urine. It therefore became possible to identify Druidic stone work and prepared surfaces for tiles, decorative stones or brickwork. Importantly it enabled us to identify tunnels that had been sealed with flint walls, possibly after back filling them with spoil from new tunnels.

More Evidence of a Stone Age Society?

On the 7th June I decided to check the tool and weapon stand in the Druid's house in the garden for what materials the tools had been used on. The theory was that traces should have been left by dirty tools. The materials left behind were ash, yew, nettle, hemp, linseed oil, bees wax, leather, bluebell sap, blood of deer, blood of sheep, blood of humans, iron, bronze, beef and oak. The human blood gave a positive response on garlic so is likely to be Roman blood. By looking at the tool stand or cupboard, it is possible to see that the Druidic stream of the population were possibly a Stone Age people with access to metals. The average person may have done most things with stone and wood tools. Bronze and iron were available and used, for example in the fireplace, as hinges and for wheels. The limited use of iron indicates that the Druids may have had limited access to it and that their manufacturing and metal treatment was a long way behind that of their continental counter parts and the Romans. Just in case the Druid who use to live in my garden was a bit behind the times, possibly similar to myself in my attitude to computers and mobile phones, I decided to check some of his neighbours. The tool/weapon stand is always in the same position in the house so easy to find. His neighbours tool stands were also without bronze or iron stains but were stained with Roman blood.

Whilst I was working on sites round the village Nigel was hard at work over at Chiswell Green. We knew that complexes the size of Henge Temples must have had metalled roads, paths and assembly areas. The surfaces could be cobbles, gravel, tiles, brick, stone or perhaps wood. Underneath the surface layer would have been foundations of some sort, but what sort? In Hertfordshire the most likely material was chalk in fact a chalk urine mortar and this is what we tended to find. In Yorkshire we had come across cobbled surfaces still visible and durable after 2000 years but did not have time to identify the foundations. Nigel had been dowsing with different coloured gravel on the Chiswell Green temple complex and although the surfaces had long gone it looked as if the temples had gravel forecourts with rings of gravel possibly graded by colour. Nigel had also found some stone tools and weapons which were analysed (Figure 10.11 and 10.12). Some of the axes had Roman blood on them. Although the first impression may be that the stone weapons became contaminated in battle there is the possibility that the odd Roman was captured and sacrificed. The blood could then have been used to anoint weapons prior to a fight.

The stone weapons both axes and spear heads are very crude. The weapons suffer from two very important disadvantages, at least when compared with metal weapons. The first is that the edge is on a wide angle so although the edge may be sharp, as it penetrates it opens up a large wound. That is it has to push aside a lot of flesh or bone. This will limit penetration compared with a mettle weapon. It will also cause a lot of tearing and mechanical trauma to blood vessels. The arteries respond to this trauma by constricting and limiting blood loss. A metal edge cuts without tearing and causes little mechanical damage or trauma. The artery therefore does not know that it has been cut and the wounded person rapidly looses blood. The Britons may have been use to a stylised combat using stone weapons in which it was possible to sustain injury without bleeding to death in a matter of one or two minutes. They may not have fully appreciated how deadly iron weapons were. The second disadvantage relates to the penetration of projectiles. Stone tipped projectiles are easily fatal as the stone tip is very sharp and the head will open up a way for the shaft of the arrow or spear through the clothing, skin and flesh. However, unlike a metal tipped spear or arrow where the spear and arrow head encases the shaft so that the shaft does not catch armour or clothing as it penetrates, the stone arrow or spear head has to be held and lashed into the shaft. The shaft and lashing therefore has a larger cross section than the stone tip and will be caught by the armour or even heavy clothing. If a stone spear penetrated armour and clothing it would be difficult to retrieve it with ease so the warrior would be without a spear, at least until he had recovered it. A fatal situation to be in. The Romans probably knew the weaknesses of stone tipped weapons and made sure that spears got caught in their shields or were quickly broken. With stone tipped weapons the Britons were at a considerably disadvantage and probably had to beat or club the Romans to death. It might be thought that the obvious thing for the Britons to do was to capture Roman weapons and use them, which they probably did. However, if the Britons acquired Roman swords it is not likely that they could put them to good use as they would not have been trained to use them. The Roman sword was only one part of a weapons system. A short stabbing sword is not of much use without a shield of suitable design preferably backed up with a helmet and armour. The use of a short Roman sword by a 'Celt' would have brought him into the position the Roman soldier required for killing him. One final point is that a weapon has a life, a limited durability. If you can hit your opponent twenty times before your weapon brakes and he can hit you thirty times before his breaks you have a problem.

Figure 10.11 Artefacts from Temple Sites
The artefacts shown include two axes, bottom right, with above them a ceremonial knife made from Druidic puddingstone cement, see chapter 16 for details. There are two lamps, bottom left and three stones marked with red ochre. Above them are three pieces of pottery the left hand one is ceramic the other two are made from druidic cement.

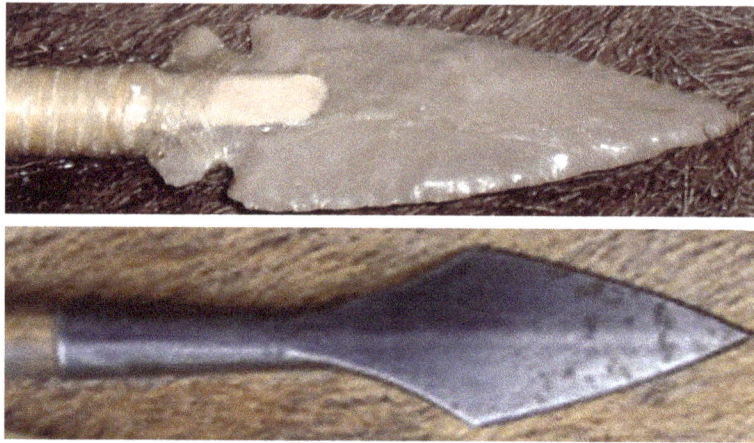

Figure 10.12 Iron and Stone Weapons
The binding of the stone blade onto the shaft would hinder penetration and withdrawal of the weapon.

Return to Yorkshire and More Discoveries

By mid June 2005 it was time for another look at the Yorkshire Stone Circle and Henge Temple. This time when we arrived the weather was hot and sunny and the long days ideal for dowsing. Since the last visit ten months earlier we had learnt a great deal more about Henge Temples. However, two hundred miles north of Hertfordshire things might have been different. The uniformity of the Druidic or Henge culture over the British Isles still had to be proved.

We were now in a position to determine the age of a Henge Temple under the broad classification of Stone Age, Copper or early Bronze Age, later Bronze Age and Iron Age. However, it was becoming apparent that the method of dividing artefacts into ages that is the Stone Age, Bronze Age and Iron Age was not very suitable for a Henge society that was beginning to look as if it might span two to three thousand years. The Danish scholar C. J. Thomsen proposed the three age system in 1836 and although widely used and useful for dealing with artefacts in museums it can be misleading when trying to follow cultures and date sites. The term 'Hengeworld' used by Mike Pitts as the title of one of his books is a much better descriptor. Hengeworld conveys the sense of a culture or civilization cut off from other cultures, possibly even from part of their own people, the secular wing of society. The secular or tribal people may have wanted to join the new culture from Southern Europe particularly the industrial and technological world of the Roman Empire. The Druids appear to have been isolated or more likely to have isolated themselves from the technical and cultural progress being made elsewhere in Europe and possibly also in the secular sections of their own society. However, if Hengeworld sounds too romantic Henge Age is another descriptor that can be used. With the exception of the final destruction of the Henge Temples, we knew it was not possible to put a date to a site either its beginning or its end. It was only possible to identify if metals were associated with a site or not and if they were, which metals. Valuable metals, just like present day family heirlooms, can be in use long after they were made. Also in conservative societies, which religious ones tend to be, tools, materials, methods, procedures can be used long after the secular society has moved onto newer and better ways of doing things. Bearing this in mind, the henges can be dated on the basis of the metals found in them and used in the construction of ceremonial chariots. Some of the tombs lacked linen so pre linen may be another descriptor.

The first metals to be identified in the Yorkshire site were those in the tomb. The bronze was made from copper and tin. The spear head came up on an iron oxide witness but not on acid cleaned iron. This was taken to indicate that the spear tip had corroded to iron oxide. The sword gave a positive response with iron oxide and a response on acid cleaned iron indicating that some iron was still present. The next items were the hinges of the gates and the torch or lamp stands. They all appeared to be bronze made from copper and tin. There were signs of iron being used in the structure of the roof. What happens is that due to weathering and wear iron oxide falls to the ground and leaves a small area of iron contamination beneath a rafter or joist. This meant that the henge was constructed in the period when bronze was being used along with iron. In archaeological terminology the Yorkshire Henge Temple was an early Iron Age structure.

The pool in the centre of the henge is a very important ceremonial structure. It is the centre piece of every Henge Temple found to date. A number of tests had been done on some pools and background levels of wool, feathers and linen have been detected in them. The pools were surrounded by a path and some pools were not circular. A straight edge could be found on some. It was reasoned that if there was an entry point to the pool the stone circle would have a gap in it. On the previous visit to Yorkshire it had been noticed that some stones were missing from the pool wall and we thought they had been removed, but alternatively, perhaps they had never been there. Using a stone witness and looking for traces of stones that had been removed the stone circle was checked. What we found was a gap where stones could not be picked up. There was no chemical stain from them. Switching to a clay witness an apron of clay was found leading to and into the pool. The clay witness was picking up large thick tiles which went into the pool to provide a footing for the priests. The edges between the tiles could be identified and they were about two feet square and set on mortar. On the outside edge of the tile apron was an area that gave a positive response with both a linen and feather witness (Figure 7.1). This is from the clothing of the priests. Facing the pool and on the left going down into the water is a path giving a positive response on wool. It is as if an initiate or a person being baptised walks down into the water on the left hand side. To his right are the priests dressed in linen robes. Round the pool is an area of wool, as if the congregation viewed proceedings from behind a low wall surrounding the pool. The clay tiles and the evidence of what clothing was being worn provided the first clue as to the use of the pool or at least one of its possible uses. The pool was in an Iron Age Henge Temple, so how far back in time do the pools go. They were there in the Bronze Age at least. Yet another question to be answered.

On the southern henges it was becoming clear that there were stone columns in or about ring three or four. The Yorkshire henge had similar structures. The temple had clear easily identifiable stone columns standing on a clay tile base. The tile base varied in size but on average was about three feet by two. The bearings of some of the stone columns are shown in (Figure 7.6). Round the chemical stain of the stone, the mortar used for the column could be detected and also the remains of the paint used to decorate them – egg white, red ochre, woad and charcoal. The charcoal could have come from burning but the others are associated with paint.

Work on the St. Albans henges had shown that the ceremonial route from the Dolmen was flanked by ceremonial chariot tracks. This was found to be the case with the Yorkshire henge. The chariots followed the tracks from the Dolmen until they came up to the western gate where they veered to the left and right. The bones were carried into the henge building.

The scale and cost of the Druidic ceremonial procedures were becoming apparent. Taking this one example the following components can be identified. First the bones are cleaned with scrapers (one has been found) then covered with red ochre. Priests with staffs walk with the bones and take them into the temple. Two chariots with their horses and drivers will have been prepared. They arrive on site and flank the priests in procession. The bones enter the temple and at a specific point are handed over to another set of priests who take the bones to the chapel and hand them over to a third set of priests. These in turn take them down the shaft to the bone depositories. This may be an underestimate of the ceremony involved but it is possible to work out some of the ceremony.

Down south another feature had been noted on the outside wall. Narrow doors had been found in the wall of the temple on the 9th circle of postholes. These were looked for and found. The bearings of the narrow doors from the centre of the tomb were – 257°, 263°, 275°, 284°, and 293°. The bearings were taken with a hand compass and it is unlikely that the Druids stood in the middle of the pool to make observations of solar events even if the doors do relate to them. At this stage of the project the use of the narrow doors was not known.

The visit to the Yorkshire farm and henge produced something unexpected. A sheep grill was being built and the trench dug for it went down on the corner of one of the stone columns of ring 3. There, as predicted, was found the large clay tile used as the base for stone columns. A great find providing a little more physical evidence that the dowsing was picking up real artefacts. At the end of the short stay the Yorkshire henge had provided new evidence confirming that the henge culture differed little between the south of the British Isles and the more northerly parts. It is also clear that the Romans viewed Druids as Druids wherever they were. Like the Dalek command 'exterminate' there was a Latin equivalent which the Roman soldier knew only too well. Outside the Yorkshire Henge Temple and to the northwest are the execution lines. Three rows with senior priests set out in front on their own. Their pools of blood, two thousand years later, show that they died holding their position. Likewise those behind them stood and died in their allotted positions. Only occasionally do we find an execution rank that has fallen back. Out in front along the boundary of the henge site can be found the pools of blood of the warriors. Those who, possibly only armed with stone weapons, decided to die fighting the Romans. The mass grave is close to where the executions took place. They are on the northwest boundary containing well over two hundred bodies. As in Hertfordshire so in Yorkshire.

While in North Yorkshire we visited a number of Druidic sites of which perhaps the best known one is the Thornborough Rings (Figure 10.13). This set of three earth rings runs on a northeast southwest line. The earthen circles are all about 240m in diameter and the distance from one end of the complex to the other is about 2300m. the northern henge is the most intact but is covered with trees and not so easy to work on. The earthworks lie between the village of West Tansfield and Thornborough which is about 8 miles north of Ripon. The central henge is possibly the most accessible from the road running between the two villages.

The rings are described as dating from the late Neolithic to the early Bronze Age and are considered a site of major archaeological interest. The rings are certainly attracting a lot of local interest as there is a proposal to extract gravel from land close to them. The rings would not be touched by this work.

You do not need to know much about Druidic Britain and in particular the Iron Age to realise that the rings may represent a Dyke Temple complex. Further south we were already picking up signs that when you have temples or temple complexes close together they may well either differ as to the time period when they were in use or they belong to the same time period but have different functions. The three earth rings were therefore of considerable interest to us. Even if only one ring was being used at any one time, the size of the temple complex indicated that it would be at the centre of a service complex. There should therefore be the 'monastic' estates for the Druids and their families. The service industries associated with chariots, horses, dogs, carpentry, metal, stone, textiles, dyeing, trading etc. And possibly signs of the secular society. In short little archaeological knowledge is required to realise that the earth rings and the nearby fields are an extremely important site for those interested in pre-Roman culture and civilisation. However, we were only to have about an hour at the northern henge. The visit started at the entrance to the henge with a nice archaeologist telling us what was known about the site and its history. His description of the inside of the henge was that there was not much there. As Nigel and I could see from the entrance where we were standing that there was a great deal there it took us by surprise but it did illustrate how little conventional archaeology knows about the 'Henge Age'. Once the talk was over Nigel and I set out on a quick reconnoitre. The inside of the earth bank had been terraced for ceremonial chariots. It was difficult dowsing on the slope so I could not determine if the roads had been covered. Moving onto the flat bottom of the henge we soon found a Henge Temple. Returning to the entrance we picked up battle blood by the gates and a little further out the execution lines. The mass grave was in the corn field. The henge was therefore in use up to Roman times and the Romans put an end to it in their normal efficient manner. There were areas of metalled road (cobbles) in and around the henge. Like the Romans the Druids built to last. Like Durrington Walls the rings probably go back to the Stone Age and there will be a succession of temples within them.

Figure 10.13 The Thornborough Rings
The rings are Temple complexes. Their size indicates that there is likely to have been a city nearby providing services.

Another North Yorkshire site of interest is Ulshaw Bridge Earthworks which is to the west of Bedale. It can be reached by going south from Leyburn along the A6108, the earthworks which run alongside the River Ure, are massive and a monument to Neolithic or Bronze Age engineers. Our main interest was in finding if it was possible to determine how the earthworks were built and what they were used for. We found the tracks of carts used for moving earth and signs of ceremonial chariots but in the limited time available no sign of other methods of moving spoil. Going north from Leyburn along the A6108 is the small village of Downholme. The village seemed to be in part at least built on a Druidic dyke system. Just south of Catterick and on the west side of the A1 is the small village of Hackforth. We were given access to some of the fields and a small stream. Like Downholme the village and farmland was on an ancient and large Druidic temple and dyke complex. There was at least one Bronze Age Temple and one Stone Age one in a single field. There were chariot tracks, wells and signs of dykes. Execution lines and mass graves showed how the Romans closed the site down. After a very quick tour we came to the conclusion that North Yorkshire was as rich in Druidic temples, dykes and landscaping as Hertfordshire.

Summary

The early months of 2005 saw progress over quite a large area of study. From the limited work done in the tropics, the stains or memory traces in the soil can be identified as readily in the tropics as in a temperate zone. It therefore appears that the archaeological history of tropical countries can be studied using science based dowsing that is biolocation techniques.

A Roman programme to exterminate the Druids from Britain appeared to be confirmed. In Berkhamsted Druidic earthworks and a barrow had been used by the Normans when they built a castle. The presence of the castle has preserved the earthworks. Within the Henge Temples a circle of structural stone columns was identified. The use of a bronze witness has provided a method of finding ceremonial roads. A large Druidic site was found at Tunbridge Wells. It has been proposed that the pre-Roman society was divided into a religious possibly monistic class and a secular or tribal class. The division between the two may have been sufficient for one to stand aside while the Romans exterminated the other.

The Dolmen started to emerge as a complex structure for excarnation by birds alone. Dyke Temples were emerging as complex large buildings designed for specific aspects of the religion in that the ceremonies in them must have differed from those in the Henge Temples.

Techniques were being developed for studying the fighting between the Romans and Druids. The pre-Roman culture of Britain or at least the Druidic section was appearing to be a Stone Age culture. They were using metals extensively but the main tools and some of their weapons at least were made from stone.

The use of witnesses was developed to the stage where it became possible to study a Druid household in detail. Both the structure of the house and the activities going on within it.

Two shafts to tomb access tunnels were discovered on a farm.

Methods of dating tombs and hence the temples were developed. Indications of Druidic artwork in the form of paintings on vertical surfaces were found.

The discovery of chalk mortar enabled Druidic construction work to be identified.

It has been proposed that the term Henge Age be used to describe the pre-Roman culture of the British Isles.

The pre-Roman terraforming in North Yorkshire is as varied and as spectacular as it is in the South of the country.

Box 10.1 Studies of the Oak's paramagnetic field (aura)

Many dowsers are aware that plants including trees have what they call an aura. The aura is the multitude of magnetic fields produced by the materials and chemicals which make up the tree. To the Druids it may have been the living essence or spirit of the tree. In the case of deciduous trees there is a marked change in these fields as the living processes shut down in the autumn and start up again in the spring. The aura of the oak is of particular interest because of the tree's association with the Druids. Early in the study of Druidic Temples we had realised that the oak's aura consisted of six rings which matched the six rings of the temple. Then three more rings were found on the temple sites which caused us to look at the oak again. When we did, it was found that, as in the temples, there was a gap and then three more rings round the oak. The overall plan of the temple matched the oak's summer ring aura even down to its size and the position of the 6 + 3 rings. Closer study of the oak revealed that round the trunk was a further element of the aura which looked like a starburst (Figure B10.1.1). This pattern had been observed on a temple site in St. Albans and at Stanton Drew late in 2005. At the time it was considered to be part of the Druidic artwork or pictograms that we were finding drawn on the ground and associated with Henge Temples and Dykes.

Figure B10.1.1 Tree auras. Round the trunk of an oak tree can be found a complex magnetic field which appears to be like square ended rays coming from the tree. The appearance is of something expanding out as a burst of energy. Hence the term 'Star Burst'.

As it seemed to us unlikely that the Druids would have manufactured out of thin air, so as to speak, a pattern matching a component of the oak's aura, the connection had to be studied. The starting point was to identify the material giving rise to the starburst and if possible what was causing the starburst. As there are not that many materials and chemicals in an oak, after a little experimentation it was discovered that a bottle of glycerine produced a nice starburst. Placed on the lawn the starburst could be plotted out and photographed (Figure B10.1.2). The design of the Stanton Drew starburst looked as if it was a reasonable match to the glycerine starburst.

Figure B10.1.2. Star bursts. If a bottle of glycerine is placed on the ground an aura can be detected round the bottle which looks very much like the 'Star Burst' aura of the oak tree. The tape indicates the north south line. The starburst is due to an incident magnetic field acting on the glycerine.

Box 10.1 contd

Another element of temple artwork is the Thunderbird design. The Thunderbird was first found in the design of the underground tomb complex, then in the main temple. The Thunderbirds were then found as part of the oak's winter aura. As with the starburst the next step was to try and construct a Thunderbird on the ground. We already knew that the eight rays from the oak were due to tannin, a material which is also found in teabags. The rays formed part of the wings. Next lignin, which is present in cardboard and wood, was tried and it produced a nice rectangle where the head of the Thunderbird should be. When a piece of an oak branch was added to a teabag and cardboard the neck was also produced. The neck was not due to tannin in the branch. When the branch was removed and replaced with glycerine the neck appeared. Figure B10.1.3 shows the Thunderbird produced by the model tree. It could be that glycerine, or a substance like it, is part of the oaks antifreeze system to protect it during the winter.

Figure B10.1.3   Thunderbird aura. If a model tree is made from cardboard, tannin (tea bag) and glycerine the complex aura of four Thunderbirds is produced. One of the Thunderbirds from the model tree has been marked out on the lawn.

After finding the Thunderbirds on the western Dolmen in Anglesey the area around a Dolmen was explored and a quite complex pattern found as is shown in figure B10.1.4. This passed the search back to the aura of the oak and it was not long before Nigel had found a complex system much of which matched the Dolmen pattern. As something new was found round the Dolmen it was looked for in the oak's aura and found. If something was found in the aura it was looked for round the Dolmen. The complexity of the picture painted on the ground round the Western Dolmen is such that it could not match the aura of the oak by chance. The Druids must have been familiar with the winter aura of the oak and used it as the basis of their Western Dolmens art design, a design which can be traced back to the Stone Age. It was used by Druids for perhaps a period of three thousand years before the arrival of the Romans. The reason for using the Thunderbirds is probably their association with the 'death' of the oak during the winter. The four Thunderbirds are the gate to the underground realm of the dead.

Figure B10.1.4   Dolmen pictogram. The pictogram round the Western Dolmen is complex. The main pigment used is manganese dioxide. Rising from the Dolmen enclosure are four Thunderbirds then a black circle with figures rising from it. The Thunderbirds are embellished with red eyes, black eyebrows and drops of blood.

Box 10.2 Dykes

In archaeology an earthworks is an artificial bank or mound of earth with or without a ditch. Britain is covered with ancient earthworks. Examples include barrows, henge monuments (circular banks and ditches) cursuses, artificial hills, hill forts and dykes. The term dyke crops up regularly in the naming of these earthworks for example Devils Dyke, Wansdyke, Car Dyke, Grimsdyke and the famous Offas Dyke. These examples are linear features that can run for miles across the countryside. (Figure B10.2.1) and where undisturbed consist of deep excavations and high mounds (Figure B10.2.2). Other dykes have the same characteristic pattern of features but are much smaller in length and depth. They are literally mini dykes. The term is also used to describe other excavated earthworks that come in a variety of different shapes.

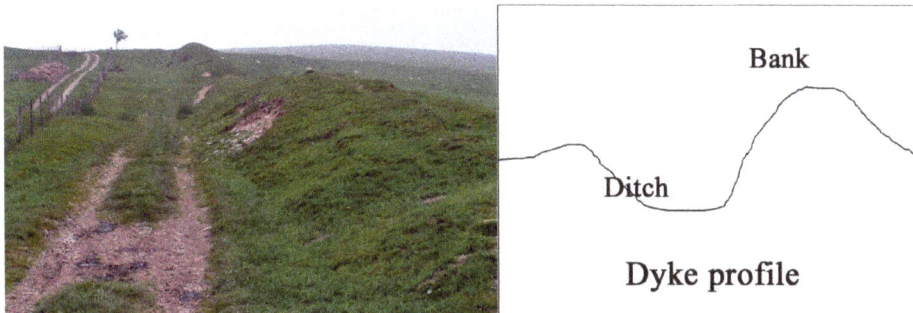

Figure B10.2.1 A section of Offas Dyke. Offa's Dyke is a linear earthworks following large sections of the current border between England and Wales. It is generally interpreted as being an Anglo Saxon defensive feature.

Figure 10.2.2 Beech Bottom Dyke, St Albans. One of many dykes in and around St Albans Beech Bottom is perhaps one of the most impressive. It is very deep, wide and terraced. It is accepted as being an ancient earthworks but its true purpose has remained elusive and conjectural. It dowses as a massive druidic roofed ceremonial structure.

The term dyke is believed by some to originate from an old English word dic meaning a ditch or trench. Archaeologists have reasonable evidence to declare the antiquity of some of these structures as being pre-Roman. Others such as Offas Dyke are considered to be post Roman defensive earthworks. Whatever evidence has been acquired their exact origins and purpose has been unknown. The scant archaeological evidence has resulted in them being attributed to numerous activities such as flint and clay mining, fortification and defence and boundary definition.

Chapter 11

Temples Provide a Time Line

A Temple Site in Stoney Lane

In 2004 I had dowsed the large expansive back lawn of a neighbour's garden and found that they had a Henge Temple almost in the middle of the lawn. It was easy to spot visually from the house. The lawn had that slight depression that shows where a temple once stood. The slight depressions are visible in the fields all over Hertfordshire and much of Britain. After finding the tomb of the Prince and the Princess, who had as usual a strong gold signal, I suggested a dig but my neighbour was not too keen on seeing his lawn disappear. However, he said that I could look at a field they had the other side of Stoney Lane which ran past their house to join Shothanger Way (Figure 11.1).

I later did this and found there was not only the telltale depression but there was a much larger expanse of poor vegetation both within and round the depression. A sure sign of human activity. A quick run round the field with rods and I knew that there had been a Henge Temple in the field. The centre of the new temple (SLT) was probably only about 300m to 400m from the temple I had studied on the back lawn. The large number of Henge Temples that we were finding presented a problem. Why so many? There was evidence that the temples burnt down and for a while we thought that the temples had a limited life after which they were deliberately burnt down. The idea was that the temple builders then moved to a new site to build another one. However, the problem of the large number of Henge Temples would have to wait.

My neighbour's offer of the field after haymaking in 2005 was just what we were looking for. The field was isolated so that we could mark out features with canes and leave them up as we gradually worked out the design of the temple and its outbuildings. The temple was centrally situated in the field so that the entire temple could be dowsed and measured up. The only thing missing was the Western Dolmen. That was way out in somebody else's woodland. However, a whole temple close to home was a gift indeed.

When starting to dowse a Henge Temple the question is where do you start? I decided to start on the central pool after first checking the tomb. The tomb we had found contained copper and arsenic bronze so it looked as if the temple was early Bronze Age.

The pegging out started with the pool. The Druidic pools were made water tight by lining them with clay. Consequently a clay witness can be used to pick up the outline of the pool. This can be followed up in the Hertfordshire area by dowsing with the chalk and urine mortar which was used by the Druids when building the flint wall round the pool.

In other parts of the country different stones are used and the mortar may differ although so far the urine component has always shown up. Once the pool wall was outlined it was clear that the pool was a large one and a break in the wall indicated an entrance to the pool (Figure 11.2). The pool entrance had been tiled as there was an area of tiles outside the pool and an area inside the pool. Having outlined with canes what was the central feature of the temple the wall was next. It appeared to be on the 9th ring of postholes, at least so we thought. In the wattle and daub wall we found narrow doors and what looked like an internal walled structure on the south west side. As I tidied up that afternoon I began to realise that for the first time in 2000 years the outline of a Druidic Henge Temple was taking shape. On previous sites we had marked out postholes, altars and tomb complexes as

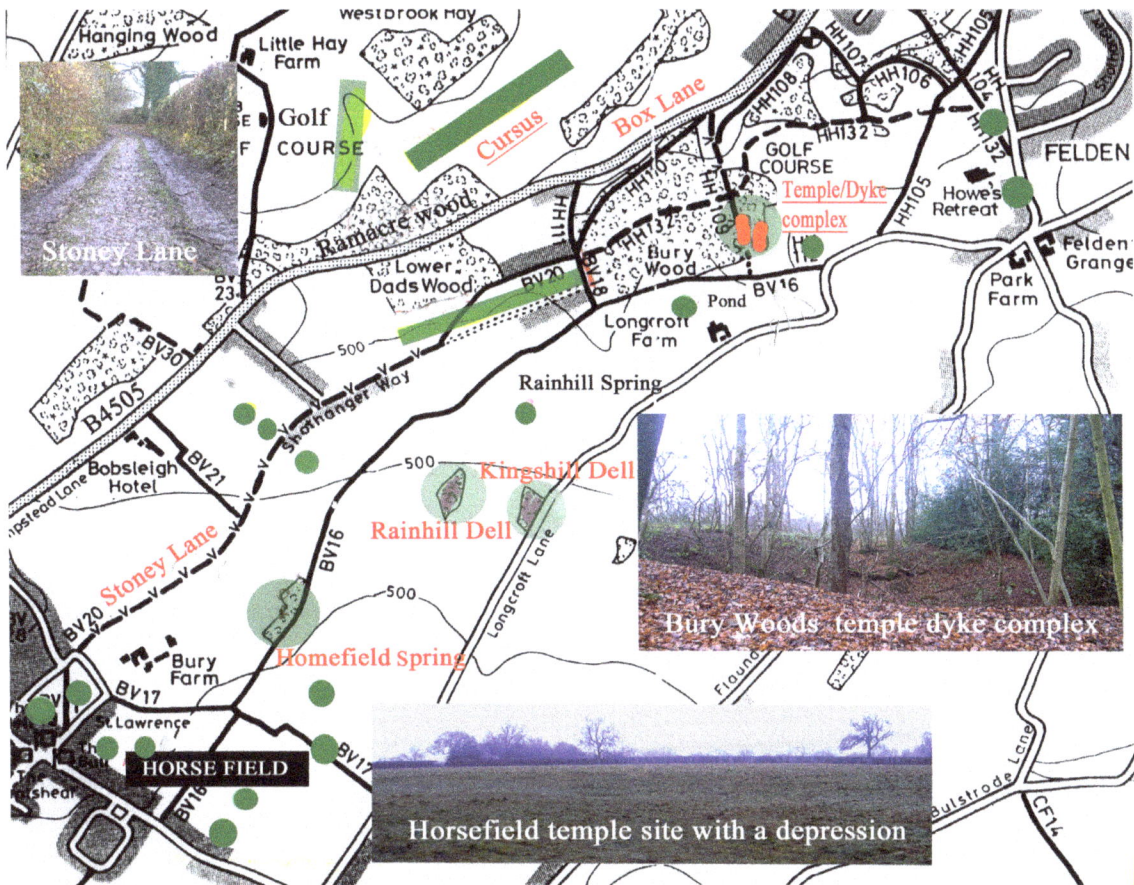

Figure 11.1 Henge Temple Location

The map shows the large numbers of Henge Temples that can be found in an area. There could be as many as twice the number shown. There are Dyke Temples in Homefield Spring, Bury Wood and Ramacre Wood. Many of the roads and lanes such as Stoney Lane and Box Lane are Druidic roads dating back at least 3000 years. Two Cursi are shown.

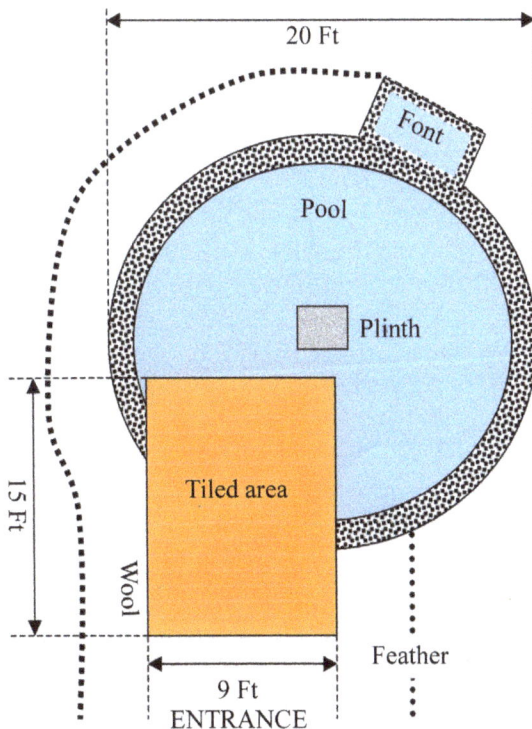

Figure 11.2 Temple Pool

One pool at the Stoney Lane Site (SLT) was studied in detail. It was probably in use in the late Bronze Age and early Iron Age. The pool was fitted with lead umbing a central pillar or obelisk and a font for the baptism of children. There was a tiled apron and steps leading down into the pool. Dead bodies were brought up to the pool on the left hand side.

139

individual components but now the canes for each component would stay and a bigger picture would take shape as each visit to the site produced more data and more canes. The Stoney Lane site I thought might even go down in the history of the Henge Age. Over the next few days work proceeded on the Henge Temple site. One group of eight Druids plus two dogs were found guarding one gate, the stone columns were being marked out, and the floor design looked as if it was 6 + 3 rings of postholes. Then when I was trying to make sense of the narrow doors and what appeared to be sighting posts a few yards from them, I realised that if I used the centre of the pool as the centre of the temple, the bearings of the doors and sighting posts for astronomical observations of the horizon did not make sense. Not only that, the position of the gates, normally on or near the four quarters, were not correct if I took bearings from the pool. This could mean only one thing. It was not a simple site with one temple but a multiple temple site and it was going to be very difficult if not near impossible to sort the different temples out. The next step was to see if the other temples could be identified by looking for their clay pools. A second pool was found and a second tomb, then a third tomb. The second tomb was Bronze Age, copper arsenic. The third tomb appeared to be a Stone Age one as no bronze or iron was found in it.

The presence of three tombs and by implication three Wood Henge Temples meant that the site was complex and was being used as a sacred site for a considerable period of time. The presence of what Nigel and I thought was a Stone Age tomb - if we were right - had very important implications. Being able to trace Druidic Temples back to the Neolithic would give Druidism a very respectable history indeed and raise questions as to its origins. Did its origin go as far back as the Mesolithic? Perhaps Druidism came in with the original settlers as the ice sheet retreated. After a few weeks work it was becoming clear that the Stoney Lane site was going to be a very difficult site to work on. We had to admit that we still had not found our ideal site which is a lone temple with only one tomb and one set of postholes.

A Stone Age Henge Temple

After the frustration of dealing with a henge site spanning a thousand years or more and finding the complexity produced by many tombs close to each other almost impossible to resolve, our luck changed. A year earlier at a congress of the British Society of Dowsers I had met a farmer who was keen on dowsing. On hearing of my interest in stone circles he had suggested that I visit his farm to look at the stone circle he had built. I kept in touch and when I met up with him again he repeated his invitation. The farm was under an hours drive away from my home and was in the depths of the Hertfordshire countryside. So, having accepted the invitation Nigel and I arrived at the farm one Saturday afternoon. Amongst acres of open Hertfordshire countryside, one could almost feel the presence of the 'Celtic' farmers and Druids in the landscape. To make sure that we made the most of our time on site I had arranged to meet up with a sensitive who, with luck, would be able to point out the most promising areas to investigate. As luck would have it, several spiritualists and sensitives were visiting the farm for a 'lodge meeting' run by a visitor from North America. So with their help we were soon aware of a number of interesting areas on the farm worthy of investigation. We found and checked out a number of Henge Temples also battle blood and lines of execution blood with the associated mass graves. The land was clearly being used for Henge Temples up to Roman times. However, the temples were in the middle of crops or difficult to get at. At the best their study would have to wait until after the harvest. Walking back towards the farmhouse across a field of young maze plants where we had come across a Henge Temple we found one of the sensitives sitting in a grassy area. The reason she gave for sitting at the particular spot where we found her was that the 'energies' were in 'balance'. In our language it meant that she had spotted something using her magnetic sensory system. A few minutes later we knew that it was a Henge Temple and it looked as if it was on its own. The sensitive had been sitting on the central tomb of the temple. A week later I was back at the farm with a group of dowsers so did not have much time to study the Henge Temple identified by the sensitive. However, I was able to see that the anterooms and bone depositories of the underground complex looked as if they were well laid out and symmetrical. This was an indicator that the Henge Temple and the tomb complex may be on their own. The underground design had not been modified to cope with previous tomb complexes. The underground tomb complex also looked as if it was Stone Age. No metal, no linen fabric. The outer wall of the Henge Temple had a radius of about thirteen metres not the usual 30m. A Stone Age Henge on its own so that it was easy to study, no complications from overlying and later henges as far as we could see. The farmer gave us permission to study it and leave canes and markers on the ground and so the first detailed study of a Stone Age Henge Temple started. Many ideas were going through our minds and being discussed in great detail but the main question was would we be able to show that Druidism, as indicated by its buildings and ceremony, could trace its ancestry back to the Stone Age. That is over two thousand years before the Romans arrived and a thousand years before the Celts are believed to have arrived in the British Isles. From what we had already learnt from the Henge Temples we were not at this stage too sure that the Celts ever did arrive in Britain, aspects of their culture may have done but we did not believe a mass migration of Celts had taken place. The 'Celtic' influence was more like Christianity. Christianity as a culture did not arrive by mass migration. The procedure was for specialists in Christianity to arrive and the indigenous people then took on board some of its elements, blended them with their own Druidic elements and from the mix made a brand of Christianity which was theirs. They owned it and with ownership came its development. Like good Christians and I suspect like good Druids, they then started exporting their brand of Christianity, with its elements of Druidism and so Druidism, through 'Celtic' Christianity would come to play its role in the development of the religion we now know as Christianity. However, at this stage ideas were only ideas and there was a long way to go before we would know how closely linked the Stone Age was to the temples. The first brief look at the site indicated a Druidic tomb complex without metal. Now we needed the detail. The first step was to mark out the tomb and underground tomb complex. As the canes were laid down on the ground above the tomb a rectangular structure appeared with outside dimensions of 2.45m x 2.10m It had well defined walls with the typical stone of the area, flint, and chalk urine mortar. The occupants were a young girl, height about 1.35m, and a young man, height about 1.50m. The young girl was a prepubescent child and the man probably not much older. The tomb was sealed with sandstone slabs and its walls plastered with a gypsum plaster. Using as witnesses egg white, egg yolk, red ochre, manganese dioxide, charcoal and red cabbage (woad) we found evidence of paintings on the side walls of the tomb. It was difficult to imagine that plaster and paintings would have survived four thousand years in the British climate but there was no sign of the paintings having fallen off and having produced a stain extending away from the wall. The 'warrior' had a flint headed ash spear, yew bow and willow arrows with a quiver and oak shield. The arrows had their flint heads at the bottom of the quiver. The position of the arrow heads in a quiver seems to vary between tombs. From the point of view of pulling the arrow out of the quiver a flint arrow head could get caught if at the bottom whilst the feather or flight end would not. For quick use it is however an advantage to hold the flight end when withdrawing the arrow. To do this without the head catching on the quiver as it is withdrawn means a suitable design for the head of the arrow or for the inside of the quiver. It is therefore possible that the way the arrows are placed in the quiver may be an indicator of the age of a tomb or perhaps alternatively an indicator of the skill of the local craftsmen. In the present case the position of the arrow heads at the bottom of the quiver could indicate that the henge dates from close to the early Bronze Age. The presence of plaster and paint in the tomb may support this date. The use of tannin for leather manufacture and linseed oil may also indicate the transition period to the early Bronze Age. The use of linseed oil on wood such as the spear is interesting as it indicates that flax was being grown for this purpose even if it was not appearing as a linen fabric in the tomb. The most likely early use of flax is to make strong threads and rope. This use of flax straw could have come in before it had been discovered how to ret the straw in water and separate the fibres in a suitable form for weaving. However, the use of linen thread to secure flint arrow heads and spear points in some Stone Age tombs indicates that the fibre had been separated from the straw to make fine threads also the use of linen in the belts of the sacrificed Druids at the gates of Stone Age Temples indicates that it was finding its way into clothing when the henge was built. Some of the Stone Age Henges may therefore date from the dawn of the 'linen age' in Britain.

There are a number of features of the tomb and the Henge Temple above it (Figure 11.3) that indicate that the structure is Stone Age.

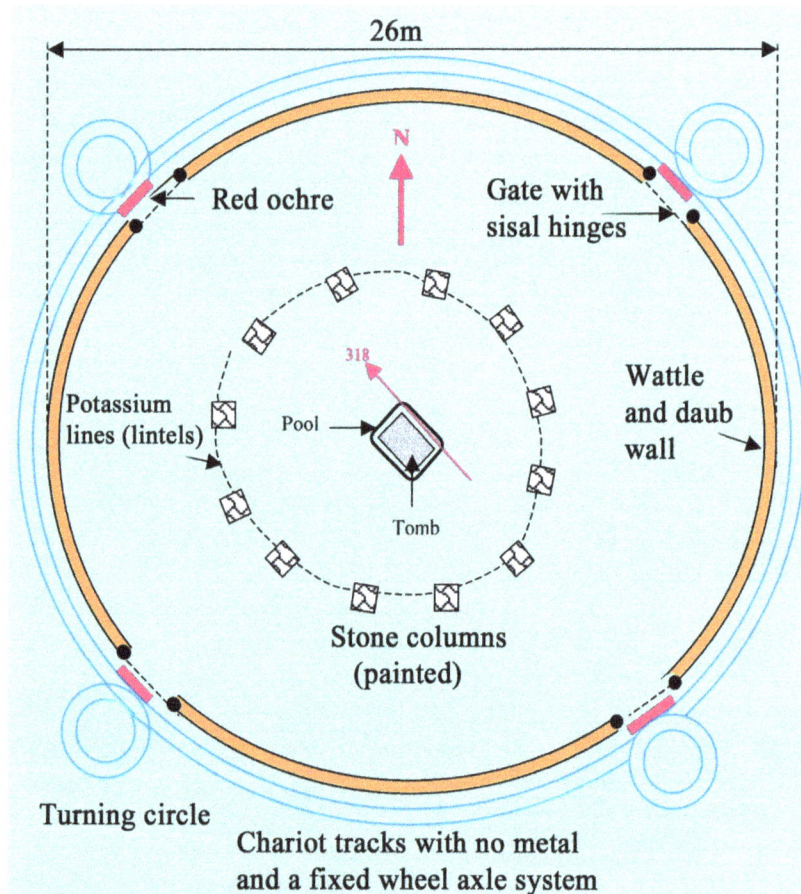

Figure 11.3 Stone Age Temple

A Stone Age Henge Temple was found on a farm. The temple appeared to have all the components of a Henge Temple but was only about 26m in diameter. The stone circle of 12 stones would have been about 13m in diameter.

For example the tomb dowsed as if it had a layer of feathers in it. The feathers were between the bodies and not on them. Feathers were found at the head and belt of both bodies. Dog hair and teeth indicated that both occupants were probably lying on wolf skin with the head of the wolf above their own. There was sable at the belt. Looking at the tomb in more detail the man had his hands by his side and the girl had her hands folded across her abdomen. Fingernails are used as a witness to work this out. The heads rested on horse hair pillows. Horse hair pillows have been found in all the tombs studied to date. The hair of the female was down below the shoulders. Again this is a constant feature through the ages. The hair of the male is normally short but does vary in how short so it looks as if male fashion varied with time. In this tomb the hair of the male was to the shoulders. The main garment being worn was made from wool with leather footwear.

From the tomb chamber there are four tunnels, the widest we had come across at the time. These went to the four anterooms of the bone depositories (Figure 11.4).

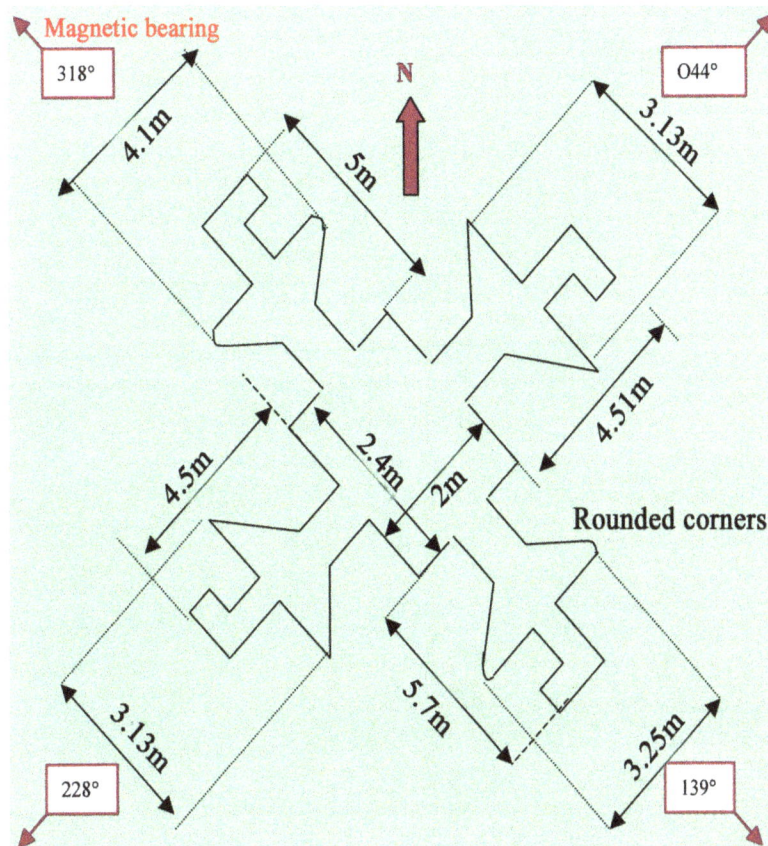

Figure 11.4 Tomb Complex

The tomb complex of a Stone Age Henge Temple is almost identical to those about 2000 years later. The 'Celtic' cross was not very convincing and later we were to discover that they were indeed Thunderbirds.

Unlike many other underground temples that had been studied the layout was ordered and precise. Soon after the outline of the underground temple had been marked out a North American Indian called into the site and after studying the outlines on the ground told us that each of the arms looked like the Indian Thunderbird with outstretched wings. The individual arms of the complex do indeed look like the Thunderbird but putting all four together the 'Celtic' cross appeared to us to be a better likeness. The bone depositories are part of the outer circle of the cross, the anterooms are the widening arms of the cross and the centre of the cross is the tomb or pool in the centre of the temple. The anterooms vary a little in size with the distant wall being three to four metres across. The side walls are 1.6m to 2.30m long. There is a little rectangular altar in each anteroom and what looks like a ceramic lamp in one corner. A passage leads from the anteroom to the bone depositories. These chambers are about 1.4m by 0.60m, not very big and not capable of holding the skeletons of hundreds of people. Using witnesses it looks as if the body was divided between the four locations. Heads to the north, legs and leather in the east, Arms, hands and Ash wood, possibly staffs, to the south, trunk to the west.

A Review of Tomb Complexes

The tomb at the centre of the underground complex is a time capsule. It is made with the techniques and materials of the time and reflects aspects of the culture of the time. It is therefore useful to review and compare what has been found with other tombs up to about mid 2005.

Since identifying the first tomb in Snowdonia (Chapter 5) many tombs had been dowsed. The common features are that they all appear to be deep in the ground and in a chamber from which there is access to four anterooms and bone depositories. The tomb is made from local stone and there is always a male and female body in it. The tomb is lined up approximately north south or north east/south west with the heads to the north. The range of deviation from the north south axis is quite large with that of the Stone Age tomb on the farm being about 45° to the east. In addition to the two bodies there appears to be a spear, bow, and quiver of arrows a shield and in later tombs a sword. Dealing with the bodies first. There is normally, but not always, a positive response with a toadstool witness and a foxglove witness indicating death by poisoning. The bodies range in age from clearly adult to, as in the Stone Age tomb, pre adolescent. The male is 1.50m and the female 1.35m. There is an indication that the walls of tombs are plastered and may have painted pictures on them. This is not the case with some Stone Age tombs.

The anterooms often show signs of having paintings on their walls. At the head end of the tomb is a horse hair pillow which has been found in all tombs to date. The female always has long hair the male short hair but sometimes extending below the shoulders. The clothing in the early tombs consists of a full length garment of dog hair (wolf) and wool garments. In later tombs a linen garment is the norm and the feathers disappear. There is no sign of a linen fabric in what appear to be Stone Age tombs. The male had his hands to his side in this tomb but this varies between tombs. The female had hers on her abdomen. Belts do vary as does the tanning process for the leather. Leather shoes are worn. The leather is tanned with tannin but urine tanned leather has been found in some tombs. Tannin indicates that the tomb could be late Neolithic or even early Bronze Age. Pig leather shoes are sometimes found on the female. Nettle fibre was used on the belt, shoes, spear and arrows. There is normally something dowsing as terracotta between the bodies and at the southern corners. This feature is observed up to the Iron Age. The corner objects responded to beef fat and may be lamps. Beeswax is not found in early tombs but appears in later ones. The shield dowsed as oak in early tombs but in later tombs the shield is ash or bronze. The arrows dowse as willow, the bow as yew and the quiver as straw. The arrow flint heads vary between being towards the feet and head. Linseed oil is on the quiver and ash shaft of the spear but there was none on the shield. In later tombs gold appears, then arsenic bronze, tin bronze and then iron.

There was the possibility that semi-precious stones and crystals were being used in the clothing or were present somewhere in the tomb. The main problem with stones and crystals is that many of them are mainly silica or some form of quartz. A positive response to a witness may therefore be misleading and only indicate that a stone of some sort may be there.

Semi-precious stones were dowsed for and quartz found on the shoes, belt and head of both bodies. Malachite was found on the head and belt of both bodies and on the shoes of the female. Amethyst was on the belt area of the male and head of the female. Sapphire dowsed on the head of the female and amber at the waist of both bodies and the head of the female. The semi-precious stones are present from the Stone Age to the Iron Age. Tree resin is dowsed on the spear and arrow heads where it is used as a glue. In some Stone Age tombs bluebell sap is used instead. Hemp was found as a binding thread in some tombs. The lack of linen fabric, hemp and gold may indicate a date very early in the Bronze Age but perhaps more likely a Neolithic origin. The presence of linseed oil indicates that flax was grown locally or the oil was being traded.

Stylised tomb complex and temple rings

Figure 11.5 The Celtic Cross
Could the 'Celtic' cross be derived from the design of the tomb complex?

A Druidic Temple Guard

Returning to the Stone Age Temple on the farm. The temple had the four sets of eight Druids sacrificed at the four quarters. Whether as part of a consecration event or deconsecration we do not know. What we did know was that at least two thousand years later the same ceremony was going on. As it was not possible to get at all four quarters only the line of Druids on bearing 256° was dowsed in detail (Figure 11.6).

The line of Druids extends from a point about 3.60m from the centre of the tomb to about 16.50m out along a radius. If the gate was near this point it would indicate a building 33m in diameter. Facing out from the tomb the females are on the left the males on the right. The dogs are on the outside of the third pair going out. The heads of the 1st, 3rd and 4th males are associated with deer antlers. This is what might be expected. What was not expected was finding antlers associated with the female Druids. There is the possibility that there is another

145

line of Druids for another temple and that the antlers belong to that group. However the tomb is Stone Age, Reindeer may have been common and female Reindeer have antlers. At the time the people may not have thought it odd to have women wearing antlers.

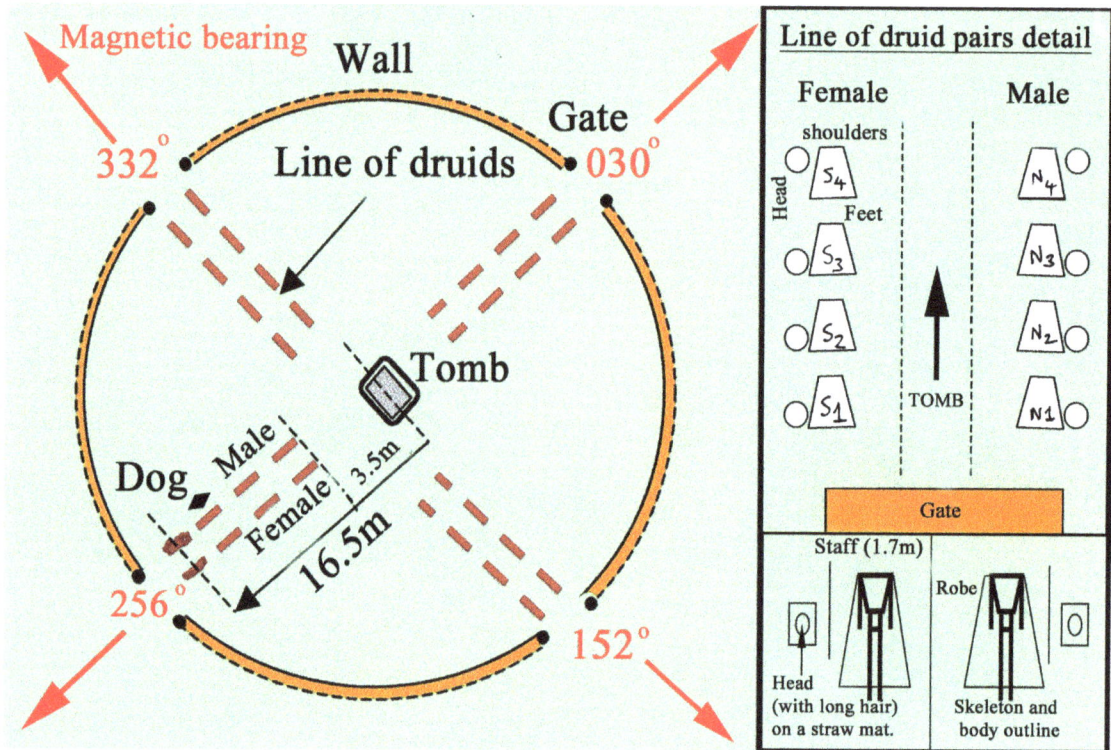

Figure 11.6 The Druidic Honour Guard

The staffs were dowsed for semi-precious stones. Quartz, amethyst, citrine, red ochre and egg white gave a positive response. There was no gold, silver, iron or bronze. The second male and female Druid did not have any stones or paint on their staffs. All the female Druids had long hair. The bodies and clothing of the Druids were tested for a range of items. The belts were leather with linen for the males but linen only for the females. Druids N2 and S2 were pre pubescent children.

At this time we did not know if the 4 x 8 Druids were left to decay on the ground, as the posture of the bodies seemed to indicate, or if a large shallow grave was dug. We knew that there was often a row of lamps along either side of the bodies.

Stone Circle

The next step was to see if the Stone Age Temple had a circle of stone columns in it. It did not take long to find them. There was a ring of twelve stone columns with potassium lines connecting them as a circle and indicating wood lintels. There were no signs of iron or bronze pinning of lintels or roof timbers. The stone columns were constructed from limestone and based on tile or brick foundations.

All but two of the columns had been painted on the two radial sides. The pigments used in the paint were: red ochre on columns 1 to 9, manganese dioxide on columns 1 to 8, woad on columns 1 to 7 and malachite on columns 1 to 7. It looks as if columns 11 and 12 may not have been painted. Column10 did have gypsum and egg white on it. The use of tiles and mortar indicates that the Stone Age Druids were already quite advanced in ceramics and construction skills. In other words they could produce bricks and tiles and secure them with a crude mortar. Trade was also going on in pigments as they were being used extensively.

At this stage we were pleased that we had found evidence that stone columns were being used in the late Stone Age or early Bronze Age as part of the temple structure. Also that they were being painted. This indicated that two sides of the column at least were dressed, possibly plastered and prepared for painting. However, the twelve stone columns did not produce a circle centred on the tomb. Although lintels connected the columns, this indicated that there may be at least two temples on the site when we had assumed that there was only one. Alternatively, the columns, or some of them, were not primarily structural. Knowing that there was possibly another tomb around we soon found one but by now it was too late to sort out the stone columns and find if there were two sets of stones or if it was one set deliberately not positioned in the form of a circle.

The Seven Altars

The Henge Temples studied to date all had seven altars. An altar is normally identified by looking for the use of mortar and the stone used for the slab and plinth. Then clay, terracotta or Druidic tile for the blood bowl or trough is tested for. Seven altars were identified and are shown in (Figure 11.7).

The altars magnetic bearings from the centre and their use were:-

034°	Horse
089°	Sheep
117°	Dog
154°	Humans
271°	Goat
307°	Pig
350°	Ox

The altar for human sacrifices was 1.89m long 0.96 wide and a blood trough 0.32m wide ran along the front. The altar for the dog was 1.30m by 0.80m with a 0.33m wide blood trough. The altar for the pig was 1.15m by 0.66m with a 0.22m blood trough. The altar size seems to match the body size of the victim.

The Clay pool

The clay pool was examined to see if there was a place for the priests to enter it. The wall of the pool was about 45cm wide with the total diameter being 4 to 5m. Quite a small pool. On the south west side there were entrance steps. On the right hand side of the steps facing the pool up the steps was an area responding to a wool witness whilst on the left side the response was to both wool and feather witnesses. Feathers and linen are taken to indicate Druidic (priest) dress.

Temple Wall and Gates

The Henge Temples that we believed to be Stone Age had a number of things in common. One was that they were smaller than the normal 60m diameter of the later temples. This Stone Age Temple was about 30 metres across. The walls were made from wattle and daub, it had

26m

Horse
034°

Cow
350°

Altars

Pig
307°

Sheep
089°

Pool
(Clay)

318

Dog
130°

Goat
207°

Tomb

Human
154°

Stone columns

Concept sketch

Pathway
to the
DOLMEN

Tomb

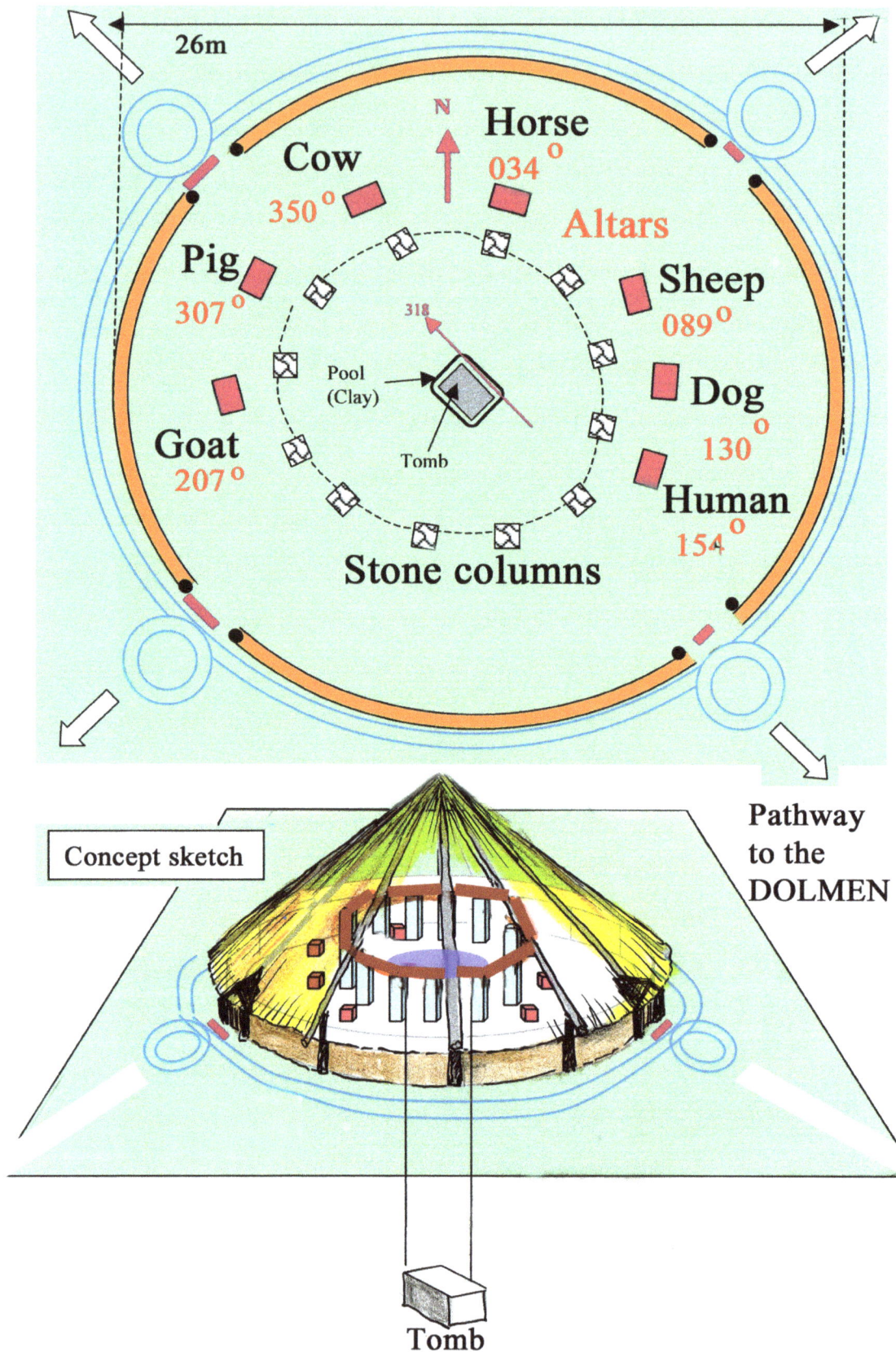

Figure 11.7 Altars
The position of the 7 altars within the temple is shown.

148

gates approximately to the north, east, south and west and the gates had hemp and nettle fibre hinges. The gates we could get at were the western and northern gates. There was little space round the eastern gate and the southern gate was in thick bushes and not accessible. The northern gate had chariots or carts going past it. The chariots stopped and, from the red ochre stain, unloaded or picked up bones. The carts had rotating axles. That is the wheels were fixed to the axles and the lubrication points for the axles were under the floor of the cart. The wheel tracks had no bronze or iron but did have wood and leather. It looked as if there was a turning circle outside the northern gate.

On the western gate there was again an indication of a chariot track and turning circle and footpaths going off to the Western Dolmen. There was also a patch of red ochre where the chariot was unloaded.

This summary of the features of what is possibly a Stone Age Henge Temple shows that in terms of basic design the temples change little over a period of at least two thousand years. Many aspects of the ceremonies remain unchanged for the same period of time. The advance of technology and the availability of materials are however revealed in the size of the temple, the clothing of the Druids, the splendour of the tomb and its occupants. The design of the tomb complex and the Henge Temple does not change. It is as if they are both locked into something permanent which sets their design.

Back in Chiswell Green Nigel had been working on Druidic art and was using as witnesses paint pigments such as gypsum, flint, sand and chalk to identify pictures drawn on the ground. The Stone Age Henge Temple was checked for such drawings and an outline of a starburst was found in a segment between two sets of eight Druids. It looked as if the artwork associated with the temples may also have remained constant over at least two thousand years.

Chapter 12

Maiden Castle – Fort or a Druidic Temple?

Maiden Castle

On the14th July 2005 I set out to visit a place called Maiden Castle, an archaeological site that is referred to as an Iron Age Hill Fort. Maiden Castle is set on a hill in the rolling landscape of Dorset just south of Dorchester, off the A354 and only about 8 miles from Weymouth on the south coast (Figure 12.1). I say set on a hill but as you approach the site from the Dorchester direction the complex looks as if it is the hill. It is said to be the greatest Iron Age fortress in Britain, a potent legacy from the early unrecorded centuries of Britain's history. According to the archaeologists Maiden Castle was in use 4000 years ago, that is early Bronze Age or late Stone Age. The archaeologists then trace the history of the site suggesting that it was occupied by late Stone Age peoples from Northern Europe for a while and then abandoned round about 1500BC. About 350BC the plateau of the hill was then reoccupied and a massive Iron Age fortress constructed. Two hundred years later sometime round 150BC the outer concentric ringed fortress was built. At least so the conventional archaeological story runs. When I read the history of Maiden Castle very little of it rang true. The pictures of the so called fortifications looked too much like a dyke system and I thought that the whole concept of hill forts needed to be examined. For example, by the time you are forced into a hill fort such as Maiden Castle you have lost the battle as there is nowhere to go and nobody was going to come to your aid. Your crops, cattle, houses, farmsteads would all be destroyed. Holed up in the fort there will be the problem of water for many people, cattle and horses. The so called stones used for slings which were found at Maiden Castle are more likely to be a store of what we have called prayer stones (Figure 10.8). If sling shot was important for your defence you want a nice flat sloping surface coming up to the 'castle wall'. You do not provide the enemy with trenches to shelter in.

The work that I had been doing with my dowsing colleague Nigel pointed to Maiden Castle being a Druidic Dyke Temple complex. It was no more a defensive position than a cathedral, monastery or church, but people might seek refuge in it if threatened. My visit in July 2005 was therefore an initial look to see if I might be right.

The car park, which acts as a good base, is near to the West Gate so I approached the site from the car parking area picking up the path that leads to a metal kissing gate. The area of the car park and the ground leading up to the first bank is reasonably even and rises slowly. It was ideal for chariots and bronze chariot wheel tracks were found going through where the gate now stands. The bronze was evidence that the place was being used during the early and later Bronze Age. Lying outside the bronze chariot tracks were carbon tracks with staff marks indicating Druids in ceremonial procession. On passing through the gate and on the right, is a bank and then a deep dyke with another bank on the far side. It looks very much like a small Dyke Temple. There is one altar on the eastern end of the bank. From the position of the altar there is a spectacular view towards the east, also to other parts of the countryside. The altar could have been inside a building but this was not checked. The altar gave a positive response with human blood and garlic. The garlic indicates that some Romans may have been sacrificed on the altar. There was some indication of a clay pool in the temple and of more altars along the top of the bank for animals. The area looked like a Dyke Temple dating from the Iron Age but a detailed survey would be required to confirm this. Returning to the path and going further in towards the West Gate I found that it is possible to climb down into the dykes. At the bottom of the first one I entered there were bronze chariot tracks, copper and arsenic, in

Figure 12.1 Maiden Castle

The name is derived from the Celtic Mai Dun meaning Great Hill. It is Europe's largest 'Hill fort' covering some 47 acres. The hill has had three dykes dug round it. The ceremonial dykes were enclosed and took on the role of imitation underground rivers. When the Romans arrived the main battle areas so far discovered were outside the East and West Gates of the Dyke Temple complex.

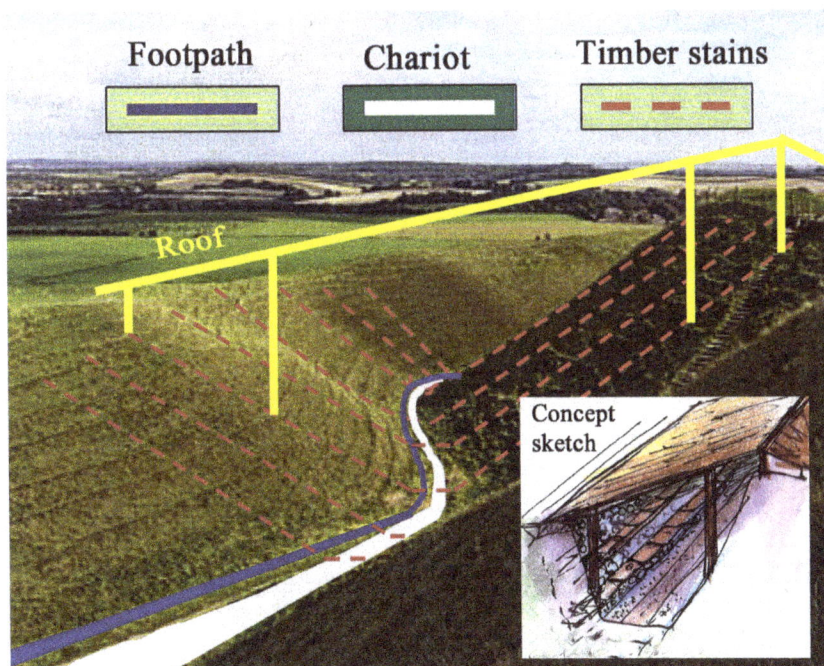

Figure 12.2 Concept Sketch of a Dyke

The large ditch dug using oxen has terraces built into the walls for priests to walk in procession. The floor is levelled and surfaced for chariots and painted with the diamond symbols of flowing water.

151

the second dyke under the main ramparts the chariot track was copper and tin. Iron tracks were not found. Up on the plateau, that is inside the fort, were Henge Temples. One tomb had linen, iron and copper in it. Along the rampart the points where spoil was lifted up from the bottom of the dyke could be identified. The lifting of spoil required special equipment and oxen to pull the 'bucket' up. After more reconnoitring I started on the return to the car park. Outside the iron gate and to the left facing the car park I found a mass grave measuring 8m by 10m. Running across the path and in front of the fence there were four lines of bloodstains. The section I looked at was 10 patches long giving at least 40 executions. After the visit I was reasonably satisfied that Maiden Castle was a Druidic Temple complex and not a fortress. Three and a half weeks later I was back with Nigel and a long list of questions that we wanted the answers to. My previous work had to be checked but importantly we had to find out if the dykes were roofed over. If they were then the fortress hypothesis would no longer be tenable and the Druidic Temple complex hypothesis would be in. On walking up to the iron gate from the car park we checked for chariot tracks. We looked at two and found the tyre marks of one to be copper arsenic bronze and the tyre marks of the other to be a copper tin bronze. This indicated that the site was in use from at least the early Bronze Age. We did not look for wood or iron wheel tracks. The next stop was the dyke system. Scrambling along the steep sides of the dykes it was possible to pick up the postholes of the posts that were supporting a roof and by using potassium as a witness, the ash shadow of the rafters going across the dyke were identified. The technique we used was to mark out posts and rafters and confirm that they were in line from one side of the dyke to the other. Where a rafter had been bedded into the bank it had left a rectangular stain. Two dykes were checked and found to be roofed in (Figure 12.2). Once the dyke is roofed in there is a need for lamps and the bronze lamp stands were identified on the outer and shallower slope of the dyke. With the bronze stain there is also a beef fat stain. The chariot tracks on the floor of the dyke were confirmed.

There was not much time to study the East Gate but chariot tracks were found following the dyke and going under the road crossing the inner dyke. This indicated that at one time either the dyke was not filled in at this point and allowed chariots to follow the inner dyke without obstruction or there was or perhaps still is a tunnel connecting the two parts of the dyke.

The area outside the East Gate is farm land and did not appear to be easy to work on so we returned to the West Gate and car park area. This time we looked at it more closely, see (Figure 12.1). The execution line was identified but this time it looked more like a battle line. There were four ranks and there was Roman blood amongst the Druidic blood. The execution lines went across to the wire fence which runs along the left hand side of the car park and up to the first bank. The execution lines may extend into the cultivated fields. In front of the four ranks are the scattered patches of blood characteristic of battle blood where 'Celt' and Roman fought to the death. In front of that area there is what appears to be a real battle line where the Romans are taking casualties. This area is about 10m in front of the car park. Just inside the wire fence and stretching across the battle line is a mass grave measuring 37 paces by 18 paces (Figure 12.3).

The Romans could have started their programme of destroying the Druids when they arrived in Britain. In other words the attack on the Druids and the execution lines at Maiden Castle could date from soon after the invasion in 43AD and then the programme progressed over the rest of Britain lasting until 62 AD or even later in the north of the country. Knowing the true sequence of the Druidic extermination by the Romans could be very important to understanding the relationship between the Romans, the secular bodies under Kings and Princes and the religious bodies i.e. the Druids.

152

```
                    Romans

         ↗                      ↖

Secular        ←───────────→        Druids
```

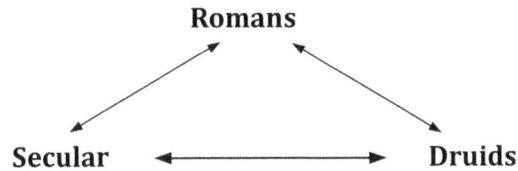

At this point all we had to go on was the idea that the genocide of the native Britons by the Romans started at the time of Boudicca's revolt. It is easy to assume that the execution lines appeared at about this time but it does not quite fit. The Roman attack on Anglesey, which took place before the revolt, could be the final phase of a process started soon after 43AD. If that is the case, Vespasion's attack on Maiden Castle would have been against Druids not against the local King. One account of the attack by the Romans on Maiden Castle describes the Romans fighting their way from 'rampart' to 'rampart'. This does not sound very convincing. The dykes had roofs and the Romans would have been no match for 'Celtic' warriors and war dogs in individual combat on 'ramparts'. The Romans almost certainly just knocked on the East Gate door and asked for it to be opened. Temple gates were wide, moved by manpower and not designed to keep attackers out. The accounts say that once in the Romans ran amok, slaughtering the inhabitants indiscriminately and burying the bodies outside. That means that there should be a lot of battle blood inside the East Gate.

If the Romans did kill everybody inside the temple complex where did all the Druids come from for the execution lines? To date we have found very limited killing in temples. With some exceptions, the killing takes place outside. At the moment it looks as if Maiden Castle fits this picture. However, because we had not looked at the East Gate there was the problem of Vespasion's battle to resolve. Was it with the secular society under their Chief or was it with the Druids. If the battle was with the secular society the written records of the date may be correct. The battle with the Druids could then have taken place much later. If the battle was with the Druids and not the local Chief, the date is still probably correct and indicates that destruction of the Druids was an important objective of the Romans as soon as they landed. Maiden Castle appeared to be the key to finding an answer. That meant one more visit to Maiden Castle.

Return to Maiden Castle

The description of Maiden Castle found in archaeological literature seemed to be based on a lot of conjecture and not very much real evidence. Even the description of the Roman attack under Vespasion did not ring true. The site, even today can be clearly seen as a Dyke Temple complex not a defensive position. A few fire arrows onto the very extensive thatched roofs and the lot would have gone up in flames. The Dyke Temple complex was not the place to seek shelter in, it would be a death trap. Any fighting would take place outside on the gradual slopes up to the temple. Nigel and I had now looked at many temple complexes and the fighting was always on the outside of the temple. Sometimes there was fighting round the gates but inside there were very few patches of blood with the exception of children's blood. The West Gate and the area round it showed the typical pattern of a battle between the Romans and Druids. On previous visits to Maiden Castle battle blood had been looked for inside the complex but none found. That does not mean that it is not there as the plateau is a large area to search. However, searches round the East Gate had failed to find any battle blood. The descriptions in the literature of Vespasion's attack on Maiden Castle are so detailed that another visit was necessary to determine if the archaeologists were right. This time to look at the East Gate and find out if there was battle blood around the gates and inside the temple complex at this point. Also to look for battle lines outside the temple, execution lines and mass graves.

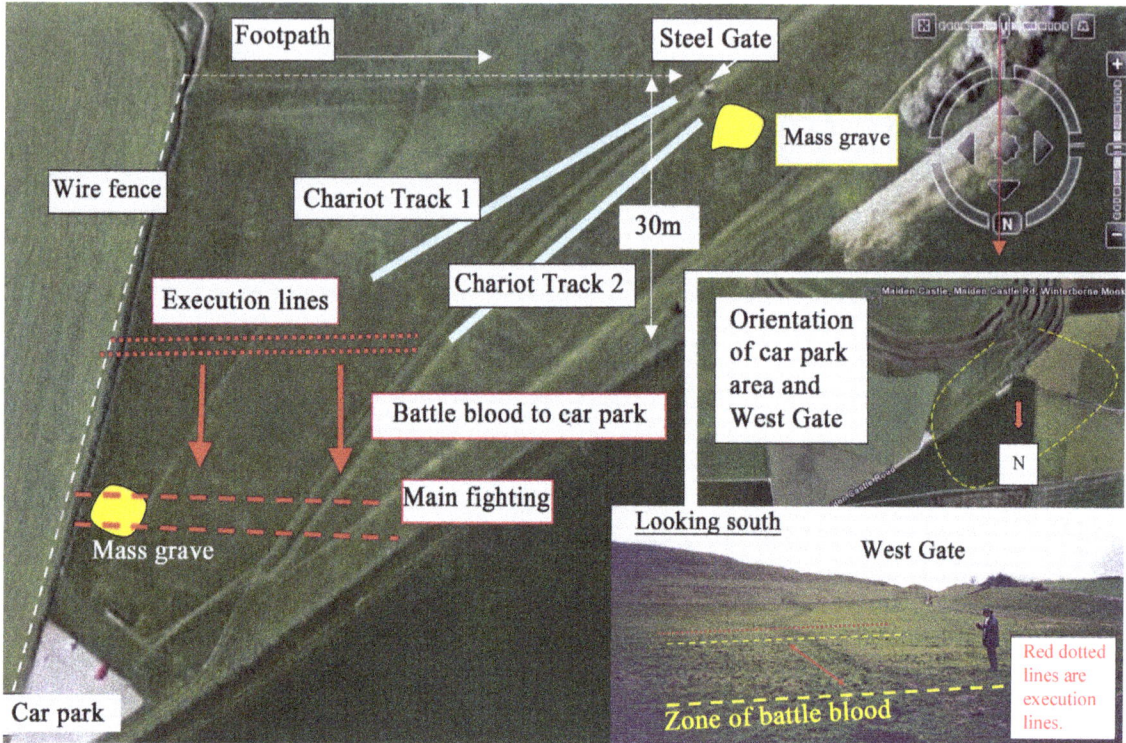

Figure 12.3 Blood Below the West Gate

The path from the car park leads uphill to a steel gate through which it continues up to the 'castle's' western complex. Two chariot tracks were located in the field beyond the car park. Chariot track 1 dowsed positive on copper and tin indicating that it was late Bronze Age. Chariot track 2 dowsed positive for copper and arsenic placing it in the early Bronze Age. Four ranks of execution lines run across the field up to the fence on the left and possibly beyond. Roman blood can also be found amongst the lines. The mass grave in the vicinity of the car park contains 37 x 18 bodies. Note: all positions are approximate.

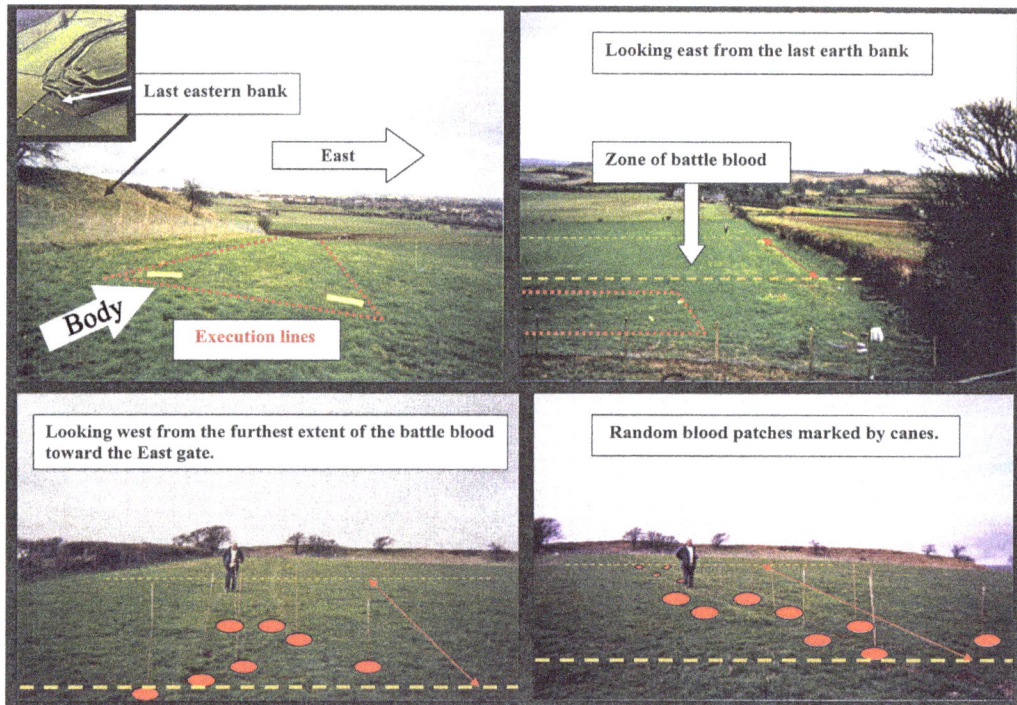

Figure 12.4 East Gate Battle Blood

The canes indicate where battle battle blood was found outside the East Gate.

154

So at 7am on the 17th February 2006 Nigel and I were on our way to Maiden Castle to see if the current archaeological view was right or if our interpretation was nearer to the truth. When we arrived we parked at the West Gate car park and then, loaded with the tools of the dowsers trade headed up the left hand side of the car park to a gate opening onto the footpath which runs along the base of the hill. The present day footpath runs along a wire fence and is on a terrace built as a ceremonial road. To confirm this we stopped along it and dowsed for the road and ceremonial chariot tracks. In the middle of the path were the 4 to 5 foot wide twin tracks from wood wheeled vehicles, ash, hemp, leather, the leather tanned with urine. The cart was pulled by a deer. We felt that these carts were Stone Age but it was still possible that they were being used in the Bronze Age for a specific ceremonial purpose. They could also be service carts repairing the road. Straddling the wood wheel tracks were the 8 to 9 foot wide tracks of bronze tyred wheels. The bronze was both arsenic and tin bronze that is early and late Bronze Age. There was no sign of iron tyred wheels but the wire fence may have suppressed the iron field so that it could not be picked up. On either side of the road there was an edging or curb. Druidic cement (Chapter 16 Puddingstone) was used to make a mortar and secure stones in place or to make a concrete as required. A chalk/urine mortar was used for the road surface and running along the centre of the road was the diamond pattern which is found on all the ceremonial roads we had studied so far. Nigel had decided to time me and it took about seven and a half minutes to obtain this snapshot of a ceremonial road. Continuing along the path and about halfway along the base of the hill we came to a recess dug out of the hillside. It looked very much like the entrance to a tunnel and a type of underground ceremonial complex that we had found in Hertfordshire. We had little time to study it but the chariot tracks went into the recess, there was the outline of the tunnel opening and a short way in on the left of the tunnel there were indications of the 'U' shaped tunnel for deer drawn carts. It is the finding of wood wheeled deer drawn carts in underground ceremonial complexes which are clearly Bronze Age and still being used in the Iron Age that makes it difficult to draw the conclusion that the use of deer as draft animals indicates the Stone Age.

Following the path round the terracing the chariot tracks go under the fence into a field and the modern footpath rises up the side of the hill. By keeping to the lower slope there is a path to a gate into a field. From the gate, looking to the right the field runs up to a hedge almost in line with the East Gate (Figure 12.1). This field and the one the other side of the hedge looks an ideal area for any battle that may have taken place between the Druids and Romans. These fields are private farmland and permission is required to enter them. If the fields are viewed from the Castle there is a gentle slope away from the Castle which is going to give defenders an advantage. On the defenders left flank the ground falls away even more steeply providing the defenders with an advantage on this flank. Because of the hedge most of our dowsing took place on the left half of the battle field (Figure 12.4).

It was an easy task to identify that a battle had taken place as there is battle blood everywhere. Following the blood down the slope the front of the battle line was found and it could be followed round to the left where the ground became steeper as the north slope of the hill is approached. The battle must have been fierce. It was only possible to mark the position of a few patches of blood (see photo) but they are everywhere and it is easy to imagine that the Druids stayed their ground and died. There is battle blood right back to the present day fence. The Romans were taking casualties and the impression is of two disciplined forces in combat. Over on the left flank there is less human blood but more dog blood. It looks as if the Druids used the war dogs to protect their flanks. The Romans won with the result that running along the top of the battle field is an execution line three rows with 22 in each row and with the chief priests in front, about 70 people. The mass grave is about 24 by 23m., there will be other mass graves but they were not found.

On the north side of the hedge where most of the dowsing was done the battle field was about 80 to 90m long and 70m deep. On the south side of the hedge the battle blood continues, there are more execution lines and mass graves but no sign of the battle extending into the East Gate area.

The picture at the East Gate is the same as at other Druidic Temples except it is on a larger scale. The Druid warriors fought a stylised battle in that none intended to survive. After the battle the older people, women and children were rounded up, had their hands tied with hessian, forced to keel down and had their heads cut off. So far it has not been possible to determine if the Romans burnt the temples down. The temples represented a lot of raw material, in particular stone and wood, they also had to be searched for bronze, silver, gold, lead, precious stones. The Romans may well have left buildings standing, at least until they had taken what they wanted. After the Romans had finished with the site the local secular society may have removed what was left. As with other temples there is some evidence that people returned and tidied up the site. The tunnel found along the footpath between the West and East Gates had been sealed off. Before or after the battle we do not know.

Maiden Castle was probably established in the late Stone Age with Woodhenge Temples and Dykes. Once established the Druids would have probably kept it going and maintained and developed it. It was too big an investment to let go. The dykes were used as ceremonial chariot ways in the Stone Age, and in the early and late Bronze Age and it was still a major Druidic centre in 43AD when the Romans arrived on the scene. With no evidence of other battles at Maiden Castle during the early years of the Roman occupation it is probably reasonable to assume that the Romans started to remove Druidry from Britain as soon as they landed. If the Romans fought the local secular society they did it elsewhere. However, after seeing what happened to the Druids the local chief and his people probably decided to get on with the Romans.

Stonehenge

After the first visit to Maiden Castle on the 14th July 2005 by mid afternoon it was time to head north towards Stonehenge. Every time we felt we had made some progress unravelling the complexity of the Henge Temples there was an urge to go and check it out at Stonehenge and at the same time to confirm results obtained on previous visits to the site. This time we wanted to look at a number of things including the structure of the roof, to see if we could pick up the equipment for raising the stones to the upright position, to see if there was a central pool and to look for HGV tracks inside and outside the henge. The following morning we were on site at 06.98 hours and walking down the tunnel towards the Stones for our two hour session. Passing the pictures showing the history of Stonehenge we realised that unbridled conjecture is the cause of a lot of inaccuracy and ideas that just do not hold water in the real world. After finding a place to leave witnesses, markers and cameras we started looking for roof rafters. We soon picked up the potassium lines of the rafters going from the outer wattle and daub wall to the linteled sarsen circle. How do we know that the rafters we were looking at were there at the same time as the linteled stone circle? The answer is that we do not. To answer that question we may have to climb up on top of the stones, an activity frowned on by English heritage and one I am now ill suited to. However, with a ring of rafters heading towards the centre it is almost certain that the sarsen, circle with its lintels is structural and provided support for the rafters carrying the weight of the roof. Next the presence of HGVs on site was confirmed. The tracks of the wheels could be followed to a stone where the outline of some structure associated with handling the stone could be traced (Figure 12.5). Beef fat which was used as a lubricant, was also present indicating that there was a mechanical device

156

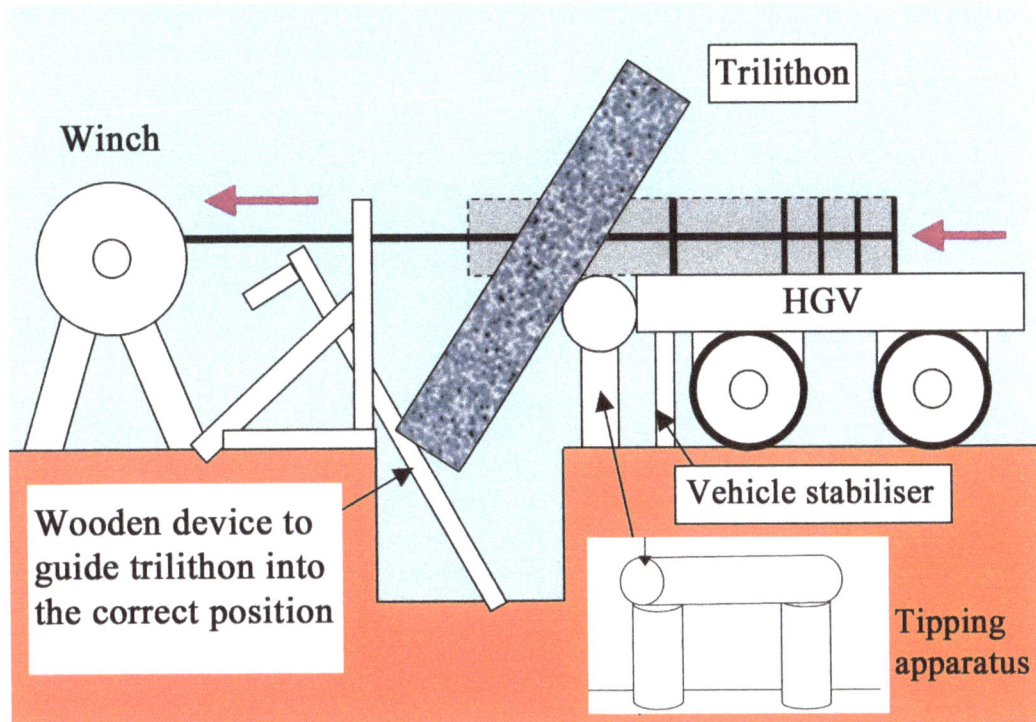

Figure 12.5a Unloading a HGV

Diagram of how a stone may have been unloaded from a HGV into a hole prepared for it. The aim of the Druidic engineer would have been to take the weight of the stone on a trestle or a robust support of some sort which would allow the stone to tip as it was pulled off the vehicle.

Figure 12.5b Tipping Apparatus

The large postholes (white) of the tipping apparatus can be seen. The blue shows the tracks of vehicles with wheels having iron tyres coming up to the tipping apparatus. Lubrication and ropes can also be located with these features (not shown).

157

for moving stones, in this case probably for tipping them up and into a prepared hole. With stones varying greatly in size and weight it is likely that more than one method of positioning stones was used.

The central pool was looked for using a clay witness (Figure 12.6). As Stonehenge is an ancient site there will have been many pools over the millennia. One pool was found and it lay within the horse shoe of the bluestones. The pool was big and would have occupied most of the central area. There were signs that a previous pool had extended out and beyond the back of the bluestone horseshoe. There was no doubt that Stonehenge was fitting into the standard picture of a Druidic Temple. It had a central pool with stones being either part of the pool structure or providing some ceremonial function a little way from the pool wall. Then there was a ring of structural stones with stone lintels and an outer wattle and daub wall. There were no HGV tracks across the ditch or dyke indicating that the ditch came after the stones were set up or a bridge was used. The presence of iron tyred wheel tracks, 6 wheels per axle, going through the Heel Stone show that the stone was probably set up after the big stones were moved on to the site. All the HGVs had iron tyred wheels, some had 4 wheels per axle some 6 wheels. The carts were pulled by oxen with some having 5 lines and some 7 lines of oxen. They were big carts, similar to those used at Avebury. They might have been the same carts that were used at Avebury as carts large enough to carry 20 to 50 ton are a big investment.

The fact that the tyres of the carts were iron confirmed that Stonehenge, at least in its final days as a temple was an Iron Age construction. It should be possible to work out when the bluestones arrived on site and how. However, whenever they did arrive they would have to fit into the design of the Druidic Temple that is either as part of the pool, as structural components, as part of the outer wall or have some ceremonial or energy engineering function. The trilithons of the sarsen horseshoe are a new feature of temple design in that we had not come across it at this time in other temples. Because of the space taken up by the trilithons other features of the temple would have to be modified or not included in the Stonehenge design. For example altars could be moved to a barrow.

The use of the term Hill Fort for the Druidic Dyke Temples that happen to be sited on hills and the association of astronomy with the remaining stone skeletons of temples is unfortunate. It is perhaps a classic example of how modern man tries to make the past fit in with his views on what should be on the top of hills or what the ancients can or cannot do. The terms Dyke Temple or Henge Age Hill Temple might be both more accurate and better terms for the masterpieces of Druidic engineering. Once it is accepted that structures such as the dykes of Maiden Castle and Stonehenge were in great buildings and were built by a civilization with competent engineers it is possible to workout how they did it and seek confirmation from the chemical stains left in the soil.

Clay pool (16m diameter)

☐ Plinth (1.2m) ▬▬ Plumbing

Figure 12.6 Sacred Pool

A pool was located on the western side of Stonehenge. The yellow broken line shows its outline. In the centre of the pool is the plinth with its associated plumbing. The lead also contains silver and the bronze on the plinth, witnesses as arsenic bronze indicating the pool to be early Bronze Age. Note: the stones are on top of the eastern edge of the pool.   *Photograph by J.J. Evendon*

Chapter 13

Henge Temple : Matters of Life and Death

Summer is a busy dowsing time so two days after the visit to Stonehenge we were on our way to the secret Yorkshire Henge Temple. The aim was to explore it in greater detail and obtain some design data and information on how it was used. Some mundane detail such as where were the toilets and washrooms but also how many Dolmen were part of the temple and where are the main bone depositories. On this visit the weather was fine and we made very good progress.

The Dolmen

The Western Dolmen had been seen as a feature of the Henge Temple from the early days of temple research. There appeared to be nothing wrong with the idea that thousands of years ago bodies had been stripped by birds (excarnation). Indeed there was not but what I had failed to appreciate was that one Dolmen could not handle the throughput of bodies. However, it was not this line of reasoning that led to the next step. The temples had four gates, there were four anterooms and four bone depositories, it might be a good idea to look for indications of more Dolmen and this is what we did in Yorkshire. We knew where the western Dolmen was so all we had to do was explore along the four points of the compass. This was not a difficult task as the phosphate halo round a Dolmen can be identified from 10m or more away using a search technique. Three more Dolmen were discovered so that there was indeed one on each quarter of the compass. Even a cursory look at the Dolmen was enough to show that each was different. The main difference was in the size and the number of bodies each could take. All the Dolmen had a wattle wall round them with a gate facing the temple. The walls, possibly the whole Dolmen, appeared to be on a raised brick plinth to prevent entry of animals. On the walls facing the temple there were lamps, the hinges of the gates were of bronze and the gates were made from ash wood. Outside the gates there were elevated areas from which ran a wooden ramp towards the temple.

The Western Dolmen

The specifics of the Western Dolmen were that it had bird perches on either side of the excarnation table where the body was laid out. The table only had room for one body at a time. When tested with hormone witnesses it was found that the occupants were women and that cats were also placed on the table. One body plus cat indicates a rather important person so the possibility arises that the Western Dolmen is for the senior female Druid or for unmarried female Druids. Figures 13.1 and 13.2 shows the general layout of the Western Dolmen which is a rectangular area about 8m by 5.9m.

There was a tiled ceremonial footpath from the Dolmen to the temple with chariot tracks on either side of it. Inside the Dolmen there were three bird perches. The excarnation table was constructed from oak posts and struts with a hazel wickerwork top. The head was to the west. The red ochre was applied from the right hand side of the table when facing the gate. Under the wicker support for the body appeared to be two lamps for burning incense. As no steps were found where the red ochre was applied to the bones it looks as if the excarnation table was at about table height. Entry and exit from the Dolmen was by an ash gate on bronze hinges. Terracotta lamps were identified on the outside wall. The posts of the wall facing the temple and the gate were painted black.

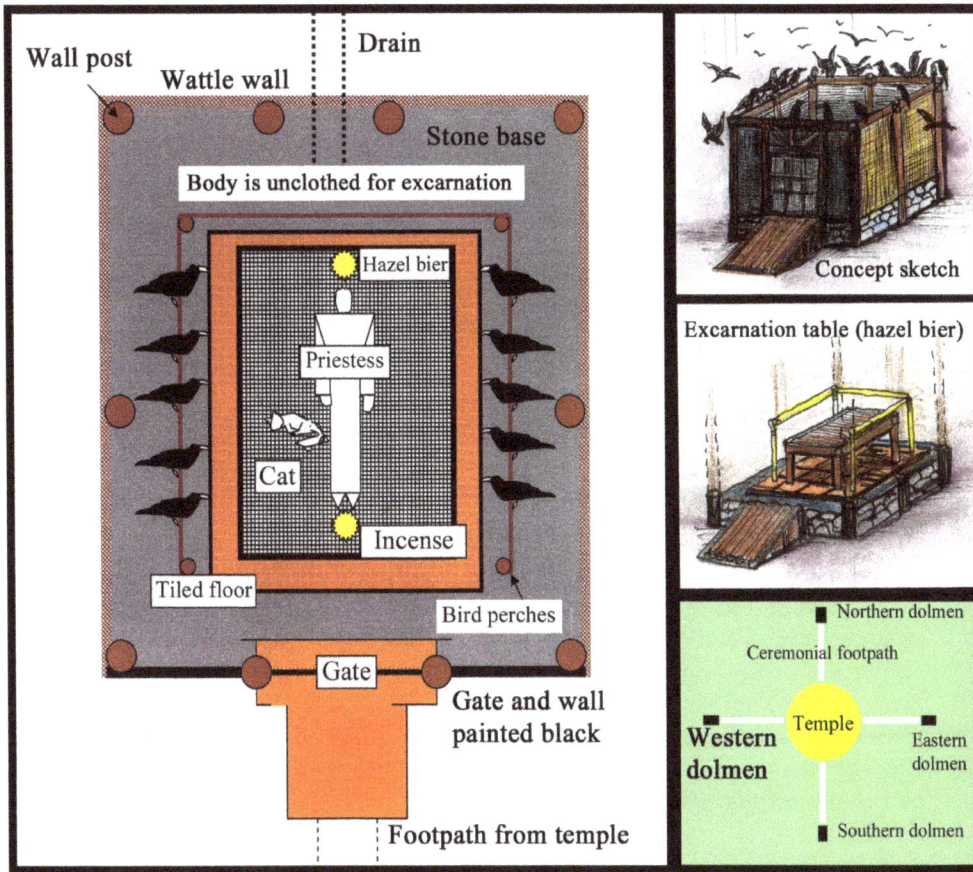

Figure 13.1 The Western Dolmen

This dolmen appears to be the most important one of the four. The body of the priestess and her cat were laid out on a hazel bier or on a flat stone. Under and round the bier there was a tiled floor, then bird perches and the enclosing wattle wall. The front wall was painted black and fitted with lamps. There was a drainage system at the back for clearing waste, rain and the water used to clean the floor. There were incense burners under the body and a patch of red ochre.

The Southern Dolmen

The Southern Dolmen (Figure 13.3) was much bigger than the one to the west. The excarnation table could take two rows of three bodies with the head to the outside. One row was for females (to the south), the north row was for the males. It looked as if the bodies had their own bier which was slid onto the table. There was a stone wall or plinth for the wicker fences to prevent animal entry. At the front and facing the temple there was a high step. Along the front wall were four terracotta lamps and bronze hinges for the ash gate. The door and posts were painted black. The floor of the excarnation area was tiled from the bird perches to the table and there was a drainage system to the rear running to a soak-away area about 16m from the Dolmen. The first 6 to 7 metres of the drain were covered with slabs of stone. The ceremonial path to the temple had postholes along either side. The Southern Dolmen with space for six bodies was much bigger than the western one and clearly intended for a different type of person. Running above and along the length of the table in the centre was a fourth bird perch.

161

Figure 13.2 The Dowsed Western Dolmen

The photograph shows the form of the various parts of the Western Dolmen marked out and identified using appropriate witnesses. The rectangle of red card is the central area of red ochre beneath the hazel bier on which the body is laid. This bench like structure is marked out in canes and red and white tape. Other markers show the positions of the tiled area, bird perches (blue card), postholes and the wall of the enclosure. Other chemicals and activities associated with its workings are easily identified and their form correlates exactly with these main features.

Figure 13.3 The Southern Dolmen

The dolmen is outlined in tape with an entrance area reached by steps. The canes to the right and left show where the edges are of the ceremonial road which leads to the temple.

162

The Eastern Dolmen

The Eastern Dolmen was similar in construction to the others (Figure 13.4). A low wall or elevated platform for the excarnation table with a wattle wall round the outside, three bird perches, tiled floor, drain at the rear, steps up to the gate area or forecourt. There were four terracotta lamps along the front wall which was painted. Egg white, gypsum, malachite, manganese dioxide and woad were identified. In front of the Dolmen it looked as if chariots may have come to pick up the bones and take them to the temple. The Eastern Dolmen's main point of interest was the excarnation table. It was long enough for two bodies. The male body lay nearest the temple, head to the west, the female body lay farthest from the temple, head to the east. Both bodies were tested using witnesses for toadstool and foxglove. The male body gave no response the female was positive on both. It therefore looks very much as if when a male Druid died his wife was sacrificed and then went through the death rituals with him before entry to the spirit world. Entry to the spirit world together was possibly considered the ultimate in sexual equality and the greatest complement that could be paid to a wife by her husband.

Figure 13.4a The Eastern Dolmen
The red and white tape shows the wattle outer wall. Within the central area are the bier (red) and the stand for the two bodies, one male, one female. There is drainage and floor tiles and in front facing the temple, a raised tiled area. A chariot comes up to the Dolmen (white card, wheel tracks) and delivers the body and takes the bones away to the temple. The front wall (blue card) and doors are painted in a number of colours.

The Northern Dolmen

The final Dolmen was the one to the north. It was in amongst crops and not so easy to study. It was, however, of the same general pattern with a walled or raised platform, bird perches, drain to the rear, platform in front of the gate, 4 lamps on the front wall and the front painted with Gypsum, Red Ochre, Malachite and Woad. The main difference was in its size. It was the largest of the Dolmen and could take eight bodies at a time, four female and four male. The bones were carried from the gate of the Dolmen a distance of about 9m to a point where a chariot or cart would pick them up to take them to the north gate of the temple.

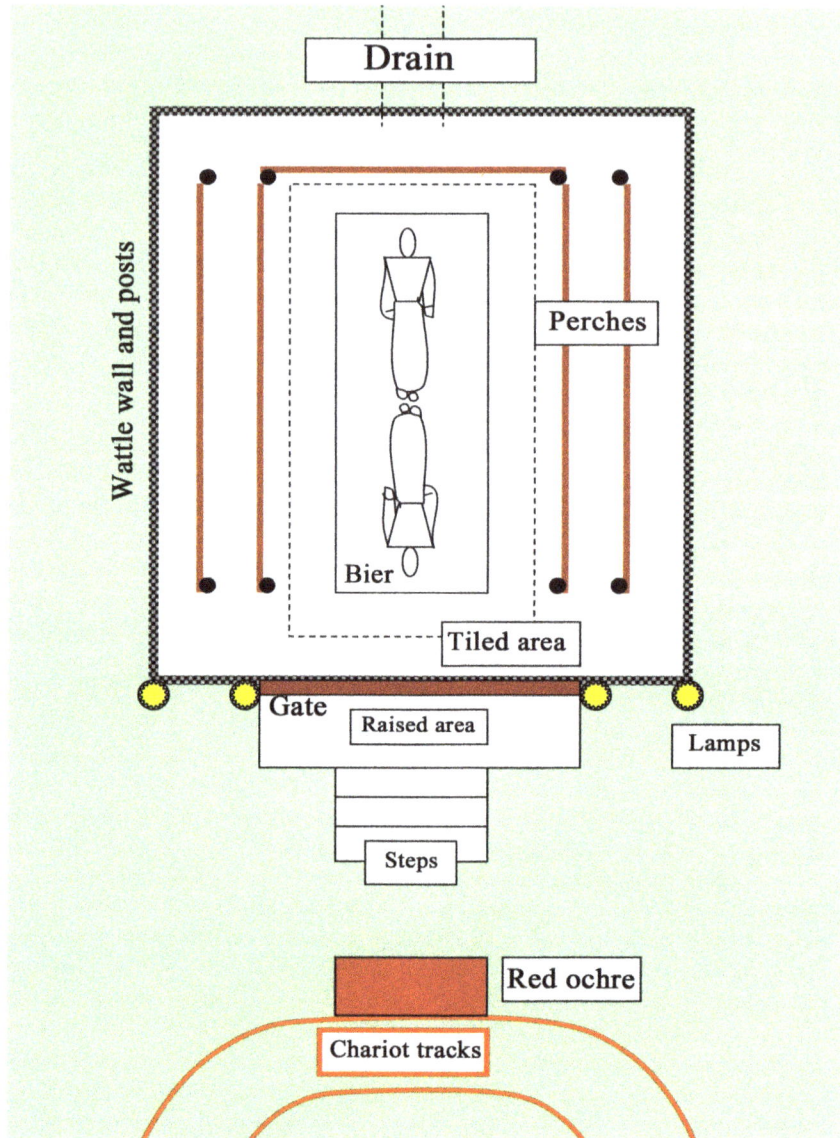

Figure 13.4b The Eastern Dolmen Main Features
The distinguishing feature of this dolmen is the double length excarnation table.

Review of the Dolmen

The Dolmen were only discovered on the 9th August 2005 and this was our first look at them. They are reasonably complex structures conforming to a basic design which protects the bodies from animals, are reasonably easy to keep clean and are designed to be bird friendly. The placing of a number of bodies out on the tables indicates that it may have been a batch process. The Druids dressing the bones would then probably wait until all the bodies on the tables were stripped. This would limit their ability to feed in new bodies. It is also possible that different species of bird specialized in different stages of the excarnation process. If this is the case it would help the birds if all bodies were at the same stage of excarnation. However people do not die in batches and they may have been placed in the Dolmen as they arrived.

The Western and Eastern Dolmen are for senior people. The Western Dolman, because it is limited to female bodies and there is no trace of foxglove or toadstool, looks as if it was

164

reserved for senior female Druids who died naturally. This indicates that at least some of the female Druids were celibate and they were associated with the setting of the sun, the world of the dead. The cat is the instrument of death and operates at night. At the moment the wall and gate of the Western Dolmen appears to be painted black as colours have not been found. Another indicator of the night. The Eastern Dolmen is for male and female and as at least some of the females were poisoned it is likely that when a Druid died his wife was also killed. The front of the Dolmen had white, blue and black but only as spots along the wall. The wall could have been painted to represent the morning sky. As the Druid couple may have had children the Eastern Dolmen probably represented birth and life. So far we have not been able to obtain enough information on the other two Dolmen but they may be for the secular society. The western and southern Dolmen have ceremonial priests with staffs. We have not found them at the other two Dolmen on this site so far. At the moment we have not found an excarnation facility for senior or important people from the secular society.

The four Dolmen extend the temple grounds over a radius of about 130m or more. The use of tiles and bricks in the construction of the Dolmen means that they would have become stained over the years with bird droppings, feathers, chitin from maggots and flies and body decomposition products. This means that even if bricks and tiles have been removed and used elsewhere they can still be identified as bricks and tiles belonging to the Dolmen. The bricks, tiles, stones and Druidic cement and concrete products do in fact turn up in the debris fields around Henge Temples and in post Roman and modern constructions. Identifying the stain confirms that certain bricks and tiles are Druidic in origin and not Roman.

Toilets

A large building such as a temple requires comfort facilities. One of the easiest facilities to find are the toilets (Figure 13.5).

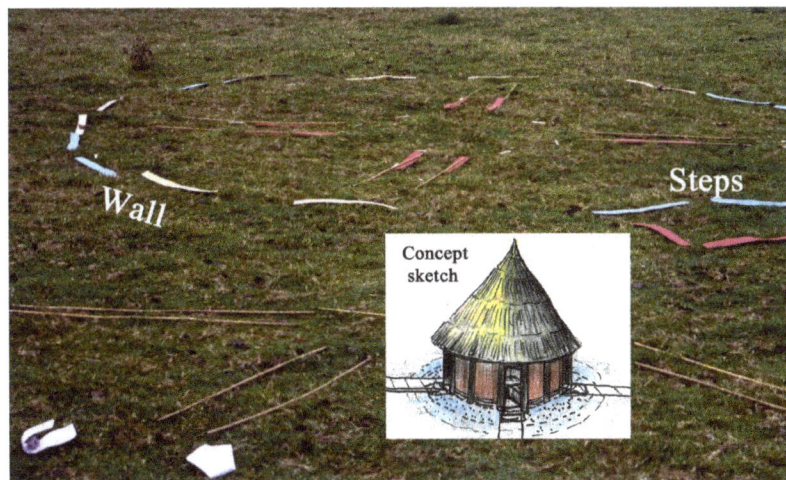

Figure 13.5 Toilets
Two toilets blocks were found one for men and one for women. They were raised with steps (indicated in blue) going up to each sector. Each quarter appeared to be a separate room or area.

Outside the south west side of the temple there were a male and a female toilet block, each with a drainage system to lead the waste away. The toilet blocks were circular and appeared to be raised. They were divided radially into four sections with access to each section being by tiled steps. The walls were wattle and daub and the floors were tiled. No lamps were found inside the toilet blocks. The doors were ash with hemp hinges. Using the contamination left by clothing the blocks could be divided into a section for Druids (linen) and the remaining three sections for the rest of the population (wool). There appeared to be a hazel wattle wall giving privacy to the Druidic section. Camomile, cedar wood and moss were used in the toilets and there was a band of camomile round the outside of the walls. It is not possible to say how the toilets were used but it does look as if there was a gully long enough for two people to use and the gulley then emptied into a central tank from which a drain ran to some distant treatment area.

The toilet blocks, like the Dolmen and Druidic houses, appear to have been well engineered and built by professionals. Bricks, tiles, stone, mortar and iron were used as required. On this visit washrooms were not found nor any kitchens or places set aside for eating.

An Iron Age Village

At a distance of about three to four hundred metres from the temple is a village with roundhouses and the first deer pen that we had found. We had been looking for signs of deer husbandry for some time as we knew that they must be reared on a large scale for their antlers. Antlers were used as tools for so many tasks that deer were a major resource and are likely to have been farmed and used as draft animals before oxen. Venison was an important component of the diet as it can be identified at spits and in stew pots. Terracing and chariot routes are visible through the village area and running across the fields. What looked like a chariot garage was found. We had first come across them in St. Albans. The garages were sunk into the ground so can be seen 2000 years after they were last used as a depression in the ground. The postholes of the walls and the roof timbers can be identified. The wheel tracks and lack of animal droppings indicate that they were reversed in or man-handled in.

The Temple Altars

After a quick look at the village, work continued on the temple. The altars were identified, see (Figure 13.6), and seven found. The altars were on the magnetic bearings of:

032°	Ox
080°	Sheep
125°	Horse
154°	Human
247°	Pig
323°	Deer
350°	Goat?

Five stone or masonry columns were then identified on the western side of the temple. There were signs that the columns had been painted on their radial sides. The columns were about 2ft by 2ft but some were smaller than this.

A search was made for large bone depositories as it was realised that the small ones could not take very many bones. Four were found, see (Figure 13.7), but there is a puzzle associated with them. The depositories where most of the bones must be stored do not have a very strong phosphate signature. The small depositories are easy to find the large ones are not. If large depositories are part of the underground complex there must be a tunnel network allowing access to them. Again this has not been easy to establish on the Yorkshire Temple site.

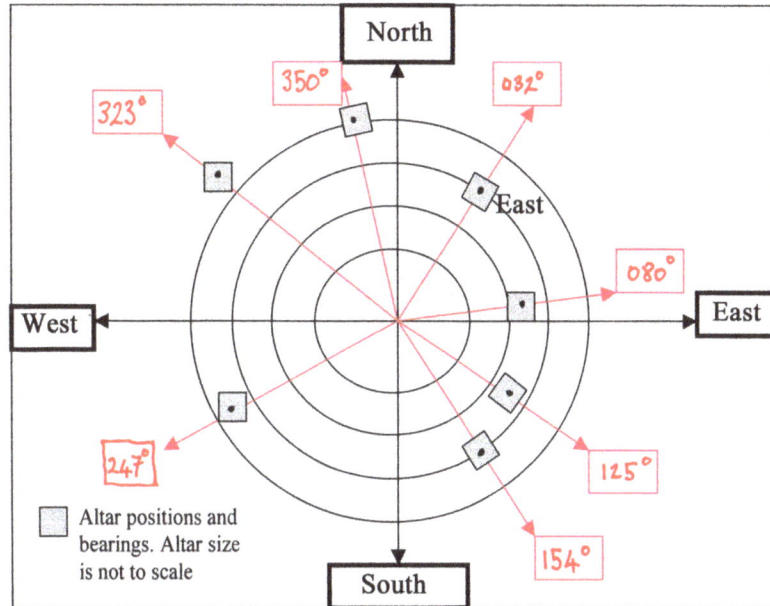

Figure 13.6 Altar Positions

The altars were not evenly spaced and their positions were
probably determined by practical aspects of temple use.

Figure 13.7 The Southern Bone Depository

The southern bone depository (cane rectangle) is large and
accessed from a triangular anteroom which has a small altar.
The floor or possibly the top of the chamber was a piece of slate.

Temple Pools

On the 15th August 2005 work started up again on the Stoney Lane site (Figure 13.8).

The temple pool is an enigma and we have made repeated attempts to find out why it was so important to the Druids. Many different witnesses had been used to try and find out what might have been put into it but little progress had been made. On the trip back from Yorkshire Nigel and I had discussed the pool at length. One idea that came up was that the Druids might have carried bodies into the water on hazel stretchers or biers. To test this idea we studied the access area to the pool. There were traces of hazel wood but not of its bark. Next the wheel tracks of a cart were identified coming in from the West Gate to the pool and back. The question was which track was coming in and which one was going out and was the cart carrying a body. As we pondered this we realised that if the body went into the water or was washed with water it would come out dripping water. Over the years the chemicals in the water would leave a trail. These chemicals included calcium and magnesium carbonate. Using them as witnesses the route leaving the pool was identified and when the other route was followed from the temple it lead to a 'Chapel of Rest'.

If dead bodies were being washed or immersed in the water it was possible that the pool could be drained. However, no drainage point could be found. There was something in the middle of the pool from which a number of rays were going to the outside wall of the temple. The narrow straight lines looked as if they were part of a ground plan for the temple and I made a mental note to return to them. On the following day we returned to the pool. This time to look more closely at the wall of the pool. Normally the pool was marked quickly but even so we always picked up the entrance and exit steps either because they were so large or, as in the present case, there was an approach to the wall and steps and then an inner tiled area for standing on. Working our way round the wall of the pool we came across an anomaly on the east, north east sector. It was an area that bulged out from the pool wall and was in fact a small pool. It was what we were looking for as we had reasoned that the Druids probably had a baptism ceremony for children. The small pool had what appeared to be gold and silver candle sticks on the wall. On the right facing the pool were traces of linen and feather from the priests clothing. On the left the wool from the garments of the parents. As a long shot we looked for the tracks of wooden wheels. We found them. The parents brought their children to be baptised in a pram. We followed the tracks back and out of the temple to see if they had a party after the event but the tracks disappeared into thick brambles. To be sure that we had discovered a font we checked with babies faeces and regurgitated milk. Positive on both. Moss was positive also blood and skin. There is therefore some evidence that the Druids may have practiced circumcision. Having found another use for the pool we moved to the altars. The reasoning was that after the sacrifice the carcass would be placed on a hand cart and wheeled out. The cart tracks were there and they led outside the temple to pairs of strong posts. Here the carcass was hung and butchered. The temple provided the abattoir for butchering animals. This raised the question of whether the Druids were following a Shamanistic ritual of honouring the spirit of the animals they used as food. Having found what happened to the animals it was necessary to find out what happened to the human sacrifice. All temples appear to have seven altars one of which was for humans. The presence of poisons from foxglove, deadly nightshade and toadstools indicated that some of the human sacrifices were possibly ceremonial in which case the body would go off to one of the Dolmen. The tracks of a cart were identified by the human altar and they were followed out of the temple. Instead of going to a Dolmen the cart went to an abattoir station, presumably to be hung, drawn and quartered. This looked as if criminals were also dealt with by the Druids and that they did not go to the Dolmen.

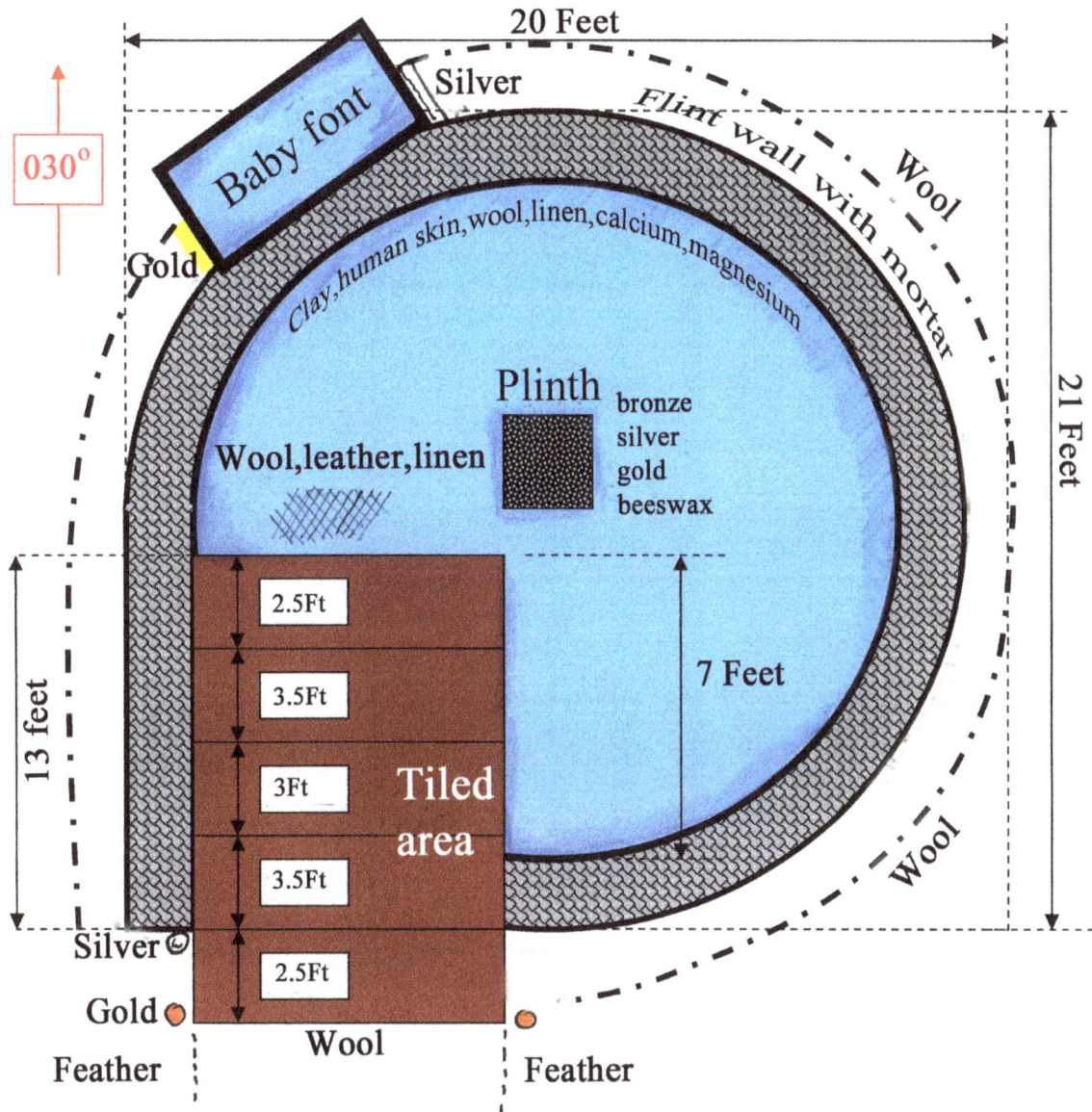

Figure 13.8 The Stoney Lane Pool

The drawing is of one of the Stoney Lane pools and is probably Iron Age as it is quite elaborate. The entrance area and font are shown. There was a plinth with a highly decorated structure on top of it. This structure may have been a fountain. The priests stood on the right facing the pool.

There were Druidic dog kennels on site. The dogs were kept in a compound with a feeding trough on one side. In the centre was a substantial and well built kennel. Dogs are in the habit of taking bones into their kennels and to the back wall to ensure nobody takes them. Along the back wall we found the stain of human blood. It therefore appears that some of the human sacrifices were fed to the dogs indicating that they were probably criminals or had stepped outside the system. Based on these observations it is possible that the human altar was used for two types of person. The criminal and the 'volunteers' who were going to take a message to the world of the Gods and Spirits. The reason for proposing that people were sacrificed as messengers is that there are 32 guardians plus the Prince and Princess associated with each temple.

Returning to the pool, there was always a rectangular area at the centre. When dowsing the area positive responses were obtained with gold, silver and red ochre. On the evening of the 23rd August 2006 I walked down to the Stoney Lane site to see if I could come up with any ideas on the pool. I used flowers to try and identify wedding ceremony activity without success. We had spent a lot of time trying to find how the Druids emptied the pool but not much time on how they filled it. I had tried to find pipes coming up to the pool without success. On the 26th August I was again working on the site when I had another attempt to find out how the water entered and left the pool. I could find no signs of terracotta or clay pipes entering or leaving the pool. This absence of evidence for a plumbing system indicated that water may have been delivered by buckets or carts or that I was not crossing the pipes at the right angle. The 'Celts' were said to have invented the barrel so water carts drawn by oxen could have been used. However, there was no sign of oxen or carts round the pool. There was still the possibility that the carts drew up at the gates and emptied the water into a temporary chute to the pool. The bucket theory was not very convincing as the late Bronze Age pool being studied was a big one and would have held several thousand gallons of water. So the mystery of how the Druids filled and emptied the pool remained. Also where did the water come from? The Chilterns are a range of chalk hills with very few streams and none for a mile or more round the site. The same problem, lack of water, applied to all the henges round the village of Bovingdon set as it was in the Chiltern Hills. Whilst pondering the problem it dawned on me that the Druids may have used lead piping so with a lead witness I started another search. With lead it is possible to do a sweep search for the target and this time I found a lead pipe coming into the pool and going to the centre where the rectangular feature had been. It looked as if Bronze Age Druids were into lead plumbing and possibly fountains. Within the temple the lead pipe had side branches to other pools. This indicated that the water supply via the lead plumbing was in use for a very long time. When one temple came to the end of its life another was built and its centre moved away from the old one by about 10m., this meant moving the lead plumbing to feed a new pool. Going back along the lead pipe it arrived at a well about 20 to 30 yards from the temple. The well was about 2m in diameter and had a large aluminium halo (blue halo using the colour wheel) which indicates that the well was in use for a long time. Having found the well it was then easy to pick up signs of a ceramic pipe. The pipe went from the well, running along a contour line to just outside the West Gate then turned left through a right angle to go into the central area of the temples. The ceramic plumbing could be traced to Stone Age pools.

On the 27th August 2006 Nigel and I did a detailed study of the well. We knew where the lead and ceramic pipes were. We now had to work out the structure of the well and well house, the way water was hauled up from the bottom of the well and what happened to it then. It was not going to be easy because many well houses would have been built over a period of 2000 years. The Bronze Age building was a substantial piece of engineering. There was a header tank about 6 to 7 foot in diameter and a lead pipe under it going off to the temple. The header tank was made from oak and bound with iron. The square postholes of the structure upon which it stood were easily identifiable. When dowsing a structure that was clearly above ground level by a number of feet the dowser looks for the drip lines and tries to confirm them by using witnesses. The signature of the header tank was a circle coming up on oak and iron. The Stone Age header tank was much smaller. It was a clay header tank with a wooden sluice gate leading to a clay gulley. It was supported on a wood frame (Figure 13.9).

The well had been in a building which, judging from the size of the postholes, had been robustly built. Using drip lines, dust lines and beef fat, which was used as lubricant, the design of the building could be worked out. It was a typical well in that there was a spindle across the well which is normally turned by hand to lower and raise a bucket except there was no sign of where people stood to do the turning. There were signs of postholes going away from the

170

well on the opposite side of the well to the header tanks. They were supporting a wattle wall. A walled path to the Well! It seemed rather odd but after checking who or what might have used it, it became clear that it was for oxen (Figure 13.10).

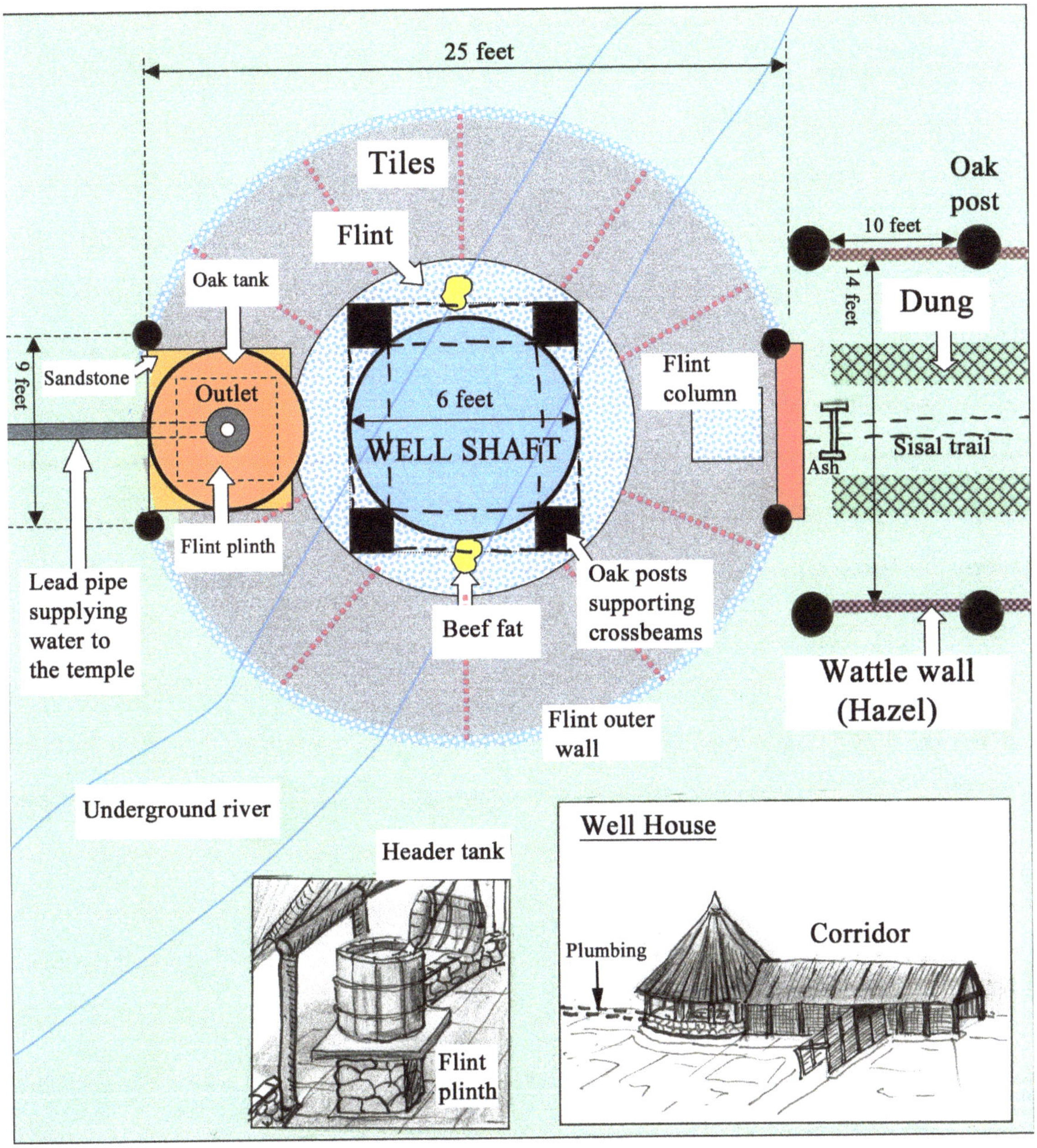

Figure 13.9 The Well and Well House

The well was a substantial piece of engineering. Its complex structure was designed to allow one or two oxen to raise water from the underground stream in a large vessel and tip it into a header tank. The illustration is a summary of the dowsing. Measurements are in feet and the sizes of all features are drawn in proportion to each other.

Well engineering form

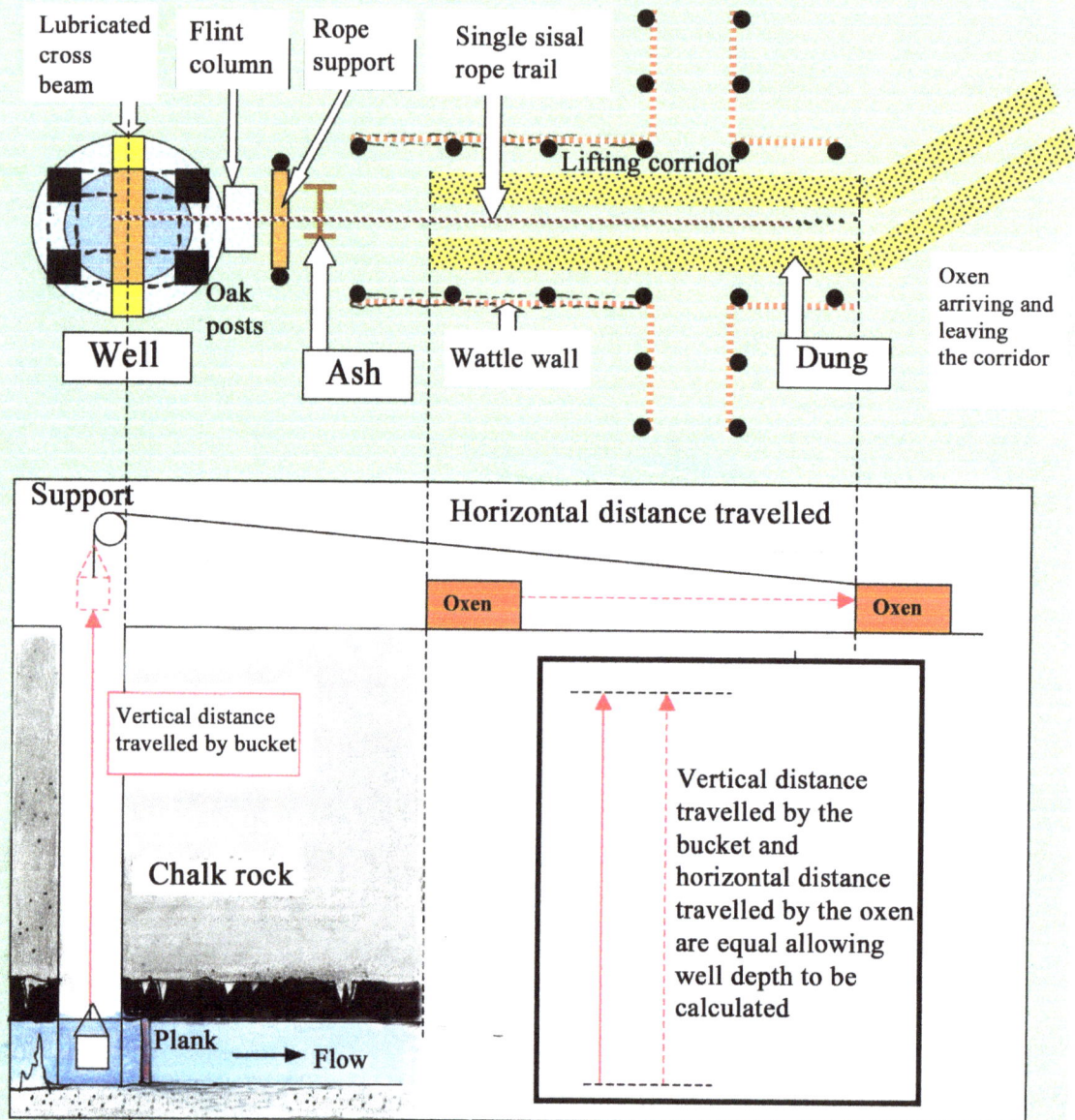

Figure 13.10 The Depth of the Well

Cattle were used to raise water from the underground river. The distance travelled by the oxen indicates the total travel of the bucket and hence the vertical distance from the river to the header tank (assuming no gearing was used).

We had stumbled on how the Druids drew water from the well. They harnessed an ox to a pulley system. The bucket was raised above the well and tipped onto a chute directing the water into the header tank. The track of the oxen went into the next field and it was possible to measure the length of rope used to pull the bucket up the well. This gave the combined depth of the well and the height to which the bucket was raised. The well was dug down onto an underground stream indicating that the Druids had the necessary dowsing skills to find underground water. The ceramic pipes sections taking water to the temple were about 2ft

172

long. The pipes were sealed where they joined each other with hemp and tree resin. The lead pipe took a direct route whilst the ceramic pipes went along the contour to a holding tank and then down hill to the temples. A hole was dug down to see if the pipes were still there but only the chemical stain remained. No attempt was made to find the lead pipe as the 'Celts' did not know how to remove the silver from the lead or so we thought at the time and probably all 'Celtic' lead was taken as booty by the Romans.

Finding the plumbing systems and the well was of considerable significance. The system provides a window on the technology of the period from the Stone Age to the late Bronze Age and Iron Age. As far back as the late Stone Age people were moving water around in ducts and pipes. They were digging wells and they were able to use oxen to raise the water from the well. The problem of how the pools in the temple were filled had been solved, but we still do not know how they were emptied. Every temple pool we looked at had a plumbing system and a well as a source of water. The well was a major feature of the temple site. The same well was used over the life of the site which may have been from the Stone Age to the Iron Age. Some of these wells still survive to the present day (Holy Well in St. Albans) and they could be 4000 years old or more.

The wells and plumbing system provide the archaeological dowser with a method of identifying how many pools there are and how many temples there might be on a site and divide them into two groups, those belonging to the 'ceramic age' and those belonging to the 'lead age'. The presence of a plumbing system is also additional evidence for a temple having been on a site at sometime in the past.

Chapter 14

The Stoney Lane Time Capsules

Many Temples on One Site

During the summer of 2005 the temple complex down Stoney Lane, a fifteen minute walk from my house, was studied in some detail (Chapter 11 & 13). This site in the Hertfordshire village of Bovingdon was beginning to present problems. Things such as the Bronze Age sighting points that we thought we had found would not line up with openings in the wattle and daub walls or with an observational point in the Henge Temple. At first this was thought to be due to identifying magnetic images instead of the true target as the sighting points. Bronze casts a multitude of images and the greatest care has to be exercised in identifying the target from amongst its images. The gates on the four quarters did not line up on the pool or the tomb. The evidence started to indicate that there may be no sighting points and that there may be more than one Henge Temple. This would mean a number of tombs for a whole series of Princes and Princesses that had given their lives over the ages to act as emissaries to the Earth Goddess or the spirits of the land. At first the depression in the ground had been assumed to be associated with just one temple then three more had been found. The four tombs were found fairly close to each other. This indicated that it was going to be very difficult to unravel all the postholes, pools, stone columns, altars, walls and processional routes etc. Looking on the bright side there were four tombs, later to be joined by more, spanning a period of time (Figure 14.1).

From work on the St. Albans site we already had evidence that woodhenges, temples and their tombs may go back to the Stone Age. If this were the case it may be possible to trace a cultural time line back for 2000 years before the Romans arrived. We already had Iron Age Henge Temples that may have been in use close to Roman times and Bronze Age Temples but we had not studied the tombs in them in any detail. With four tombs close together on the same site it appeared that it might be possible to obtain cultural information covering a significant period of time. The Stone Age tomb at the farm (SAF) (Chapter 11) has been included in the comparison. Eventually we had six tombs which were called Stone Age 1, 2 and 3 (SA1, SA2 and SA3) and Bronze Age 1, 2 and 3 (BA1, BA2 and BA3) plus the one at the farm (SAF). At this time it was not known which was the oldest. The tombs had one common feature in that they were constructed using flint and mortar and had lime stone covers. A review of the tomb designated SA1 should help to set the scene.

Stone Age Tomb 1 (SA1)

This tomb has been allocated to the Stone Age because no metal was detected in it. According to the dowsing it is deep but remember that could be the present day position of the stain from a surface structure now long gone. The Prince and Princess appear to be dressed in wool down to their feet. There is no linen present but a red sable witness gives a positive response on the head of the man and woman. The man has dog (wolf) hair and skin over the whole body. The woman does not. Both of them appear to be covered with or lying on feathers. There is leather at the feet and belt area of both which gives a positive response with urine but not tannin. Red ochre appears to be spread the full length of the bodies. Amber was detected in the belt area of the man and on the head and belt area of the woman. Amber is a tree resin so this does not mean that actual amber is present. Amethyst and citrine were not detected but white quartz was. Malachite was found on the body of the man but not that of the woman. The hair of the woman was down to the waist, that of the man to the shoulders. There was a flint headed spear bound with nettle fibre, bluebell glue and resin, a yew bow almost as long as the tomb, flint headed arrows and a leather quiver with the arrows flint heads up.

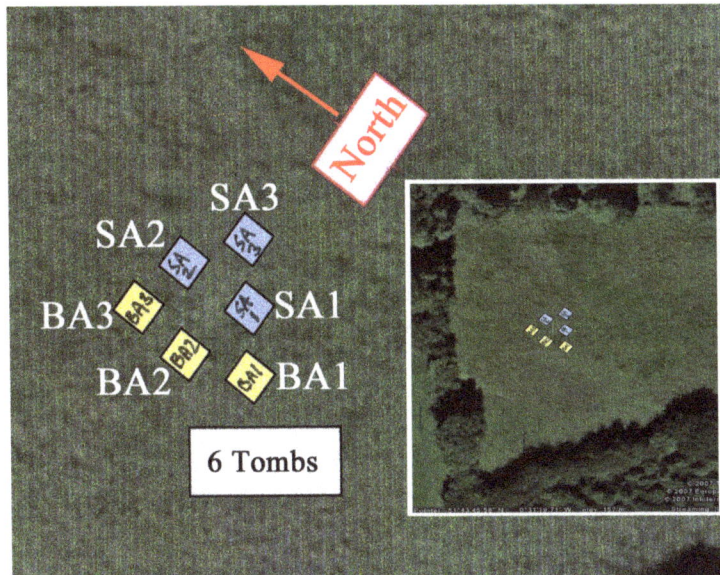

Figure 14.1 Stoney Lane Tombs

The Stoney Lane site was home to many tombs. The first six
to be identified were within a few meters of each other. The
proximity of the tombs indicates that the tomb temple sites
were back filled when a new one was dug. The closeness of the
tombs explains why the antechamber and bone depositories
of the underground temple are not always symmetrical. The
temples could be at different depths.

The contents of the tomb look very Stone Age but religions tend to be conservative and just
because there is no metal in a tomb does not mean that the society did not have metal and use
it. The absence of linen is probably indicative of the age as the Prince and Princes would have
been sent to the Earth Goddess dressed in the best clothing available. The clay pool that may
have been a contemporary of the tomb was small and simple with wooden steps into and out
of it. Because the Stoney Lane site contained a series of tombs originating in the Stone Age and
Bronze Age it provided an opportunity to compare tombs. The layout of the tombs is shown in
(Figure 14.1) and the detail of SA1 in (Figure 14.2).

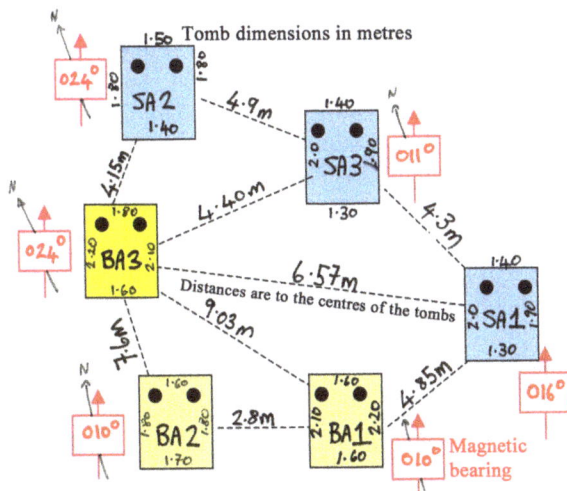

Figure 14.2 A Stone Age Tombs

Tomb SA1 is on a bearing of 16°. The
bearings of the other tombs range from 10° to
24° indicating that the builders were not using
true north. Magnetic north and hence the
dowsers north does vary with time and it is
possible that the dowsers north was more
important than true north. The fact that the
dowsers north moved about would support the
idea of a spiritual being moving around. The
dowser can determine north when underground.

A Review of Tombs

Box 14.1 contains details of the tombs. In table 1 the main features of the tombs are given that is size, corrected bearing and whether they contained metal or not. The tombs have been labelled Bronze Age if they contain bronze and Stone Age if there is no bronze present. One of the Stone Age tombs (SA2) contains gold and silver on the belts of the two occupants. The belt was not tested for lead which would have contained silver.

In table 2 the use of linen identifies two of the Bronze Age tombs, BA1 and BA3, with BA2 having wool and no linen. The bodies in BA2 were those of children and both bodies were covered with feathers, which is more in common with the SA2 and SA3. It therefore looks as if BA2 could be a very early Bronze Age tomb.

To determine if the Prince and Princess were poisoned a foxglove and toadstool witness were used. The occupants of BA1, BA2, SA1, and SA3 had been fed toadstool those of BA3 and SA2 had not. The occupants of both these tombs could have been children. Foxglove had been given to the occupants of all tombs. Foxglove is not a nice poison as it induces vomiting. Other poisons may have been given with it to reduce side effects. The poison appears to have been given with honey as pieces of chalice have been found on a temple debris field contaminated with toadstool, foxglove, deadly nightshade and honey.

The following is another description of a Stone Age tomb which was designated SA3 as no metal was detected in it.

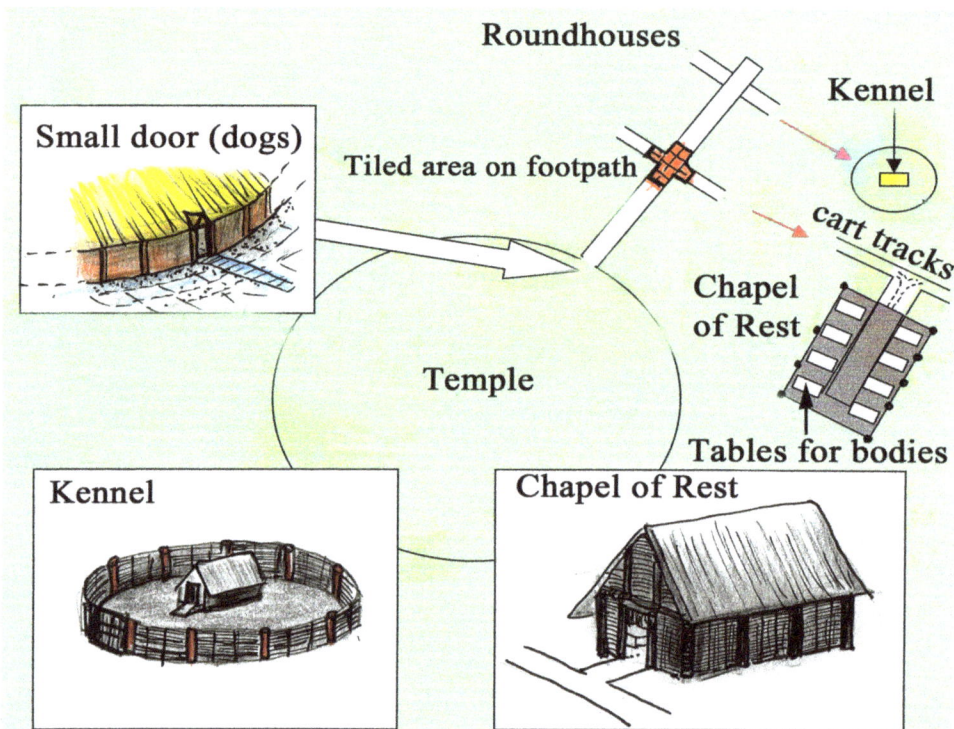

Figure 14.3 Temple Doors
The ceremonial path from a temple door follows a straight path for over a hundred meters. At intervals there are tiled crossroad. The approximate site of the Chapel of rest and the dog kennels are shown. Some houses were found on the left of the path.

Stone Age Tomb (SA3)

The tomb dowsed as if it was constructed in the following way and is a typical Stone Age tomb. It is a walled structure, outside dimensions 1.70m x 2.74m internal dimensions 1.30m x 2.30m, with two people in it. The tomb lies along the magnetic bearing 16° or when corrected 11° east of true north. The walls of the tomb are covered with a gypsum plaster with egg white detectable on the side walls indicating that they may be painted. The tomb is covered with three slabs of limestone, the gap between slabs being sealed with chalk mortar.

Slabs of stone are found on the tops of all tombs. The slabs are large as only three appear to be used. In the present case that means the slabs are about 930mm by 1700mm. To be sufficiently strong they would have to be 100mm thick or more. Such a slab of stone could weigh in at over 300kg, say 700 to 800 pounds. A piece of stone for the tomb may require between six and ten men to move it on the surface. Underground it would present many problems particularly moving it round corners. This is a piece of evidence suggesting that the tombs were surface or sub surface features which, in this case three to four thousand years ago, left a chemical stain that has moved deeper and deeper into the soil, clay and chalk. Returning to the contents of the tomb, there is no metal in the tomb, no linen and the shield is made of oak. However, the feathers are selectively used on the belt and headdress along with red sable. The use of a dog/wolf pelt is probably a status symbol in an age when not much was available to indicate importance or position in society. The burial of the whole body with a shield, spear and bow at a time when the normal procedure was to allow birds to strip the body and then place the bones in an underground vault indicates that the 'Prince and Princes' were intended to go somewhere to represent the group from which they came, envoys to the Goddess of the Earth. The use of poison, digitalis from the foxglove, could ensure that they were not physically injured and the hallucinogenic toadstools and fermented drink, probably beer, helped them to pass over into the spirit world. The two bodies at about 1.60m (5'2") are small and so could possibly be young teenagers but they were not children. The 'Prince and Princes' tend to be young, sometimes children but also older people, for example the Prince in SA1 is bearded. The hair style of the Princess is constant with hair down to the waist. That of the Prince shows some variation from shoulder length to short. The Prince in SA3 had short hair. The clothing was made from wool and leather, the tannin process being based on tannin. This indicates that SA1 is older than SA3. Tanning replaced urine as the method of preparing leather and it was also used on pigskin to produce leather which is sometimes found on the feet of the Princess.

No linen was found in the tomb. A red sable fur witness was used to try and identify the use of sable and a positive response was always found. However it does not mean that red sable was used, only that the fur of a relative may have been used. The same applies to the dog hair witness. The hair and skins used could be that of a wolf. The use of bird feathers varies. In SA3 the feathers appear to be limited to the belt and head area which is where the red sable is also being used. In two Stone Age tombs SA1 and SAF and one Bronze Age BA2, the feathers extend over the full length of the body. In BA2 the fact that the feathers extend the full length of the body could be taken as evidence that BA2 is very early Bronze Age. Indicators that this may be the case are: the bronze is copper/arsenic, no linen was found, the tomb did not have a tiled floor. However, the tomb needs to be studied in more detail if it is to be linked to the early Bronze Age. A better comparison can be made between BA3 and SA1. BA3 is defiantly late Bronze Age as the bronze is copper/tin, linen garments are present with no wool or dog hair being detected. Feathers are restricted to the head and waist. Linseed oil was used, also hemp and importantly it looks as if red wine and olive oil may have been used. These later two items would only have been available from well established trade with Europe. If this tomb is compared with SA1 clear differences emerge. In SA1 there is no olive oil or red wine, no metals, no tiles, dog hair or skins are used, the leather is tanned with urine and the feathers extend over the body. The fibre used for binding in SA1 is nettle in BA3 it is hemp. The arrows are held in the quiver flint head down in BA3 and flint head up in SA1. The technology of the age is clearly demonstrated when tombs that may be a thousand years or more apart are compared.

Temple Walls

The complexity of the Stoney Lane site was such that it was not possible to relate features to each other. Temple walls were there but they could not be related to a pool or tomb. Stone columns and altars could not be related to other features. As a result selected aspects were studied. One feature that had been discovered in the walls of temples were narrow doors. These were present in some of the walls at the Stoney Lane site. One that faced the kennels was looked at. The track running from the door gave positive responses when using witnesses for leather, tannin, linen and feathers. There was no response on wool. Positive responses were also obtained on human skin and dog hair. The path had two carbon tracks with staff marks on the outside of them. The dog hair was also on the outside of the carbon footpath. As the path was followed towards the kennels and buildings, every so often it would disappear and be replaced by tiles in the form of a tile crossroads. The path was about 100m long (Figure 14.3). From this it looks as if some of the small side doors do have a ceremonial purpose. In this case for taking dogs into and out of the temple. The tracks from foot traffic indicate that priests were both barefoot and leather shod and appeared to be in ceremonial procession with dogs. The absence of wool fibres may indicate that the path was only used for ceremony when everybody wore linen whether it was summer or winter or alternatively, it may be that linen was worn all the year round for all duties, summer and winter with no wool garments. However, the presence of wool in the Druid's houses indicates that wool was worn by the Druids family and by the Druid some of the time at least. If this is the case then the path was only used by people when they were dressed in linen and so it was a ceremonial path.

On the St. Albans site chariot tracks had been followed round the outside of temples. Some tracks were clearly ceremonial with the chariots acting as an escorting guard. Other carts were smaller and carried the bones. On the eastern side of the Stoney Lane site a gate was identified to enable chariot tracks to be studied. The gate had oak posts, an ash door and hemp rope hinges. The swing of the doors inwards can be followed and their size measured. When closed the gates overlap and do not meet up in the manner of modern gates. From tracks identified by the East Gate there was a cart, drawn by a horse, which had oak wheels and no metal tyres. The route taken by the cart was into the temple to a point where the bones were unloaded as indicated by a patch of red ochre. Dowsing where the horses head is likely to have been revealed an area of hemp and ash wood which indicated that they were used for the bridle and bit of the horse. Knowing where the head of the horse was also indicates the direction the cart was going. Along the track there are red ochre stains, which are on the right hand side of the cart as it went in and on the left hand side when the cart was leaving the temple. We have not yet been able to work out what this means (Figure 14.4).

At the North Gate a more complex situation was found. There appeared to be the tracks of a horse drawn cart mixed with those of a hand cart. The picture of the horse drawn cart was similar to that found at the East Gate. That is a cart turning into the gate, going down to an unloading point indicated by a rectangular patch of red ochre. The cart had rotating wheels with an axle length of about 1m. On the right hand side when the cart was going into the temple there is a red ochre stain which, unlike the East Gate, remains on the right side when the cart exits the temple. The hand cart enters the temple along the same track as the horse drawn vehicle but quickly departs from it and then heads further into the temple and its own rectangular red ochre stain.

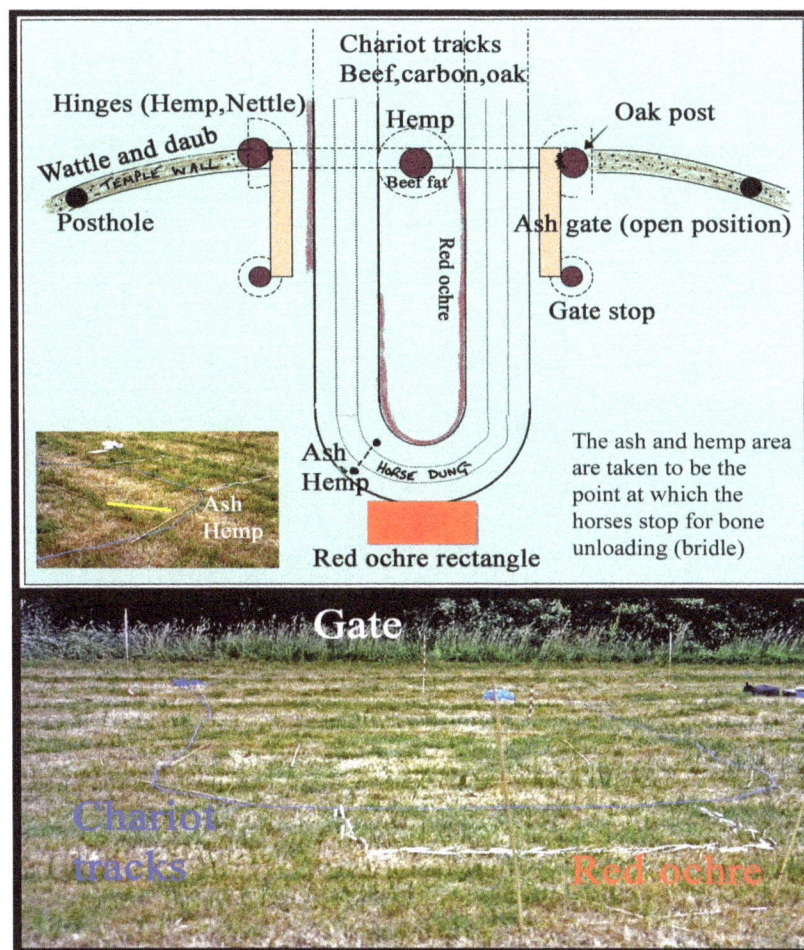

Figure 14.4 The East Gate and its Chariot Route
The design of the temple gates follow that of the Eastern Gate. The two ash gates open inwards onto a post and close onto a central post. The hinges are made from fibres such as hemp and nettle. The red ochre stain is rectangular which indicates that bones were carried in a crate which may have been unloaded onto a stand.

The hand cart then exits joining the horse drawn vehicle track near the gate. Both vehicles had oak wood and leather wheels. It is not possible to say if the two sets of tracks belong to the same temple. The use of a hand cart is more likely to be associated with an earlier age than a horse drawn vehicle. Because both carts are ceremonial it is not possible to say if they belong to the Stone, Bronze or Iron Ages. However, the gate hinges are rope and made from hemp so they could be late Stone Age or early Bronze Age (Figure 14.5).

The Chapel of Rest (Figure 14.3)

Having picked up the trolley tracks going towards and away from the pool (Chapter 13) they were traced back about 100m to a building. This building had a door about 1.6m wide on bronze hinges and was about 5.5m wide and 3.7m long. There was a corridor down the centre with a wood framework for four bodies on either side. It looked very much as if bodies were wheeled in on their bier and moved onto the 'shelving'. Camomile and Cedar Wood oil witnesses gave positive responses in the building. The wheels of the trolley appeared to be on a rotating axle.

Dog Kennels (Figure 14.3)

The dog's kennels were a bit further on from the chapel of rest and were in two rows of four. The kennel compound had a diameter of about 4.5m with a rectangular kennel about 1.5m by 2m.

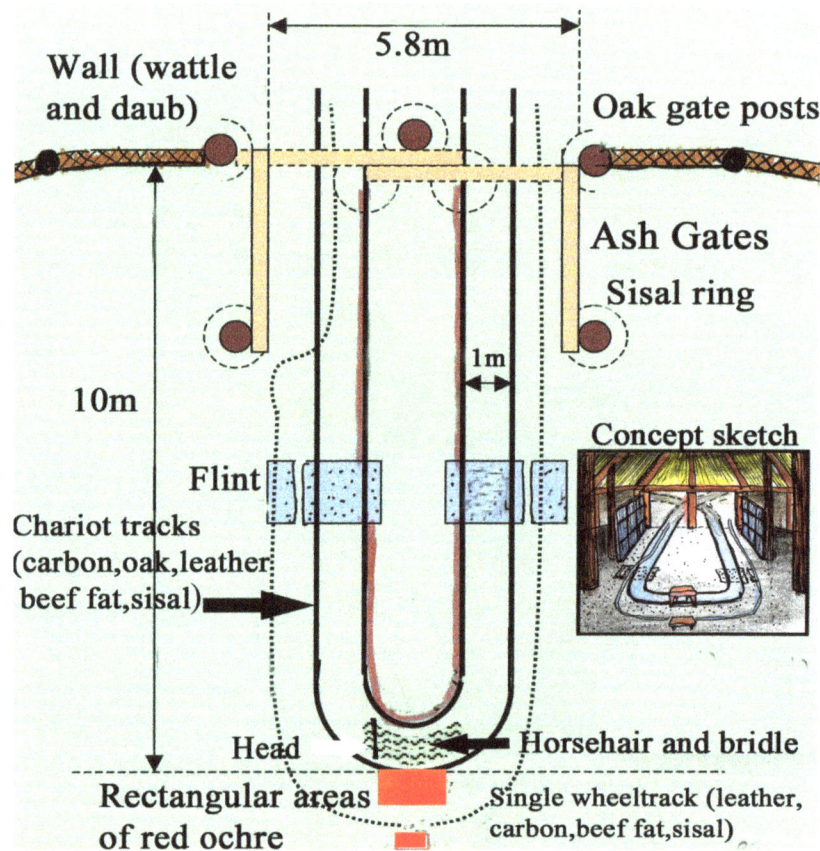

Figure 14.5 The North Gate and its Chariot Route

The illustration shows a concept sketch of the North Gate of a Stone Age Temple at the Stoney Lane site. The gate was about 5.8m wide. Two vehicles are involved in bringing bones to the temple, a small possibly wheel barrow sized one and a small horse drawn chariot. Both vehicles enter the temple through the gate and go to an unloading point about 10m inside the temple and identified by rectangular patches of red ochre. The position of the horse at the unloading point is indicated by an area of horse hair from the body and an area of hemp and ash wood from the bridle indicates the position of the head.

A Temple Floor Plan

We had checked the rectangular object at the centre of the main pool and were convinced that at the centre of each pool there was a plinth or some feature standing out of the water. It had been painted with pigments such as red ochre and had gold objects or gold leaf on it. (Figure 13.8) There was beeswax which could be for protection or used for candles. Radiating from the plinth area were lines. When we first found them we put them on the back boiler with the intention of returning to them when we had finished with the site, little realising what was going to emerge. On the 25th September 2005 after some site clearing we eventually got down to looking at the rays expecting them to be part of a construction plan for the temple. I picked up a ray at the centre and followed it out. It finished at a temple wall. I then checked to the right, there was nothing but on the left the line continued. The thought that we might be dealing with an image of the oaks rays and the triangles of the 'Celtic' Cross came to mind. Did the Druids draw out the oaks rays and triangles and use them as a plan. I returned to the centre without following the line and said to Nigel that I thought we had the rays of the oaks aura and that we had better mark it out. The mapping started, I dowsed, Nigel placed the canes. The line out to the wall was marked then I followed the line to the left but it only went about 1.7m before turning left towards the centre and then left again. There was a rectangle at the end of the ray not the top of a triangle as I had expected. A line then returned to the centre. Looking back from the centre we got the impression that we were looking at a Thunderbird. If it were it would have been painted on the ground possibly using manganese dioxide or charcoal and red ochre. We checked and found that the lines we had followed were painted with manganese dioxide and that the rectangle had a red ochre eye with a manganese dioxide eyebrow. Below the head were three drops of blood in red ochre. Walking back down the neck there was a line coming out of the neck. By following this, the wings of the Thunderbird took shape. We started to map out the Thunderbirds and found that there were fifteen. It was clear they belonged to different sets and our bet was that there should be four in a set. To prove this we needed a lone temple so the hunt for one was on. The Thunderbird was the first very clear and undisputable artwork that we had found. The stone circle at Stanton Drew had shown us that the Druids were into drawing pictures on the ground. But where did the Druids get the Thunderbird from. It is known in North America and possibly other parts of the world. Nigel later rechecked the aura of the oak. We had for some reason assumed that the aura finished with the triangles but this was not the case. Above the triangle was a rectangle joined to the triangle by a 'neck'. I confirmed Nigel's findings. We now had a direct link between the temple and the aura of the oak. A link that could not be by chance. The next step was to find the source of the different part of the oak's aura (Box 10.1). A task that would take a few weeks. The skill and knowledge of the Druidic dowsers had been demonstrated once again, the relationship of the living world, in this case the forest trees, to their religion had also been confirmed. We also had another time line marker to trace Druidism from the Stone Age to when the Romans arrived. A period possibly exceeding 2000 years.

Summary

The Stoney Lane temple site was yielding valuable information about the design of the Henge Temple and the ceremonies conducted in and around it. The site covered a 2000 year time period and possibly much more. The first clear evidence of the Druids painting figures on the ground derived from the magnetic field of the oak was found. Dating tombs was not easy within the Stone Age or Bronze Age. This might be due to the wealth available at the time the tomb was made and the rank of the individuals or that technical progress was slow. The transition between the Stone Age, the early and late Bronze Age and the Iron Age is easily identified.

Box 14.1 Dating the Stoney Lane Tombs

Tombs without metal
SA1
SA3
SAF

Tombs with metal					
Silver	Gold	Arsenic/Copper Bronze	Tin/Copper Bronze	*Fabric*	
SA2	SA2	X	X	Wool	
X	X	**BA1**	X	Linen	
X	BA2	BA2	X	Wool	
BA3	BA3	X	**BA3**	Linen	

The presence of wool and linen are indicated. Wool was available before linen. Like metals, fabric may have chronological significance. The silver may be silver in lead artefacts and not the pure metal. Lead was in use before bronze. Gold may well have been available in the Stone Age.

Stone Age	----	----	No Bronze or Iron
Early Bronze Age	----	----	Arsenic/Copper Bronze
Late Bronze Age	----	----	Tin/Copper Bronze
Iron Age	----	----	Iron

Wool
Linen

Possible chronology based on tomb analysis

Table 14.1 Chronology indicated by metals and fabrics found in the tombs

Tombs

MATERIAL (NON METAL)	SA1	SA2	SA3	SAF	BA1	BA2	BA3
Wool	√	√	√	√	X	√	X
Linen	X	X	X	X	√	X	√
Dog hair ♂ (body)	√	√	√	√	√	X	X
Dog hair ♀	X	X	X	X	X	X	X
Urine Leather	√	X	X	X	X	X	X
Tannin Leather	X	√	√	√	√	√	√
Feathers general	√	X	X	√	X	√	√
Feathers Head/Belt	X	√	√	-	√	X	√
Plaster (Gypsum)	X	√	√	√	√	√	√

METAL

	SA1	SA2	SA3	SAF	BA1	BA2	BA3
Silver	X	√	X	X	X	X	√
Gold	X	√	X	X	X	√	√
Arsenic bronze	X	X	X	X	√	√♀	X
Tin bronze	X	X	X	X	X	X	√
Iron	X	X	X	X	X	X	X
Bearing (degrees)	16	24	16	51	1	10	24
Dimensions (m)	2.00	1.80	2.74	-	2.10	1.80	2.20
	x	x	x	-	x	x	x
	1.40	1.50	1.70	-	1.60	1.60	1.80

Table 14.2 Tomb Chronology Possible chronology indicated by the materials used in the tombs. The order of the Bronze Age and Stone Age tombs does not indicate their age.

Chapter 15

Tunbridge Wells and Pontypridd – Unlikely Twins

Calverley Grounds Park

On the 28th August 2005 Nigel and I met up with two dowsing friends who were studying the archaeology of Calverley Grounds Park in Tunbridge Wells, Kent (Chapter 10). This was a return visit aimed at a more detailed study of the Druidic temples in the park. The park has a valley running through it. The slopes on either side extend up from an old stream bed to higher ground with the slopes showing the typical terracing associated with Druidic ceremonial roads. The first stop was to look at one of the terraces on the left hand side on entering the park (Figure 15.1). The terrace and the slope up to it is one of those somewhat secluded places where young people lay down to talk and pass the time. As a result small change falls out of pockets to be lost in the grass. Having a sharp eye for coins I soon had enough of them to pay for my lunch and a coffee for Nigel. Running along just inside the boundary hedge is the flatter surface of the terrace. I found the bronze tracks of ceremonial chariots along this hedge and also lower down the slope. The chariots had a 7ft axle length and the bronze stain was copper arsenic. I found no tin at the higher level but there was a copper/tin track lower down. There was also the double footpath of Druids with their ash staffs. The path looked as if it was heading for the temple site which lies on a plateau tucked behind the entrance gate on the left hand side. On the far side of the temple and outside the park boundary are buildings and a car park. The temple site was checked for a pool, which it had and the presence of a tomb which was also present. The rings of postholes were then found, 6 + 3, and the imprint of a circle of twelve stone columns. There was a wattle and daub wall round the temple on the 9th ring of postholes. The temple had been burnt down and the potassium lines ran from the centre to a stone column and then to the wall. Bronze pins were used in the roof, the bronze being copper arsenic. There was a plinth in the pool and a lead pipe ran from the plinth and out of the temple towards the floor of the valley. The plinth could be the base for an obelisk but there was no evidence for that at this stage. Finding lead with arsenical bronze indicates that lead was in use in the early Bronze Age. Also of great interest was the fact that there is a considerable difference in hydrostatic head between the pool on the plateau and the source of water on the floor of the valley. I have come across these phenomena on at least two other sites and it raises the question as to how the Druids moved water uphill. In all the cases I have come across so far, header tanks at the well do not appear to be practical because of the height difference. It therefore looks as if the Druids had a method of pumping the water at least to a water tower. The answer to this mystery may be locked in the soil of Calverley Grounds Park. According to our guide, until a few years ago, there used to be a ring of stones on the plateau. The stones were removed and used to make walls and garden features in the park. If this is the case, it may be that after being left undisturbed for over 2000 years as a sentinel to a past age, in the 20th Century the stone circle was in the way of garden designers and it was destroyed.

Further back in the corner and next to the car park there used to be a lone trilithon. Almost certainly part of another temple which had a linteled stone circle. From old photographs it could have been mistaken for a Victorian garden feature but to find out our guide and Nigel went to have a look at where the trilithon used to stand whilst I looked for mass graves and execution lines. The trilithon hunt was difficult due to undergrowth but it was not long before the stains left by the stones were found. The uprights were only 1.2 to 1.5m apart. This made us realise that the Druids did not trust the stones to span more than about 1.5m. This can be seen at Stonehenge where the lintel stones only span short distances. Wood lintels span much greater distances and enable a much more open space to be created in the temple. The question

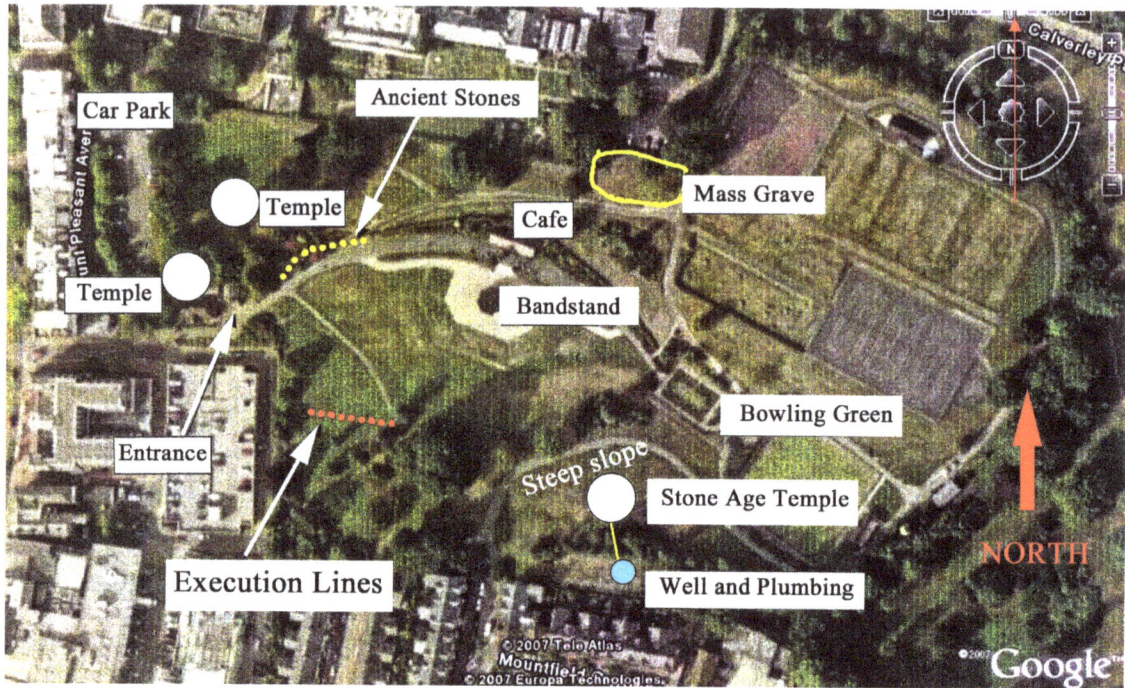

Figure 15.1 The Tunbridge Wells Temples

Two Henge Temples are indicated on the map and the approximate position of execution lines. A third and possibly the last Henge Temple stood where the modern car park is on the left on entering the Park. The yellow line indicates the position of stones that may have come from the last temple.

Figure 15.2 Stone Columns

Some of the stones from the circle of stone columns in Calverley Grounds Park are now used as ornamental landscape features in the park.

184

therefore arises as why the Druids used stone lintels. The answer may lie with energy engineering, see Box 19.1. The dowsable outline of the stone circle was later followed into the car park by our guide. Our guide confirmed that this was where the centre of a temple use to be. It is interesting that part of a temple 2000 years or more old survived into modern times and somewhere under the car park a Prince and Princes still lie.

On entering Calverley Grounds Park there is some low ground on the right and then the ground gently rises. This area contains at least one mass grave, battle blood and execution lines. Going further into the park on the right hand side are terraces with ceremonial chariot tracks. On both sides of the valley the copper arsenic and copper tin tracks appear to be in different places. It looks as if the tracks from the early Bronze Age do not mix with those from the late Bronze Age. Still further into the park, towards the Bowling Green, on the right is an elevated plateau (Figure 15.3). This was the site of a Stone Age Temple. The underground tomb complex was well formed in that adjustments do not appear to have been made to the basic design to accommodate underground neighbours. It looked as if children were in the tomb. The occupants of the tomb had wool garments but no linen, the male had a dog or wolf cloak. There was no gold or silver in the tomb. A ceramic pipe led from the clay pool to a well the other side of the boundary hedge. The presence of a Stone Age tomb and temple indicates that the valley was a Druidic site for possibly 2000 to 3000 years before the Romans arrived in Britain.

Our guide then took us on a tour of the park and pointed out stonework in the garden. The stones used could have come from the Druidic Temples and buildings. Using witnesses it is easy to test a stone for contamination which might indicate its origin. For example if the stone has been painted, egg white and pigments will show up. If the stone is from a Dolmen, bird faeces and feather will indicate its origin. A number of stones were tested and there were quite a few which were contaminated by something which indicated that they had originally been brought on site by the Druids and used in the temple complexes. Some of the big squarish stones found in walls near the entrance and below the site of the temple were contaminated by egg white and paint pigments on two sides only. This indicates that they were part of the stone columns in the temples, that is they are from the stone circle that stood there until recent times.

On the floor of the valley near the Bandstand are chariot tracks and turning circles. A bit closer to the exit is a well, possibly with the answers to how the Druids pumped water. The park has a wealth of archaeological data. It has been used for at least 4000 years if not much longer and every use that has been made of it will have left some trace. It is only a matter of finding these traces and interpreting them.

One of the reasons for visiting Tunbridge Wells was to find if the general picture we were building up of Druidic life and work fitted what could be found in different parts of the country. One theory that we had built up was that the British Isles were under a 'Pax Druidica'. That is the Druids had enough authority to see that conflicts were resolved in a certain way, there were rules of conflict. This is probably why Boudicca for example does not appear to have lived in a fort or have had a palace guard. It is also why after 3000 years there are few if any fortifications from the Henge Age. All the so called hill forts that we have looked at are Druidic Temple complexes and not defensive positions. This lack of fortifications is quite different to the succeeding 2000 years after the Romans arrived. This period produced fortifications over the whole land on a grand scale. Weapons were freely available in pre-Roman times. Anybody could make their own flint headed spear, flint axe or bow and arrows, so shortage of weapons is not the reason why the society appears to have been peaceful.

185

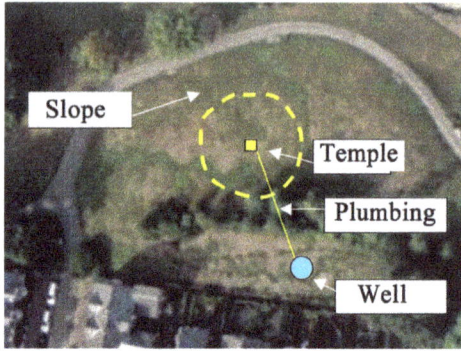

Figure 15.3 A Stone Age Temple

At the far end of Calverley Grounds Park on high ground overlooking the bowling green there use to be a Stone Age Henge Temple. The pool was connected via a ceramic pipe to a nearby well.

Figure 15.4a The Rocking Stone, Stone Circle at Pontypridd

Dr Price tried to repair the circle by replacing stones he thought were missing. The new stones were not magnetically aligned with the originals and can easily be identified by dowsing the magnetic fields between the stones. The massive rocking stone was surrounded by a pool. The blue rectangle was the ornamental plinth and the line of the pipe bringing in water is indicated.

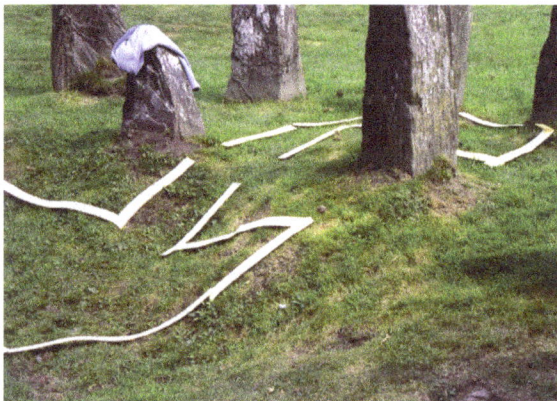

Figure 15.4b The Underground Temple of Pontypridd Rocking Stone

An anteroom and bone depository are shown. The tunnel to the depository goes under and between two portal stones. The outer stones may have been structural.

186

Pontypridd

To extend our territorial coverage Nigel started to take advantage of his visits back home to see his parents. They live in Pontypridd, South Wales, and a place I had some familiarity with as I use to visit the University there from time to time in my university days. One of Pontypridd's claims to fame and also of the nearby town of Llantrisant is their association with a certain gentleman by the name of Dr William Price. One of Dr Price's achievements is that when he died at the age of 93 and was cremated, the twenty or more pubs in Llantrisant ran dry during the festivities. Not many of us will be able to claim that as an achievement. Dr Price was an eccentric, possibly one of the most eccentric characters in Welsh history. His life is described in a book written by Dean Powell called 'Eccentric the Life of Dr William Price'. His main claim to fame is that he caused Parliament to pass the Cremation Act in 1902 which, with a much more religious society than today, had an effect throughout the country. How he did it is an exciting narrative in its own right but for my story it is Dr Price's involvement with ancient stone circles and neodruidry that is important. Dr Price was almost certainly not a dowser and responded only to the visual impression of the ancient landscape. One part of his local landscape was the Rocking Stone (y Maen Chwyf) and stone circle on a hillside overlooking the valley and town of Pontypridd. Nigel started to investigate the circle and other areas of the park during trips to Pontypridd. He found ample evidence of Druidic activity with terracing, roads and temples. Wales was of particular interest to us for a number of reasons. Snowdonia was where I had found the first Prince and Princess. The Romans had not been able to clear the land of locals in the way they had in lowland Britain. Importantly the subsoil rock in most of Wales is hard rock. This raised the question of were the Druids able to tunnel into such rock and create the underground tomb complexes we had found in lowland Britain and Yorkshire. If they could, were the temples built to the same plan round Pontypridd as in the St. Albans area. Finally, it was on record that Dr Price had added stones to the circle to fill in what he thought were missing stones. The new stones would not have been magnetically aligned and should be identifiable using science based dowsing methods. After Nigel had done quite a lot of work to make sure that there was a Druidic Temple at the Rocking Stone site I accompanied him to Pontypridd on the 3rd September 2005.

The Rocking Stone and the surrounding stone circle have modern residential buildings on one side and the old workings of a quarry on the other. There was however sufficient space round the stones to map out the greater part of the temple. The first thing was to confirm the tomb. This surprisingly could be done by walking on the Rocking Stone (Figure 15.4). Magnetic fields associated with the tomb penetrated the stone with no difficulty. The tomb had arsenical bronze, linen and gold over the chest of the woman and at the waist and head of the man. There was an ash spear and yew bow. The two occupants were adults and had taken toadstool and foxglove as their last drink. The tomb complex followed the usual pattern of four anterooms leading to four bone depositories. The access shaft was found on the edge of the quarry with the tunnel running into the tomb. The tunnel and underground complex were dug using bronze tools. Other materials identified in the tunnels were carbon, oak, ash and potassium but no iron. The design of the underground part of the temple complex followed the pattern found elsewhere.

The pool in the centre of the temple was identified using a red clay witness and the wall round it was marked by the present day inner ring of stones which, apart from the stones used by Dr Price to replace missing ones, had the diamagnetic north at the top and were linked to each other with a north pole facing a south pole. The outer face of the inner ring of stones was a north pole. There was a ceramic pipe running into the pool from the north, the joints of which were sealed with hemp and tree resin. The pipe finished at a rectangular plinth which had left in the ground stains of gold, arsenic, copper, silver, beeswax, amber and red ochre. On the eastern and western sides of the pool, paths came up to the pool wall. Inside the pool wall and on each side was a tiled apron about 5 by 6 feet in size, presumably for the priests to stand on after entering the pool for ceremonial purposes. Four to five feet away from the pool wall was an outer circle of stones (Figure 15.4) which had portal stones on the north, south, east and west. Paths came up to these portals but only the West Gate path was looked at. This path had stains from linen, feather and wool and it went in the direction of a huge stone Dolmen. The stone Dolmen had been in an enclosure the door of which was hung on bronze hinges. There were bronze lamps on the wall facing the temple. The ground was stained with bird droppings and feathers. No foxglove or toadstool was found and the bodies placed there were female. This was more evidence that the Western Dolmen was for female celibate priestesses.

Returning to the temple, we found the 6 + 3 rings of postholes but the stone columns were between ring 6 and 7. This places the main support for the rafters well away from their mid point. This may indicate that the Rocking Stone in the centre was used as the footing for a rafter support system at the centre of the temple or some other method of support was used. On moving out to ring 9 a thirteen foot wide East Gate was identified. The ground round the south and the west of the temple has been disturbed by mining activity so much of the outer parts of the temple can not now be identified. Walking west from the temple into the park or common land there is a Cyst Tomb. The tomb is interesting in that the stones making up the four walls are all correctly aligned magnetically. The stones contain phosphate which prevents the dowser identifying the phosphate signal of the stain left by the body. There was a horsehair pillow, woollen clothing and the skin, hair and blood could also be identified. There was no bronze, toadstool or foxglove stain.

After a days work on Dr Price's stone circle it can be confirmed that Dr Price got it right. The stone circle was part of an early Druidic Temple that followed the design pattern found in other parts of the country. It is an interesting site because the underground tomb complex is in hard rock. As a result it should be possible to excavate it and find out what is there. Because it is early Bronze Age the Romans may not have known that it was there and hence may not have looted it. Dr Price was not aware that Druids were dowsers and practiced what is called energy engineering when setting up stones. The stones he used to fill in the gaps can therefore easily be identified because they do not link magnetically with their neighbours. From this site we learnt that the structural middle size stone circle can be on different circles or radii particularly if a central support for the rafters is available. Later work in the area by Nigel has identified ceremonial roads running along the Welsh hillsides and there is every evidence that the Druidic civil engineers were just as active in Wales as elsewhere. What we do not yet know is whether the Druidic and secular societies had the same relationship in Wales as in England. The Welsh secular society should have been able to hide and protect the Druids amongst the hills and mountains – if they had wanted to. At the moment it appears that they did not do this.

Stonehenge

Things were moving fast with new discoveries which had to be tested out on Stonehenge. So on the 17th September 2005 and with fine weather we were back at Stonehenge looking for plumbing systems and wells. We found the ceramic plumbing system with its hemp and tree resin seals and the lead one. The lead system fed two pools. The main pool and possibly the last one to be used faces the summer equinox and had an impressive central plinth and possibly a fountain or obelisk. The stain of the plinth contained gold, silver and bronze. The lead pipe contained silver.

We marked out one pool to see what it looked like (Figure 15.5 and 15.6).

188

Figure 15.5 The Stonehenge Ceremonial Pool

At the centre of Stonehenge was a pool. Over the millennia as the temple was rebuilt new pools in a slightly different position were dug. The outline of one pool, possibly the last, is shown. It is a large pool with a central plinth to support a fountain or obelisk.

Figure 15.6a The Pool Plumbing

From the central plinth of the pool a pipe makes its way out of the temple and to the road.

Figure 15.6b Stonehenge Plumbing

Outside the circle the ceramic pipe has been divided into individual pipe lengths. The cross markers show the position of the hemp and wood resin seals.

189

After making a note of where the water supply pipes went under the perimeter fence so that we could find the well we rechecked the execution lines and confirmed our previous results

As with other temples the battle blood was outside the temple wall (Figure 15.7).

Once outside the perimeter fence we went round to pick up the pipe lines from the pools in Stonehenge (Figure 15.8) confirming on our way that the HGVs had iron tyres on their wheels. Following the pipes we eventually arrived at the well. The engineering associated with the well was the same as that at the Stoney Lane site. A header tank and windlass system with the bucket being raised and lowered by oxen. The length of the oxen track again gave the combined depth and height to the header tank. As at Stoney Lane the Druids had dug their well on an underground stream.

Having found more evidence that Stonehenge was a standard Druidic Temple, an upmarket one perhaps, one at the cutting edge of energy engineering technology with its lintels and five great trilithons but still a standard Druidic Temple centred on a mystic pool of water held in a large clay grail. The large stones so beloved of those trying to relate them to astronomical events were inside a large dark building with access through its gates to the sun and moon as they rose and set but no more. This may have been all the Druids required as the gold and metal work of the fountain or obelisk would have picked up the sun and moon on the horizon and made a spectacular feature. The arrangement of the stones in the temple and their size almost certainly relates to the energy engineering of the Druids although they may have been set up on astronomical or magnetic bearings. All one can say at this point is that "we are working on it".

Figure 15.7 Stonehenge Execution Lines

Round Stonehenge can be identified the execution lines where the Druids and their families were executed by the Romans. The markers indicate a pool of blood. The typical three ranks move from left to right. The mass grave can be found by following the drag line of the bodies.
(Aerial photo by JJ Evendon)

Figure 15.8 Stonehenge Water Supply

On the far side of the road to Stonehenge the ceramic and lead pipes can be found and traced to the well.

Chapter 16

Giant Strides Start with Small Steps

Dykes Start to Take Shape

After the visit to Stonehenge to find its plumbing system things progressed slowly for a while. Laboratory work continued and I discovered how to make synthetic oak auras, both the 'Starburst' and the 'Thunderbird' and the physical mechanisms that produced them became clearer (Box 8.3 and 10.1). On the 26th November 2005 Nigel took me over to a new dyke he had found at Wheathampstead. Once you are aware of dykes they appear to be everywhere and must be one of the major features of the Druidic landscape. Some of them are so impressive that it is extremely unlikely that they were built by large numbers of people armed with buckets, deer antlers and oxen shoulder blades. They were built by engineers using such tools as part of their tool kits just as picks and shovels are found on modern civil engineering sites. But just as today there is the power of mechanised equipment behind the construction team, so in the Henge Age there was the power of oxen teams.

Dowsing down in the dyke I was able to confirm Nigel's findings. The large straight running dyke was still 5m to 6m deep in places after over 2000 years. The bottom in the area I visited was still wide enough to take the 7ft axle of a ceremonial chariot. Horses and deer were used as draft animals with the deer being associated with small carts which appeared to have a rotating axle and wooden wheels. The sides of the dyke had been terraced and there were posts for supporting a roof. The rafters and joists of the roof could be identified. We found the stains from bronze lamps and beef fat and there were recesses in the side of the dyke. The recesses looked as if they had been at one time the entrances to tunnels or small dykes now filled in. At this time we were still thinking of tunnels to bone depositories and did not realise what form the tunnels might take. We also visited an open field site in the St. Albans area. Standing on one edge of the site it was possible to see many depressions in the fields showing that over the thousands of years of the Henge Age the fields had been home to many temples. The temples had plumbing either ceramic or lead, the Dolmen were there and at one Western Dolmen the cats had joined the Priestesses on the bier. We were always finding deer associated with small carts and although they were being used in the Bronze and Iron Age they appear very early on and are associated with what could be Stone Age Temples. The idea that deer may have been used as draft animals, possibly before oxen and horses, was beginning to seem a real possibility. Another thought that kept recurring was that the scale of civil engineering was so great that records and plans of some sort must have been kept. If the Druids worked in a monastery type of system then taxation and its accompanying records might not have been necessary. However, the wealth for buying all the gold, bronze, lead, red ochre and other pigments could have been generated by selling goods and services to the secular society. This would undoubtedly have required records. The temples and tomb complexes would also have to be planned and architectural plans drawn up even if only very rudimentary ones.

A Stone Age Temple

On the 3rd of December 2005 we had a look at a Stone Age Temple which Nigel had found on some playing fields near his home. The temple was on its own and would clear up a few questions for us. It had a small clay lined pool complete with ceramic plumbing. The tomb contents were Stone Age. The four Thunderbirds were there also rings around the tomb painted with manganese dioxide, the rays from the centre were in blue woad. The temple was smaller than the later ones but the 6 + 3 ring pattern was there. The Stone Age Temple confirmed that the basic elements of Druidic Temples where there 2000 or even possibly 3000 years before the Romans arrived. It also says that the 6 + 3 ring system, Thunderbirds and tomb design go back even further in time. The system is too complex to have materialised out of thin air and it must have a history of development.

Subsidence into Druidic Tunnels

A few days later on the 8th December Nigel phoned to say that he had found out that since 2002 the Local Authority at Hatfield had been aware of subsidence under a road in Hatfield, a town north of St. Albans. A survey had been done for the Council of the area which said that there were extensive chalk mine workings under the houses, road and school. The surveyors had drilled and sent a camera down to obtain data. Nigel called in on the site and it dowsed very much as if it was a Druidic underground complex. He also heard that the 'chalk mines' were going to be looked at in more detail. If the 'chalk mines' were to be looked at somebody might confirm our dowsing which would be great. They might also steal some of our thunder as I was way behind with writing the book. There was only one thing to do – go and dowse the site to see what a survey of the mines would find. On the morning of the 10th December 2005 Nigel and I were on the site of the subsidence in Hatfield. There was an empty house in front of us with cracks in the walls indicating that the corner of the house was sinking. It looked very much like two other properties I knew which were on the shafts to underground tomb complexes. A little further up the road, outside the school gates, there was a Druidic Temple. We started by confirming the tunnel which went from the road up to the corner of the house where the shaft appeared to be located. The presence of an access tunnel to a tomb complex was not saying that there was a chalk mine there so the next step was to start looking for other tunnels and mine galleries round the house. There was an edge of something then at a distance of 8 to 10 feet another edge running parallel with it. If this was a mining gallery it should have supporting columns to hold the roof up. We found what looked like the columns, 2ft square. So far so good. We then explored and found that on walking away from the house the gallery divided into two with a smaller one going off at an angle to the left. The 'Y' junction did not look like the mine workings I was familiar with so we checked for Druidic tunnels. Bronze chariot tracks, horsehair, beef fat lubrication, bronze lamps and clay or terracotta lamps with beef fat were looked for. The rectangles found earlier were decorative painted diamonds and not roof supports. The 'mine', or at least the part of it near the house, was in fact a Druidic ceremonial complex of a type that we had not come across before. All the tunnels studied to date had been small, just wide enough for people to move along. The tunnel we had just found was large enough for a horse drawn chariot. The dykes were make believe tunnels but did the Druids construct real tunnels to take the ceremonial chariots? If they did where did they start and finish, how were they ventilated? What ceremonies were going on in them? Fortunately there were clues to help us answer such questions.

Dells

One of the puzzles about Dells was that whilst some fitted the model of the Druidic Temple in that they were shallow bowls or depressions some did not. They were steep sided and deep. Early on I had thought they might be where people had dug down on the gold signal from the Princess. They had postholes round them and perhaps they had been partly filled in or dug down on for clay or chalk. They were an enigma and so had been left on the back burner. Now the thought arose that they might have something to do with a tunnel system for chariots. Perhaps a point at which the spoil was removed, or to provide ventilation. If there was a connection the easy way to find out was to visit one and dowse for tunnels. After lunch on the 10th December 2005 I walked from my house to a deep steep sided Dell in a nearby field. I reasoned that any tunnel coming from it or passing through it could be picked up by looking for tunnel walls leaving the Dell at right angles, this strategy paid off. There were at least five tunnels converging on the Dell. One came in from the direction of a Druidic Temple about 150m away. This tunnel appeared to continue through the Dell and out the other side. I followed this one and after about 20m the tunnel started to curve into a spiral. It would be a difficult job to sort out how much of the spiral was tunnel and how much a chamber in which the chariot turned round. Being December, time on site was limited but I had the distinct feeling that at least part of the mystery of the carvings on the stone outside New Grange Cairn and also those inside might soon be solved (Figure 16.1).

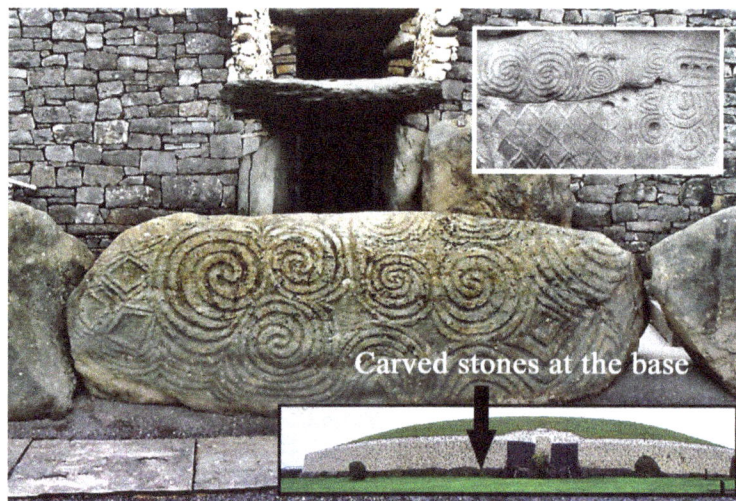

Figure 16.1 Henge Art
The photograph shows the spirals, diamonds, 'U' shaped carvings found in Henge stone carvings at New Grange. It is suggested that the spirals and diamonds characteristic of Henge artwork are derived from the fields associated with flowing underground water. They are found as an earth energy grid.

The carvings might be depicting the spiral chariot route round the breasts of the Earth Goddess. Or so I thought at the time. On the 11th I was back over the fields. My way took me along a path that gave an excellent view of the smoke plume rising from the Buncefield Oil Depot. It was big, rising high into the sky before being carried by the wind. The media were full of the size of the explosion and fire. After taking a few photographs I continued on my way to the Dell and its tunnels. The Dell has a number of tunnels leaving it so I started to see where the tunnels went. There were three tunnels going out into spiral loops and one tunnel doing an inverted 'U' or horseshoe, leaving and returning to the Dell (Figure 16.2).

The horseshoe shaped tunnel had deer as the draft animal pulling a small cart. The three spirals or chambers showed that horses were the draft animal. I tested the centre of the chambers with a Druidic painted pebble and obtained a response. Perhaps the chariots took prayer stones into the chambers. It looked as if the three chambers could represent the three breasts of the Earth Goddess. This speculation was later to be proved quite wrong. What the 'U' or horseshoe tunnel represented was not clear as it is a loop in the ground. The answer may lie with the use of a deer to pull the cart. Having dowsed the chambers and loop I followed the tunnel back from the Dell towards the temple and about halfway between the Dell and Temple found the entrance to the tunnel. The temple and the tunnel complex were clearly part of the

194

same temple. This started to make sense of another type of Dell. The temples on the sides of hills had always been a bit of a mystery. A hillside appeared to be an odd place to build a temple even if you do level part of the hillside for the main part of the temple. There was a fine example of such a Dell and Temple about two miles from my house so on the 12th I visited it to find out why it was built on the side of a hill. Looking at the Dell with new insight it was obvious that the side of the hill had been dug into. It had been dug as the splayed entrance to a tunnel. I knew the temple had been in use up to Roman times because in front of it were the execution lines and a mass grave. Walking up onto the slope above the temple I was able to pick up the tunnel coming from the entrance. The splayed stone dressed front of the tunnel could also be identified. Following the tunnel up the hill the three spirals were there and also the tunnel loop with its deer tracks. It was beginning to look as if the subsidence at Hatfield had led to new discoveries. There were not only ceremonial chariot tracks running over the countryside and in dyke systems but they were also running underground. The Druidic engineers believed in serving their goddess both above and below ground. The dyke system was also looking as if it might be an integral part of an underground system as well. The scale of the Druidic engineering just kept on growing as did the complexity of the temples and their associated ceremonies.

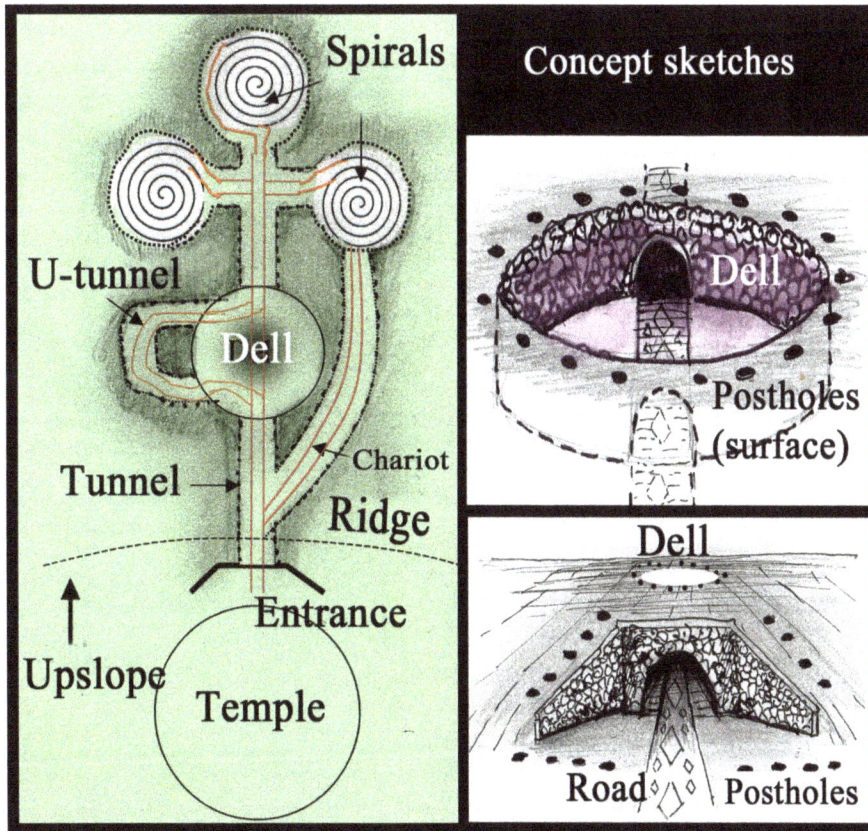

Figure 16.2 Dell Complex

The drawing shows the Henge Temple with chariot tracks entering the opening of a tunnel. The tunnel goes to a deep steep sided Dell from which other tunnels can be traced. They include a 'U' shaped tunnel used by deer drawn carts. From the Dell a tunnel goes out for about 10m and then divides into three. The three tunnels terminate in small rooms arranged in a cloverleaf where the chariots appear to go round in a circle before exiting.

Hertfordshire Puddingstone

For some reason, on the afternoon the underground spiral tunnels were confirmed my attention was attracted to a small piece of puddingstone I had picked up in a temple debris field. It was stained red so may have been in contact with red ochre. Hertfordshire puddingstone is a mysterious stone that looks like modern day concrete. The stone has attracted the attention of geologists who have declared that it is 20 million years old and was a type of conglomerate possibly formed in rivers. Conglomerates are made from deposits of pebbles embedded in a fine grained matrix such as silt which then hardens by natural cementation. Some blocks of Hertfordshire puddingstone are said to be 6 metres across and one metre thick. The stone is sufficiently unusual to have its own Webb site.

However, I decided to analyse a piece of puddingstone in fact 4 pieces (Figure 16.3). The first piece gave a positive response with red ochre, manganese dioxide, red cabbage (woad), egg white, egg yoke, urine, blood, carbon, potassium, calcium, magnesium. This is a sure indicator that the piece of stone had been painted at sometime. The remaining three pieces did not give a response with paint pigments. They did, however, give a positive response with witnesses of urine, human blood, phosphate, deer antler and oak. The Druids used urine for a number of things and could easily have used it to make a concrete. Their tools included deer antler and mixing may have been in an oak vessel, both of which left traces. Continuing the analysis, wood ash gave a positive response, possibly indicating that something had been fired. Lime, camomile and cedar wood also gave positive responses. On a later occasion the analysis was extended and positive responses were obtained with gold, bronze, human hair, sand and blue clay. When the analysis was extended to more pieces of puddingstone there were variations between stones but the common element such as urine, antler, oak, human hair and skin, bronze, wood ash and clay remained. When analysing lumps of concrete none of the contaminants listed above were found but iron was.

Figure 16.3 Puddingstone

The photograph shows typical examples of Hertfordshire Puddingstone. The cement of the pudding stone is so hard that the stones of the aggregate split when the pudding stone is broken. From left to right the main pieces are associated with the spirits of the fish, deer and horse. The small pieces are associated by dowsing with ox, lamb and bird.

196

The significance of the findings could be very important. Either the Druids had spotted the artistic value of puddingstone a long time ago and had gilded and painted many pieces and I just happen to have four pieces contaminated with gold, bronze, deer antler and oak. Or the Druids had developed cement and concrete a long time before the Romans arrived and were using it in their temples and civil engineering. I told Nigel that I thought we might be onto something quite important but that we had to make sure we were right. Nigel immediately started searching for specimens and then found that there was a Web Site recording the location of notable pieces of puddingstone *(www.megalithic.co.uk)* – Hertfordshire pudding stone trail). In the meantime I had to find ways of making sure that the analysis was right. The theory was that the Druids or somebody before them had discovered how to make a type of cement. The cement might be derived from lime, blue clay and possibly other things. These might be heated to form a clinker. The clinker would then have to be ground into a fine powder using, perhaps, similar tools and methods to those used for making flour. Iron was not available so contamination by iron tools would not be expected. Modern concretes and cements and Roman ones are contaminated with iron and so can be identified. To make a cement or concrete mix a bronze tool and or gold tool might be used to 'bless' the product. If this were so, concrete destined for the temple might contain gold, copper arsenic bronze or copper tin bronze. From the mixing process dear antler and oak wood would be incorporated as contaminants. The liquid being used would be urine with its load of contaminants. If this story holds it should be possible to find puddingstone with copper arsenic bronze and copper tin bronze. Examples of both sorts were soon found. There may be some puddingstone for low grade uses which lack the gold. This has now turned up.

Having realized that puddingstone may be a Druidic artefact, if it is used as a witness it should pick up all the areas in the temple where it was used. This was quite a breakthrough as puddingstone witnesses immediately picked up the plinth in every pool. Hardened areas such as steps and edges of roads could be identified.

To check our theory about the origin of puddingstone we started visiting people's front lawns to dowse their pieces of puddingstone and concrete. Concrete always had iron in it and no gold or any of the other things. Puddingstone pieces were often angular and looked as if they had been part of something. It then occurred to me to see if platinum and nickel were present as these may accompany some golds and not others. I found a sample in which there was platinum and nickel and another where it was absent. The evidence that Hertfordshire Puddingstone is a Druidic concrete is good and it can be traced back to the early Bronze Age. It certainly makes a valuable witness for studying Druidic artefacts. It picks up steps, kerbs, places where puddingstone artefacts have stood. Later work showed that the Druids made small objects such as bowls, cups and knives from the cement,(Figure 10.11).

Henge Age Art: The Origin of Spirals, Diamonds and Zigzags

During February 2006 Nigel was working on the river complex in St. Albans and on the river running along side the Stanborough Lakes in Welwyn Garden City which lies to the north of St. Albans. His main aim was to identify the Druidic civil engineering associated with rivers and lakes. However, as Nigel knew, if there were Druidic buildings then there could be paintings on the ground. It was whilst looking for signs of paintings alongside the rivers that Nigel came across some natural magnetic field patterns. The patterns are so important to the understanding of Henge art that I asked Nigel to write an account of how he made the discoveries. His story which follows shows biolocation in action and the possible origin of two if not three of the best known motifs associated with the Henge people's art.

I was going to dowse the clear open area leading from the car park I was using in Welwyn

Garden City, a town north of St. Albans, down to the river bank and look for artwork and any other signs of Druidic activity.

I did a search pattern using a 5 metres wide band round the waters edge, initially without witnesses, to focus on detecting magnetic fields. Immediately I picked up a response which I followed. The shape that started to emerge was not the usual rectangle, square or circle. When walking along the bank the rods flicked open and shut regularly in a band a few metres wide running parallel with the river. I recognised this sort of response as typical of a detailed pattern with magnetic field lines changing direction and shape many times. I adapted my search pattern accordingly to try and follow the fields and my first impression was that there was a line of concentric circles running along the river bank. I tried my usual pigment witnesses but they gave negative responses. I tried quite a range of different witnesses, again with no response. Without witnesses I located strong responses but with witnesses there was no response. By now I had an idea that the responses were occurring in distinct circular patterns about 1.5 metres in diameter with possible rings inside them. At the edge of one of the areas I carefully inch by inch dowsed and followed the line to map out the shape. After about a quarter of a circumference following the arc of a circle I lost the responses. I varied my intercept angle looking to regain the responses if the line had changed direction. Indeed it had and I picked it up curving in at a sharper angle compared to the original line. Again a short distance (less than a metre) along the arc and I lost it before relocating it as it curved inward on an even tighter arc. I had now found the pattern and I was following a spiral path into the centre of what I had thought was a circle.

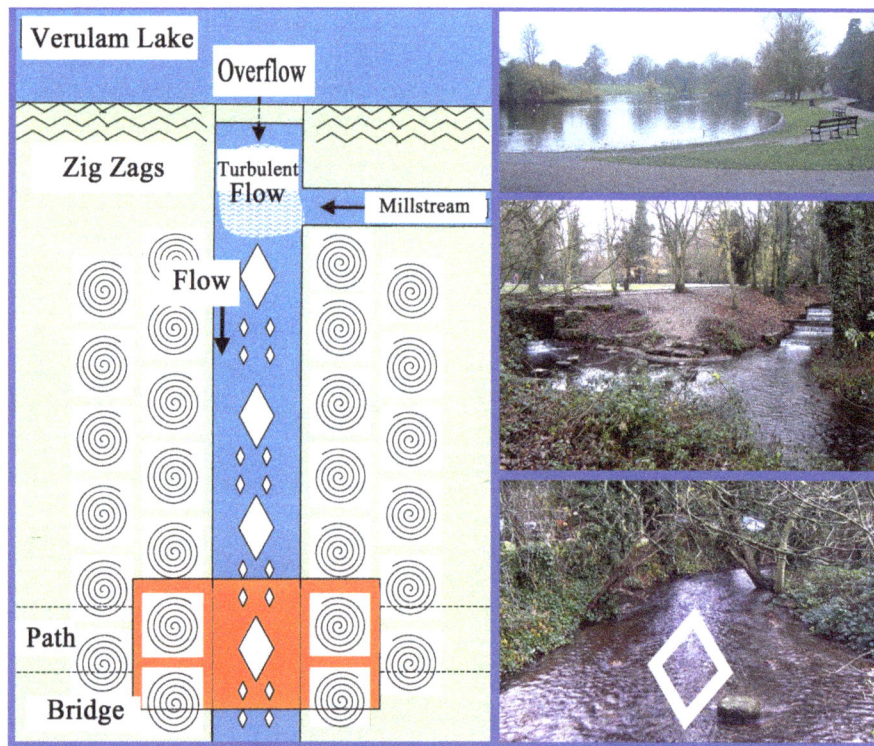

Figure 16.4 The Magnetic Patterns of Flowing Water
*The diagram shows the spirals and diamonds associated with running water.
The water must be flowing freely to generate the pattern. If the river or
stream runs under a bridge the diamond can be found on it making it easy
to dowse. This is particularly useful for dowsing the complex colour pattern
inside the diamond. On the banks of the lake a zigzag pattern can be found.
Although these are common patterns on other features they would have been
associated with the land.*

198

I remembered Geoff once saying to me 'I am always suspicious of spirals' when dowsing complex forms it is quite easy to jump from one line to another and think you have a pattern that actually isn't there. I checked it again inch by inch – it was a spiral. I moved along to the next one and that was a spiral as was the next. So there, running about a metre and a half from the waters edge, was a line of spirals. What they were I had no idea as they didn't respond to any of the usual witnesses. Then of course I realised there was one witness in particular I hadn't tried. Having no luck with other witnesses and now with the light fading I rushed back to my car to get a bottle of water to use as a witness. With the sun now close to its nightly liaison with the horizon I had just enough time to carry out some checks. I used the water as a witness by blanking out the plastic bottle by holding carbon in my other hand and sure enough the water witness responded to the spirals. I wondered if the spirals were caused by water seeping out of the river underground but I couldn't see how it could produce these regular geometric spiroglyphics. I had dowsed leaks on the side of the local canal and water pipes and hoses and they didn't have this sort of activity. It then occurred to me that the spirals may be generated by the river itself. This would be easy to check. I placed the bottle of water between the river and a spiral and the spiral disappeared. This confirmed they were part of the energy system of the river. I just had enough daylight time to check the opposite bank of the river and sure enough there was a corresponding line of spirals on the far side. Whilst on the far side I checked a similar sized area around the lakes edge. I was unable to locate any spirals. but did find a zigzag line of responses about 1 metre from the lake edge. I drove away from the river preoccupied with the idea of how similar the spirals were to the spirals in prehistoric rock art. Had we solved the mystery of some of this artwork and its symbology? It was a very interesting phone call to Geoff that evening, the plot was thickening I had never imagined river dowsing would produce such developments.

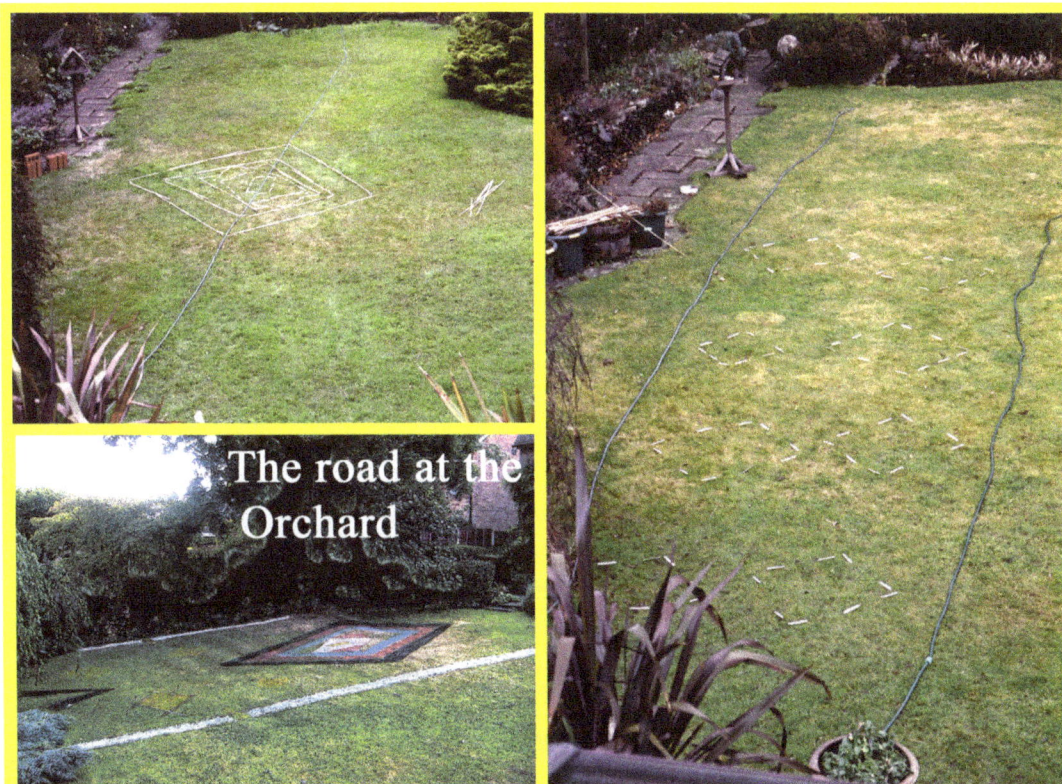

Figure 16.5 The Structure of a Diamond

One of the diamonds which are generated by the flow of water through a hose has been dowsed in detail. The different sections of the diamond have different colours which match the colours used by the Druids for painting them on the ground. When two hoses are used to represent the river two rows of smaller diamonds are created.

The weekend was approaching and this was going to give me chance to check stretches of the River Ver in St. Albans. The weather was forecast to deteriorate over the weekend with the worst arriving on Sunday. On Saturday I rose early and was down on the river bank at the foot of the Cathedral opposite the Westminster Lodge Leisure Centre. Here the river meanders across a small flood plain/water meadow. On one side the bank is flat and accessible over large stretches. The river is also shallow with a firm gravel bed. The weather was perfect and I quickly located a line of spirals in a similar position to those I had found at Welwyn. I also found evidence of a second line of spirals running adjacent to and outside the original spirals. The second set of spirals seemed to poke out from between the first. I followed both upstream heading north for several hundred metres to a large pedestrian embankment which acts as a dam for the ornamental lake in Verulamium Park. Periodically I checked the spirals by blocking them with a bottle of water as at Welwyn and the water witness had the same effect i.e. they vanished. As I approached the embankment the river slows and becomes a confluence between an overflow from the lake and the river emerging out of a watermill complex on the right. This caused the spirals to fade and become unreliable when dowsed. I checked the lake edge which has a good clear tarmaced pedestrian walkway round it and again confirmed there were no spirals and located the same zigzag line of responses as at Welwyn. The Ver flows down the north eastern side of the lake where its course has been engineered as part of the lake complex. Spirals were easily located on the smooth tarmac running between the river and the lake. This was all I could do for the time being as I had to return to family duties.

A few days later I found myself back at my starting point and this time I was in the river Ver. I wanted to check some other responses I had found on Saturday when I had detected a pattern of something in the middle of the river. I had been unable to identify it using witnesses. My immediate interpretation was that I had found posts or pillars of an old bridge or crossing point. If I could identify what they were made of and find a pattern typical of a structure used for crossing the river it would show once again that stains can be stable features in a river bed. What I was hoping for however was that the structures would prove to be Druidic. I exhausted my witness base and still had not identified the energies. I climbed back up onto the river bank and checked the old river bed. The energy patterns were not there. I checked a length of some 10 meters and found no response. It then occurred to me that I had better do the same in the river. As I again walked upstream I obtained similar responses to the original ones at regular intervals. The responses ran along the whole stretch and like the spirals faded and became indeterminate in the area of the confluence. This provided a clue to what might be going on in the river. The dowsing was very similar to that of the spirals i.e. no witnesses and a complex pattern of responses which seem to switch on and off. There was no option but to fine dowse the shape of the energy pattern. Dowsing form in a river is not easy as it can be difficult to place markers into the river bed. I returned to my starting point as this was shallow and had a clear space. By now the first specks of rain were falling and the wind started to gust. Luckily the far side of the bank was lined by several mature trees that protected the river from the wind and acted like leaky umbrellas keeping the worst of the rain off me. I followed the lines of energy marking their changes in direction with a vertical cane stuck into the river bed. Over a stretch of some 5 metres I had four groups of canes each marking a distinct energy system. I joined them up with string and the pattern revealed itself to me. They were diamonds, the river had diamond shaped fields running along its middle. I continued my search pattern either side of them as I had also found a number of regular responses here. Running either side of the large central diamonds was a line of smaller ones, the river was completely filled with diamonds. I looked at the pattern of the canes with a combination of exhilaration and relief. The sky was now dark grey and the rain was falling heavily. The discovery of diamonds in a water course was a convenient point to finish the days work.

After Nigel phoned me about his discoveries I quickly laid out two hosepipes on the lawn, one for each side of the river and dowsed the space in between. The response I obtained was two rows of diamonds. I then repeated the experiment with one hosepipe and obtained a single row of larger diamonds (Figure 16.5). I was also able to confirm the spirals generated by the flow of water. It was now clear that Nigel had managed to solve one of the mysteries associated with ancient 'Celtic' art. For generations people have pondered on the significance of zigzags, spirals and diamonds without any clear interpretation emerging. We now knew their probable origin and what they might mean. The symbols represent the land (zigzag) and water. The combined symbols as in (Figure 16.1) probably stand for 'Life' or something close to it. When placed at the entrance to a tomb the symbols probably mean that there is life after death. The Henge Age equivalent of the Christian cross.

Summary

The small advances described above are only small in the sense that individually they took little time. They were however quite important steps forward. The dykes and the discovery of tunnels sufficiently large to take horse drawn chariots reemphasised the civil engineering capability of the Druids and the immense investment in religious buildings both above and below the ground. The origin of Druidism had now been pushed back into the Stone Age, 2000BC plus. The Henge civilization had cement and concrete available long before Rome appeared on the scene. The dowsing skill of the Druids has again been illustrated by their knowledge of the magnetic fields associated with running water.

The year 2005 would end with a direct comparison of standard archaeological methods and conjecture with biolocation based archaeology.

Chapter 17

Following the Footsteps of the Archaeologist

The Destruction of Archaeological Information

One evening in November 2005 I decided to watch an archaeological programme on TV. The programme was about a place I had never visited although I had passed by the site many times over the years. The archaeological site providing the focus for the film was Durrington Walls. It is said to be one of the biggest henges in Britain. It is however now much eroded with little to see and little known about it although conjecture abounds. Because of this the site does not attract as many visitors as the nearby 'Woodhenge' Also part of the site is in private hands and permission is required to visit it. Durrington Walls is said by archaeologists to date from the Neolithic, about 2500 BC. The site covers 30 acres (12 hectare) and there are two posthole ring systems which archaeologists have identified within it (Figure 17.1). The one to the south and near the river Avon has six rings of postholes and the one to the north has two rings.

The ditch is estimated to have been 16m wide and 6m deep (see page 44 of Circles and Standing Stones by Evan Hadingham) I knew from the description given early in the programme that the site was a Druidic Henge Temple complex and that the six ring circle was a 6 + 3 ring temple. The ditch, as they called it, would have been an enclosed dyke, a ceremonial chariot track. The programme had not been on for long before I realised that the archaeologists were doing (in fact had done as they had long left the site) an enormous amount of damage. Tons of soil were being moved, postholes were being cleaned out, for what reason I could not see, what they thought was a road but was in fact the foyer of the temple looked as if it had been destroyed. There was no attempt to identify the tracks of ceremonial vehicles or parades of people or the patterns in the ceremonial foyer and concourse surfaces. Aghast at what I had seen I rang up Nigel, told him what had happened to the site in the name of archaeology and said 'we have to go down to see how much damage the site has sustained'. It was also an excuse to look at what could be an interesting site. The two large posts reported by the archaeologists appeared to be on ring 4 or 5. They were therefore not the gateposts of the temple which should be on the outer ninth ring. They therefore had some other function which we had not come across before. The list of objectives for the visit to Durrington Walls began to build up. They were:

- find out how much damage had been done to the site

- obtain some information on how the posts were erected

- to look at the ceremonial area in front of the large posts and referred to by the archaeologists as a 'Neolithic road'

- identify the ceremonial chariot track in the dyke and obtain details of its construction

- to look at the construction of the dyke roof.

There were many more things that could be added to the list but there is a limit to what can be done in one day.

The Durrington Walls' Dyke

With a reasonably clear idea of what we wanted to do we arrived at the car park next to Woodhenge on a fine sunny day in the middle of December 2005. It took a little while to grasp which part of the landscape was the dyke and Henge system of Durrington Walls, partly because of its size and partly because the dyke was nearly filled in and not all that clear until the eye tuned in. There is also a road, the A345 on a raised causeway, running through the site and splitting it into two unequal halves. The target area was the smaller half because it contained the temple and was where the archaeologists had been working. To enter this part of the site requires the permission of the land owners. To reach it we had to walk down a slope and across the road. After arriving on the site the first dowsing survey was on a small length of dyke close to the road. Like all ceremonial dykes it was complex. At the bottom of the dyke was a road with clear well defined edges, a paved surface of some sort with a row of blunt ended diamonds running down the centre. The diamonds are characteristic of ceremonial roads. When they became part of the Druidic ceremonial art is difficult to say but they are on roads that almost certainly date back to the end of the Stone Age. The diamond is shown in figure 17.2 with its colours.

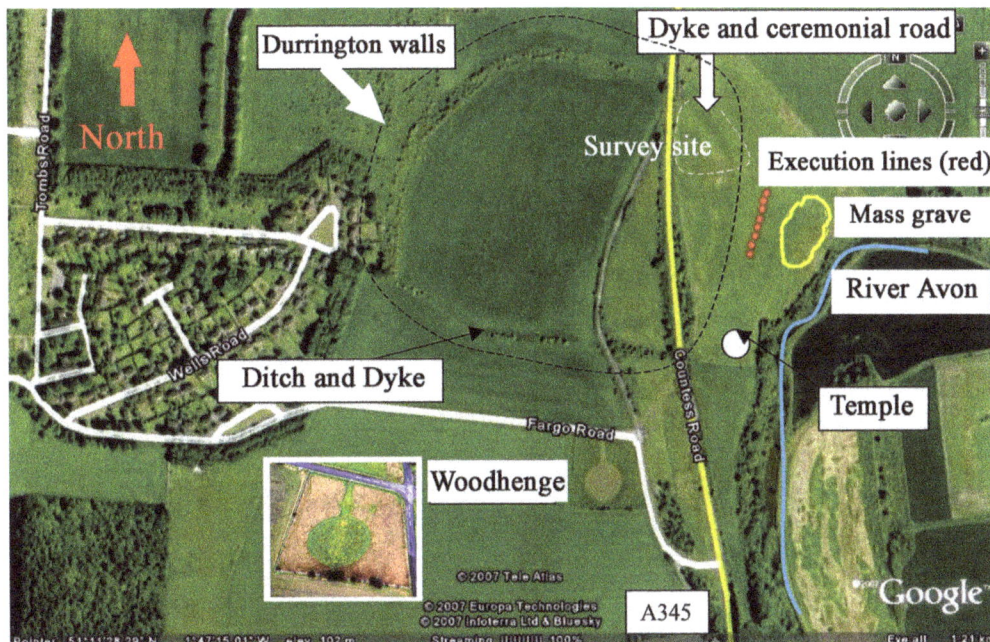

Figure 17.1 Survey Sites at Durrington Walls
A section of dyke was studied close to the main road (A345). At this point the ground is reasonably flat and easy to work on. The River Avon comes close to the Eastern Gate of the temple then bends away leaving an area of ground which was used for execution lines and mass graves.

Three diamonds were dowsed in detail. The pigments which were almost certainly used by the artists were manganese dioxide for black, red ochre (red oxide), woad for blue, gypsum for white and flint. Sometimes charcoal is used for the black pigment and henna for yellow but they were not found in the three diamonds dowsed. The edging of the road was strengthened by the use of a cement and brick. The main road surface comes up on chalk and urine witnesses which indicate the mortar that was used by the Druids. At the time of writing we do not know if there was a metalled surface on top of the mortar. However, the archaeologists did not draw

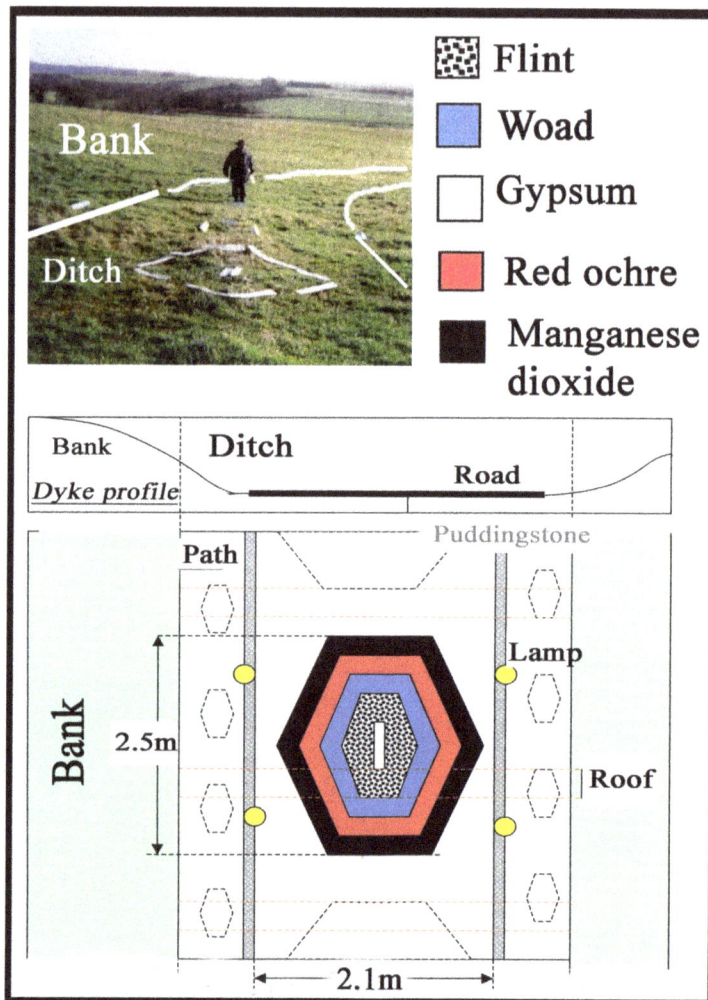

Figure 17.2 Painted Road Diamonds

The diamonds painted on the ceremonial road at the bottom of the dyke were blunt ended but in other respects very similar to the diamonds found in the St. Albans area. There were smaller diamonds on the footpath. The straw bands indicate the position of the rafters.

attention to such a surface in the entrance area and foyer of the temple. The road surface shows the tracks of iron and bronze wheeled vehicles drawn by horses and deer and the bronze tracks contained copper, arsenic and tin. The arsenic indicates that the origin of the site goes back to the early Bronze Age. The iron chariot tracks were the first to be found on a ceremonial road and indicated the use of iron tyres for ceremonial purposes in the Iron Age. Wooden wheeled vehicles with tyres of leather and hemp also used the road but they could have been service vehicles of any age and do not necessarily indicate that the dyke had a pre Bronze Age origin. About halfway up the outer side of the dyke there is a footpath. The carbon tracks left by feet and footwear and the regular imprint of ash staffs which is so typical of Druidic ceremonial processions were present. The tracks indicate that people were walking round that section of the dyke at least. The path was engineered with a surface and edging and had the coloured diamonds running along the centre. The same colours, design and pigments were used as on the road but the diamonds were much smaller. As the dyke was roofed over lighting was required. The lamps and their position in the dyke can be identified from the stains left by the beef fat and the herbs used in the incense. The material of the lamp such as

204

pottery, bronze and iron can also be identified. On the outer side of the dyke (Figure 17.3), there had been bronze lamps possibly on a cement base between the path and road.

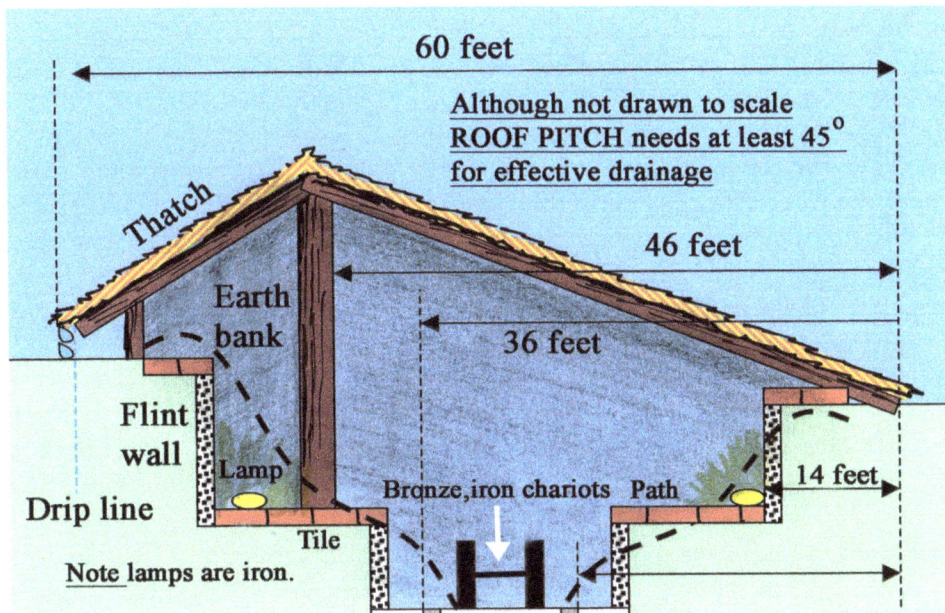

Figure 17.3 Concept Sketch of the Structure of a Dyke
Dykes were not a simple ditch or trench with banks of earth heaped alongside them. The dykes were highly engineered features contained within a building. At Durrington Walls the roof was supported by two rows of posts. Two tiled pathways on terraces flanked the highly decorated road. The whole structure symbolised an underground river.

Iron lamps were found on the inner bank mounted on a cement base. This indicated that lamps were not placed on the inner bank in the section of the dyke being looked at until the Iron Age as no bronze and ceramic lamps were found. On the outer bank ceramic, bronze (copper and arsenic) and iron lamps were found, indicating that the dyke was used for a period spanning at least 2000 years. The lamps were placed on a ceramic tile. The bronze lamps were on the outside of the ceremonial walkway. The lamps along the walkway used beef and olive oil, the iron lamps along the inside of the road were not placed on a ceramic tile and only burnt olive oil. The use of olive oil for lighting a track nearly a mile long indicates its use in considerable quantities in Druidic Temples. There must therefore have been a considerable import trade in olive oil at this time. However, it must be remembered that witnesses are not infallible and the olive oil witness may be picking up something else.

This was the first time we had been able to look at a dyke in any detail. The reason was that the dyke at Durrington Walls is nearly filled in and we had a near flat surface to work on. Round St. Albans where our home sites are situated, the sides of the dykes are steep and the dowser needs the skills of a mountain goat. It was now clear that the best dykes for investigation are those that have been filled in by erosion over the millennia and not the well defined deep ones. After dowsing the road and path the next stage was to identify the holes of the posts that had supported the roof. The roof spanned 16m and it could be either a single roof sloping inward or a ridged roof. To achieve the desired slope on a thatched roof it makes sense to use a ridge with the thatch sloping away on either side. A 16m single slope thatch roof would require a wall on one side at least 8m higher than the lower wall. We knew from previous work on dyke roofs that there would be large beams spanning the dyke supported by posts. We would be able to identify the posts and the inner and outer walls if present, also the drip line from the roof. For identifying the wood structure there are (Dec 2005) two techniques which pick up the pattern of beams, rafters, joists and roof supports. The first is the pattern of the straw dust being generated by friction between the straw roof and the wood members supporting it. The straw dust falls to the ground creating the pattern of the roof timbers. The second method is to identify the potassium pattern created by the ash falling from burning timbers. The section of roof studied is shown in figure 17.3. Three large beams were identified on the inner side of the dyke. These went across the dyke supported by two posts each, one on the outside of the road and one in the row of lamps on the outside of the path. There was some evidence that there was a roof ridge along the inner row of posts so that the roof would have sloped down from this point to the outer wall and beyond. The dyke at Durrington Walls had been easy to study and the first detailed cross section of a dyke had been obtained. During the survey we had met up with the farmer and he had given us a lot of useful information about the site.

The dyke had been constructed in the early Bronze Age if not before and so somewhere on the site there were going to be early Bronze Age tombs and underground temples. The presence of Stone Age tombs would indicate that the dyke may have started life in the late Stone Age or Neolithic, 2000BC or earlier.

The South Temple

From the dyke where we had been working it was a downhill walk to the temple that had attracted the archaeologists earlier in the year. Henge Temples can be identified by scanning for the postholes. Once these are identified the pattern of the posts indicates where the centre of the temple with its pool, pool wall, font and plumbing are to be found. The central area and position of the East Gate was also quickly identified. On first dowsing the central temple area it came up as a dowsing desert. No pool or pool wall, even postholes seemed to be missing. At first it appeared as if the archaeologists had scraped the place clean. Even the so called road or temple foyer and concourse yielded nothing at first. This could have been due to poor dowsing conditions and not entirely the result of over zealous archaeological activity. However, on the basis that the holes dug by the archaeologists should show up a careful search was started. By this time we knew that the temple had the usual 6 + 3 rings of postholes with the wall and its gates on the outer ring. On a long established site there will be the footprint of a number of temples. As the one identified first seemed to stand out quite well the others were not looked for. It is not possible to relate a given tomb to a particular temple without some very careful measurements and study of ceremonial tracks. A long and difficult process, so this was not attempted. One tomb had been quickly located and it was early Bronze Age. No attempt was made to relate it to the postholes that were being identified.

The two large postholes were soon found and they could be identified by the iron left on the side of the hole by the archaeologist's tools. The cotton from the clothing also left a dowsable trace. There was however nothing left of the post and what might have been on it such as paint or fixtures. We could not determine if the posts above ground were round or square, whether there was a gate with hinges, whether they were painted or not. Fortunately for us, the zest of the archaeologists for burrowing five feet into the ground does wane and the area between the postholes had only a modest depth of soil removed. As a result the stain of the lintel between the two posts was still there. It had been painted on the outward facing side with gold, red ochre, gypsum and manganese dioxide. Woad was not tested for but it could have been used. The potassium from the burning of the lintel was present and could be traced between the two posts. Discovering that some chemicals had moved sufficiently deep into the soil to still leave a trace meant that we might be able to find out what sort of activities had taken place in the foyer. Bronze wheeled ceremonial chariots had left a track which came in or out on one side of the gate, it went up to the ceremonial wood trilithon where the bones were unloaded. This activity left a stain of red ochre in the soil. The ceremonial vehicle then proceeded out of the gate. Just inside the red ochre stain, almost under the lintel, a diamond had been painted on the ground (Figure 17.4).

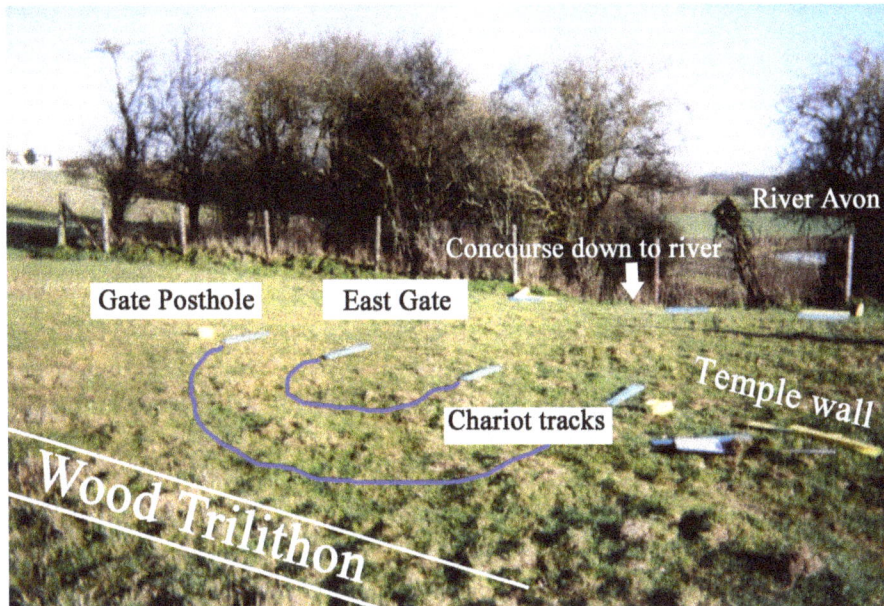

Figure 17.4 Temple Entrance
The wood trilithon is set back in the temple. The chariot delivering the bones makes a sweep round in the foyer and concourse to unload the bones at the trilithon.

Having found the temple wall on the 9th ring it could be followed round until the Eastern Gate was identified. There was a bronze hinge on the left hand post as one faces the temple. The right hand post should also have a hinge but one was not found. The forecourt of the temple could be traced as it ran towards the river Avon. Just in front of the wire fence and hedge a band of chalk and urine was found, the mortar used by the Druids, then there was a gap of a few inches and a band of something giving a positive response on a witness of Hertfordshire puddingstone matrix (Chapter 16). The material the puddingstone identified we interpreted as Druidic cement. The presence of cement indicates that as the land sloped down to the river a much stronger and more solid surface had been required (Figure 17.5).

It was not possible to work on the river side of the fence as permission had not been obtained. Apart from steps down to the river there could be a pumping station for the water required by the pool in the temple. If a pumping station was not to be found but instead a well was found to be the water supply for the temples, it would indicate that the Druids required pure clean water for the pool. Contaminated water such as river water may not have been considered good enough for the temple pool. If this proves to be the case there would have been a reason for why the river water was not good enough. Later work back in the lab would show that pond water was inferior to clean water when used in energy engineering. If river water is also inferior the temple at Durrington Walls may have a well. After dowsing the temple foyer and the concourse between the river and East Gate, attention moved to the West Gate. Outside the zone of destruction in the centre of the temple the double ceremonial path was found by dowsing on carbon, ash wood and feather. This path would be expected to go off to the Dolmen of the priestess which could be a hundred metres or more away. Back in the centre of the temple, the one thing the spades of the archaeologists could not destroy was the tombs of the Prince and Princess. One tomb had been found early on and it had a bronze shield and a bronze head dress on both occupants. The bronze contained arsenic not tin. The Princes wore a gold dress down to the waist. A further three tombs were found including a Stone Age one. The tombs were the centre pieces of the usual underground temple complex. The site had therefore possibly been used from the late Stone Age or Neolithic to the Iron Age (Box 17.1) . Before moving outside the temple it was decided to find out how the giant five ton posts of the trilithon had been moved. Contrary to popular belief amongst archaeologists and the media the ancient engineers used the methods any sensible person would use. For example they could bring the posts in on a wheeled wagon pulled by oxen and then use a tipping bench to take the weight of the post. As all lorry drivers know, you cannot put the full weight of a load onto the rear (or front) axle. When tipping the posts it is likely that boards on the far side of the hole would be used to direct the base of the post down as an 'A' frame system lifted up the far end. If required, a stop would be erected to prevent the stone or post going too far and if necessary stops on either side. To see if the system used at Durrington Walls was anything like this, the first thing to look for were the wheel tracks and signs of oxen (hoof and faeces). These were found and could be identified at some distance from the postholes. Next what might be a tipping bench was found, big solid wood posts about 70cm square with a grease line between them.

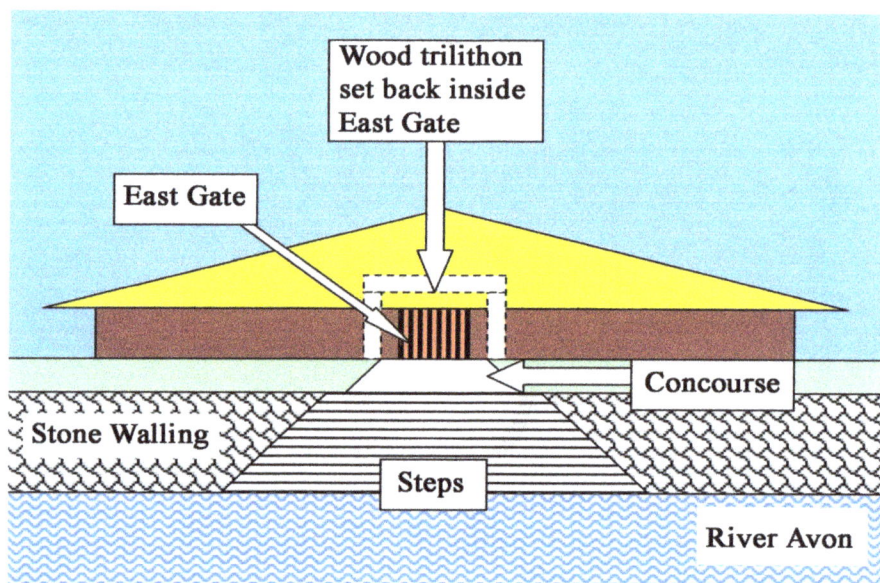

Figure 17.5 The East Gate
The concourse outside the East Gate of the temple goes to the bank of the river. There is a steep slope from the concourse to the river and steps would be required. They are indicated by the presence of Druidic cement.

The three lines of oxen can be picked up on the far side of the posthole so it looks as if the oxen had gone across it. This may indicate that the load came off the rear of the wagon although the front is a possibility. How the engineers used the cart and tipping bench would be determined by how much room they had in the temple. Once one post is up it will restrict options for the second. Time was getting short so we did not look for the presence of 'A' frames or other equipment that may have been used to lift the posts and lintel. It was more important to determine the date the temple ceased to be used. Much to my amazement the removal of earth had not removed all the signs of battle blood, it could still be identified in the foyer. The eastern foyer and concourse had patches of blood all over it, some as far back as the wood trilithon. All round the eastern and northern sides of the temple there was battle blood. Execution lines were found running to the north of the eastern entrance along the river. Behind the execution lines there was a large mass grave. The search for battle blood, execution lines and mass graves only covered part of the site (Figure 17.2). The full picture will have to wait for another visit but the days of the temple as a Druidic site came to an end when the Roman General Paulinus paid it a visit sometime between 43 AD and 60AD.

After a days dowsing we had an idea of what the Durrington Walls site had been. It was a large Druidic ceremonial dyke and temple complex dating from possibly the late Neolithic which would make it about 4500 years old. The full scale of the engineering is difficult to envisage but the circular dyke was probably mostly covered in by a timber and thatch roof making it a very large enclosed space. This in itself would have been a massive structure as the dyke is about 1.5 km long. Road surfaces, lamps, roofing would all have to be maintained. There would have been within the circular dyke at least one temple which with the standard 6 + 3 ring structure is about 60m in diameter. In the vicinity of the temple there would have been buildings such as washrooms, toilets, dog kennels, garages for chariots and all the other services and structures. The Dolmen to the four quarters, the processional ways for chariots and priests. There would have been painted surfaces with red, white, blue and black at least. The river frontage would have had some importance but not necessarily a connection with Stonehenge. Durrington Walls was a big and important site in its own right. If bones did come in by river it may be possible to pick up the red ochre stains on the bank. The henge culture of the Druids on this site may have spanned two and a half thousand years at least. In fact the elements of Druidism may have been practised long before the first Prince and Princess were laid to rest in underground temples. It may be that Durrington Walls, as a sacred site, has a history going back beyond 4500 years ago and has many secrets to reveal. Lets hope that the archaeologist in their zest for digging do not destroy the memory of the land first. It is the chemical traces in the soil that hold the information and digging is not the way to get at it. Some keyhole archaeology may be justified but not the removal of soil by the ton. To read the secrets of the Durrington Walls Henge Age Dyke Temple requires archaeology based on biolocation.

Box 17.1 Tomb Evaluation at Durrington Wells

The Henge Society is characterised by the sacrifice and burial in a tomb of what I have referred to as a Prince and Princess. The society of the time probably looked upon them as envoys to the spiritual world and the Earth Goddess. The basic design of the tomb and the underground complex remains unchanged from the early Bronze Age, possibly Neolithic, to the Iron Age. The contents of the tomb however change and indicate its age. On a site such as Durrington Walls the tombs indicate the age and history of the site. Although they were not looked for it is possible that there are other temple sites within the dyke system. The tombs found were all within the South Temple Five tombs were found in a small area of the temple and their contents analysed.

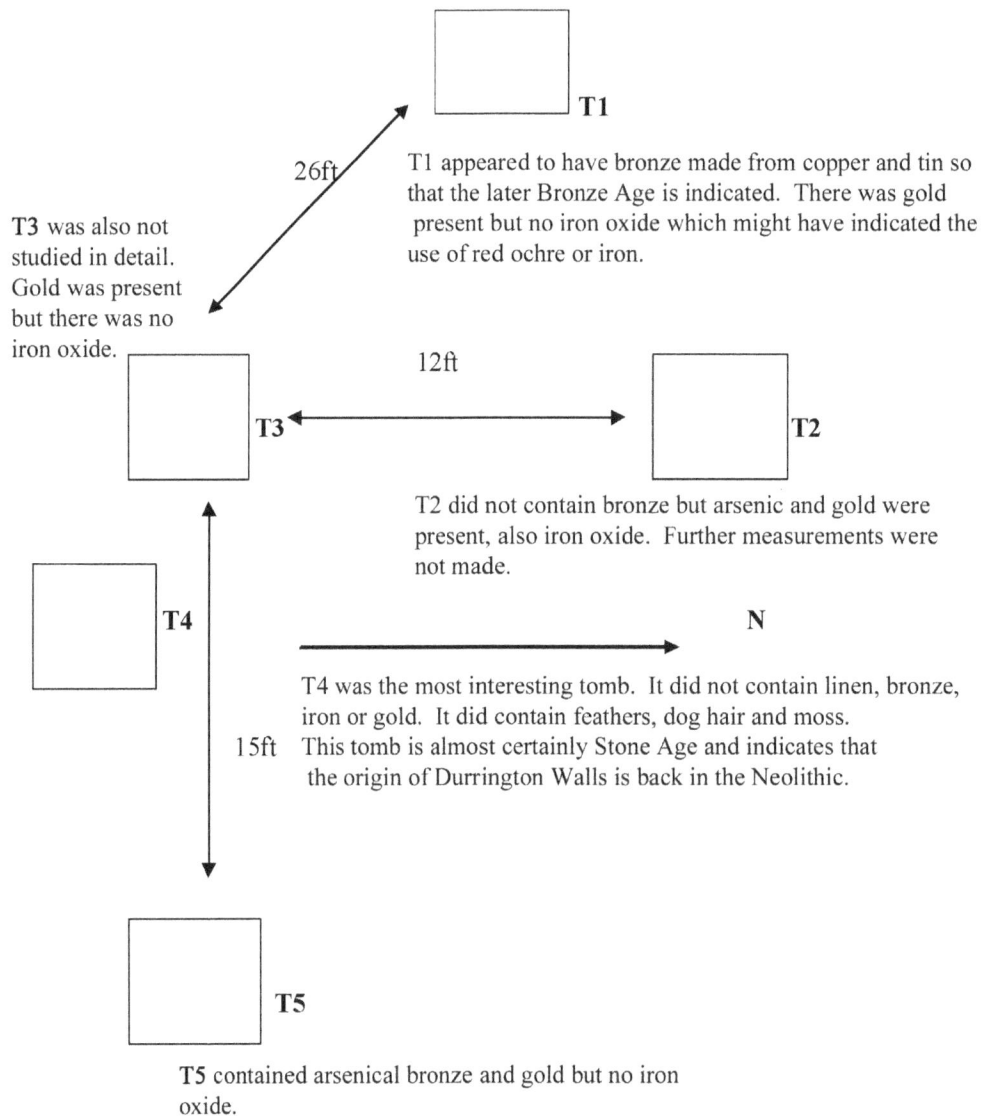

T1

26ft

T3 was also not studied in detail. Gold was present but there was no iron oxide.

T1 appeared to have bronze made from copper and tin so that the later Bronze Age is indicated. There was gold present but no iron oxide which might have indicated the use of red ochre or iron.

12ft

T3

T2

T2 did not contain bronze but arsenic and gold were present, also iron oxide. Further measurements were not made.

T4

N

T4 was the most interesting tomb. It did not contain linen, bronze, iron or gold. It did contain feathers, dog hair and moss. This tomb is almost certainly Stone Age and indicates that the origin of Durrington Walls is back in the Neolithic.

15ft

T5

T5 contained arsenical bronze and gold but no iron oxide.

Figure B17.1.1 The approximate relative positions of the five tombs which lie within the boundary of the temple destroyed by the Romans. Each tomb probably represents a temple and pool which could have had a life measured in hundreds of years. The temple site was in use during the Iron Age so there should be an Iron Age tomb somewhere.

210

Chapter 18

Epitaph to a People

The Death of a Civilization

During the Christmas break 2005 work in the laboratory continued. We wanted to know if anything dowsing wise happened at the winter solstice which the Druids might have picked up. Possibly something that indicated what time of year it was. Also we were still investigating the winter aura of the oak (Box 8.3, 10.1). However the main dowsing activity between Christmas and the New Year involved a visit to Anglesey to see what impact the Romans had on the Island. The scale of the programme of genocide being carried out by the Romans in what is now England was becoming very clear to us. Execution Lines and mass graves around temples were everywhere, north, south, east and west. What was not clear was whether it was Roman policy to systematically work their way across the country, killing the Druids and their families as they went. If it was, the invasion of Anglesey was perhaps the culmination of this programme on the west coast. The alternative but less likely theory was that the genocide followed Boudicca's revolt and the invasion of Anglesey predated the mass killings in the rest of the country. Yet a third possibility is that after the removal of the Druids the revolt led by Boudicca set off a second wave of killing in which the non Druidic part of the population was the target. The use of execution lines by the Romans with, as it were, 'officers' in front of the three lines makes it look as if the executions were for the benefit of an audience so somebody must have been left to act as an audience and bury the dead. The scale of killing, however, indicates that they were few in number.

There was one idea that had been developing in our minds as more and more temples were found with execution lines and as we played with the different pieces of the puzzle. The idea was that the population was divided into a religious Druidic and monastic group and a secular group with its Nobles, Princes and Kings. There is said to be historical evidence for this division of the pre-Roman society. The evidence for a secular division of society is used by most writers covering the pre-Roman period and the Romans conquest. For example in John Peddie's book 'Conquest The Roman Invasion of Britain' little if any reference is made to the Druidic or spiritual section of British Society. The book is all about Tribes, Chiefs, Princes and Kings that is the secular part of British Society. The split in numbers between the two groups is not known. The mass ceremonial killing of the Druids would make sense if its purpose was to demonstrate to the remaining population that Druidism had come to an end. The mass killing of Druids also makes sense of why so many temples, both large complexes and small ones, have so many people lined up outside them to be killed and so few signs, with some notable exceptions of big battles with the Romans outside and inside the temple complexes. A religious monastic group would not have many warriors but would have plenty of artisans and priests who could be captured and tied up ready for execution. The Romans, just like Henry VIII, realised that 'monastic' lands were of great value. In addition there was the wealth in the form of bronze, silver, gold, lead and semiprecious stones. Confiscating Druidic land and the systematic destruction of the Druids by the Romans makes sense both on material grounds and political grounds. It is not difficult to understand the material aspects. A society that has been accumulating gold for hundreds if not thousands of years will have a great deal. The Druids had wealth passed from one generation to the next. Again, if that society was trying to keep its treasures away from the enemy they would be moved west and eventually land up in Anglesey. Such a hoard would provide a big pull for a Roman General which, when combined with political objectives, could be irresistible. The indication is that by the time Paulinus arrived off Anglesey much of mainland Britain had been cleared of Druids. Also, provided the

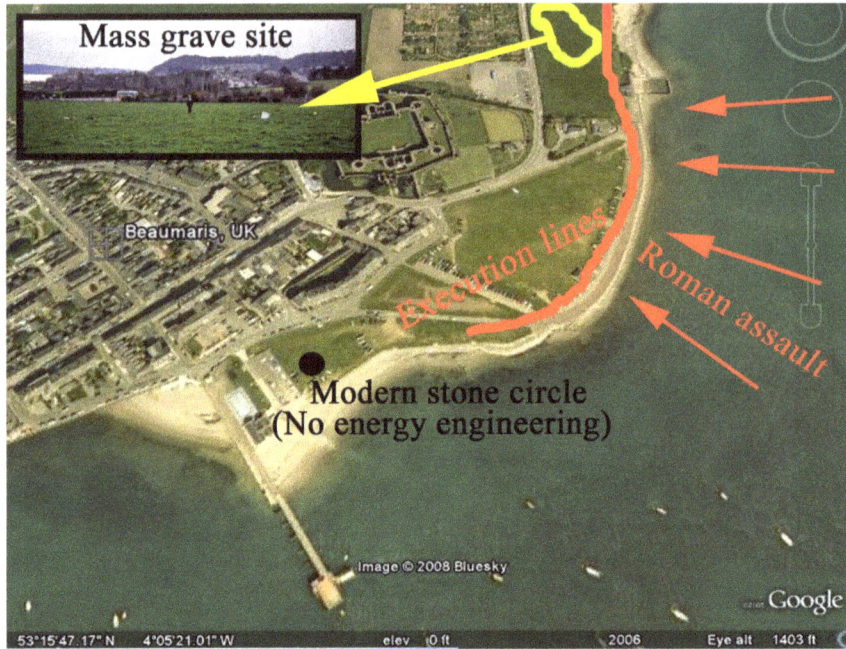

Figure 18.1 Beaumaris Town and Survey Site
The diagram of the Beaumaris area of Anglesey shows where the Romans may have landed and the approximate positions of the mass graves and execution lines.

Figure 18.2 Beaumaris Beach and Seafront
The beach in front of the car park at low water showed where the mass graves were, where the horses came ashore and where the jetties and houses may have been.

Romans were only dealing with Druids the rest of the population may have allowed them to travel freely along the road network created by the Druids. To illustrate this understanding between the Romans and the secular society it is worth noting what John Peddie says in his book 'Conquest, the Roman Invasion of Britain' (pg. 112). He says that the Romans may have maintained 130 small forts scattered over the country. Small forts are easily dealt with by the locals if they have a mind to and could have only been maintained in terms of supplies and military effectiveness re law and order with the cooperation of the local people. The local Princes or secular Chiefs could therefore be standing aside from the Roman conflict with the Druids for their own reasons and it was only when they were dragged in by Roman mismanagement such as an insult to a Queen, Boudicca, that they would turn on the Romans.

The invasion of Anglesey was therefore likely to have been a Roman v Druid affair. Because the Romans were not facing a field army of trained troops the attack on the Druids would not count as a battle for the records and so does not enter the annals of Roman military history. The event was going to be the slaughter of a Stone Age community, something even a Roman army could not be proud of and two thousand years later little would be known about the invasion. Even the landing areas are not known but educated guesses by experts tend to home in on Beaumaris. What accounts there are of the resistance on the beaches indicate that no disciplined regiments of warriors awaited the Romans, but this could be wrong.

The pull of Anglesey as an area for archaeological dowsing had been growing as it looked as if it may provide us with some answers to a number of questions. The Ordinance Survey Map of the island showed that it was full of ancient henge sites. Nigel had done some research on possible landing sites for the Romans and so armed with this information, on Wednesday the 28th December 2005, early in the morning, we set out for Anglesey. Over the Christmas period traffic is light which combined with good weather led to a speedy journey. We arrived in Beaumaris early in the afternoon (Figure 18.1) .

The weather was clear with a biting wind and by good fortune the tide was out, exposing a large expanse of foreshore. At low water it is easy to appreciate why the Romans might have selected Beaumaris for their landing. On the Beaumaris side the tide goes out well over a hundred metres exposing a firm stable beach with a gentle slope. On the far shore wide expanses of beach are also exposed bringing the two coasts close together. Along the present day promenade behind the sea wall, Beaumaris boasts a large grassed area with ample car parking spaces. The area has been reclaimed from the sea in recent times so possibly three metres of land fill are now on top of the original beach or fields. The land fill does not appear to affect any magnetic fields coming up from below so dowsing was not affected.

On arrival the plan was to first walk along the coast and identify the most plausible landing place for a legion of the Roman army. Six to ten thousand men could be involved with their horses and transport. A mile or more of coast might be required for such a landing. Looking along the present day shoreline it was clear that such a stretch of beach was available with the present day Beaumaris being about in the middle. After parking the car the 'L' rods and human blood witnesses came out for a quick 'look see' in case there were any signs of a battle. Much to our surprise patches of human blood started to turn up straight away. The patches resolved into the sporadic patches of battle blood. The idea being that their randomness indicated that they were the result of combat. Sometimes the bloodstains can be identified as either Roman blood or 'Celtic' blood but on this occasion there was not time to sort the patches out. Then again much to our surprise we identified an execution line and traced it from the Pier and lifeboat station at the south end to the north end of the Green and car park. The distance is about 600m and at least a thousand people were put to death in the 600m. One of the puzzling features about the situation was why so many prisoners were captured and tied up. Soldiers coming ashore and trying to establish a beachhead are not going to be taking prisoners and tie their hands behind their backs. This problem had to wait as there was a large expanse of foreshore to check and a battle line had not been identified. Before leaving the car park a search was made for mass graves. It is possible for a dowser to pick up mass graves at a distance, sometimes at well over a 100m. Searches from the promenade drew a blank on the car park and Green but indicated at least three mass graves on the foreshore (Figure 18.2). Once down on the beach the first of the graves was identified and its outline determined. It contained about 170 bodies or rather their chemical imprint.

As at high water the mass grave was well out to sea the question arose as to whether it was a grave or just where the bodies were burnt. A potassium witness did not pick up any sign of wood being used for burning but wool, iron, leather and urine and footwear all indicated a mass grave dug in the ground for Roman dead. It looked as if the present day foreshore had been fields at the time of the invasion. The bodies would be buried and there would have been hundreds of years for the chemicals to move deeper into the soil. The chemical stains became so deep that they have survived being inundated by the sea. The presence of about 170 dead Romans indicated that fighting had been hard. However 170 in a force of thousands is a minor loss. We did not have time to check the occupants of the other two mass graves on the stretch of beach in front of the car park. On the southern side of the pier (Figure 18.1) there were more mass graves and from one of them could be traced the smear of blood from an execution line. It is possible that some of the graves were for the executed and some for battle casualties. Before the tide came in it was necessary to try and find evidence of a landing force. The present land exposed by the tide was clearly dry land two thousand years ago so the landing must have taken place further out to sea. Even so there should be signs of fighting close to the beach and it was not long before patches of battle blood were being identified down to the waters edge. Next came the identification of horse tracks, four horses abreast, coming up from the waters edge and the direction in which the beach of 2000 years ago lay. Also found were the postholes of one roundhouse and the postholes of what could be two jetties, one higher up the beach than the other. Also found coming up the beach from the direction of the mainland were tracks of a wooden wheeled vehicle and the tracks of a bronze tyred vehicle which is likely to be a ceremonial chariot. As the afternoon wore on the cold wind was dulling the enthusiasm for dowsing. To the south could be seen the dark outline of a weather front which was moving closer and by the late afternoon flurries of snow were sweeping across the beach. The warmth of the hotel called and later that evening over a glass of wine, the afternoons dowsing was reviewed and the following days work planned.

By the following morning it was high tide so with no beach to explore it was decided to determine the extent of the execution lines. The frost and snow were disappearing in the sunshine and it promised to be a good day for dowsing. The first step was to explore the execution line to the north of the Beaumaris car park. The line ran into the old disused open air swimming pools. This area of the coast is being eroded and it is not possible to trace the lines any further. Inland and behind the old pools is a hill, the seaward side of which has been cut into by the sea. The spoil from collapsed soil and rock is on the rocky shore below the hill. It was easy to see that the hill makes an ideal defensive position. Because of this there was the possibility of finding a real battle scene as legionnaires pushed the 'Celtic' warriors up the hill and off the cliff edge at the top. After entering through the iron gate at the bottom of the hill, as the ground started to rise, patches of battle blood could be found. One cluster of patches proved to be Roman blood so the defenders put up a fight and could get the better of the legionnaires. It did not take long to find the mass graves and we looked at one of the large ones which was about halfway up the hill in a little detail. It had about 240 Druids in it and about 20 Romans. A little higher up was a small grave for 15 warriors. It was an unusual grave, the first of its type that we had found. The occupants were Druids, as judged by the linen and the use of beer and toadstool tea. They had their heads on, unlike the victims of the execution lines. They also had iron swords or at least the scabbards. The leather in the grave was tanned with tannin, unlike Roman leather. They were also deliberately set in a small grave on their own. Because of this it occurred to us that the Romans if they were doing the burying were unlikely to have conferred such an honour on Druidic warriors. The question therefore arose as to who buried them. Until this moment we had always assumed that Roman camp followers equipped with iron tools would have done the burying. It seemed to fit the pattern of the Roman dead and Druidic dead being buried in the same grave with the Roman dead having a position of privilege at the north end and having full clothing and sometimes complete with bronze items and a sword or scabbard. We were use to dowsing the walls of tunnels and ditches for traces of the tools used in their construction and could always pick up the deer antler, bronze or iron left by the tools being used. At Durrington Walls it had been possible to identify the iron tools and cotton clothing used by the modern day archaeologists. When we tested the mass graves on Anglesey we found that they were dug using antler picks. This indicated that the burial was done by the locals and they had sufficient freedom to be able to put selected warriors into their own grave.

Continuing the investigation of the hill we found an execution line running up to the top and down the other side towards the rocky beach. Once on the beach we doubled back and walked along the shore under the cliff which was formed by the hill where the fighting had taken place. A phosphate area was identified in the scree at the bottom of the cliff. Perhaps a mass grave that has come down with the erosion or maybe where bodies were thrown over the cliff edge at the time of the battle. Returning to the coast road and going north along the coast are found more execution lines. The associated mass graves were not found as it was not possible to access the fields on the island side of the road or the beach on the other due to high tide. We then retraced our steps to the car park and took a break for coffee and to discuss the next step. We decided to investigate the road and accessible areas to the south between the pier and Gallows Point. Again we found evidence of execution lines along the coast but not all the way to Gallows Point. The total length of coast checked was at least 3km, nearly 2 miles, most of it with an execution line running parallel with the coast. With 3 executions every 3m and one priest every 6m there were about 4,000 executions plus the battle casualties. Looking at the Island as a whole it may be possible to estimate how many people were executed. The Island has a coastline of about 125 miles or 200km. If Anglesey is ringed by execution lines for only 10% of its coast something like 24,000 people could have been executed round the coast alone. This is about 250 people per square mile as Anglesey has an area of 85 square miles (221 square kilometres). If this figure is added to those killed in execution lines round the temples in the interior and battle casualties it looks as if the Romans wiped out most of the islands population. Back in Hertfordshire the killing was on a similar scale but it is not easy to quantify. In Anglesey, because it is a well defined geographical area for which the population can be estimated, it is possible to quantify the scale of the killing in relation to the local populations. It could be that only a very small percentage of the population survived. The presence and use of small Roman forts on the island after the invasion indicates that not much of a population was left for the Romans to worry about.

The following morning was spent looking at an inland site. If the B5109 is followed out of Beaumaris, about one and a half miles out of town on the right is an unmade road to a farmstead called 'Cremlyn' and just beyond the farm buildings there are two standing stones, one in each of two large fields. I had looked at enough lone standing stones to know that they normally meant one thing, that is the remains of a western Dolmen. The Dolmen that was set aside for the priestesses and their cats. Why this should be I do not know. It may be that I have not looked at enough standing stones and perhaps the next series could be from the eastern Dolmen. However, one possible explanation is that the priestesses were sufficiently important to warrant a special Dolmen and as only one appears to have died at a time perhaps an extra special Dolmen could be justified. Although today there is often only one stone left standing, the 'foot print' of the other stones and the phosphate shadow of the flat stone on top which received the body are easily identified. Whether in Aberdeenshire or Devon it is always the same picture. The Ordinance Survey Maps showing the two quite separate standing stones thus indicated that Druidic Temples were also there. So on a fine late December morning we set off along one of Anglesey's narrow country roads climbing onto higher ground with low hills and valleys. It was a short journey to the unmade road leading to the stones and then a quick word with the wife of one of the farmers cleared the way for our visit. The fields opposite the farmhouse were typically Druidic with mounds and terraces. Very much like the farm in North Yorkshire where we had found the structure of a Druidic village not far from a temple. Visions of a week or more in ideal summer weather sorting the site out floated before my eyes. However, the here and now was cold wind and mud, lots of it as Nigel and I walked up to the fields where the stones stood isolated and alone in a sea of mud. The first thing we found was what appeared to be a natural rock table top running west from the path and projecting into the field then dropping several feet to the sloping ground that went down into the valley and to one of the standing stones. The 'table top' was an early ceremonial site but it did not belong to the standing stone. After identifying its Prince and Princess we walked down to the standing stone. The stone was about seven foot high, decaying and with pieces of stone around its base. The first step was to identify the footprint of the other stones and the phosphate halo and shadow. This was not quite as easy as it sounds as Nigel and I were working in thick mud. Whilst I was confirming the trappings of a Dolmen including the ceremonial route between it and its temple, Nigel was checking for signs of paint using a manganese dioxide witness. He soon found a rectangle round the Dolmen. This was not too unexpected as walls could have been painted and the paint would have been washed down into the soil. However, this did not fit as we had not picked up walls and the line was about two feet thick. It looked like a deliberate rectangular mark on the soil round the Dolmen. The next step was to walk along one of the sides to pick up any line leaving it at right angles. This process revealed lines coming out from the sides of the rectangle which quickly resolved into Thunderbirds. We had discovered that Dolmen have religious artwork round them.

As this was the first time we had found such artwork round a Dolmen, mainly because we had not looked for it, on returning to Hertfordshire we quickly checked and confirmed the presence of artwork round other western Dolmen. The coloured pigments used were also identified and (Figure 18.3) shows the artwork as it may have appeared over two thousand years ago. After confirming that the standing stone was the remains of a Dolmen the ceremonial path was followed to the temple, mainly to identify where it was. The temple had the typical 6 + 3 ring construction but was it in use when the Romans arrived? If it was, battle blood might be in and around the temple. After some searching a few patches of blood were found in the temple but most patches of blood were outside in the form of execution lines stretching round the northern sector. Outside the three lines were a few patches of battle blood as single warriors took on the Roman soldiers or some poor individual did not move fast enough. Round the temple was a debris field from which some shards of pottery were obtained. On returning to the dirt track and the first sacred site we had found we dowsed another execution line running for a short distance parallel with the track. The Romans had certainly visited the site and killed all those they found. Although we did not check for men, women and children we knew from previous sites that all were put to death. The site would have been left as a ghost town, not necessarily burnt down but perhaps allowed to rot and collapse. Finding out what the fate of the Cremlyn site was, fire or decay, will have to wait for the next visit to Anglesey.

By Friday it was time to start the journey back to St. Albans and review what had been achieved. We knew that the Romans had arrived in typical Roman fashion, not that that had

216

been in doubt. Beaumaris had probably been one of or possibly the only landing site. There had been fighting at the landing site and the Romans had sustained casualties. It would be possible, given time, to check all the mass graves and determine Roman casualties. Druidic casualties might be more difficult as they could have been thrown into the sea. The local population had been rounded up and executed. Some of them had dug the graves for those executed. The fighting had extended up the hill behind the old swimming pool. Amongst the Druids were warriors with iron weapons. We should have looked for bronze weapons but did not think of it at the time. These warriors received a special burial. We now knew that on a stable beach the stains remain in place and can still be dowsed. This promises to open up a whole new land area for archaeological study. The western Dolmen has artwork round it which is based on the winter aura of the oak and probably symbolises death. Perhaps of most importance, Anglesey probably offers the best location for the study of the Roman attack on a section of the indigenous population of the British Isles. Also for a study of the aftermath of such a meeting between Rome and a less technically developed people. Anglesey must have had most of its population killed within a period of a few days. The landscape would have reverted to scrub and forest and then in due course a new population would have moved in. When did they move in, who were they and where did they come from? Did the Romans keep the island free of any new people trying to move in? Any population pressure there might have been on the land in mainland Wales was of course removed by the Romans in the way it had been removed in Anglesey so there would have been little need for people from the mainland to resettle the island.

Anglesey contains many treasures waiting to be discovered including possibly gold, silver and bronze artefacts hidden from the Romans, the Anglesey Druidic Hoard!

Figure 18.3 The Cremlyn Dolmen
This single standing stone in a field at Cremlyn Farm is the remaining one of a pair. The other stone is missing but its magnetic stain still remains. The stones were part of a dolmen and had four thunderbirds round it based on a black rectangle encompassing the stones.

217

Glastonbury Tor

The visit to Anglesey provided evidence that the Roman attack on the Island was not just an attack on a main centre of Druidry with the idea that once destroyed the rest of Druidry would collapse. The attack was part of a programme of genocide covering the whole of Britain. The visit also had an archaeological output in that it identified at least one place, possibly the only place, where the Romans landed on Anglesey.

The Anglesey invasion may demonstrate the force and effort the Romans were prepared to bring to bear on any perceived resistance to their authority. Perhaps it also demonstrates the vigour and determination with which the Romans were pursuing their genocide policy. The Druids were not being pursued to break their power; they were being pursed with extermination in mind. There was something about the Druids that the Romans did not understand or feared in some way.

Another centre of Druidic power was what is now known as Glastonbury Tor and the surrounding countryside (Figure 18.4).

Figure 18.4 Glastonbury Tor   *Aerial photo by JJ Evendon*
The main survey sites including the terracing are shown.

Glastonbury and its famous Tor is situated in Somerset and is within easy reach of London along the A303 and B3151 and from the Bath, Bristol and Wells direction along the A39.

Glastonbury is on high ground in what use to be a wet and marshy area called the Somerset Levels. The Tor was one of a number of islands rising from the marshland. Glastonbury Tor is a 500ft high natural hill which has attracted religious activity over a period of at least 5000 years. In 1993 Nicholas R Mann wrote and published a book called "Glastonbury Tor" 'A guide to the history and legends'. To help set the background as to why the Tor is important to a wide range of people including archaeological dowsers I quote from his book -
"...there is evidence of enormous labour upon the Tor's slopes, that if done solely for agriculture was easier done elsewhere, and if for some ceremonial purpose – a labyrinth? A processional way? – was likely done in a time so remote that the significance is hard for us to understand or even imagine".

The visual evidence of enormous labour and the association of the Tor with myth, legend and early Christianity make the Tor and the tower of St. Michaels' Church which caps it a place of pilgrimage. It is a sacred spiritual site in a way that few other places are.

The book by Nicholas Mann contains details of the geography and history of the Tor and has a chapter on the terraces round the Tor. It was these terraces and the way the Romans treated the place and its people that attracted us to Glastonbury on the 18th February 2006.

According to the model of Druidic ritual that we were working with the terraces should be ceremonial roads built by the Druids. The official view is that the terraces round the Tor are likely to be Medieval strip-lynchets built for agricultural reasons. Terracing of hillsides is very common in Britain, for example they can be seen clearly on hillsides when driving along the M4 between Swindon and the River Seven. Nigel had also found them in Wales on the higher slopes of hills near Pontypridd and minor terracing can be seen in the flatter countryside of Hertfordshire. The Tor has attracted a lot of archaeological work over the years, work that has perhaps produced more conjecture than facts. However, importantly there are written records of what the Tor has been used for since Roman times which supplement archaeological findings. So our working hypothesis for the visit was that the Tor is a Druidic Temple with a spiral ceremonial chariot road climbing up to the temple complex on the top. From the literature there appears to be a case for the terracing to be seen as a labyrinth.

However it did not appear to be a very convincing case. We were confident that the roads had not been destroyed by archaeologists but we knew a lot of digging had taken place at the top by two groups of people. The first group were builders preparing foundations for buildings such as towers and churches. The second group were archaeologists trying to find the foundations of buildings such as towers and churches. It was possible, and indeed what we were hoping for, that after the destruction of the Druidic Temple there was enough time for chemical stains to be taken deep into the soil and be protected from both builders and archaeologists.

From the car park in the town we walked up and along the north side of the Tor so that we could approach it from the north east. From the road running past the back of the Tor there is a public footpath which crosses a field to concrete steps that take a winding route up the steep eastern end of the Tor. We climbed the steps as far as the second terrace which looked a good place to start our study. Then walking about 30m away from the steps we prepared to solve one of the mysteries of the Tor (Figure 18.5). I did the dowsing and Nigel took the notes. The first check was for bronze. I found two bronze tyred wheel tracks about 7 to 8 ft. apart, the tracks were about 1 foot wide. There was a positive response on both arsenic and tin which indicated that the terrace was in use in the early and late Bronze Age. There were no iron wheel tracks which is a common finding on ceremonial roads. Iron tyred wheels are used in ceremonial vehicles but they may also indicate commercial vehicles which are drawn by oxen. Two horses were pulling the chariots. The next step was to find the edges of the road. Just like today's engineers the Druids put a kerb or edging along their roads and this would have been important for the stability of the outer edge of a road which was on a sloping hillside. Using a witness for Druidic cement a kerb was found along the outer edge of the road. The road surface gave a positive response on chalk mortar indicating that it may have been metalled in some way, possibly gravel. The presence of gravel has not been reported by archaeologists so a compacted chalk surface may have been used. The bronze tracks are well defined and can easily be identified by a dowser using a bronze coin as a witness. There was no doubt that the terracing had been built to take a ceremonial road and that it was in use during the bronze Age. If this was so then the diamond marks should be along the centre of the road. It did not take long to find a row of diamonds the outer edges of which had been marked our using red ochre. The next question was 'when was the road built?'

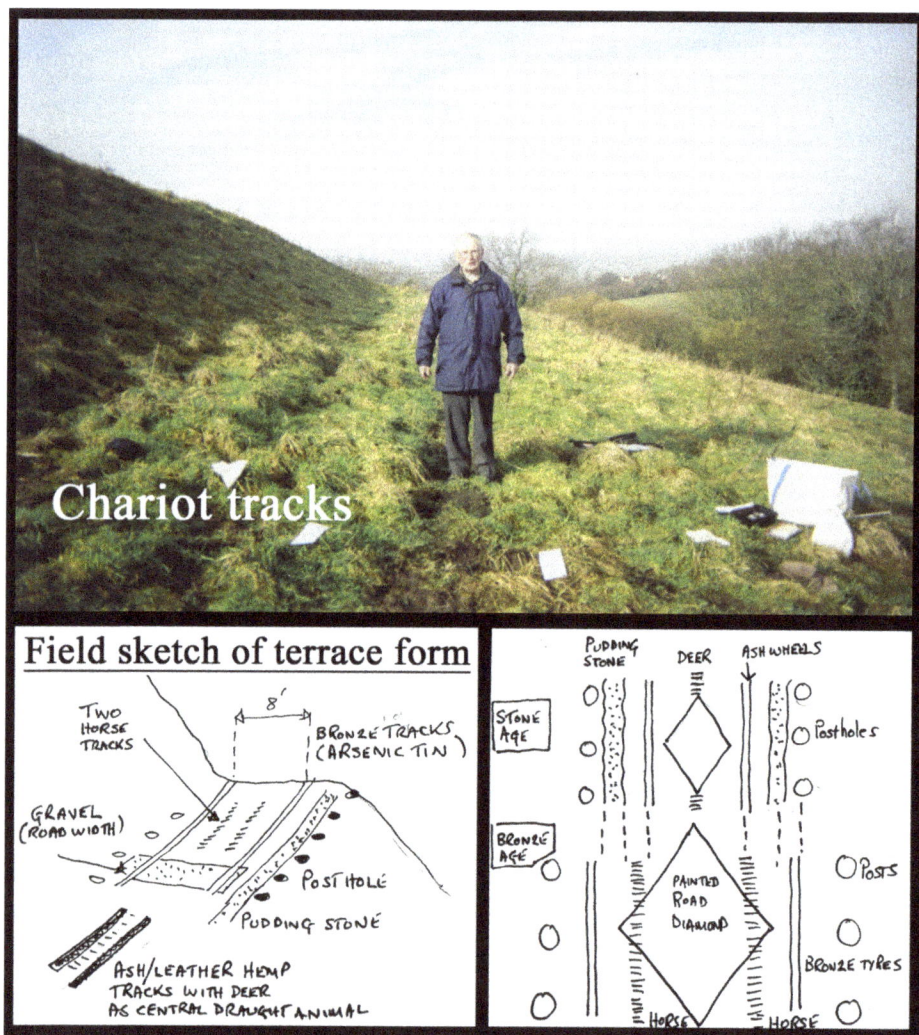

Figure 18.5 Tor Road Terracing

On the terrace it was possible to identify the diamonds and the chariot wheel tracks. The outer edges of the terrace (road) were hardened.

220

We had not been able to answer that question when studying a ceremonial road at Maiden Castle the previous day. However the first thing was to see if wooden wheeled vehicles had used the road. We soon identified wheel tracks four feet apart using ash wood, hemp, leather and urine witnesses. The cart was pulled by a deer and on either side of the cart were the footpaths of Druids with their ash staffs. There had been flint kerbs on either side of the road which had been about seven to eight feet wide. We often found evidence of these small deer drawn carts on ceremonial roads, for example at Durrington Walls and other sites. There was evidence that they were being used into the Iron Age so they were not proof that a road was being used in the Stone Age. Then I came up with the idea that if the small vehicles were Stone Age and the ceremonial road dated from that time the diamonds should be smaller and perhaps in a different place, that is out of phase with the large ones. There were indeed small diamonds that could be out of phase. However, this was not good evidence. The diamonds are multilayered like an onion and perhaps I was picking up the central area where the outer layers had been lost to erosion. Perhaps the location of the diamonds changed over time and there was some evidence for that. However, we did not feel that the presence of small diamonds took us that much further forward. Then Nigel came up with what could be the clincher. We knew that there were postholes along the outside of ceremonial roads and in some cases there was a wattle wall. All we have had to do, said Nigel, is to identify any postholes that might be associated with the wood wheeled vehicle tracks. If the deer and wood wheeled cart belong to the Stone Age it will have its own postholes. It did not take long to find them and they were more or less where the bronze wheel tracks were. The posts and the chariots could not have been there at the same time. This is good evidence that the terracing on the Tor was first constructed in the Stone Age. Not conclusive evidence as religious systems tend to be conservative and the Druids may not have moved to bronze until the Bronze Age was well advanced. However, we now knew the answer to the mystery of the terracing. The terraces are ceremonial roads. We did not have time to determine if they were all enclosed and roofed over but they were probably first established in the Stone Age. This means that the surveying, engineering logistic and mathematical skills were available at that time. I say engineering skills deliberately; do not imagine for one moment that the Druids moved earth and rock only using buckets and shovels.

So the terracing of the Tor may have started about four to five thousand years ago. When did the use of the ceremonial roads end? Looking back down from the terrace to the field far below I said to Nigel 'We forgot to check for battle blood and execution lines before we came up', then I added that no battle was likely on these steep slopes so we can give it a miss. 'Better not' said Nigel 'remember what the Romans did on Anglesey. There could be a ring of execution lines round the Tor'. There was no answer to that logic so with that we started the climb down. On reaching the field I took the top and Nigel the lower part which was nearer the road. There was some scattered stains of battle blood but the main finding was four sets of execution lines each line consisting of three ranks. It was clear that a very large number of people had been executed on the north east approach to the Tor. They had been lined up, as elsewhere, for an audience to witness the executions. We had never come across four sets of execution lines, one behind the other. As the ground was rising the Romans could have started on the front ranks first and worked their way back. Those in the rear ranks would have seen the executioners gradually approaching them as they dispatched line after line of their friends and neighbours. Two thousand years later tourists walk through the field quite oblivious of what happened there in the past. I often wonder how many of those people who claim to feel the spirit and mystery of a place like the Tor do pick up the main events that happened there. However, we now knew that the Druidic complex on and round the Tor was a very big one in terms of both numbers of Druids and the area covered and that it came to an end sometime between 43AD and 60AD.

221

If the Tor had been a Druidic Temple then at the top should be the tomb of a Prince and Princess so with a few rests on the way to check terraces we eventually made it to the plateau and St. Michael's Tower. After two perhaps three thousand years as A Druidic Temple site the Christians are said to have adopted it somewhere about 500AD. Anything that was pagan would have been removed or used to construct new buildings. The building and rebuilding activity of the Christian era and the digging of the archaeological era will have destroyed much of the 'memory' of the sacred area at the top of the Tor but it should not have affected the tomb complex. The chambers may have collapsed under the weight of buildings above them but it should still be possible to pick them up. At least that was our hope.

Whilst Nigel looked for altars I looked for the Prince and Princess. I found them under the tower but to one side of the centre. The bronze was copper tin, that is late Bronze Age; the Princess was dressed in a gold jacket. Having found one tomb, there could be more, I looked for signs of the clay pool. Using red clay as a witness I found a 40ft diameter pool with the Tower in the centre. This is an incredibly large pool to have on the top of a hill. The Druids were water engineers but filling a pool that size when it is on the top of a hill would present problems. In addition, a pool at the top of a hill is likely to have a different ceremonial role to one down below on the surrounding flat ground. As the Tor was almost certainly seen by the Druids as representing part of the Earth Goddesses anatomy it was not difficult to identify what offering was made. As with the underground shrines to the Earth Goddess it was likely that milk had been used as an offering. This proved to be the case, and it filled the 40ft diameter pool. At this point it may be as well to remember what the dowser is responding to. It is not an actual pool but the stain left in the ground by what was there at a time in the past. A small clay pool may have been at the top of the Tor but with all the construction and archaeological activity the clay may have been spread out over a wider and wider area with its load of adsorbed chemicals including the milk.

Altars were found round the central pool. One was for sheep another for deer a third was for human sacrifice. The postholes could be traced out but only as far as the slope of the land would allow. It looked as if the Druids had modified the circular design of the temple to take account of the topography of the hill. Chariot tracks came up to a flat area which is just outside the temple on the south west side. The surprising thing was that so much was still dowsable. The evidence we were gathering indicated that there is a lot of information about the early and Druidic phases of occupancy of the Tor still locked in the Tor's magnetic memory banks (Figure 18.6).

We descended the Tor by following the path to the south west. On the way we found evidence of oxen being used on the slopes. It is not a good idea to use oxen to pull carts up steep slopes and the Druids would have known that. The Druids were however experienced in using oxen to operate lifts such as drawing water from a well, hauling earth and rock up slopes or in this case supplies and building materials to the temple. We soon found one suitable site for a lift which looked as if it had been in use during the Bronze Age, perhaps earlier as rope based on nettle fibre had been used at one stage.

Nearing the bottom of the hill there is a field (Fair field) which is to the right of the path and it slopes up to the first terrace. There are execution lines running parallel with the first terrace and in front of them a thirty foot wide band of battle blood.

After half a days work it was clear that the Tor was a major Druidic ceremonial site that had been in use for 2000 years if not much longer. On and around the Tor will be the history, the social and technical development of the Druids and of the Henge people. This information is now accessible using biolocation techniques based on a scientific approach.

On the way back to Hertfordshire from Glastonbury it seemed as if every hill showed examples of terraced ceremonial roads. How they were used, why they are where they are, what they were used for is a story that is slowly being revealed. The ceremonial roads may fall into two groups, the permanent ones and the one offs to bless or sanctify an area of land. The latter group is proposed because even given 3000 years there are an awful lot of terraced roads and roads do not make the best agricultural land. Perhaps a terrace is to be preferred to a dyke but why does Maiden Castle have dykes and Glastonbury Tor terracing? Why seven terraces and why three dykes? To answer questions like this, one must understand the religion and beliefs of the Druids, and that means understanding their temples, both those above and below the ground. The best known one above ground is Stonehenge. Understanding what and how Stonehenge works would be a step forward.

Figure 18.6 Tor Temple

On the Tor summit a temple was identified with a Bronze Age Prince and Princess under the tower. There was a central pool. The temple wall extended to the edge of the summit plateau on the east. Ceremonial chariots came up the terraced ceremonial roads of the Tor and they finished their journey at a turning circle.

Chapter 19

The Power of an Ancient Landscape

Stonehenge: The Trilithon Powerhouse

At first sight the design of the stonework in Stonehenge does not appear to fit easily with the concept of energy engineering (Chapter 2). For example, if the sarsen trilithons were energy amplifiers how did the magnetic fields they generated pass through the linteled circle of sarsens? The stones in the circle would be expected to act as witnesses and block the field from the sarsen horseshoe and cast an energy shadow. That is unless the witness effect of the stones in the linteled circle was neutralised in some way so that the field could pass through them. Did the Druids know enough to do this?

The first step was to see if the trilithon design resulted in an amplification of the field from the stones of which it was made. House bricks were used for the experiment. Bricks may appear to be an odd choice for sarsens but it is the magnetic fields that are important not the material from which the object is made. Two bricks were stood on the lawn with a north and south pole of the same magnetic axis facing each other and with the diamagnetic north at the top. In other words two bricks were set up as if they were going to be part of a stone circle. A third brick was then laid across the two bricks as a lintel. Using the length of the brick as the axis round which to rotate it there are eight different ways in which the lintel brick can be placed on the two uprights. However, before placing the lintel on the two uprights the size of the paramagnetic field from first one brick then two bricks alone was measured by walking in towards them and waiting for the rods to cross. Once this had been done a number of times to be certain of the field's size, the lintel brick was placed on the two upright bricks. The extent of the field from the bricks was again measured and measurements repeated as the position of the lintel brick was changed through its eight positions. The effect of the lintel was either to suppress the field, make no difference or, in one position to magnify the field from the two upright bricks, see figure 19.1.

The reason for a lintel stone was now clear. It was a field amplifier and the brick experiment was saying that there was probably only one way to place the lintel on the two uprights to achieve amplification, at least with the bricks and on the side of the trilithon I was using. The next step was to make four more trilithons and construct a horseshoe from them. The reason the Druids had used the horseshoe design was almost certainly to focus the energy in some way. The obvious thing to do was therefore to dowse across the open end of the horseshoe looking for an energy beam. The five trilithons were set up magnetically as if they were part of a stone circle but physically in the shape of a horseshoe and then tested. There was no field coming out of the cavity. This was not a set back as I already had an idea of how the stones should be arranged. The next step was to reverse the position of the 3rd trilithon (Figure 19.2).

This change produced a strong field emerging from the cavity. The field was tested for colour and it gave a positive response with blue and with an aluminium witness.

It was now clear that the horseshoe cavity, provided it is set up correctly, is an amplification system for the blue line, the line which according to mythology is used for communication. It is also said that spirits move along it. When I placed a brick in the beam the beam was blocked, as would be expected. Using bricks I was therefore able to show that provided the stones of the horseshoe were set up in a particular way the Druids may have constructed a field

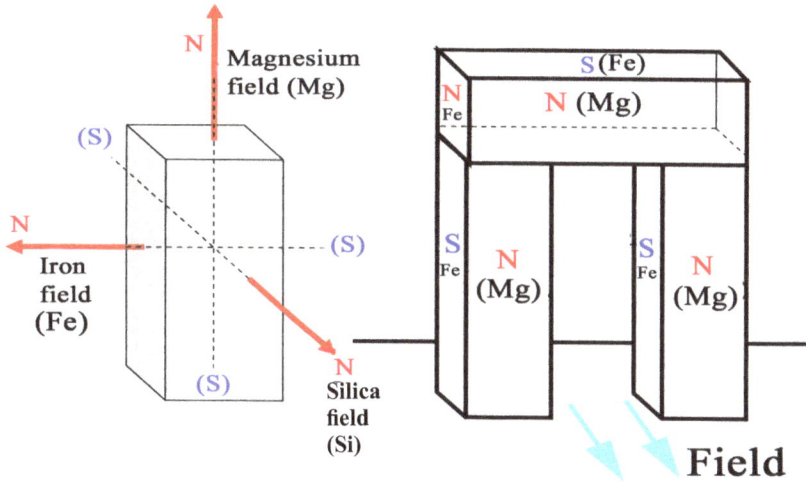

Figure 19.1 Linteled Bricks

When a lintel brick is correctly placed across two standing bricks a field emerges as shown from one face of the trilithon. The field appears as a line going off into the distance. The bricks or stones used will determine the composition of the fields. In this case the bricks used for the experiments produced a blue line containing fields due to aluminium, iron and silica.

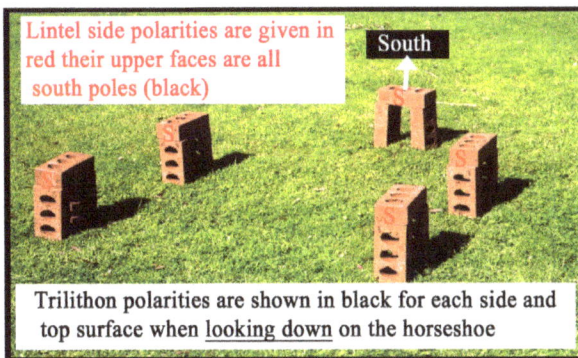

Figure 19.2 Trilithon Horseshoe

A model of the trilithon horseshoe was made using bricks.

The uprights of the trilithons were assembled as if they were part of a stone circle with the north of one stone facing the south of the next.

The lintel stones were placed so that a field was projected from the north inner face.

Set up with all trilithons facing in the system does not work.

If the middle trilithon is reversed the enclosed space acts as a resonant chamber and a field emerges from the open end of the horseshoe.

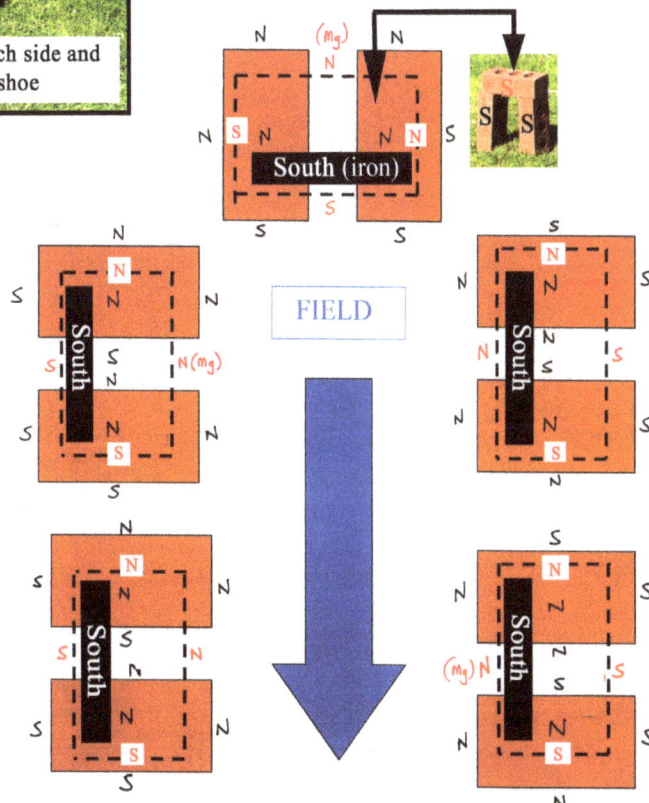

amplifier. If they had, the linteled circle also had to be properly constructed to allow the blue field through. The next step was therefore to construct a linteled circle or at least a section of it from bricks. The upright bricks alone as was expected, blocked the beam from the horseshoe but by placing a lintel brick on the uprights and rotating the lintel bricks about their long axis a position was found which removed the field of the brick circle and allowed the beam to pass through (Figure 19.3).

Did the Druids know how to do this? We would know when we next visited Stonehenge. Then finally the Heel Stone, if set at the right angle it would allow the beam to pass down the Avenue and at the same time split the beam to either side. It did not take long to find the angle at which a brick would split the beam from the horseshoe (Figure 19.4).

Based on experiments with bricks on the lawn it was now clear that a magnetic field amplifier could be made and a linteled circle designed to allow a field to pass through it. The question was did the Druids know this and if they did would we be able to show that they did. Stonehenge is now in such a state of disrepair that it seemed unlikely that traces of the energy engineering in terms of blue lines could be picked up. However, just in case, a model of what remains of the horseshoe and linteled circle was set up in the garden (Figure 19.5).

The model of the horseshoe was based on two uprights and two lintels missing. For the circle there were lone uprights on either side of four linteled bricks. When dowsed, a blue line could still be identified. This was evidence that there could be some residual activity from a damaged horseshoe and meant that we had to visit Stonehenge. A visit was organised for the 5th April 2006. Unfortunately I caught flue a few days before and the visit had to be postponed until the 16th April 2006. A long wait when you are wondering if your predictions are right or not. At last the 16th arrived and Nigel and I set off not knowing what we would find but with a quiet bet that the Druids did know what they were doing. And if they did! What did they do with such a powerhouse? We stopped at a roadside café about a mile from Stonehenge for a quick lunch and a review of our action plan, a list of all the things we wanted to check. We knew that if the Druids had built Stonehenge on energy engineering design principles the Heel Stone should be a beam splitter and provided it had not been moved by more than a few degrees it should still be splitting any beam coming from the sarsen horseshoe. There could therefore be a beam in an easterly direction which we would not be able to access as it would be inside the perimeter fence. Then a beam down the Avenue and a third beam crossing the A344 and heading towards the car park and the western end of the Cursus. We finished lunch and headed for the car park at Stonehenge. Having paid our £2 we found a parking spot and with our dowsing gear stepped across the temporary fence of the car park and were on our way towards the Avenue and Heel Stone. Once clear of the car park there was enough space to look for the blue line that might or might not be coming from the Heel Stone. Walking so as to cross the line at 90° and with the colour wheel at blue series 4, I started searching with Nigel watching closely for any signs of the rods crossing. After what appeared to be ages the rods crossed. I had found one edge of a 20ft wide blue line. This made it look as if the Druids almost certainly built Stonehenge on energy engineering principles. Confirmation would however have to wait until the following morning when we had permission for a two hour session with the stones. Although we had found the blue line it was not possible at this stage to say that it had its origin in the sarsen horseshoe as many stones have their own intrinsic blue fields. This was the case with the blue stone at the Westport site, Killadangan (Chapter 2). We continued walking towards the Avenue and on reaching it checked to see if there was a blue line going down the Avenue, there was. This energy line was heading off in the direction of Woodhenge and Durrington Walls. Woodhenge is about one and a half miles from Stonehenge as the crow flies. So far so good, it looked as if the Heel Stone was splitting a beam (Figure 19.6).

Figure 19.3 The Lintelled Stone Circle

The lintelled circle of stones is important to the energy engineering. If any of the stones are misaligned there is a chance that the system will not work. The figure shows the model of the linteled circle set up in the garden. There is at least one other arrangement of stones that will allow the field through but the one shown is close to that used by the Druids at Stonehenge.

Figure 19.4 The Heel Stone Beam Splitter

The model of the trilithons was extended to include the Heel Stone as a beam splitter.

The blue line is indicated by the canes. It first penetrates the linteled circle then splits at the Heel Stone so that three blue lines are created.

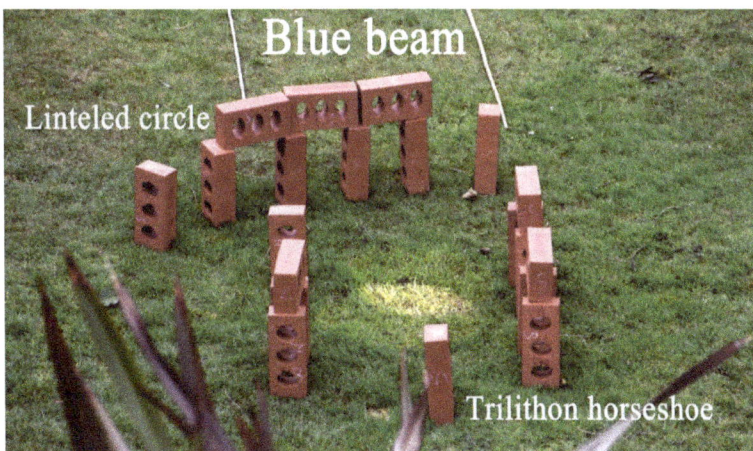

Figure 19.5 The Model of Present Day Stonehenge

The trilithon horseshoe was recreated modelled on the present standing stones.

The trilithons still produced a blue line that penetrated what remains of the linteled circle.

227

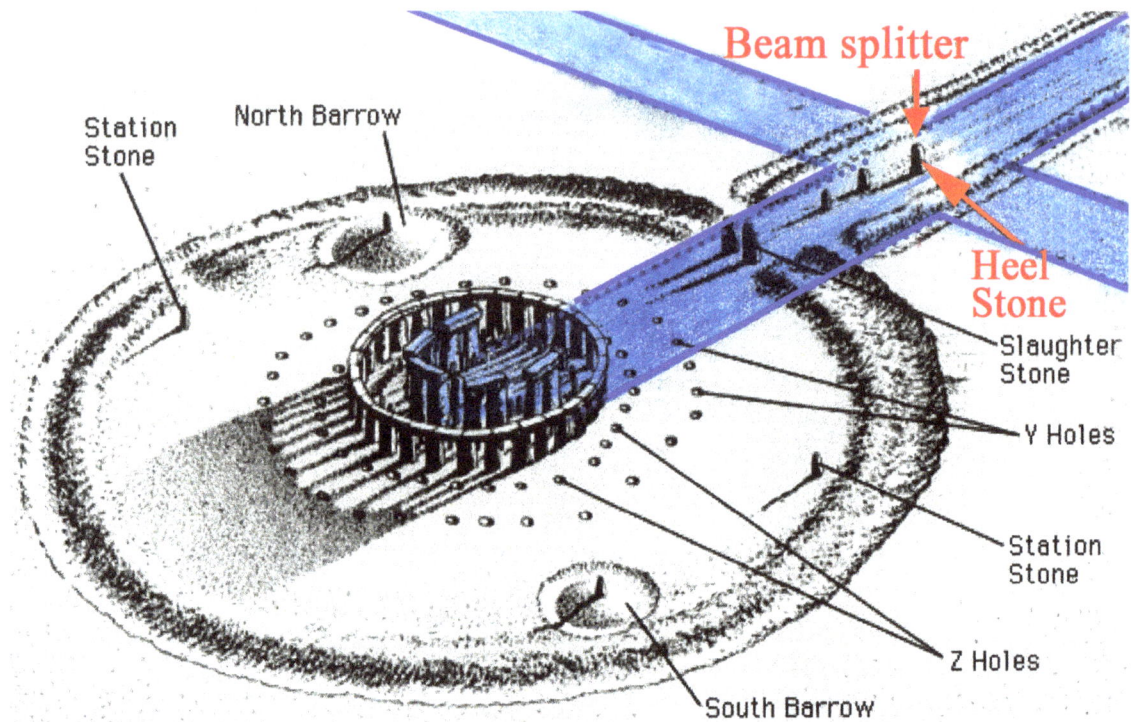

Figure 19.6 Stonehenge Blue Beam Generation

The diagram shows the position of the blue beam system generated by the trilithon horseshoe and the Heel Stone splitting it.

On previous visits to Stonehenge we had not checked for signs of Dolmen round the henge so on this occasion we looked for the northern one as we were in the right position relative to the henge. We found several small Dolmen able to take one or two bodies at about 48m from the fence. Then a little further away from the fence, 75m, we found a large one able to accommodate at least eight bodies at a time. This confirmed that Stonehenge, at least for part of its life, was a working temple complete with Dolmen.

Cursi 1

The next item on our agenda that afternoon was the Cursus. To the archaeologists Cursi are enigmatic Neolithic monuments. The book by Alistair Barclay et al 'Lines in the Landscape' gives an insight to the archaeologists approach to Cursi, their methodologies and interpretations. Why archaeologists refer to them as monuments is not clear. To reach the Cursus we had to walk back past the car parks to the unmade dirt road running from the A344, across the Cursus towards a Military Camp in the distance. We checked the road when we were about 30m from the Cursus ditch and found that it was a Druidic road. There were two sets of bronze wheel tracks with an axel length of about 2m (7ft). Between them was a gap with dog hair and tracks from wood wheeled vehicles pulled by deer. The bronze was both arsenic and tin bronze indicating early and late Bronze Age. There were also iron wheel tracks with axle lengths of about 1.2m. These iron tracks went straight across the Cursus whilst the bronze tracks curved and entered the ditch or dyke which encircles the Cursus. The term ditch is not a good one but it is commonly used. The so called ditch is a road designed to take chariots and possibly other vehicles and foot traffic. Dyke is therefore the preferred term. The iron tyred vehicles crossing the Cursus almost certainly postdate the active use of the Cursus.

228

One of the reasons for visiting the Cursus was to see if it might have been the site for the stylized or ceremonial battles that we had predicted. A working hypothesis had been developed to explain the absence of fortified buildings and the population's lack of military skills. The hypothesis was that disputes were settled by battles fought to a strict set of rules. If we were right one of the places to look for the ceremonial battle fields was a Cursus. The Cursus was and even today is a large flat area. In the Henge Age it was used for dealing with cattle, deer and probably other animals coming in from the outlying farms and ranches. We already knew a certain amount about the Stonehenge Cursus (Chapter 6, 9 and 15), the drover's routes coming in or going out at the western end, the fencing, animal pens and roadway round it. If there was a ritual battle site on the Cursus it could be anywhere. We therefore used the dirt road crossing the Cursus as a dividing line and Nigel took the west section and I the east section. Much to my surprise I was soon picking up the stains of blood within about 15m of the road. Exploring further east I found more bloodstains and then they came to an end. Some checks with toadstool and garlic witnesses indicated that the stains were from Druidic blood. There was no trace of Roman blood. On his side of the road Nigel had not found blood on the Cursus but had found execution lines on the outside of the sunken roadway or dyke. At one point the execution blocks were six ranks deep then a space and another block in front, again six ranks deep. It looked as if people had been brought out of the city and executed possibly in their thousands along the length of the Cursus. There were execution lines on the north side of the Cursus and on the south side. The ones Nigel had identified extended to the west as far as the Cursus Barrows. We did not follow them to the eastern end of the Cursus. A Cursus is an agricultural, social, commercial and sporting facility with probably other uses as well including religious functions. Its size possibly reflects the size of the city or community it serves as well as having to provide the minimum dimensions for functions. A bit like a sports field. Walking west along the cursus we found occasional indications of battle blood until the line of three fenced barrows on the south side was reached. There is also another barrow so small that I had missed it and it was Nigel who picked up the shallow ditch and slight rise in the ground. The ditch gave a positive response on a human blood witness. It looked as if there had been so much blood available that the whole ditch was stained. There were also cart tracks in the ditch. Walking round the barrow on the inside of the ditch, seven drainage gullies were found, each one from an altar. That was the last observation we had time for as the car park closed at 6pm. We also wanted to visit Woodhenge before it was dark. Further investigation of the Cursus was left until the following day. Once we reached Woodhenge we soon found a blue energy line running through the site but we were not able to say that it came from Stonehenge. However, the main thing we were after was evidence that the Woodhenge site was in use when the Romans arrived. This evidence was provided when we found execution lines round the temple and near the car park. There seemed to be no doubt that the site was occupied and being used by Druids at the time of the Roman invasion.

That evening we retired to the Wheatsheaf at Lower Woodford, a village not far from Stonehenge, for a meal, a glass of wine and to discuss the day's events and plans for the following day. Prior to our visit to Stonehenge I had constructed a model in my garden of the present broken down sarsen horseshoe and the sarsen linteled circle. I had found that the horseshoe produced a blue line that penetrated the linteled circle and that the line could be split by a brick (a beam splitter) placed where the Heel Stone stood The linteled stones had to be correctly set up magnetically otherwise the field was blocked. However, I had two worries. One was that the fallen stones might neutralise the fields or if they did not, that the field could be so strong because of the size of the stones that it would saturate the sensory system and blind the dowser. This could make it impossible to work on the stones.

The following morning at 06.45 we were once more amongst possibly the most famous stones in the world and hoping that by the time our 2 hours were up we would know if the henge was designed as a power house. We started work on the fallen lintel stone and fallen upright of the trilithon at the base of the horseshoe. We were measuring the polarity of each of the six faces of the stones that is whether a face was a north or a south pole. The fallen lintel stone has the mortise recesses on what would have been the underside of the stone. On the other and upper side a mortise recesses had been started then stopped. Why? My bet is that the Druids had identified the wrong side of the stone as a magnetic north. The lintel would not work if incorrectly positioned magnetically so they had to start again. The magnetic poles of the stones are shown in (Figure 19.7).

Figure 19.7 The Polarity of the Stones

Polarity of the stone's surfaces was determined using a long cane and string. When the string comes into contact with the stone surface the string adopts the polarity of that surface so it can be dowsed on the ground.The results showed that the stones were deliberately positioned with an ordered polarity to create a magnetic field amplification system. This order of polarity is unlikely to have been achieved by chance.

The results we obtained indicate that the Druids positioned the trilithon stones magnetically so that they would generate a large and powerful focused field heading in the direction of Woodhenge and Durrington Walls. The linteled circle also had its stones magnetically positioned so that the field could penetrate them. I tried to identify the blue field within the linteled circle but could not. It was not until I was nearing the Heel Stone that I was able to pick it up. This indicates that the field from the damaged sarsen horseshoe is still very powerful. In fact the field is sufficiently powerful to saturate the magnetic sensory system. If the horseshoe were intact the field would probably be several times stronger still.

By the end of our two hours we had shown that the sarsen horseshoe was a magnetic field amplifier, but why build such a powerful device? Why split the beam? Why send it onto Durrington Walls and Woodhenge? We would come back to these questions but there were two other things to look at. The first was the ditch. There had to be a reason for a ditch round Stonehenge and the most likely was that it was not a ditch but a dyke that was now mostly filled in. We already knew that there were chariot tracks in the ditch or dyke so it was a matter of confirming it and this is what we were able to do. Bronze and iron tracks and ceramic torch stands were in the ditch so, as we had thought, we were dealing with a dyke and not a ditch. The dyke probably links up to a larger dyke system or to ceremonial roads. Whilst I was looking at the dyke Nigel was looking for signs of his latest discovery, the heptagon tunnel system that runs round the outside of the tomb complex. He found signs that it might be present as part of the Stonehenge underground complex. We then moved onto the eastern barrow which is down by the road junction. It is a ceremonial barrow with seven altars for human sacrifice. There were seven drainage gullies for the blood and the ditch round it was an almost complete human blood stain. The use of the barrow for human sacrifice does not preclude its use before or after a period of human sacrifice for some other purpose. However, we did not pick up any sign of other uses. We had come across seven raised altars for human sacrifice before. Where it was possible to trace the removal of the bodies from the raised altar barrow they could be followed to a Dolmen which was used for those who fought the ceremonial battles. The raised altars at Stonehenge indicates that there is an arena for battles close by. The South Barrow was used for animal sacrifice and we did not pick up any sign of human blood.

After identifying that the sarsen horseshoe was a magnetic field amplifier and that at one stage Stonehenge had had a dyke associated with the temple we drove back to our B & B for breakfast and a rest before returning to the Cursus. Cursi are widespread in Britain and as already mentioned those in the Upper Thames Valley have been described in a book by Alistair Barclay and co-authors. A Cursus is often referred to as a monument and dated to the Neolithic and early Bronze Age. If a Cursus is a monument it would need a lot of justification in any age because of the large amount of land that it occupies. A much more logical explanation is that a Cursus is part of the working fabric of a town or city and was used on an almost daily basis for a wide range of activities. This was our view as a result of earlier visits to the Cursus and on the previous afternoon we had picked up information backing this view. By early morning we were back on the Cursus with the first task to sort out the battle blood that we had found. The main area of battle blood was in a restricted area with well defined boundaries (Figure 19.8).

The bloodstains were checked with toadstool for 'Celtic' blood and garlic for Roman blood and were found to be 'Celtic' with no Roman blood. The area where the blood was found either had a wall round it or it was defined by markings in some way. Flint and mortar witnesses indicated a wall. Iron tyred vehicles also came up to the eastern side and then swung away as if there was a barrier of some sort. We did not have time to study the area in detail but it would be nice to know where the gates are, did teams of warriors march up to the gates, how were the dead taken away, are there any signs of metal weapons? Were there stands for an audience? The size of the arena indicated that it could take from two combatants up to possibly two rows of 30 warriors facing each other. On the previous day I had found more patches of blood lying outside the rectangle and closer to the dirt road. This time I checked the boundaries and again the blood was in a defined walled area. This time the area was somewhat smaller at about 15 to 17m square and it was separated from the larger arena by a ceremonial road complete with diamonds and bronze tyre marks. Just in case the small arena was for women warriors the blood was checked and found to be that of women. Some more evidence will be required but it looks as if the 'Celtic' women could be warriors as well as the men. We now had the first clear evidence that ceremonial battles were fought on the Cursus. What the role of the battles were remains to be discovered and many more Cursi will need to be checked to see if battles took place on them. To study the arena at Stonehenge in detail could easily take a day or more of dowsing time which we did not have so we had to move on. Walking down the Cursus towards the western end we picked up patches of bloodstains. One area contained both Roman and 'Celtic' blood, another Roman blood and dog hair. The Roman and 'Celtic' bloodstains appeared to be in an enclosure as if there had been a staged fight

231

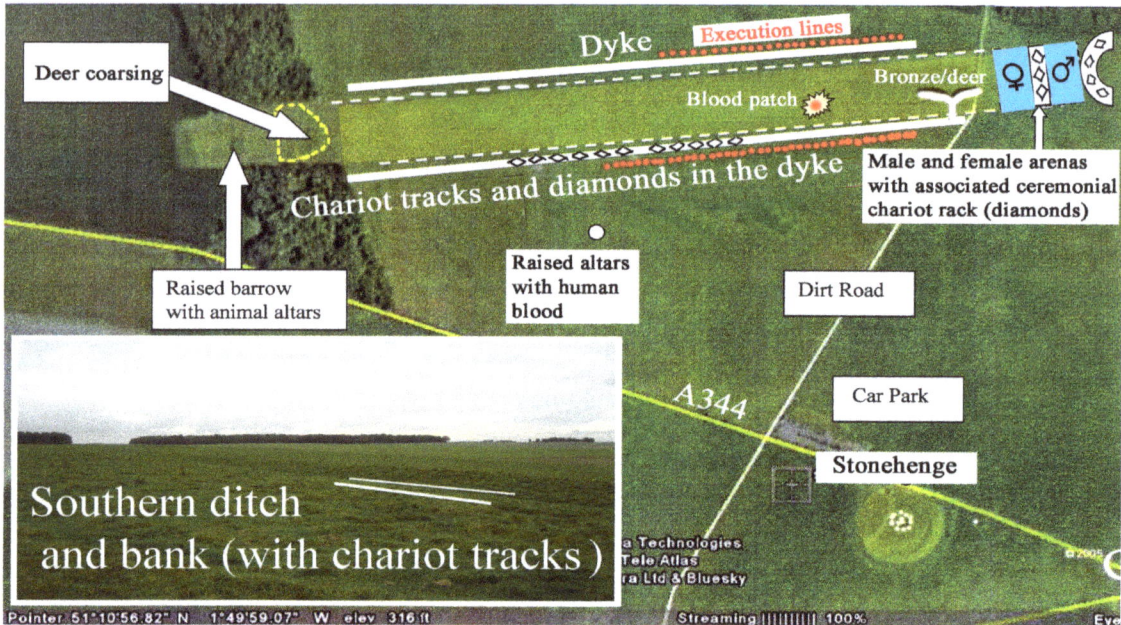

Figure 19.8 Stonehenge Cursus

The Cursus is about 1.75 miles long and a 100 yards wide. Just east of the dirt road crossing the Cursus are the two arenas separated by a road marked with diamonds and chariot tracks. To the west, animal pens and other structures have been found. At the far western end there is a large barrow with altars for animal sacrifice. From the postholes it looks as if the barrow was in a building. The size of the Cursus and the numbers of people executed by the Romans indicates that there was a large town or city nearby.

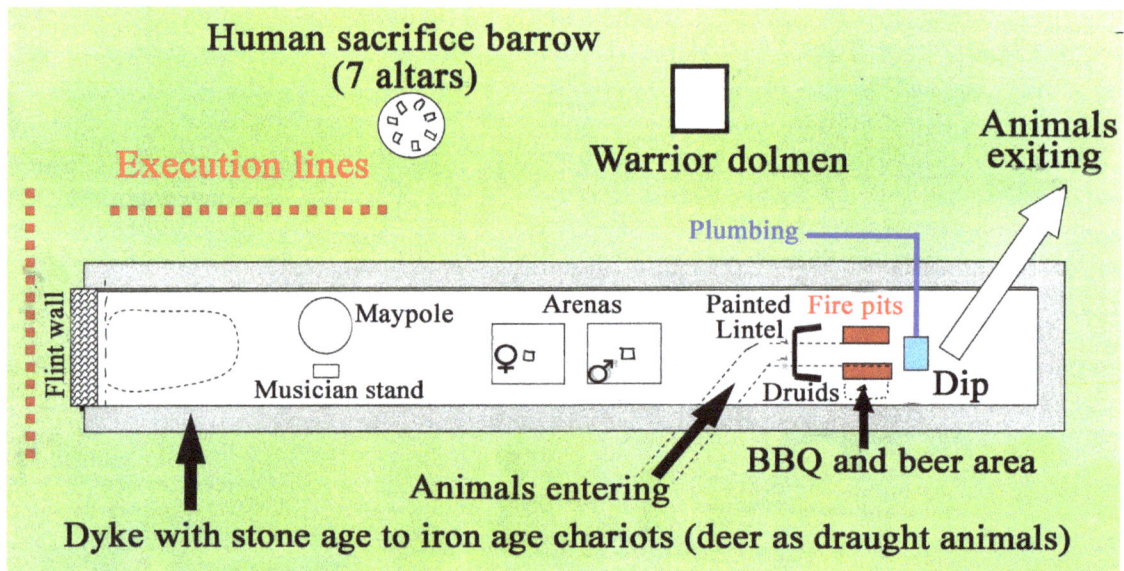

Figure 19.9 The Little Hay Cursus

The Cursus at Little Hay had ben studied in more detail by mid-2006. It is very similar to the Stonehenge Cursus with a surrounding dyke or ceremonial track. The wheel tracks indicate that the Cursus was established in the late Stone Age and was used until the Romans arrived, a period of about 2000 years.

232

between Romans and 'Celts'. This enclosure is just before the first fenced barrow is reached.

The next stop was the low barrow with its seven altars for human sacrifice. We were interested in how the victims arrived and how they were taken away. There was the possibility that they were the survivors of the staged battles. Were they treated as criminals or with some degree of reverence? The use of seven altars indicated that human sacrifice was on a large scale. There was at least one other raised altar barrow for human sacrifice less than a mile away at Stonehenge so they were a reasonably common feature of the landscape. One of the fenced barrows also had a trench round it with human bloodstains. However, there was no way of telling if the two barrows were in use at the same time for human sacrifice. We looked for indications of how the victims arrived at the barrow but did not find any signs. This might be expected if they were the victors and still able to walk. Identifying their mode of departure was easier. An iron tyred vehicle entered the ditch and the bodies were loaded onto it. The cart came out of the ditch on the Cursus side and then circled round the barrow until it headed for the A344. The track eventually went across the road so it was not followed to see what happened to the bodies. If they arrived at a Dolmen it would look as if they were being treated as warriors. If they landed up in a pit then perhaps they had been on the wrong side or were not respected citizens. One of the gullies on the small barrow had Roman blood in it so the facility may also have been used for prisoners.

Our next objective was to see if we could identify deer coursing. The idea was that if the course was from east to west we might be able to pick up deer blood as we approached the western end. In fact as we approached the end, patches of deer bloodstains started to appear with a halo of dog hair round them. We also found signs of Roman blood associated with dog hair which may indicate that Roman prisoners might have been dispatched by dogs. Eventually we came to the barrow at the western end of the Cursus. It soon became clear that this barrow was for animals. There were seven altars with gullies draining into the ditch. There was no sign of human blood.

There was one final thing to do before heading home and that was to try and determine the age of the Cursus. The method was the one we had used at Glastonbury. The tracks of the wooden wheeled carts in the road surrounding the Cursus were identified. Next the postholes and walls either side of the cart tracks. Then the Bronze Age chariot wheel tracks were identified and their position compared with the walls either side of the Stone Age cart track. The bronze tracks were on top of the walls and postholes and indicated that Stone Age people had built the Cursus or at least started its construction. Like Stonehenge the history of the Cursus starts back in the Stone Age.

On leaving the Stonehenge area we realised that the scale of killing by the Romans was such that they had gone into the city and local towns and villages and rounded up the whole population and executed them. There appears to have been little fighting going on. If this can be confirmed it indicates that a major battle may have taken place somewhere in the Stonehenge area when most of the fighting men had been killed. A possible alternative is that each temple had its own Temple Guard who died defending their own temple as they had done at Maiden Castle. If this were the case the Romans could pick off the fighting men temple by temple and never had to face a Druidic army.

The significance of the Cursi is that they may provide a window on the activities of the secular population. When studied in detail the Cursi will reveal a great deal about what was going on outside the temple complex. The detailed study of Cursi was not part of the project which was focusing on The Stones. However, they proved to be irresistible and have been included. The Cursus is still a sacred area with the ceremonial chariot track round it. Its use and manner of use was almost certainly governed by the Druids. The full story of the Cursi is going to take a long time to unravel but when it is it may well match the 3000 year story of the temples in duration, detail and drama.

The following is an account of some initial studies of Cursi:

Cursi 2

After working on the Cursus at Stonehenge we now knew what to look for and how to identify them. It was therefore not very long before three Cursi had been identified back at base followed later by one in Wales. One of the Cursi was at Chiswell Green, St. Albans another was along Stoney Lane, a lane going out of Bovingdon towards Hemel Hempstead. Both of these Cursi had Stone Age Temples on them which must have predated the Cursus. The temple on the Stoney Lane Cursus differed from previous Stone Age Temples that had been found in that it was a full size temple at about 60m in diameter. The central tomb did not appear to contain anything more advanced than binding thread made from nettle, urine tanned leather, wool, dog hair and feathers. It did however have two side walls plastered with gypsum. This along with the size of the temple is taken to indicate that the tomb is likely to be late Stone Age and that the Cursus dates from then or the early Bronze Age. The second Cursus near Bovingdon is east of the Little Hay golf Course on Boxmore Trust Land and is more interesting. It is nearly half a mile long and is in many ways similar to the much larger Stonehenge Cursus. The ceremonial track round the Cursus has chariots from probably the Stone Age. The wheels are wood with leather (urine), nettle and hemp binding. The axle length is about 1.80m (6ft) so slightly smaller than the bronze wheeled chariots. Interestingly the chariot is pulled by two deer. This was the first time we had found two deer in harness and indicates that deer may have been the first draft animals used in the British Isles. At the moment we cannot distinguish between deer in general and Reindeer. However, because Reindeer feed on lichen which is only found further north we had assumed at this point in time that the Druids were using deer other than Reindeer. Deer are commonly found to be the draft animal with pre-Bronze Age vehicles which are being used for ceremonial purposes. The Bronze Age chariot tracks are about 2m wide and the stains in the ground is from arsenical bronze and tin bronze. There are also iron tyred chariot tracks. Extensive lengths of the Cursus and its northern end have execution lines along them. The Cursus was therefore in use up to Roman times. Its use could span at least 2000 years and possibly include the Neolithic making 3000 years of use. The Little Hay Cursus is on high ground overlooking the valley of the River Bulbourne through which the A41, Grand Union Canal and the main line railway between London and the north runs. From this valley a spur valley runs west in the direction of Chesham (B4505) and on either side of this small valley there are Dyke Temples. They are in Bury Wood to the south and in the Ramacre Wood on the Little Hay side. This results in an extraordinary concentration of Druidic ritual landscape engineering (see map). One important aspect of Cursi as already mentioned is that they are part of the fabric of any town or even large village. They are therefore going to be very useful archaeological sites, provided archaeologists do not go and dig them up.

Having learnt how to identify Cursi in the landscape and realising how much information they contained they started to attract dowsing time. If they were 'home' so as to speak to many activities then they would eventually tell us a lot about life in pre-Roman Britain. One obvious thing to look for were signs of the Beltane Fires between which animals were said to be driven. Wood fires leave a deposit of ash rich in potassium. The potassium would have been taken down into the soil and leave a stain for the dowser to find. Early on the evening of May 9th 2006 I walked down Stoney Lane to the Bury Wood Cursus, a walk of about half a mile, climbed over the style into the field and approached the Cursus. Once I was on it I did a circular sweep search using a potassium witness. The strength of the response indicated something big. There was no doubt that there was a strong source of potassium towards the western end of the Cursus. I set off towards where I was hoping the fires would have been but it was about 150m before I reached them. The significance of this is that Beltane Fires are going to be easy to identify on a Cursus and can be used for finding a cursus.

The stains of the fires were in rough scrub ground, not easy to map out. However, it was clear that there were two rectangular fire pits. Extending from one end of each of the fire pits, the eastern end, there were walls on puddingstone foundations behind which Druids stood. The linen and feather of the Druids clothing could be identified in a sort of Druids enclosure. The animals had to pass between the Druids before going between the fire pits. The animals going between the fires included cattle, horses, sheep, and deer. Adjacent to the Druids enclosure there was a large painted rectangular area for reception of the animals. There was also a blue energy line running along the length of the Cursus and going into the gate of the Beltane complex. By the time I had discovered all this a large red sun was approaching the western horizon so I packed up and walked home along a lane laid down by the Druids possibly 4000 years or more ago and still carrying their paintings including the diamonds. I had discovered the Beltane Fire Pits on one Cursus but as they were in a scrub area they were not easy to work on. Across the valley, the Little Hay Cursus did not run into rough ground so that was the next place to look for the Beltane Fire Pits. The following afternoon I was on my way to the Little Hay Cursus carrying witnesses and 30 canes to mark out the fire complex. Using a search technique I soon found the fire pits. They were the same size and shape as the ones on the other side of the valley, neat rectangular structures. The results of the survey are shown. (Figure 19.10). The fire pits (14ft by 4ft) and the Druids enclosure, extending about 16 feet from the fires, were the same size as the one found on the Bury Wood Cursus. In front of the entrance was a large rectangular area (about 60ft by 32ft) marked out with paints of different colours. Moving in towards the fire pits there is indications of a painted oak lintel across the entrance.

The Little Hay Cursus was used for deer coursing with the deer killing zone at the northern end in front of a flint wall and what looks like a ditch. Behind the wall there may be a viewing stand. Moving away from the northern end and towards the southern end of the Cursus there are two arenas. These are walled enclosures designed for ceremonial battles involving up to 28 warriors. The figure of 28 is based on the capacity of the Dolmen which was used for the dead warriors. The arena are going to require much careful work as they contain a lot of detail. At the moment it looks as if the warriors marched in to the arena and formed up in two double ranks facing each other. Between them was a rectangular plinth based on what is called puddingstone. Bronze, gold, toadstool tea and yeast (beer) stains are in the puddingstone rectangle and there are two drip lines from the rectangle, one to each of the apposing ranks of warriors (Figure 19.9).

The warriors were barefooted and where they were standing is stained with toadstool tea and beer and skin from their feet. At the moment we have not worked out if the plinth or what was on it was mobile or permanent but the most likely option is that it was brought in for the ceremony and then removed. The feathers of the priests are all round the battle area but not where most of the fighting was taking place. Whether they were just part of the ceremony or umpires is at the moment not known. The plinth and bowl for the drink was placed at the centre of a Thunderbird display. Four manganese dioxide and red ochre Thunderbirds going out from a central rectangle with a black circle round them. (Figure 19.10).

The tracks of the carts that collected the bodies after a battle and the blood drip line from the cart can be followed from the arena to the warriors Dolmen. The cart from the Barrow where humans are sacrificed also joins up with the arena carts and both go to the Dolmen. It is going to take some time to work out what is going on but one possibility is that the survivors of the battle were sacrificed. In passing, it is perhaps worth noting that if the local ravens and crows were to deal with 28 bodies in one go there would have to be a lot of them with the required nesting and roosting places. The use of excarnation by the Druids would have had a profound effect on local bird ecology and the avian scene in the countryside would have been quite different in Druidic times to what it is today. The Cursus was still in use up to the Roman period but this may not apply to the arena.

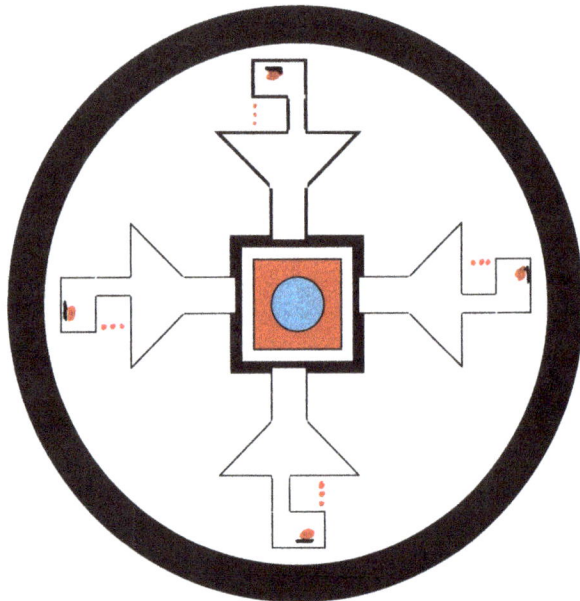

Figure 19.10 The Arena Symbology
The arenas had at their centre a plinth or some stone structure.
Round it was a black rectangle, four Thunderbirds and a black circle.
As all the combatants were due to die either in the battle or on the altar the symbol has been interpreted as indicating the gate to the spirit world.

A Welsh Cursus

On the 12th May 2006 I was back at Plas Tan y Bwlch, the Environmental studies Centre in the Snowdonia National Park where I first found the Prince and Princess at the centre of a Henge Temple. The Centre's buildings are about halfway up the northern slopes of a valley and from them fine views are to be had of the River winding its way to the sea and of mountains in the distance. In the extensive gardens below the Centre there are flowering shrubs including in full bloom during my visit the largest rhododendrons in the country. The steeply sloping lawns descend from the house almost as a cascade with each slope stopping for a path or wooded slopes. The paths wind back and forth to deal with the steep slope and eventually reach the valley floor and a flat lawn area which shared the Henge Temple site with a final area of woodland before the main road and flood plain was reached. The Prince and Princess were still where they were laid to rest over two thousand year ago. I checked the temple, as at the time of my last visit to the Centre I did not know much about their structure. It was a typical 6 + 3 Henge Temple above ground. Below ground it had the four anterooms and bone depositories and the encircling heptagonal tunnel round the tomb complex. I checked for execution lines and found them round the temple. The following morning I set out to find the Cursus. The reasoning was that a temple along with signs of houses and roads indicated that the community was large enough to have its own Cursus and the most likely place for it was the flat ground running alongside the main road just above the flood plain. Cursi can be identified in a number of ways. The more we learn about them the more indicators of Cursi we have available. However, the best method for the site was to find the Beltane Fire Pits

236

as the potassium could be picked up at over 150m. When I reached the flat ground down by the entrance gate to the Centre I found fields on each side of the access road providing a reasonably flat area that looked as if it was large enough for a small Cursus and which could have been selected as a site for a Cursus. I did a sweep search using a potassium witness and obtained a response from the field which ran back towards the temple and was overlooked by the buildings of the Environmental Studies Centre. On entering the field I soon found the two fire pits then a little further on the first of two arena. Round the outside of the Cursus were bronze chariot tracks. The Plas Tan y Bwlch Cursus, although small, appeared to be identical in design to other Cursi. The presence of execution lines round it showed that it was in use up to at least 43AD.

A Ceremonial Tunnel

The buildings of the Environmental Studies Centre are perched on the side of the valley. Looking at them from the garden, on the left of the centre there is a row of study and lecture rooms with a car park in front of them. The main building is on the right. Below the car park the ground starts to drop steeply. For years water diviners and dowsers had homed in on what they called an underground stream running across the car park to the gardens. The stream looked as if it came out of the side of the mountain, ran under the buildings then under the car park. The stream then passed under the lawns and headed for the valley. On previous visits to the Centre I had accepted that it was an underground stream but had always been suspicious because of its straight parallel sides. On this visit I decided to challenge the underground stream hypothesis. It was not long before I had identified that the stream was a Druidic ceremonial tunnel (This does not exclude the possibility of a stream being under or near to the tunnel).There were chariot wheels with tyres made of leather, bronze (arsenic and tin) and iron. I could dowse paints from the chariots which were pulled by deer. I could follow the tunnel towards the mountain but could not get at the end of the tunnel to see what was there. Was it a 'Hall' deep within the mountain? Tunnels have entrances so I followed the tunnel under the steeply sloping lawns until it reached some trees and a vertical drop. I went round the side so as to speak and found myself on a garden path which crossed the route of the tunnel. I could find no sign of a tunnel under the path but there was a vertical rock face back amongst the trees towards which the tunnel had been heading. Recessed into the rock face was the blocked entrance to the tunnel. A large block of stone was over the tunnel entrance. On the right of the rock was a clear rock face running back to the tunnel entrance. On the left was a curved wall of dry stone walling which ran behind the blocking stone and was clearly there before the stone. Running up to the stone were the tracks of all the vehicles identified in the tunnel (Figure 19.11).

Dry stone walling was used extensively by Druids in Wales. In studies of dry stone walls in the Pontypridd area it had been found that in Druidic walls the stones were laid according to their magnetic axes. This creates a field in front of the stones much like that found round a stone circle. I tested the polarity of the outside surface of the stones and they were all north poles. In modern dry stone walling the polarity of the outer surface of the stones is randomly distributed. The dry stone wall of the tunnel entrance is therefore Druidic and at least 2000 years old and possibly much older.

Behind the Environmental Studies Centre the paths climbing the mountainside are mainly Druidic and so are of a similar age to the tunnel, possibly even three or four thousand years old. This is indicated by the stonework supporting the paths which has a north pole on the outer surface of the stones. The age and origin of the paths or roads is revealed by the diamond markings on the paths and by the traces of linen and feather from the Druids clothing. Along the side of the hillside road there are occasional recesses cut back into the side of the mountain. At the moment our view is that they are the entrances into tunnels, as tunnels have been detected behind some of the alcoves. However, the stone for the paths had to come from somewhere and the recesses in the sides of the roads or paths may have been quarried to provide that stone. The scale of the Druidic terraforming work round the Environmental Studies Centre is impressive. A large community dedicated to ceremony and the Earth Goddess must have lived in the area and over the centuries built an extensive road and path system going up the side of the mountain and constructed tunnels into the side of the mountain. Perhaps the Halls of the Mountain Kings are waiting to be discovered in Snowdonia.

Figure 19.11 Entrance to a Hall of a Mountain King
The stone is blocking the entrance to a tunnel dug by the Druids. The dry stone walling is to the left. The chariot tracks go up to the stone.

Caerleon

On the way home from Wales I stopped off at Carleon. Carleon was a Roman town of some importance. The amphitheatre survives to this day and I wanted to see if it was constructed on a Druidic temple site. As luck would have it, when I arrived there was half a legion of school children being educated in Roman military procedures and commands. However, I managed to identify that there were a number of tombs of Princes and Princesses in the centre and also the postholes of temples. The outlines of Thunderbirds could also be traced. The Romans had built the amphitheatre on top of a Druidic temple complex and round it were the execution lines. The Romans had destroyed the Druidic centre, executed everybody and then possibly after a gap of a few years built their own 'sacred' building on the site. The ground of the amphitheatre was covered with bloodstains. People in their hundreds had died in the arena. There were blue lines running across the amphitheatre with the ones I found arising from one section of wall. There was no sign of how they were formed. I mention this because the Romans, like the Druids, appear to have been keen on blue lines and built them into their shrines and amphitheatres. At least this is what dowsing mythology holds to be the case but it has yet to be proved.

238

The Maypole

Returning to the Cursus, there was one more item that might be on a Cursus and that was the Maypole. There was not time to look for it in Wales so the search took place back in Hertfordshire. A large lone posthole with flint supporting stones had been found on the Cursus at Chiswell Green. This looked a likely candidate for a Maypole. The most likely position for a Maypole on the Little Hay Cursus was to the north of the arena. It was therefore near the arena that I started the search for the Maypole and worked my way north. After a while I found a large posthole that had held an ash pole. Round it, at a distance of about 14ft, there was a 10ft wide ring that gave positive responses with both a human skin witness and a wool witness. There was no leather so the dancers were barefoot. Between the ash pole and the skin zone, nettle fibre and hemp fibre could be identified. The nettle fibre only reached the skin zone, the hemp extended to the outside of the zone. My interpretation of this was that the Maypole site was used from possibly early Bronze Age if not late Stone Age when nettle was the main fibre for twine and rope. The Cursus was still in use up to the Roman period. The Cursus, in common with some temple sites, had a very long period of use. At least 2000 years but possibly much longer as it was on a designated area of land that could not be burnt down or destroyed. Like the temples, what went on in the Cursus area, such as the ceremonies, remained stable over a long period of time. The animals were driven between two fire pits into a dip, there were ceremonial battles, Maypole dancing, and many more activities which went on year after year through millennia. I could not find any indication of an area for an audience round the Maypole but that does not mean that there was not one. Having failed to find evidence for an audience area I pondered on how to identify the 'Bandstand'. The thought occurred to me that one thing that might not have changed over the past 4000 years, or more, is that musicians play much better if beer is available, so that is what I looked for. It did not take long to identify the 'Bandstand' using yeast. The 'Bandstand' was to one side 32ft from the pole and occupied an area 11ft by 26ft. The musicians were Druids and traces of beeswax and tree resins (rosin) were present in the enclosure. There was no trace of bronze left by metal musical instruments. Later, signs of horse hair and 'cat' gut, that is collagen from sheep's intestine, were found in the area indicating the use of stringed instruments.

Chapter 20

Things Get Spooky: Quantum Entanglement

Splitting the Blue Beam

Since my early work at Killadangan I had known that the stone circle builders knew how to build beam splitters. At the time I thought that the builders had just hit on a method that enabled them to produce twice as many energy beams using stones as 'mirrors' reflecting the beams in some way. It did not occur to me to see if a split blue beam or line differed in any way from a normal unsplit one. In fact at the time they appeared to be the same in that they could be detected using blue on the colour wheel and also with an aluminium witness. The blue lines or beam I had tested could also be blocked with an aluminium rod. There was nothing to rouse any suspicions that split lines might differ from unsplit ones and this was the state of play when I visited Stonehenge on the 16 April 2006. However, subsequent laboratory research showed that the 'reflection' of magnetic fields from surfaces was not a normal reflection process. In retrospect it is perhaps obvious that magnetic fields are not going to reflect from surfaces and that it might involve an emission process or perhaps field lines reflecting off field lines. Another alternative is that it could possibly be more like thermal radiation with the incident radiation being absorbed by a black surface which then heats up and emits its own thermal radiation. I knew that a simple process was not involved as I had gathered some evidence that the blue lines being emitted by stones in response to an incident blue field were not likely to be the incoming field just having its direction altered. It was in fact likely to be a new field generated by the stone. However, these suspicions did not enable me to identify any difference between the applied incident field and the emitted one. The usual tests said that they were the same and there it rested until I was writing up the work we had done at Stonehenge on the 16 and 17th April 2006. It was at this stage that I realised that the Druids would not have built such a large energy device just to split a blue beam if all they got out was two of the same. For the effort they had put in they could have produced as many blue beams as they wanted using much smaller stones. The Druids must have seen the split lines as differing from the normal ones in some way and wanted the biggest and best they could get. This sent me back to the model of Stonehenge on the lawn (Figure 20.1).

Everything worked and all the fields did what they should do. The brick acting as the Heel Stone split the incident beam and sent one off on either side into the distance. Now I did know that physicists use beam splitters of various types and that one type was a crystal. The crystal has to be correctly aligned but when it is, it can be used for producing photons that are, to use their terminology 'Quantum Entangled' (QE). What this means is that when one photon of light comes into the crystal two photons of light go off in opposite directions, one to the left and one to the right. Now, and this is where it gets exciting, if you do something to the photon speeding off to the left at the speed of light the one going off to the right at the speed of light seems to know what you have done to its partner and behaves as if it has also been affected in the same way. In other words two quantum particles, such as photons, can affect each other however far apart they are if they are quantum entangled. So if you absorb the left hand photon and it disappears the one to the right should also disappear at the same time even if they are travelling away from each other at the speed of light. At least this is my interpretation of quantum entanglement. Having thought of the possibility of split beams being entangled the next step was easy. I had out on the lawn the split beams and the normal beams and I could easily block any one of them using a cork board or an aluminium rod. Blocking the normal beam going to the beam splitter at 'B' (Figure 20.2) just blocked the down stream beam as it always had done in many experiments with field lines.

Figure 20.1 The Stonehenge Trilithon Energy Model

The bricks represent the five Stonehenge trilithons. They are energy engineered precisely to emit a magnetic field in the form of a beam. The emitted field hits the brick representing the Heel Stone and is split into a left and right beam whilst the primary beam continues on its way.

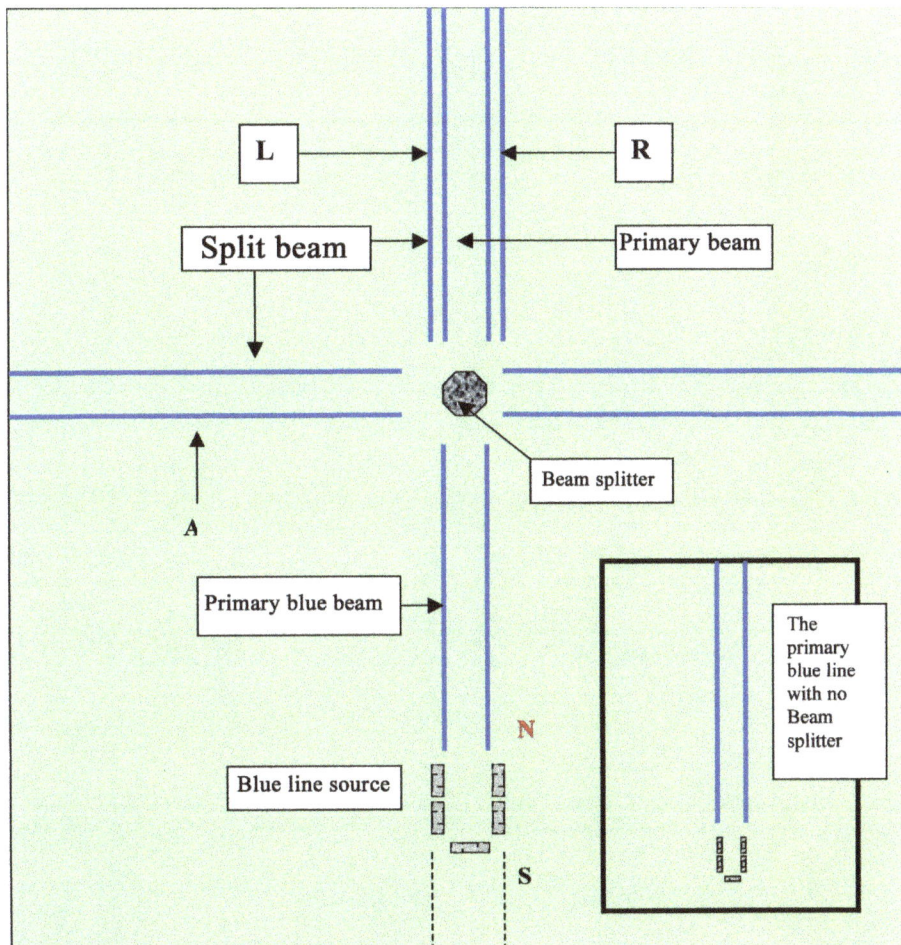

Figure 20.2 The Beam Splitter

The diagram shows the primary blue beam from the trilithon horseshoe being split by a suitably angled stone, the beam splitter. When this happens a left and right beam are produced and the primary beam acquires a split beam on each side. If the left or right beam is interrupted for example at 'A', all the split beams disappear and only the primary is left. If the split beam is blocked at 'L' or 'R' then only the beams on the left or right close down.

As a result the split beams disappeared as would be expected. Nothing else happened. Now for the next experiment. I already had a gut feeling about what was going to happen next. Something at the back of my mind said that however unlikely it was, it was going to happen. The Druids knew what they were doing. I selected a spot out on the left arm of the split beam at 'A'. Normally the cork board would block the beam so that it would disappear downstream, which is what happened. But the beam had also disappeared upstream towards the beam splitter. Not only that the right hand split beam had gone as well. All of the split beam system had gone. It was as if by absorbing one field line all field lines had been absorbed. I then checked the normal beams before it was split. It was still there but the beam continuing through the beam splitter had changed. It was narrower and I soon established that it consisted of a central normal blue beam with a quantum entangled beam on each side which disappeared when I blocked the split beam at 'A' (Figure 20.2). The side beams therefore appeared to belong to the split beam system. It did not take long to show that wherever a split beam was blocked it would result in the disappearance of the whole system. Thirty metres along one beam and the effect was the same. The idea of magnetic fields being entangled at a quantum level seemed a bit odd so I decided to up the stakes a bit by placing a second beam splitter in the system on the left split beam (Figure 20.3).

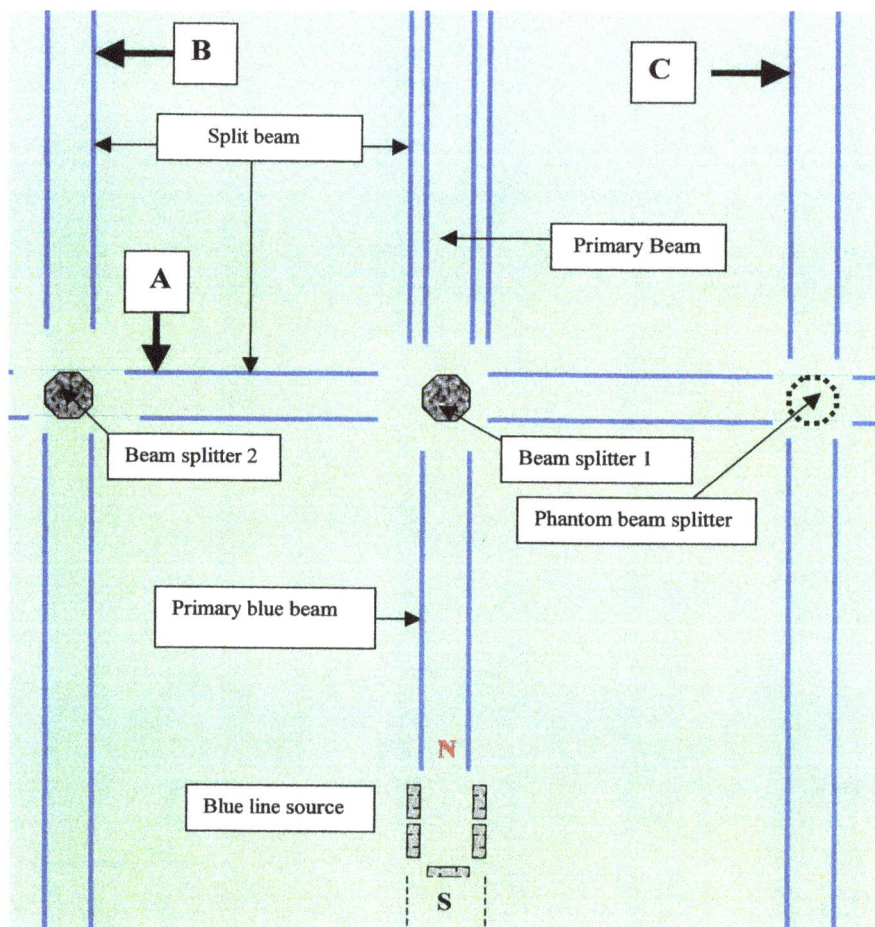

Figure 20.3 The Second Beam Splitter
If a second beam splitter is placed in the path of the split beam coming from beam splitter 1 the beam is split at beam splitter '2' and a phantom beam splitter appears as shown. If the beam is blocked at 'B' or at 'C' all the split beams disappear. The phantom beam splitter is a hologram of beam splitter 2 in that magnetically it appears to be identical to it.

The left hand beam split but so did the right hand beam which had no beam splitter. It was as if there was a phantom beam splitter in the right hand beam. If I moved the second beam splitter on the left hand beam nearer or further away from the first beam splitter the right hand phantom beam splitter followed it. If I blocked one of the beams coming from the second beam splitter the whole lot closed down. If I blocked a beam from the phantom beam splitter on the right the whole lot closed down. The evidence for quantum entanglement was growing but I was not convinced and a few more experiments were lined up. Returning to the Druids for a moment, one of the theories attached to the blue lines is that they are used for communication. At its simplest level this could be a signal such as the line is there or it is not there. This simple signalling could be done with the primary beam designated blue 1 (Box 20.1) but it would be a one way system as the target is down stream.

The sender would not know if the message had been received unless the recipient had their own blue 1 line pointed at the sender and they could signal back. Also you could only point your blue 1 line to one target at a time. Compare this with a blue 3 line (Figure 20.1) which is a two way system. The sender can block it and the recipient can block it. Also by adjusting the angle of the beam splitter you have a certain amount of control over the direction of the beam. You can also split and resplit the beam so that a number of targets receive the message and each target can send a message back and to everybody else. It was now apparent that the Druids may have had a good reason for investing in a system that produced quantum entangled blue 3 lines powerful enough to travel miles across the countryside. But did they? Were the blue lines from the Heel Stone quantum entangled and did the blue 1 line to

---

Box 20.1 Blue Energy Lines

Dowsers often refer to an energy or magnetic field as being of a particular colour. This use of colour to describe a field comes from a French dowser Henri Mager who in the early years of the 20th century designed a segmented Rosette or colour disc to test the purity of water (Box 1.4). The Rosette was later modified by a British dowser Clive Beaton. The modification was to draw eight equally spaced circles on the disc to represent the series of a field. The colour disc acts as a general wide spectrum witness and in the present context it is the blue segment of the disc that is used to identify the magnetic fields of interest. As already mentioned the pigments used to produce the colour blue can influence the dowsers' response to a field.

The most common magnetic field to give a positive response with the blue segment held in the centre (series 4) is that produced by aluminium. The aluminium may be in the form of metal (elemental) or part of a chemical compound incorporated into rock or other materials. This field is referred to as blue 1. A blue 1 field is blocked by an aluminium witness.

When the blue line from trees, upright poles and mirrors was discovered only certain blue pigments would give a positive response with it. The colour blue or an aluminium witness would not block the field. This field was referred to as a blue 2 field and is thought to come from protons in water. The blue 2 field later became called the Chronon field as it is almost certainly the field discovered by Russian dowsers.

The third blue line (blue 3) was found when a blue 1 line was split. The split line gives a positive response with blue on the colour disc and it is blocked by an aluminium witness. The blocking is, however, unusual in that the field disappears both upstream and down streams and extends to other fields with which it is quantum entangled. The term quantum entangled is used loosely and is not meant in the true physical sense.

Box 20.2 Polarized fields and quantum entanglement

From laboratory work I knew that magnetic fields were polarised. They behaved as if they were much thinner in say a vertical direction than a horizontal one. This could easily be shown by slit experiments in which a field could pass through a vertical slit but not a horizontal one. Some plastics such as polystyrene acted as polarizers passing the field if the plastic sheet is held in one position and blocking the field if it was rotated through 90?. The plastic sheet was acting in the same way as a piece of Polaroid does in relation to light. Returning to the model on the lawn (Figure 20.3). When I placed a sheet of plastic in a blue 3 beam (split beam) either the left or right half of the system would close down but not both. If the plastic sheet was then turned through 90? the closed down half appears and the other half disappears. This may indicate that there could be two fields involved with their axes at right angles. Using the garden model (Figure B20.2.1), a plastic sheet can be used on the left side to suppress say the left hand half of the system so that the right hand half is still intact and the blue 3 lines easily identified. Dowsing behind the plastic sheet on the left hand side no dowsable field should be detected or in front of it. If a board is now placed on the suppressed blue 3 line distal to the plastic sheet the right hand half of the blue 3 lines will now disappear. In other words the plastic sheet may cause the dowsable line to go but there is still something passing through it that is blocked by the board or by an aluminium rod. That is something is allowed to pass through the plastic. In the laboratory I had a split beam model set up on the dowson bench. It gave me two quantum entangled beams going off at right angles and a beam going down the length of the bench. The side beams from the splitter went through the windows of the lounge. These beams could be detected the length of the garden and if one was blocked 30m away from the beam splitter the blue 3 lines disappeared. I then set up a blue 3 line which passed through a wall. I blocked the blue 3 line so that it disappeared by placing an aluminium rod on the wall where the split beam went through it. There was now no dowsable field between the beam splitter and wall or on the other side of the wall. The next step was to 'push' on where the blue 3 line should be by using a diamagnetic field from a salt crystal. When the missing or invisible line had been pushed past the aluminium rod and it could no longer be affected by it, the dowsable blue 3 line appeared. This experiment again indicated that there is a non dowsable component in the blue 3 line system. Perhaps it is the spooky force linking quantum entangled photons!

Figure B20. 2.1 Polarised magnetic fields. The diagram shows how a plastic sheet that acts as a polarizer, can selectively block part of the split beam if placed in a Blue 3 beam at 'A'. The dowser's sensitivity to the fields does not appear to be affected by the direction of polarization of the field. This is also the case with vision and light. When an aluminium rod is placed on the wall at 'B' it removes the incident primary and the split beams back to the source. The dowser cannot detect a magnetic field but a field of some sort is still present. The diamagnetic field from the salt crystal will push this unknown field away from the aluminium rod to 'C' when the dowsable primary and split beam fields reappears.

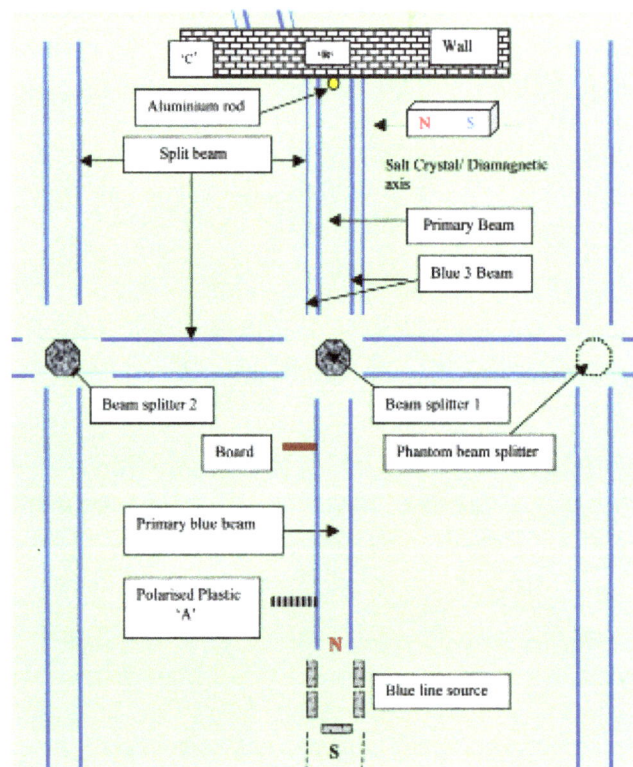

Woodhenge have blue 3 lines on either side of it. It was now time to ring Nigel, which I did. 'Nigel, I said 'you are not going to believe this so you have to come over and see it for yourself'. 'I have a set of quantum entangled magnetic field lines on the lawn', Nigel duly arrived and was suitably impressed. He then confirmed my findings on the magnetic model of Stonehenge. We both knew that we now had to make yet another visit to Stonehenge. It had to be demonstrated that the split lines from the Heel Stone were blue 3 lines. If they were it should be possible to close the system down from a number of points such as Woodhenge, from near the western end of the Cursus or at any point along one of the lines. The date fixed for the experiment was the 7th May 2006. That gave me time to do some more experiments as I was not satisfied that quantum entanglement was involved. On the basis of the Russian Chronon theory it was possible for signals to pass up and down a field line faster than light. There was also no doubting that if you did something in one part of the system the rest of the system knew, even if it was tens of metres away. But there was something else going on. One of the early observations I had made was that once the blue 3 fields had been closed down by blocking a blue 3 line with cork or wood, if the block was left in place the blue 3 lines re-established themselves. As there was nothing dowsable left to penetrate the barrier the indication was that something that was not dowsable was doing the penetrating and then re-establishing the dowsable system (Box 20.2).

The Secret of the Heel Stone

Sunday May 7th at last arrived. An early start, fine weather, little traffic and Nigel and I were once more at Stonehenge with a new assignment. I parked the car close to the dust road going to the Cursus and right down by the field so that we were overlooking both the ghost city of the Druids and the Cursus. I selected the spot because I had worked out that the blue 3 line from the Heel Stone should be crossing the car park nearby. As the sun was shinning Nigel decided to go and buy himself a hat from the Stonehenge shop. While he was doing that I started looking for the split energy beam from the Heel Stone. I soon found it and marked both sides of the beam, which was about 20ft wide, with canes. There was no doubt that it was a blue line and repeated tests showed that it was there and on its way to the western end of the Cursus. Next, the test to determine if it was a quantum entangled energy beam. I brought the aluminium cane into the beam and placed it against the fence. A retest and the beam had gone on the upstream side towards the Heel Stone. A quick climb over the fence to check and it was also gone on the down stream side. Repeating the test a few times it looked as if the split beam coming from the Heel Stone at Stonehenge could be a quantum entangled beam (Figure. 20.5a). Nigel returned from the shop, almost unrecognisable in his new Stonehenge hat, 'Nigel' I said you are the second person on the planet, indeed in the Solar system, in fact in this sector of the Galaxy to know that the Druids used quantum entangled magnetic fields'. Perhaps it was a bit premature as there were more tests to be done but again I had confidence that the Druids knew what they were doing and were still well ahead of us in their knowledge of the fields. I checked the effectiveness of an ash staff on the field and like the aluminium cane it closed the field down. We then walked from the car park towards the Heel Stone. I lined up the Heel Stone with the western end of the Cursus and identified where the beam should come across the road (A344). We found the beam and placed four canes in the ground to indicate position and direction. Next the beam coming from the Heel stone and going down The Avenue was identified. This beam has a central blue 1 section and then on either side a blue 3 section. The blue 3s are the entangled beams. The sections of the blue line were marked with canes so we knew exactly where the blue 1 and blue 3 energy beams were (Figure. 20.5c). We started work and it was not long before we realised that every time a car went past the Heel Stone and through the west arm of the split beam as it crossed the road, that the entangled beams closed down and disappeared. This meant that we had to work during gaps in the traffic flow. This we did and had soon demonstrated that if you placed an aluminium rod or an ash staff in any one of the blue 3 beams they all closed down. Nigel then went 300 paces down The Avenue and was able to close down the split beam going to the car park and Cursus using an aluminium cane. Stonehenge was behaving just as its model behaved in the garden.

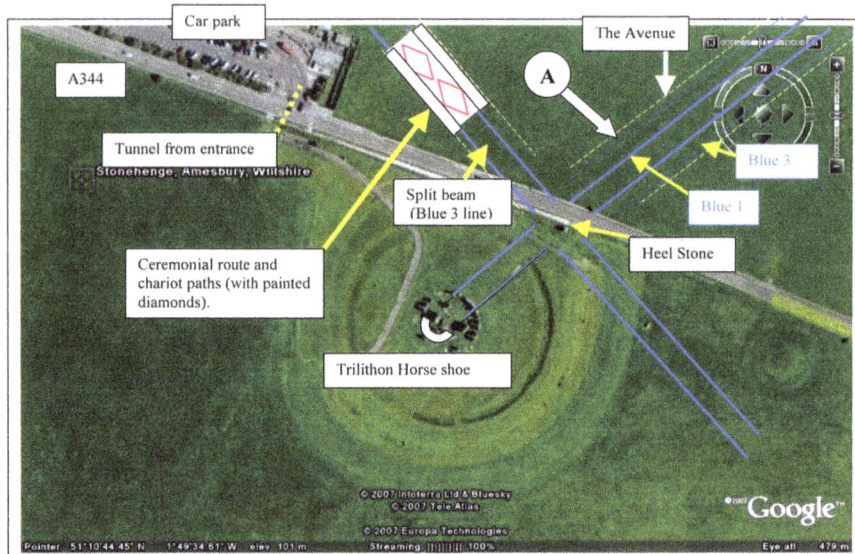

Figure 20.4 The Stonehenge Blue Beam System
The Heel Stone splits the magnetic field (primary blue line) coming from the trilithon horseshoe. The primary beam continues down the Avenue which is a ceremonial route with split beams on either side of it. Only the left hand split beam from the Heel Stone has been investigated. The left hand split beam goes past the car park and following it there is a ceremonial route from near the Heel Stone. If the left hand split beam is blocked all the split beams disappear. If the split beam is blocked at 'A' only the left hand beam closes down.

We had planned for one of us to be at Woodhenge or the western end of the Cursus and to close the system down from points which were about one and two miles away but it was clear that the traffic made that impossible. However, there were other jobs to do not least of which was to determine if the arm of the split beam going to the car park was also a processional way. If it was it would be good supporting evidence that the Druids did know that one arm of the split beam was there. We soon found that the car park arm was a ceremonial way with the central chariot road being marked with large rectangles or diamonds. The diamond was not pointing along the road but it was close. On either side of the chariot tracks were footpaths each with small diamonds (Figure 20.4). The footpaths were for the Druids and gave a positive response to a feather witness. The Druids held ash staffs and the staff marks showed that after walking along the path they stopped and turned in to face the road with the staff at their side. The chariot tracks were very interesting. There was no bronze only iron. In other words when the Heel Stone was installed to split the beams from the trilithon horseshoe it was the Iron Age. This was additional confirmation that the Stonehenge we see today was constructed in the Iron Age. It also indicates that the Druids may either not have known enough about energy engineering to construct Stonehenge until the Iron Age or only acquired the engineering skills to risk undertaking such a mega project in the Iron Age. It should be possible to check on their knowledge because if they did know how to use split beams, the Druids may have constructed smaller systems with a Heel Stone closer to the circle. If they did, it will be possible to identify it and the ceremonial roads following the two sides of the split beam. One further point about the ceremonial way. It also had the tracks of a wooden wheeled vehicle pulled by a deer. It is therefore possible that this element of ceremony was still there 2000 years, possibly much more, after it was first identified in the late Stone Age and early Bronze Age.

The ability to split beams and hence to have some understanding of them and possibly need for them goes back to before the last rise in sea level as the Killadangan circle shows with its beam splitter. Before leaving for Woodhenge we confirmed that The Avenue had its diamonds, two rows of them and that the paths on either side also had small diamonds. The position of the Druids on the footpath was also interesting. They were on the edge of the blue 3/blue1 beams. By standing facing the blue 1 beam the Druids could place their staffs in blue 1 and so it would be outside the quantum entangled blue 3 beam. If the Druid brought the staffs back into the blue 3 beam they could close the entangled lines down. The Druids were therefore standing in a position which would enable them to close and open the entangled blue 3 lines and hence possibly send a message. However, that is conjecture, the next stage was to visit Woodhenge to see if it could be dialled from Stonehenge.

Woodhenge

From the previous visit to Woodhenge we knew that there was a blue beam running through the temple. When we arrived the beam was soon identified and marked out with canes. It had a central blue 1 beam with two blue 3 beams flanking it. The blue 3 beams were turning on and off as would happen if traffic were acting on them. For a while we thought that we had identified the energy beam from Stonehenge. Then Nigel noted that the beam was coming from the wrong direction. I took a bearing (166°) and we then went off to consult the map. We soon found that we had got it wrong. Stonehenge was on a bearing of about 254° from Woodhenge not 166° and the blue beam should be coming across the field to the west of Woodhenge. Knowing where to look we started the search and soon found an energy beam coming from the direction of Stonehenge. It had the central blue 1 beam and the two flanking blue 3 beams which spent most of their time off due to the traffic. We then checked both beams for signs of associated ceremonial roads. They both had them. The energy line at 166° had two rows of diamonds on the chariot way and two sets of chariot wheels which were iron tyred and on axles about 4ft long. Between them was dog hair. There was no sign of bronze tyred wheels but there was one set of leather tyred wood wheels. The picture on the energy beam from Stonehenge was similar. No bronze tyred wheels. The blue energy beams have to be looked at in detail but it appears that they were in use in the Iron Age and were clearly known to the Druids. How they were used, apart from being a basis for ceremony, and their significance to the Druids remains to be discovered.

Druids and the Blue Beam

Although the energy beams have been referred to as quantum entangled there is almost certainly something simpler going on. However, when something that is happening to one beam is transferred miles to all the other beams it is a spooky effect as they say whether it relates to QE or not. The clarity and crispness of the response is what you need of a signalling system but did the Druids use it for such a purpose. Can it be used for such a purpose? If they did and could communicate over miles if not tens or even hundreds of miles it would certainly have caused great concern to the Romans who only had visual systems. As it happens, blue 3 beams are easy to generate so it may not be too long before somebody has the answer to the question 'can a signal be sent?' However, we may never know if the Druids used it for such a purpose. There is one myth relating to blue lines that does need an answer and that is the one that says spirits move along them. When a beam is split, it is not the brick or crystal that does the beam splitting it is the magnetic field of the brick, stone or crystal that is doing it. In fact you do not need a physical object to split the beam it can be done with a magnetic field alone. However, as all physical objects have magnetic fields including people the question arises as to whether it is possible to use a person instead of a brick to split a field? As you may have guessed, it is. Place a person in the beam rotate them so the magnetic field of the body is at the

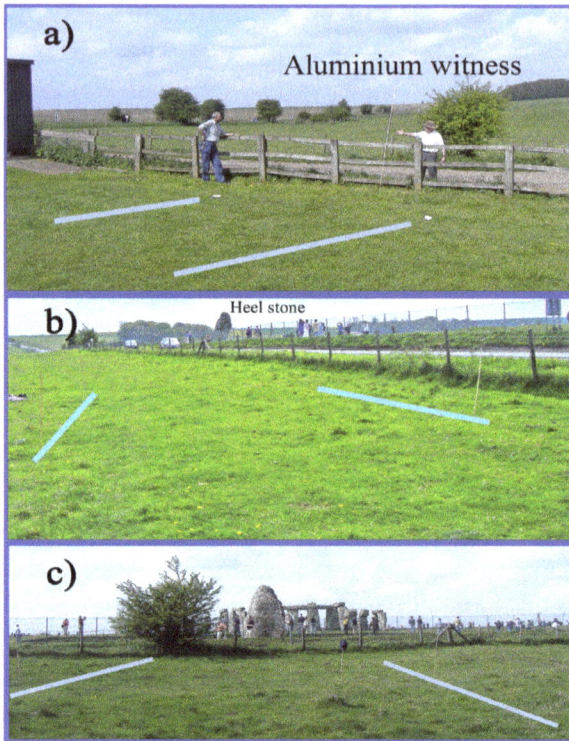

Figure 20.5 The Left Split Beam (Blue 3)

a) The split blue beam (blue 3) from the Heel Stone can be identified crossing the dirt road on its way to the cursus. The white markers show the edges of the beam. When the aluminium cane is placed in the beam it closes the beam down both upstream to the Heel Stone and down stream on the far side of the fence.

b) Looking towards the Heel Stone the edges of the split beam have been marked with canes.

c) Looking along the processional way towards the Heel stone the primary beam and the flanking split beam on the right can be identified.

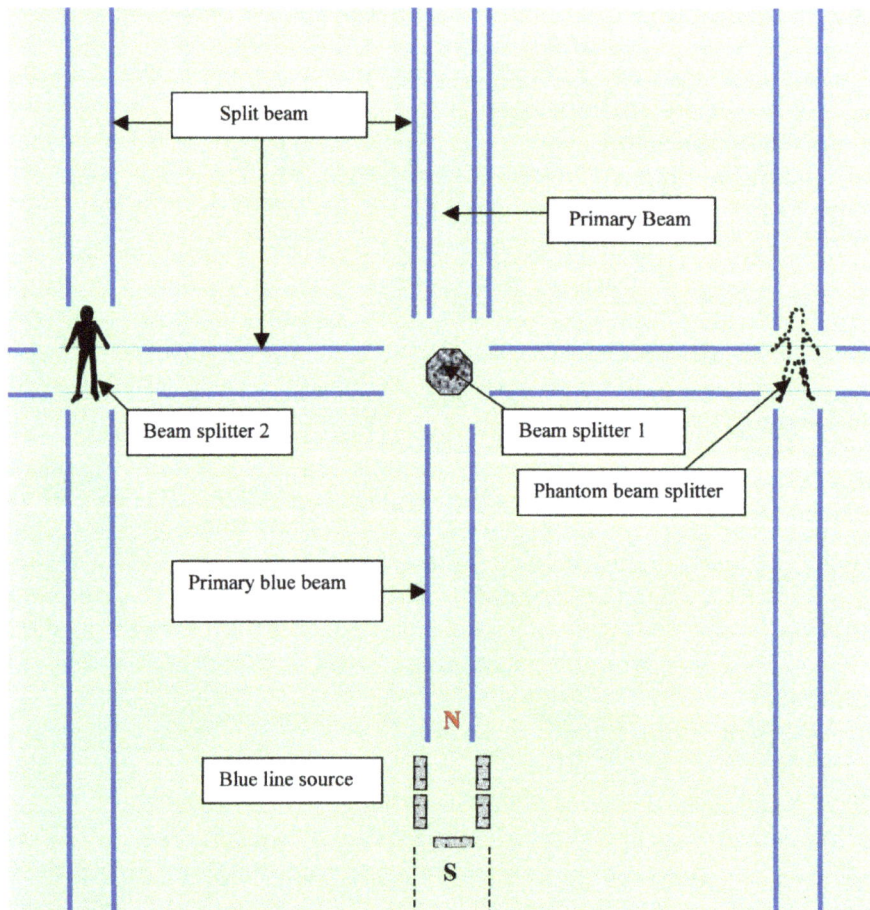

Figure 20.6 A Phantom Person

When a person is used as the second beam splitter a magnetic hologram of that person appears as the phantom beam splitter.

right angle to the beam and bingo you have a split beam. If you place that person in a split beam at point B in figure 20.5, a phantom beam splitter appears at E (Figure 20.5)just as it does if a brick is being used. If the aura of the phantom beam splitter is now dowsed for glucose, calcium, phosphate, nicotine, sweeteners and other markers it is easy to demonstrate that the phantom is the person standing at B. The person's magnetic field 'spirit' as the ancients may have called it, has moved along the blue line to become the phantom beam splitter. The magnetic fields at 'B' and 'E' are so entangled that the fields at 'B' are reproduced at 'E' in some detail. A whole lot of information has been transferred from 'B' to 'E' sufficient to produce a magnetic hologram of a person. A person who is very sensitive to magnetic fields may be able to identify the information content of the phantom at 'E' and perhaps they may even be able to 'see' the person as a phantom figure. We now know that the Druids were master dowsers and that we still have not caught up with them so perhaps some of them could see the phantoms and understood what the phantom was saying. We are working on it as they say.

Summary

When considering the magnetic fields generated by placing stones in the form of a circle it is tempting to think that the discovery was not all that significant. The stone circles have provided generations of painters, photographers, chatterers, seekers after that something the rest of the planet lacks with some physical objects to satisfy their needs. It has given people an excuse to walk to some desolate spot and withdraw momentarily from 'life'. Given enough 'monkeys' and plenty of time somebody is going to find out how to place the stones for the greatest effect. However, when considering stone circles it is easy to overlook the fact that they are but part of a system. There are outlying stones which appear to interact with the fields of the stone circle. The fields so produced were created for a reason and not just to look 'pretty' as they say. In mainland Britain not only have most stone circles been removed but those that have survived have often had their outlying stones removed. Stonehenge is an exception in that some of its outlying stones are still in place. The stone circle at Killadangan is another exception. At the time of writing they are in fact the only two circles I know of with an existing beam splitter still in place. There must be others around and even if the beam splitter has gone it may be possible to show where it might have once stood. The significance of the beam splitter is that it is unlikely that the 'monkeys' playing with stones are going to discover it and work out what to do with it. It would have taken very sensitive dowsers, able to 'see' the fields so as to speak, to work out what was going on. Or, an understanding of the mechanism of how the fields interacted. It is too early to say if the Druids used the blue lines for communication but my bet is that if it is possible, they did use them. Such an understanding of how to use magnetic fields would place the Druids way way ahead of present day dowsers in their knowledge of the phenomena and in their practical skills.

Chapter 21

Closing the Circle

Vision is the art of seeing what's invisible to others.
Jonathan Swift, Class of 1686

The Reality of Dowsing

The work recounted in the previous chapters is based on the human magnetic sensory system detecting magnetic fields and the differences in their intensity. The sources giving rise to the fields are then identified using a range of techniques. Whatever term is applied to the detection of these magnetic fields whether it is dowsing, divining or biolocation the fact remains that a clear single demonstration of the reality of the phenomena acceptable to most people does not exist. Those who do dowse of course know of its reality but those who have to make decisions and allocate resources would like to see some unequivocal evidence as to the reality, accuracy and reliability of the human magnetic sense. The reasons why this proof is not available at the time of writing are given in Appendix 4.

The work reported in the preceding chapters and those that follow opens up a whole new area of archaeology and introduces archaeologists to a very powerful new investigative technology. Because of this it might be as well to challenge the reliability of the magnetic field surveys. One premise on which science based biolocation is based is that results can be checked using standard scientific methods. For example, to build a stone circle as the ancients built them you have to line up the magnetic fields of the stones. It should be possible for geophysicists to determine the diamagnetic and paramagnetic axes of stones as laid out in ancient and modern stone circles. In fact the diamagnetic axis normally follows the ferromagnetic axis so that only leaves the paramagnetic axes to be found. The chemical stains in the soil and on stone tools, bricks, tiles, ceramics and Druidic cement artefacts can all be analysed using conventional chemical analytical methods. The results can then be used to check the dowser who is using the magnetic fields of the chemicals to identify them. The magnetic fields of postholes as found by dowsers can be checked with the archaeologists magnetometry techniques. The archaeologist's magnetic methods may only identify postholes with supporting stones round them. This should be no problem as the dowser can identify the stones that support the posts, any plinths and the carbon remains of posts without supporting stones.

For most people the preferred and most telling proof that the results of the biolocation techniques described here are linked to what is in the ground will be visual. For example the depressions in the ground, clay linings of pools, altar tops, raised earth banks, terracing on hillsides, pits, dykes and Dyke Temples, excavations in the sides of hills indicating tunnel entrances, Druidic cement and concrete, temple debris fields with pottery shards, bricks, tiles, flint tools and weapons, stone circles fitting into the design of Woodhenge Temples. Even many present day roads still show the cuttings, embankments and levelling done by the Druids. The fact that the original engineering was Druidic is confirmed by the paintings they left on the road and the bordering dry stone walling. All of this helps to support the reality of the dowsing results. The Druidic art work and temple design are linked to the form of tree auras and the auras of streams. The pictures that can be drawn on the ground round an oak tree using the aura or magnetic fields of the oak fit the temple design, 6 + 3, and the picture of the Thunder Birds that can be found round the Dolmen and inside the temple. The match between the temple design and the paintings on the floor with the aura of the oak is so precise that there is no possibility of it being by chance. The ceremonial roads are looked upon as make believe underground rivers and have the rectangle or diamond, which is found with the real streams, painted on them. The painted diamonds are irregular and have geometric and colour detail on them only found on the real diamond above a river. They are repeated at close intervals along a road and are painted using a horse hair brush. Painting them must have taken a considerable amount of labour and materials so it would not have been done unless it was seen as very important. The paintings are spirit paintings, that is they have something of what they represent in the paint and when the images change, as they did at intervals during the 3000 years or more of the Henge Age, the material creating the spirit also changes.

In the New Scientist 2nd August 2008 page 50-51 Robin McInnes is quoted as saying in relation to landscape 'There is nothing that enables us to look back centuries – except perhaps artwork'. The quotation is in relation to changing shorelines but it probably applies much more generally. It is I suspect the Druidic artwork, their spirit paintings that will help us understand them and their 3000+ year old culture. A barrow is a barrow until you discover the artwork, the paintings on and around it. Unravel the artwork and the barrow is no longer just a mound of earth in the landscape.

However, there is still one vital question. If there are so many temples around why is it that the underground chambers and passageways have not already been found? The answer to that is that they have. It is just that they have not been recognised as Druidic underground temple systems. Any cavity in the chalk tends to be referred to as a chalk or flint mine and filled up with hardcore and concrete. If the Druidic tunnels and chambers collapse and cause subsidence it is called a Swallow Hole and filled in. If the 'mine' or cavity looks like a cave or grotto with commercial potential then it may be cleaned up, renovated, a history of the site developed and opened to the public.

This is the case with Scott's Grotto and the Hell-Fire Caves.

Hell-Fire Caves

Scott's Grotto in Hertfordshire was the first cave system to be visited. The site was Druidic with a large and quite deep excavation going down to the entrance. Inside were tunnels and chambers. Chariot tracks, linen and feather trails revealed that the site was Druidic. Our visit confirmed that it was possible to identify tunnels and chambers as Druidic and not to confuse them with modern excavations. After the discoveries at Scott's Grotto the exploration of so called chalk mines and follies took on a new urgency. Many of the explanations of 'mines', tunnels and follies did not ring true. It was only in a religious ceremonial context that costs could be justified and that these underground workings made any sense at all. They were not like any mine I had visited. The dowsing we had been doing had indicated for a long time the presence of underground passages and chambers. More recently we had been finding tunnels large enough to take chariots and what appeared to be large chambers into which chariots were driven and could circle round. These tunnels and chambers were sufficiently large to be a possible cause of the subsidences that are not unusual in Hertfordshire and the mysterious swallow holes that suddenly appear and are feared by builders. After the finds at Scott's Grotto and their clear Druidic links the thought that, for the price of an entry fee, it may be possible to walk into a number of Druidic underground temple complexes which were 2000 to 4000 years old and wander round the passageways and tunnels we were dowsing from the surface caught our imagination. The next cave system on our list was the Hell-Fire Caves. On the 1st June 2006 Nigel and I headed for The Hell-Fire Caves at West Wycombe in Buckinghamshire. The caves are a popular tourist attraction and are said to have been excavated on the site of an ancient quarry in the 1750's by a local wealthy gentleman, Sir Francis Dashwood. There is a

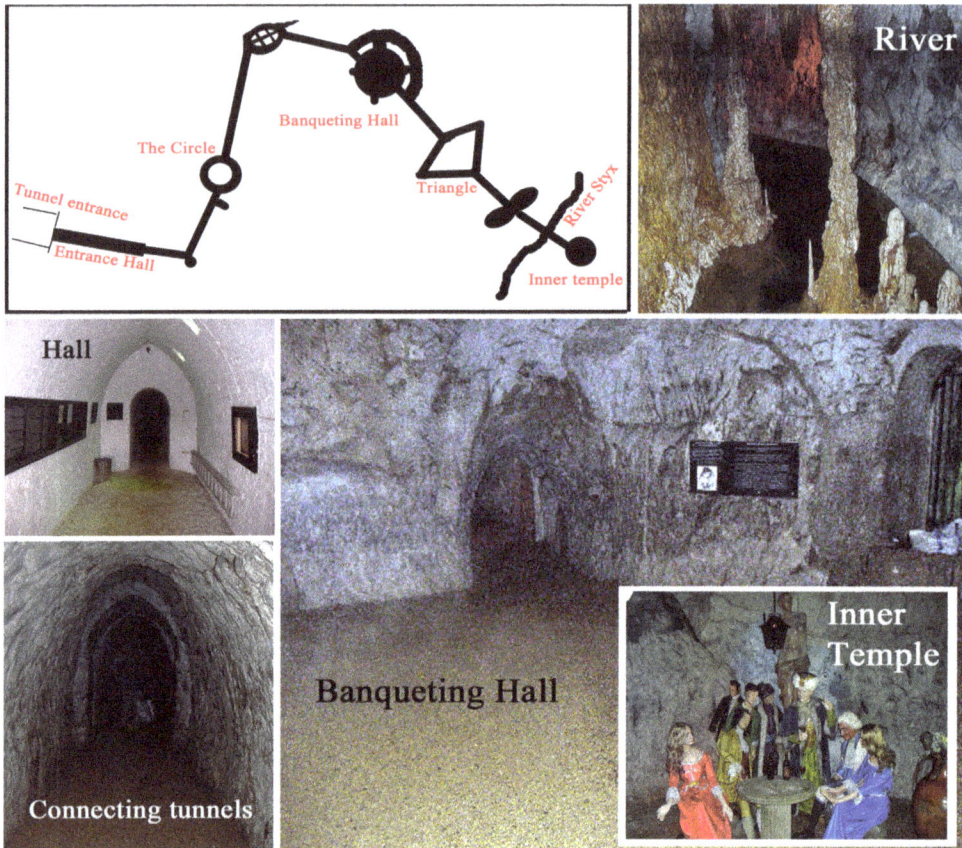

Figure 21.1 Photomap of the Hell-Fire Caves

The map of the Hell-Fire Caves shows their complexity. The main ceremonial areas were the circular tunnel, the Banqueting Hall and the shrine by the river Styx referred to as the Inner Temple. The caves have been developed in recent times and made safe. Some of the tunnels and rooms may therefore not be Druidic.

Figure 21.2 The Entrance to the Caves

The entrance was the site of the battle with the Romans. The ground was saturated with blood the stains of which indicate that the entrance is about the same size today as it was 2000 years ago.

story behind the digging of the caves and the reason for digging them, see *www.hellfirecaves. co.uk*. One part of the story is that Sir Francis Dashwood dug the caves to provide work for unemployed farm workers. Once dug the caves are said to have been the meeting place of the Hell-Fire Club which was founded by Sir Francis.

The history of the caves has been written up in an entertaining booklet by Sir Francis Dashwood, Bt. Called 'West Wycombe Caves' which can be obtained by visiting the web site. On page 3 of the booklet are an interesting two sentences which I quote 'Why he (Sir Francis) chose to have a long winding tunnel dug a quarter of a mile into the hill with all sorts of chambers and divided passages instead of just enlarging the quarry is still, however, a mystery'. The design is obviously symbolic and is thought to have something to do with the Eleusinian mysteries of ancient Greece. After walking through the caves and seeing the river Styx the thought occurred to me that the Greeks may have pinched the Druids ideas.

Figure 21.1 shows in outline the tunnel complex. It is said to be over a quarter of a mile long, say about 400m or more, so it was not dug in a few weeks or for fun. The tunnel is a sizable engineering project and it quickly becomes obvious on walking into it that it was not dug to mine chalk or flint. The chalk from widening existing tunnels and passages and making tunnels and passages safe may well have been used in construction work nearby. However, the design of what was there or was being constructed was not modified to facilitate chalk mining as a commercial operation. In the parts open to the public I found no visible evidence of working faces, supporting columns or track ways for removing spoil as would be expected in a mine. The rounded vaults of the tunnel roofs indicate a care in construction not normally associated with mining. Vaulted tunnel roofs help to protect against roof falls and aid ventilation. However that is jumping ahead so back to what we found as we found it.

The entrance to the 'caves' is cut back into the hillside to form a forecourt in front of the tunnel entrance, figure 21.2. The first step was to check for signs of chariots, priests, dogs and deer.

There were two sets of chariot tracks running up to the tunnel with dog hair and dog pad skin between them. The wheel tracks gave positive responses on copper, arsenic, tin, iron, and leather. Ash staff marks were identified. The ancient roadway had hardened edges or curbs, as the curbs could be identified with puddingstone (Druidic cement). From the initial survey it looked as if the entrance to the caves had been in use from the early Bronze Age at least. The forecourt was then examined for blood and it soon became apparent that there had been a ferocious and bloody fight. So much blood had been spilt that individual patches were not easy to see. It was possible to identify Roman blood as patches and also dog blood, but Druidic blood was everywhere. The initial point of contact between the defenders and the attacking Romans is normally indicated by a line of beeswax and paint pigments. This is possibly due to the defenders attacking the shield wall and flakes of paint landing on the ground. The inferior weapons of the Druids would soon break and the Romans would move forward but not without loss. The entrance area or forecourt to the 'caves' looked as if it was a typical entrance to a temple. All the signs of ceremonial chariots and priests in procession were present. The only draft animal detected was the deer. No signs of horse or cattle were found. The chariot tracks entered the tunnel and finished just before the tunnel became a much smaller pedestrian only passage. There was no turning circle for the chariots so it appeared to be a dead end for them. This did not look right so we went back to the entrance and started to examine the tunnel walls for signs of side chambers. We found two sealed entrances on each side of the tunnel. If the chariot tracks were followed closely they could be seen to bend in towards the wall and the sealed entrances. The Druids used a chalk and urine mortar which when spread over chalk bricks or flint makes them appear very much like the parent chalk surfaces around them. There was blood in the tunnel so fighting had continued into the underground temple.

After the entrance tunnel there is a passageway leading to the Steward's Cave. The caves or side chambers are behind iron gates so it is not possible to check their use. However, the Steward's Cave dowsed as if it had phosphate and red ochre in it so it may have been used at onetime for bone storage. Following the passage into the hill the next feature is The Circle. A circular tunnel has been dug which encloses a pillar of chalk. The tunnel is wide in parts allowing several people to gather as a group. If the left hand tunnel is followed an area of wall on the right can be seen to be of chalk blocks and it looks as if it is making good a wall and roof collapse or possibly it is sealing off an area. The walled area gives a response on witnesses of human milk and terracotta. A circle with a central area giving a positive response on human milk has been found a number of times when dowsing in the fields round Bovingdon. They occur in threes and are smaller than the one in the Hell-Fire Caves. Continuing on there then comes an extensive length of tunnel leading to Franklin Cave and then onto the 'Banqueting Hall'. On entering the Hall we instinctively knew where we were. I said to Nigel that we were probably the first people in 2000 years to stand where we were standing and know that we were in a temple to the Earth Goddess. The Hall is large and appeared on quick examination to be surrounded by a circular service tunnel from which there were four access points into the Hall. They enter the Hall about 1m above the gravel floor. To check that our first impressions were correct the chariot track round the outside and the human milk in the centre were confirmed. The 'chariot' track could however be the wheels of modern equipment used in the Hall during functions and this will have to be checked. There were indications of a central pool and plinth. The pool could not be confirmed but the puddingstone plinth was there as were traces of the bronze, gold, silver and red ochre used to decorate it. Whatever it was that had stood in the centre of the Hall it had been richly decorated. From the central rectangular plinth there were supports on each of its four sides coloured with manganese dioxide and cadmium selenium red, at least that is what it dowsed as. This pigment has been found in some art work in which red ochre was being used. It looked as if most of the pattern had been painted in red ochre which then ran out. The last part of the artwork was then completed using the cadmium selenium red. From this it is concluded that the red ochre and the pigment giving a positive response with cadmium selenium red must be similar in colour.

The Caves were open to the public so it was not possible to do a detailed survey and measure everything up. What we did find is shown in figure 21.3. We were now use to Druidic art but the floor of the Earth Goddesses Hall was something new. The central plinth with its gold, silver, bronze and red ochre was set in a background of white gypsum. Then came a circle of lamps followed by areas of black, red and blue pigments. Next came what could have been a wheeled cart track with the diamonds symbolising water and the river. We could not be certain that the cart tracks were ancient since the Hell-Fire Club could have wheeled their feasting supplies in on bronze tyred trolleys in. A more detailed study of the tracks was required before it could be said that they were due to Druidic vehicles. The diamonds were in white gypsum, red oxide and black manganese dioxide. The Hall was going to be full of memories and it would take a lot of very careful work to read them.

A compass was placed in the centre of the floor and it was clear that the design of the floor painting was lined up with magnetic bearings. The entrances from the circular service tunnel were on the four quarters, North, South, East and West. The plinth was lined up with South West, North West, North East and South East. It was evident that the Druids could not only navigate underground but when underground knew where the points of the compass were. The Druids underground navigation skills are illustrated by the next section of tunnel. This tunnel heads straight for the River Styx both in terms of direction on the horizontal plane and in terms of the vertical, the slope of the tunnel. They were navigating in three dimensions so that they hit the underground stream both at right angles and at the correct level or so it appears. An alternative explanation is that they dowsed the position of the underground

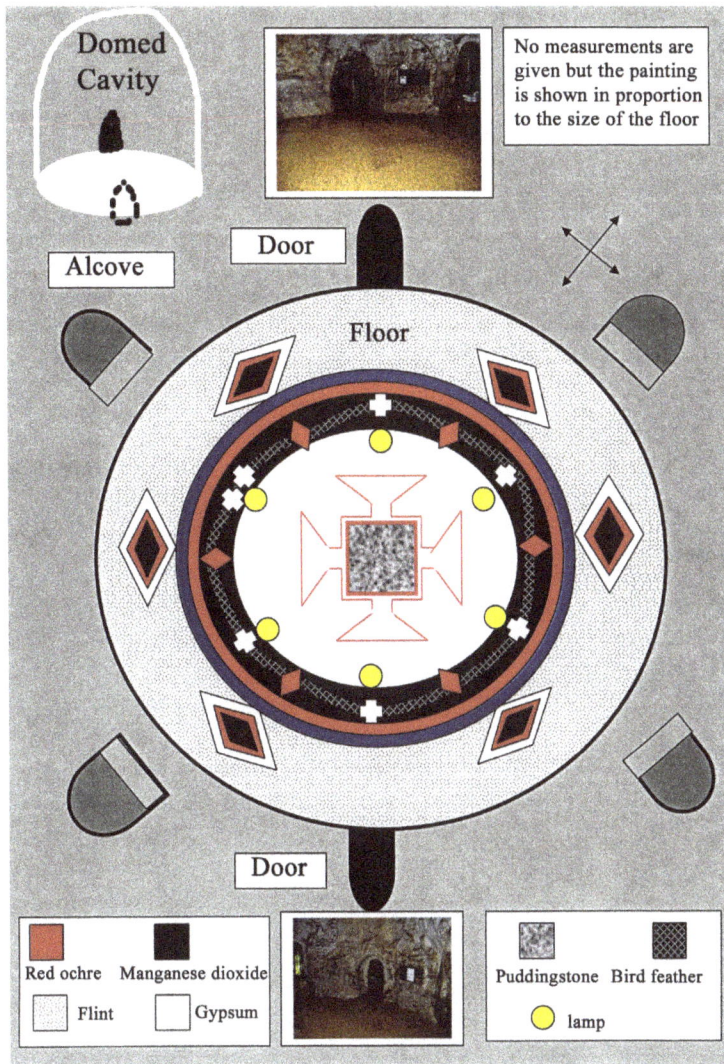

Domed Cavity

Alcove

Door

No measurements are given but the painting is shown in proportion to the size of the floor

Floor

Door

Red ochre Manganese dioxide

Flint Gypsum

Puddingstone Bird feather

○ lamp

Figure 21.3 The Banqueting Hall Floor

This painting was dowsed on the floor (now protected by the chalk and gravel of the present day floor).

The hall has a high domed roof and is referred to as a bell chamber.

Human blood can be found all over and dowses as druidic.

Figure 21.4 Tunnel Entrance

Despite the pile of rocks blocking the tunnel its entrance is easily identified on the beach at Inner Hope.

The figure is standing above where the tunnel was dowsed to be.

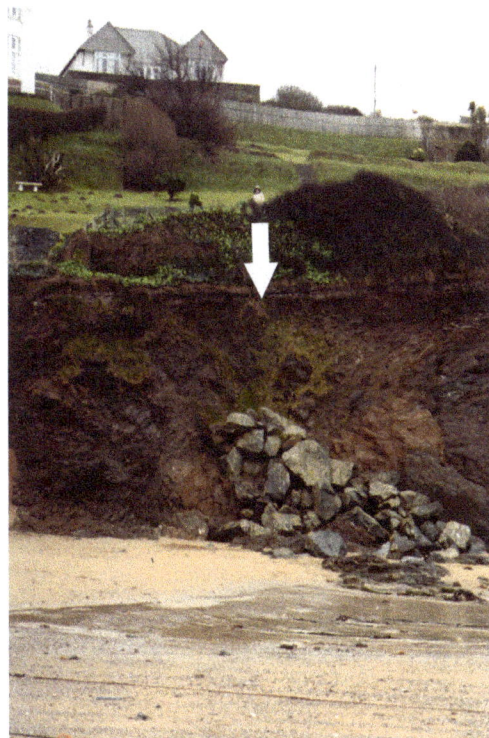

255

stream from the surface and sank a well down to it. Once in the stream it would be possible to move along it and select a position from which to dig out to the Banqueting Hall. Either way the underground navigation skill is not far short of that of present day engineers. How the Druids did it is another mystery waiting to be solved.

Part of the value of visiting and seeing an underground Druidic Temple complex was that we now knew in detail what to look for. To date we had figuratively speaking been feeling our way along underground systems trying to interpret the dowsing responses and form images of what might be there many feet below us. We had for example assumed that underground streams were water percolating through rock or chalk. It was not until we saw the River Styx that we realised that in chalk, as in limestone, tunnel systems can be formed by the running water complete with stalactites and stalagmites. Nigel had found dowsing evidence of such caves at Chiswell Green and had identified what appeared to be Druidic tunnels approaching and crossing the stream by a wooden bridge. At the time we were not convinced that cave systems and streams were present in chalk. To see such a system for real increases confidence in ones dowsing and ability to interpret findings. Once we knew what to look for and which witnesses to use, similar systems to the Hell-Fire Caves were identified at Chiswell Green and close to the Little Hay Golf Course near Bovingdon. The Hell-Fire Caves are not only Druidic tunnels and caves into which one can walk and marvel at the Druids skill but the caves and tunnels can also be dowsed on the hill above. A dowser can therefore draw on the ground the pattern of the underground tunnels and chambers and then walk round the underground system to check the dowsing.

The Stoney Lane Temple Site

Back during the summer of 2005 when we were working on the Stoney Lane temple site a well had been found which supplied water to the temples. The wellhead structure had been a substantial construction and oxen were used to pull large buckets of water up the well and empty them into a header tank. A similar system had been found at Stonehenge. At the time I had accepted the findings but had some reservations. My reservations were due to the fact that in most wells the water seeps in from the surrounding strata and so at any one time there is only a limited amount of water in the well. A 2m diameter well with water a metre deep is going to be quickly emptied by an ox drawing water. If too much water is withdrawn the sides of the well start moving into the well with the percolating water. This would not please those Druids who had to go down and clean it out. However, if the Druids were to sink their wells into an underground cave system with a stream running through it they could lower a bucket into a stream and if need be dam the stream to provide the necessary depth of water. Large quantities of water could then be removed at any one time as the tunnel acts as a reservoir. Having realised this as I was writing up my notes on the Hell-Fire Caves I decided to take a break, collected my dowsing equipment and set out for the Stoney Lane temple site. Using the new witness for finding caves I found that the well went down to a cave system and that I could trace the system through the field. At one end of the stream was the well and towards the other end the stream had tunnels approaching it and an oak bridge had been built across it. On one side of the bridge was the chamber referred to as the Inner Temple in the Hell-Fire Caves. On the other side was the tunnel going to the large hall (Bell Chamber) or temple to the Earth Goddess. The tunnel disappeared through a hedge and into somebody's garden so I could not follow it. Back at the well there were two tunnels coming up to the well but no sign of a bridge.

With an increasing number of underground temple systems being identified it became apparent that the Druids had evolved a complex system of subterranean temples with a range of different styles.

An evolution of styles is to be anticipated as the Henge Age extended over at least two thousand years and possibly four thousand. With the tombs and above ground temples an increase in complexity is observed on moving from the Stone Age through to the Iron Age. The complexity of the temples was revealed stage by stage as work progressed and we learnt more and more about them. It started with Woodhenge Temples consisting of rings of postholes. Then as discoveries were made, the temples revealed their complex design and the supporting buildings and structures around them. Then the dykes were recognised as part of temple complexes followed by the so called Hill Forts and earth rings. Dykes were roofed in and could extend for hundreds of metres and even for miles.

The underground temple complexes were following a similar progression becoming more complex with time. First a tomb then anterooms and bone depositories were found. Then access tunnels and circular tunnels round the tomb complex. Tunnel complexes consisting of three chambers and a side tunnel for a deer drawn chariot were then found linked to surface Henge Temples. This was then followed by the discovery of the big chambers to the Earth Goddess with a passage way to an underground stream and ancillary chambers. It is not possible to identify a time sequence for these developments or the presence of parallel developments at the moment. However, it does look as if the big chambers, the Bell Chambers, were the last stage of development as many if not most were sealed sometime in the late Iron Age.

The biolocation therefore indicated variety in the design of underground complexes and the presence of a very large number of them in the Hertfordshire area. The ceremonial tunnels were found to be present in Wales, both North and South, but up to June 2006 it had not been possible to find details of rooms and Halls because of the difficult terrain above them. Then Nigel remembered a tunnel in the hills above Pontypridd where he played as a boy. In June 2006 he went to find the tunnel and eventually rediscovered it. On June 18th 2006 Nigel found the battle blood outside the entrance to the tunnel along with chariot wheel tracks, the feathers and the linen of Druidic clothing. Inside the entrance a veritable Hall of a Mountain King extending back and down. Druidic dry stone walling, straight and vertical sides to the hall, positions for lamps with traces of beef fat. Two thousand years after it was last used the cave is no longer a safe place so details of its construction and what is there will have to wait. Suffice to say that the first Druidic Mountain Temple has been identified.

Walking on the sands at Hope Cove and looking towards the hotels of Inner Hope a pile of large unusual stones and boulders can be seen. They look as if they have just been tipped over the edge of the cliff above them. I had not paid much attention to them thinking that they were the remnants of some beach work or were to control wave action. However, when dowsing for fissures and tunnels on the cliff above the cove I found a tunnel. I followed the tunnel towards the cliff and eventually traced it to near the cliff edge. This indicated that there should be an opening in the cliff face. I then got my wife to stand over the tunnel and went down to the beach. There was no tunnel entrance visible. The pile of rocks I had noticed had been placed over it, see figure 21.4, presumably to prevent access.

Summary

When the quest for the Secret of the Stone Circles started the only relevant data available came from dowsing or biolocation. Support for the dowsing evidence came from modelling the dowsed systems in stone and confirming that they produced the same pattern of magnetic fields as observed on site. The next stage saw the collection of dowsing evidence which supported theories or ideas of what might be in the ground. Pieces of a large jigsaw started to fit together and as more data was gathered there was a convergence of evidence identifying what was there, increasing the probability that a particular idea or interpretation was at least near the reality. Then came the physical evidence culminating in walking through battle zones and into the underground Druidic Temples to the Earth Goddess. The ornamentation and artefacts have long gone but the chemical stains in the ground tell the story of the Druids. The large size and extent of the Druidic tunnels has led many people to suggest chalk mining as the reasons for their presence. Dates of digging are normally given as being the Middle Ages or even later. However the imprint of chariot wheels, the footprints, clothing, staff marks and the passage of deer reveal their true origin.

So far I have not found any record of the tunnels and chambers so common throughout Britain being recognised as Druidic and over 2000 years old with histories possibly going back 4000 years. There also appears to be no recognition that the above ground Henge Temples and Dyke Temples are linked physically and ceremonially to the underground temple systems. The total scale of the engineering involved and deployed by the Druids in building the above and below ground systems must rank them as one of the great builders of antiquity. They are the giants of mythology who's works are not only all a round us but are under us as well. In the mountains there are many Halls hewn for the Goddess or some other deity waiting to be discovered.

Since this chapter was originally drafted many more physical links confirming the dowsing have been identified. A Bell Chamber being the cause of subsidence in a local field. A chamber in rock strata which was too near the surface with the result that the top of the chamber collapsed to reveal the outline of the chamber and its entrance. The entrances to tunnels and chambers are now being identified on a regular basis.

Chapter 22

How Ancient are the Druids?

Avebury

By August 2006 we had learnt so much about the Henge Culture and the Druids that a lot had to be checked out on sites in other parts of the country. Part of our operating thesis was that there was a common culture over the whole of the British Isles. Regional differences yes, but there should be a common core to the Henge culture from Lands End to John O'Groats and beyond and from the West of Ireland to the East Coast of England. In addition to the geographical spread of the culture there was also a temporal spread of at least 3000 years and we were beginning to wonder if it went back to the retreat of the ice sheet from the British Isles. The first site to be revisited was Avebury Stone Circle. I drew up a list of small investigations and they included:

- To confirm that the dykes were ceremonial chariot ways and to check for their use by Stone Age, early Bronze Age, late Bronze Age and Iron Age chariots.

- To find the entrances into and the exits from the dyke.

- To find the underground stream or streams that might be associated with the Avebury Stone Circle complex.

- If we found an underground river, were there tunnels from it to Bell Chambers?

- Were the West Kennet and Beckhampton Avenues enclosed ceremonial ways following the course of an underground stream?

- Were there any sign of decorative banding, 6 + 3, along the outside of the Avenues? These might have been marked out in stone, gravel and or paint.

If the six studies confirmed the presence of ceremonial dykes, enclosed processional routes, underground streams and Bell Chambers then the Avebury Stone Circle complex would fit the design of temple complexes found elsewhere in the country. The Avebury site might be bigger and grander than most other sites but in general design it would follow the pattern found in the fields round Bovingdon and elsewhere in the country. In addition to the above we also wanted to look at The Sanctuary.

What was The Sanctuary? Nobody seemed to know. The archaeologists had dug large holes on the site but had not learnt a great deal. From some of the photographs of the archaeological digs it looked as if nothing of the original site remained so deep and widespread had the digging been. We kept our fingers crossed and hoped that the chemical stains that biolocation relied on had been taken down by thousands of years of rain into the soil beyond the depth of the excavations. There were reports that there had been two circles of stones (sarsen megaliths) present on the site in 1663. In 1930 it was discovered that there had been five rings of wood posts at The Sanctuary. There had also been two megaliths in the outer ring which appeared to be the final pair of stones belonging to the West Kennet Avenue. If this were the case the West Kennet Avenue would have linked The Sanctuary to the main Avebury Stone Circle complex. There was no evidence as to why The Sanctuary was where it was, why it was designed the way it was, how it was constructed and what its function was or indeed if it had more than one function. Silbury Hill was also on our list. Like Everest it is there and

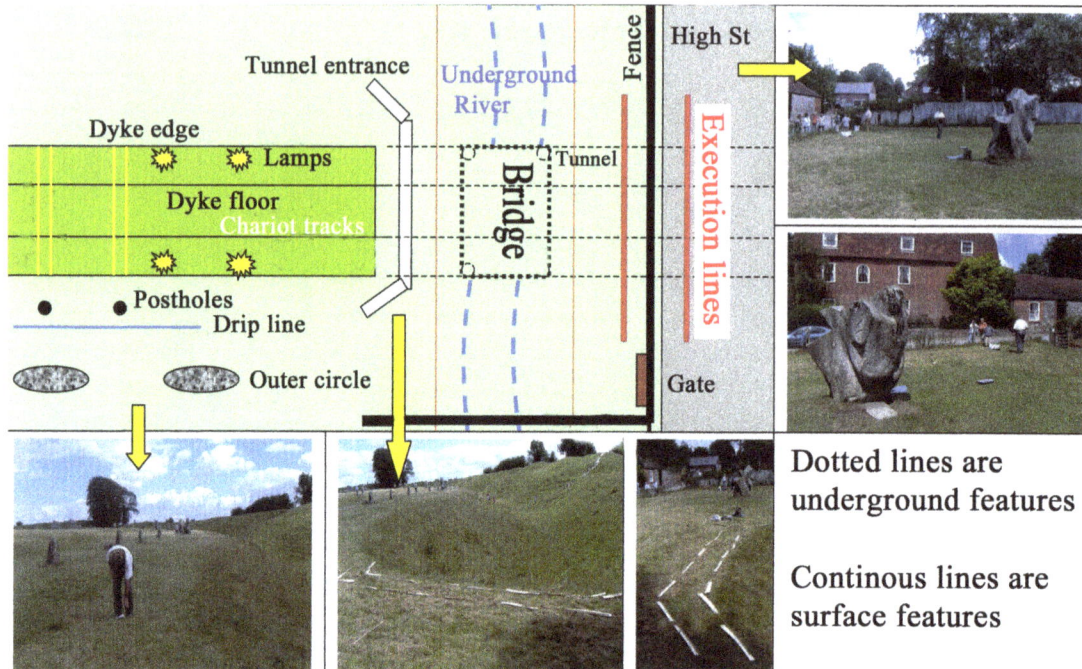

Figure 22.1 A Tunnel Entrance in the Dyke

The diagram shows the point where the Beckhampton Avenue and the village High Street cross the dyke. There is a subterranean stream under and following the Avenue along the High Street. Where the dyke meets the Avenue there was at one time a tunnel to allow chariots access to the other side. The entrance to the tunnel is now sealed and covered over. Along the sides of the dyke there are postholes and between them, going across the dyke, are the potassium lines from the rafters. Outside these postholes is a line of standing stones and concrete blocks representing missing stones. There are execution lines in front of these stones and along the High Street.

Figure 22.2 No Turning Circle for Chariots

The entrance to the tunnel from the dyke is now sealed and covered over but the lack of a turning circle indicates that the chariots went straight on under the Avenue.

260

as one Guide Book says "the original purpose of Silbury Hill remains one of the great unsolved mysteries of pre-history". Could biolocation solve the mystery? It was worth a go and even if we could not solve the mystery antiquarians had laboured over for centuries in an afternoon we might find some pointers for our next visit. Finally there was Wansdyke. A long dyke, stretching from the Avon Valley south of Bristol to the Savernake Forest near Marlborough. According to some authorities the dyke comprises a number of sections so is not one continual dyke for about 40 miles. However some sections are ten miles long or more and provide impressive examples of civil engineering. The question was whose civil engineering skills built it. The dyke is said to date from the Dark Ages 400-700AD. We knew however that there was only one group of people who would have the reason and resources to build a dyke on the scale of Wansdyke and our money was on them and they were the Druids.

Beckhampton Avenue

We arrived at Avebury just after lunch time on the second of August 2006 and decided to start on the Beckhampton Avenue. This Avenue runs parallel to and just to the left of the village High Street. Part of the Avenue is accessible via a gate almost opposite the lane leading from the High Street to the museum, National Trust Shop and café. The gate leads on to a section of the outer stone circle and behind the stones as you approach them is the dyke (Figure 22.1). The dyke stops about 20m short of the High Street and begins again on its far side. The Beckhampton Avenue along with the High Street crosses the line of the dyke at this point. It looks as if a section of the dyke had deliberately not been dug to allow the Avenue access to the circle. If this were the case the chariots in the dyke would need space to turn. As no turning space was provided, at least none can be seen, the most likely explanation for the sudden end of the dyke is that the chariot road running along its bottom had been made to pass through a tunnel under the Avenue.

The next task was to determine if there was a stream under the Beckhampton Avenue, Nigel had developed the hypothesis that Druidic ceremonial roads such as the Avenue would often follow underground streams. It did not take long to find the stream and at this point along the Avenue it was clearly following the route the Avenue was taking. Where the stream cut across the line of the dyke an oak bridge was detected. It looked as if the chariot tracks in the dyke went into a tunnel and crossed the stream using what appeared to be a wood bridge. This conjecture might be true but the oak bridge and stream could be at a lower level and many feet below the chariot way. The dyke may therefore pass into a tunnel many feet above the stream. The tunnel entrance was identified, marked out and photographed, see figure 22.2.

 It followed the same design as all the tunnel entrances that we had studied. The stones of the tunnel entrance may have long gone but they have left a chemical stain that shows what was there at one time in the past. The survey findings are shown in figure 22.1. The subterranean stream dowses as if it is about 11 ft (3.3m) wide and possibly at a depth of 45 to 50ft (13 to 15m) below the surface. The age of the tunnel is indicated by the materials used for the tyres of the chariots. The presence of nettle and hemp fibre and leather indicate a pre Bronze Age chariot. Unfortunately such chariots were used for ceremonial purposes in the Bronze and Iron Ages so they cannot be used on their own as evidence for use in the Stone Age although it is likely. Arsenic and tin were identified with copper and iron stains indicating that the dyke and tunnel were in use from the early Bronze Age to the Iron Age. The only draft animal that had left traces in the tunnel was the deer.

As the High Street was more or less on top of the western ceremonial route from the stone circles it appeared to be an obvious place to look for execution lines. The three lines of blood pools characteristic of an execution line were found running along the present day road and grass verge. This provided the evidence that the temple complex at Avebury was in use until the Romans destroyed it round about 50 AD or earlier.

Walking along the dyke to the south of the High Street we identified that it had postholes along the top edge (Figure 22.1) and there were also potassium lines from timber beams running across the dyke to its far side. The dyke was therefore roofed over at one stage. There was a drip line outside the postholes supporting this view. On the sloping sides of the dyke we found torch points. The design of the Avebury dyke followed the pattern of dykes found elsewhere in the country. It was a deep excavation with a road for foot and wheeled traffic at the bottom. There was a thatched roof over the dyke supported by wood posts and on the outside, signs of a hazel wicker wall between the posts. The wall on the far side of the dyke was not investigated.

On the inner side of the dyke there is a row of standing stones running parallel with the dyke (Figure 22.2). Many of the original stones have gone and been replaced by concrete blocks. The stones were transported to their present position on iron tyred vehicles. This indicates that they were erected in their present positions when iron was available to the Druids. If stones have been painted at anytime the paint produces a stain in the soil round the base of the stone or where the stone once stood if it is no longer there. Paint pigments were found to be present on three faces of the stones. The two sides and the side of the stone facing the dyke. The presence of paint indicates that the stones are likely to have been sheltered from the weather. To check this idea potassium lines from roof rafters and joists were looked for and found. Our initial thoughts were that the stones were intended to be structural first and decorative second. However, the potassium lines did not go to the stones but to one side of them. This in itself would not rule out a structural role for the stones. There could have been a wood lintel running between the stones. But again the potassium line did not fit such a theory and was just outside the line of the stones. That meant that we had to look for postholes. It was not long before we realised that there were no postholes in the right positions. This raised the question, if the roof was not held up by the painted stones or by wood posts what was holding it up? We knew that the Druids had mortar, cement, bricks, tiles, ceramic pipes, concrete, dressed stone and flint available long before the Iron Age and used all quite liberally. The obvious thing for them to do was build stone or brick supporting pillars and this is what they had done. Using a witness of Druidic cement the position of the supporting stone pillars could be identified. They were just behind and to one side of the painted stones. In front of the painted stones there were lamps. The standing stones are therefore likely to be all that remains of a processional way (Figures 22.3a and b). Who used it and for what are questions that will have to wait for another visit. It is likely that the processional way indicated by the stones was there when the Romans called as there is an execution line on the side facing the central stone circles.

To provide a date for the digging of the dyke we required evidence from the road at the bottom of the dyke. Having studied the tracks of wheeled vehicles, at the moment all that can be said is that bronze was available for chariot wheels. However, the dyke does cut through a number of temple sites and in theory the date of the temples, as indicated by the age of their tombs, should provide information on when the dyke was dug. Unfortunately that is a task which is going to have to wait for another visit, perhaps many more visits as the dyke is both wide and long.

The Subterranean Stream

One problem with the Beckhampton Avenue is that most of it is under houses and their gardens or on cultivated farmland and so is not accessible without permission from the owners. The West Kennet Avenue or at least extensive parts of it is by contrast on pasture land and accessible to the public. So it was to the West Kennet Avenue that we turned in order to answer some further questions and to confirm answers already obtained. The first step was to try and trace the subterranean stream to and through the main circle. This was not easy due to buildings and fences but the stream identified in the south eastern quadrant, where the main circle is situated, is almost certainly the same as the Beckhampton Avenue one. Under the stone circle the stream breaks up into a number of small streams before they rejoin as one main stream which then approaches the start of the West Kennet Avenue. It is possible that the breaking up of a subterranean stream into a number of streams attracted the Druids and caused them to select such sites as the centre of their temples and temple complexes. The number of dowsable diamonds associated with running water increases, and they reduce in size, above multi channel subterranean sites. The multiple channels of an underground stream therefore have a very characteristic dowsing signature which the Druids would have been able to identify. If the multiple water channels can be identified under other temples the reason for the Avebury Stone Circles being where they are may have been discovered (see Silbury Hill and Dragon Hill).

Moving up stream and crossing the road from the stone circle complex to the West Kennet Avenue the underground stream can be identified as a single channel and followed as it runs between the two rows of stones, only a few of which now remain. The Avenue follows the stream and although the whole length of the Avenue was not checked to make sure the stream was under it, it looks as if the shape of both of the two Avenues and the position of the stone circles may have been determined by a subterranean stream flowing from the area of the Sanctuary to the end of the Beckhampton Avenue. There is no record of a second Sanctuary at the end of the Beckhampton Avenue but there is almost certainly something similar there waiting to be discovered, hopefully not by archaeologists digging.

The West Kennet Avenue

Returning to the stones of the West Kennet Avenue (Figure 22.4). The stones were painted on their sides and on the inward face. The outer face of the stone, which was against a wicker wall was not painted. The pigments used included red ochre, manganese dioxide, woad, gypsum and henna. The bonding agents included egg white and egg yolk with egg white being found on all the stones checked. The stones were moved into position on iron tyred vehicles. There were lamps in front of the stones burning beef fat and according to an olive oil witness, olive oil. Olive oil would have been rather expensive so the witness may be picking up something else. The stone avenue is within a building and postholes can be identified running along either side of the Avenue. As the Avenue may have been in use for a thousand if not thousands of years the covered way will have been replaced many times. This means that some careful work will be required to identify the actual covered way in which the present day stones stood. However, we found evidence of stone pillars lying just outside the standing stones and these we suspect were almost certainly part of the last covered way to be erected along the West Kennet Avenue. The road built along the Avenue followed the underground river and it had hardened curbs and painted diamonds along its length. There were potassium lines between the stone pillars indicating that they supported roof rafters. Straw dust lines confirmed the role and position of the rafters and drip lines showed the extent of the roof. With the procession of priests and chariots being inside a covered way which ended either in a large building, a covered way or dyke the question arises as to where were the spectators. At one time we thought the spectators sat on the elevated banks running round the dykes but this did not make sense when we found that the dykes were covered. So far we have found no sign of spectator areas. This could indicate that religious ceremonies involved most of the

Figure 22.3a Painted Stones

The stones shown in the photograph and those that use to be there were painted. This indicates that they were protected from the weather and were in a building of some sort.

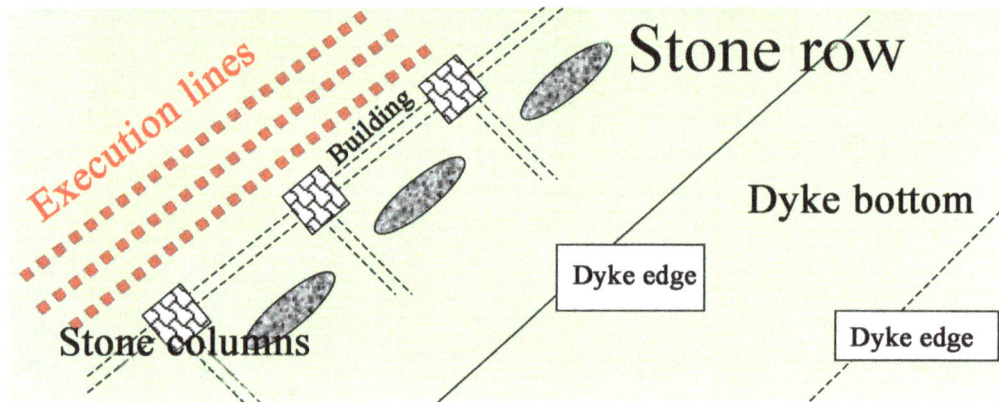

Figure 22.3b Stone Pillars

A diagram showing the supporting stone pillars for the roof which protected the stone row.

Figure 22.4 Dowsing the West Kennet Avenue

Most of the stones in the West Kennet Avenue have gone but some are now replaced by concrete blocks. The Avenue was almost certainly used for a thousand years if not thousands. The diagram represents the enclosure and the tracks as they were in the Iron Age. The use of painted stones, stone pillars and potassium lines show that the Avenue was roofed in.

population and that few were left to be spectators. We have looked for spectator stands at a number of sites but to date have only found what might be viewing stands on a cursi.

The road along the Avenue has two chariot tracks and the tracks of priests on foot. Illumination was provided by lamps in front of the stones. The hazel wattle walls would let some light in and it is possible that they could be removed when required to let more daylight in. However, the working hypothesis is that roads and in this case the Avenue, were make believe underground rivers. The energy pattern of the river, that is the diamonds, was painted on the prepared and metalled surface of the road, the road followed the meanderings of the underground stream and the cave through which the stream flowed was represented by the covered way. This for the moment is the most likely explanation for why the Druids constructed their roads in the way they did. There is, however, another possible reason for roads running in covered ways and that is it keeps the rain and snow off the road. The roads would then remain open in the winter, and vehicles would not need to be weatherproof making them much lighter. The people of the Henge Age were experts when it came to working with wood. They may not have been expert at laying hard road surfaces with foundations. The main draft animal early on was the reindeer and or the deer which have a soft foot pad, not a hard hoof which is more likely to churn up the road surface. To be able to travel independent of the weather conditions, including summer heat, would have had considerable benefits. A covered way approach to road building would have had many advantages over an all weather road surface. It is therefore possible that it is the approach the Henge Age road engineers used in preference to hardened weatherproof surfaces. Lamps would enable the roads to be used at night. As traffic developed, horses and oxen came to be used, and wheel loadings increased. As this change took place resources would eventually have to be concentrated on the road surface itself.

The presence of extensive mileages of covered roads indicates a well ordered and peaceful society. Roman society was violent and they lacked the civil discipline required for covered roads to be an option.

Wansdyke

The following day, the 3rd August 2006, was to start with a study of The Sanctuary and the far end of the West Kennet Avenue. We wanted some evidence to show that the final section of the Avenue was following an underground stream and that the processional road was still in a covered way. Then there was the Sanctuary itself with years of archaeological digging having only deepened the mysteries surrounding it. However, on our way from our overnight accommodation to The Sanctuary on the A361 we crossed Wansdyke. Wansdyke is one of those massive earth works that crisscross the British landscape. As we drove towards Avebury after one of those large farmhouse breakfasts we found that the road crossed Wansdyke with the dyke visible on our left hand side and farm buildings to the right (Figure 22.5).

Beyond the buildings the dyke climbed a slope and disappeared over a hill. We knew that if the dyke had been filled in when the A361 was built it would still be possible to dowse the chariot tracks, footpaths, postholes in fact probably everything associated with the dyke. They should be dowsable when walking alongside the modern road as the fill would have preserved the dyke. Having parked the car I started dowsing on the path which ran alongside the road and Nigel walked down to the dyke and along it to find a suitable place to investigate the structure of the dyke. The dyke is not a simple cutting in the ground. The actual dyke is the centre of three ancient roads. Looking to the north, on the left (Figure 22.6) there was a road (A) at the same level as the field next to it. This road had bronze chariot wheel marks and painted diamonds. Then on its right was the bank (B) and then below this bank the bottom of the dyke (C), a climb back up to field level (D) and then another road (E). Whilst road (A) seemed to be Bronze Age on road (E) the short axle length of the vehicles and the use of wooden wheels on road (E) indicated that it was perhaps of Stone Age origin. Down in the dyke (C) the diamonds, lamps, bronze and iron chariot tracks indicated that it was a ceremonial dyke spanning at least the Bronze and Iron Ages. The dyke appeared to be following a stream of water but with such a big excavation in the ground the water may be following the excavation not the other way round. After about thirty minutes work it was clear that Wansdyke was built during the Henge Age. It was also probably in use as an excavated ceremonial road from the early Bronze Age if not the Stone Age. The length of the dyke indicates that it is too long for it to be following a single subterranean water course. The reason for it being where it is and why it is so long still has to be discovered. The reason may be a very simple one such as joining up separate underground streams with above ground 'streams'. To show if this is the case or not would involve an extensive dowsing survey but it should be possible. Having shown that Wansdyke is a Henge Age ceremonial dyke and predates the Dark Ages by a few thousand years the next dyke on our list was Offa's Dyke. The same reasoning applied. The only people with the skills, resources and reason to build such a structure were the Druids. We had got it right with Wansdyke. Would we be right with Offa's Dyke. However Offa's Duke would have to wait. The Sanctuary a few miles down the road was our next stop.

Figure 22.5 The Wansdyke
The photograph shows a section of Wansdyke as it is today. The excavated dyke is the central part of a three lane road system. All three may not have been operational at the same time. It is easy to see that if the dyke fell into disrepair vehicles would have developed a road running alongside it.

The Sanctuary

To most people looking at a map of Avebury and seeing the Avenues leading to the stone circle complex, it looks very much as if the Avenues are processional ways. If they are, then we need to start looking for viewing areas where the audience and spectators stood, or do we?

266

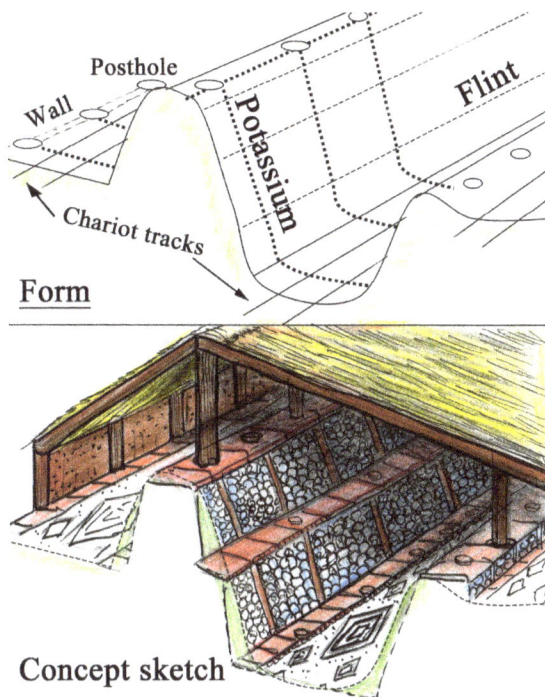

Figure 22.6 Concept Sketch of Wansdyke

The precise detail of a section of the dyke was not dowsed due to time limitations.

However from the form, as with many other dykes studies, (and in particular a detailed study of the huge dyke Beech Bottom in St. Albans) it is clear that the dyke had a substantial roof structure as part of a building covering the roads.

Inside, the walls of the dyke were clad in stone, highly painted and decorated and possibly terraced.

Table 22.1 The Analysis of a Tomb at The Sanctuary

The tomb contents indicate a Stone Age Temple.

	Female	Male
Dog Hair	x	√
Wool Garment	x	√
Linen Garment	x	x
Cotton Garment	x	x
Nettle	√	x
Hemp (Body)	√	x
Hemp (Bow)	x	√
Yew (Bow)	x	√
Feathers (Body)	x	√
Leather (Belt)	x	√
Leather (Shoes)	x	√
Urine Tanning (belt)	x	√
Urine Tanning (shoes)	√	√
Pig Skin Shoes	√	x
Human Hair	Long	Short
Deer Antler Head	x	√
Cat	√	x
Henna	Torso	Torso
Egg White	x	x
Red Oxide	From the Chest Up	From the Chest Up
Tree Resin	Body	Body
Linseed Oil	Body	Body
Linen Thread	x	Small patches
Gold	Chest	x
Silver	Head	x
Bronze	x	x

This idea that a procession must have spectators is most certainly the result of a modern mind set. The chances are that two or more thousand years ago everybody was a participant in one way or another. The fact that covered ways stopped the procession being seen was not likely to be of any great consequence. Viewing stands would probably not have been required as everybody was involved. What would have been required are assembly areas, places to harness deer, horses, line up priests, toilet and wash areas, robing areas, water and possibly even areas to provide food. In looking at the map of Avebury, The Sanctuary looks as if it is a contender as an assembly point for the start of processions down the West Kennet Avenue. There was only one problem. It could have been in use for 3000 years and we would have to separate the different layers built up through time.

I had visited The Sanctuary on a previous occasion and knew that it had postholes in a pattern characteristic of a temple. On arriving at the site the first step was to determine if at one stage in its history the Sanctuary had been a temple. There was one set of dowsable postholes with the characteristic 6 + 3 pattern. At the centre was a tomb. Tombs are time capsules and so we spent a while determining what was in it. The occupants, a male and a female, were small and almost certainly children. There were deer antlers in the tomb and a bow. The materials used in the tomb are listed in Table 22.1. The contents are characterised by a lack of bronze and linen. The use of children as sacrificial victims is not unusual and not characteristic of a time period as far as we know. No shield was detected. The structure of the underground temple had the Celtic Cross or Thunderbird form characteristic of the Henge culture. We concluded from the tomb contents that the site had been used for a temple during the Stone Age. To confirm the sites use as a temple we checked for and found the clay of the central pool. The next step was the source of water for the pool. To me it appeared that the obvious thing for the Druids to do would be to dig down to the stream that the Avenue was following. However, this was not the case and I found that a ceramic pipe had run from the pool to the east. I followed it outside the limits of the temple, across a path running past the site, possibly the Ridgeway, and into the next field. Here I found the well and the underground stream supplying it. The well confirmed that we had found a typical temple dating from possibly the late Stone Age or early Bronze Age. Quite clearly the temple predated the present processional way, the stones for which were moved in the Iron Age. In addition to the temple there were signs that the site had been used as a deer pen, possibly for harnessing deer to chariots. Later, the site was used as the covered end of the processional way going down the West Kennet Avenue. The position of stones and postholes leading to the Avenue were identified.

Fortunately, the excavations over the years at The Sanctuary had not destroyed many of the stains in the soil. It is possible that the stain penetrates the 'bed rock' which normally defines the working depth for archaeologists. If this is the case it may be possible for dowsers to investigate graves where bodies have been removed and still identify sex, clothing and artefacts that may have been buried at the same time. Using the chemical stains in the soil it was possible to show that The Sanctuary had been a religious site possibly as early as the Stone Age. It was also a religious site in the Iron Age and the chariot tracts indicate that it was part of a ceremonial system in the Bronze Age. The uses to which the site was put over possibly 3000 years varied from a stand alone so as to speak Henge Temple to the starting or finishing point for processions centred on Avebury. That is the site became just one component of a very large ceremonial complex following one or more subterranean streams. The site may have been selected because it is above an underground stream or its source. There is also another underground stream close by which had been identified in the Stone Age and was used as a source of water for the temple and possibly for the people living round the temple. As this stream was capable of delivering significant quantities of water it may have been the one that determined the siteing of the temple. Three thousand years is a long time so The Sanctuary may be hiding many more uses to which it was put over that time period. The final use of The

Sanctuary was as part of the covered way system associated with the West Kennet Avenue and it was almost certainly destroyed by the Romans.

Silbury Hill

On the afternoon of 3rd August we moved the site of our investigations' to Silbury Hill (Figure 22.7). There is a car park from which the hill can be viewed and between the car park and the hill a large expanse of thistles which ensures that most people stay in the viewing area by the car park. We, however, had to find out what was round the base of the hill and that meant walking through hundreds of yards of thistles. As we approached the base of the hill we started looking for chariot tracks round the base. We soon found them. They included early and late Bronze Age and Iron Age tracks. The chariot tracks from the west came in and headed for the hill and disappeared into it. This made it look as if there should be a tunnel entrance in the hillside or that the tracks predated the hill. Above the tracks there was indeed a depression in the side of the hill as if there had been the opening of a tunnel. However, we kept an open mind on tunnels into the hill. We knew that one reason why the archaeological digs had produced so little information was because there was little if anything in the hill, it could all be below it or it may have been on top of it. The hill could have been built to reach into the sky from some ancient sacred area. If this were the case the secret of the hill lay below it or at its top and so far archaeologists appear to have neglected both by going for the middle. Walking clockwise round the base of the hill, chariot tracks could be followed. At approximately the north and east quarters the tracks went through the bank surrounding the hill (Figure 22.7). This is what would be expected with a temple. The chariots could not go all the way round the hill which is joined by a causeway to the elevated land along which the A4 runs. Following the clockwise sweep of the chariot tracks towards the road (the bank leading up to the A4) the chariot way expands into a large turning area (Figure 22.7). The chariot tracks can be followed and they head for the left of the turning area and the elevated carriage way of the A4. The tracks do not turn but go straight into the sloping side of the bank leading up to the A4. It is possible to climb up to the top of the bank which I did. Once level with the road I found bronze tracks leading to the A4 and under it. From the top of the bank it is also possible to pick up the splayed tunnel entrance. The tunnel has been sealed and its entrance is no longer obvious, except to the trained eye. The tunnel was not followed under the A4 and its return to the hill not identified. Having found that ceremonial chariots of the Bronze and Iron Age were being used round Silbury Hill the next stage was to see if the site was being used when the Romans arrived. The presence of extensive execution lines answered this question in the affirmative. Execution lines were found round the north side of the bank and although they were not looked for they may also be round the west and east gates.

There seemed to be plenty of evidence indicating that round the base of the hill there was a large ceremonial area which had been in use from possibly the Stone Age to the arrival of the Romans.

The actual hill itself is fenced in to discourage access although a few people brave the thistles to find the footpath up the hill. From a distance, a hint of a terrace system can be seen on the sides of the hill but the hill is not sufficiently large for a spiral terrace to take chariots up to the top. The small lengths of terrace or alcoves in the side of the hill could be the result of digging for materials, archaeological work or they could be an original feature of the hill. To find out we would have to climb up and look at them in detail. We followed the path to the top and decided to start our study there. The top is an area about 60m in diameter so it was large enough to build a temple of some sort. We set to work to find postholes, tomb, pool, gates, altars and it was not long before we were able to piece together a temple. A temple 40m above ground level!

Figure 22.7 The Silbury Enigma

From the air Silbury Hill looks enigmatic and the mind struggles with possible purposes and uses for such a structure. The engineers of the time would have used cattle as motive power and lifts and scrapers for handling spoil. The task is unlikely to have involved the armies of people beloved by archaeologists. The flat ceremonial area round the hill and the Ceremonial Courtyard can be seen. The chariot tracks going round the hill appear to start with Stone Age ones then develop into Bronze Age and Iron Age tracks. From the Ceremonial Courtyard chariot tracks enter a tunnel under the A4 road. The short lengths of terracing on the side of the hill provided support for stone pillars and posts supporting structures built on the hill.

Aerial photo by JJ Evendon

270

If the ancients were capable of building a 40m high hill a temple would not be too difficult. However, the clay pipe plumbing we had found to the pool did not look as if it was intended for a hydrostatic head of 40m. There was however, just outside the outer wall, signs of lead plumbing. Another odd feature was that the gates on the edge of the plateau were opening into space, unless there were wood steps from them to the ground. The hinges of the gates were hemp and nettle that is they were typically Stone Age. The tomb was in the centre of the temple which is where it should be. As this was close to where the subsidence had taken place if not on it, we began to wonder what we were looking at. However, as it was now late we decided to call it a day and return to our bed and breakfast accommodation. A quick clean up and then onto a near by pub for some supper. A glass of local beer helped to ease the fatigue of hours of dowsing and we started to think through what we had found at the top of Silbury Hill. The dowsing was clear and precise. The tomb was about the right size. The plumbing to the pool was there, it could be traced down the side of the hill. The temple fitted the top of the hill like a glove. But surely the tomb would have been discovered by the archaeologists. If not the tomb then an antechamber at least. The other possibility was that the tomb was below the hill. Possibly 45m down, about 146ft. Could a temple still show up with 146ft of chalk above it? The idea that the magnetic fields from chemical stains dating from over 2000 years ago could penetrate such a distance took some believing. If it was true, we had found what was below the hill and possibly one of the reasons why it was where it was. It was clear that the tomb had to be dated as the hill must postdate it. By now we were sure that our survey of the top of Silbury Hill had been a survey of what was below it and that meant another survey. This time, after dating the tomb, we were going to ignore the temple and look for something quite different. The hypothesis we were working on was that the hill had been built as a communication centre. When working round the base we had looked for blue lines and had found one coming in from the west but the few blue lines we had picked up as we worked our way round the hill did not indicate a communication centre. On the top of the hill it might be different. Our tasks for the following morning were clear: date the tomb, identify the blue lines and then see if there was a built structure on the top of the hill or on its sides and if still time, look at the underground stream to see if it split into many channels under the hill. If it did, that could be the reason for the ancients selecting the site for the hill. After a second glass of beer we joined local people and visitors to watch Morris dancers and a visiting German folk dancing group perform in the village High Street. We did wonder if Morris dancing went back to the Henge culture, but if it did, how could we prove it. Perhaps the answer could be found in another glass of beer.

The following morning we were back on the hill. The tomb contents matched the one at The Sanctuary so was likely to be Stone Age. The ancillary parts of the underground temple such as anterooms were present. This meant that we had found what the hill had been built on. The precision with which the top of the hill fitted the outer wall of the temple indicated that the hill was precisely placed by engineers who could work to fine tolerances. By the time the construction workers had reached the top of the hill, the temple would have been lost sight of. The engineers would have to know the location of the temples outer wall under the hill and how to position the rim of the plateau at the top of the hill. An alternative to surveying methods is to dowse the temple from the top of the hill. As we could do it the Druids could almost certainly have done it. Next we checked the blue lines. Unlike the situation at ground level the top of the hill is a veritable cross roads of the blue lines. The theory is that blue lines are used for communications, for spirits to travel along them. If this is the case Silbury Hill could be an important communication centre. We just have to work out how the Druids might have used it and where the blue lines are coming from and going to. This would require a major study. All that can be said from a short visit is that Silbury Hill may have been a communication centre.

The next step was to look for signs of buildings at the top and on the sides of the hill. The most important indicator of building activity is provided by the stains left by Druidic cement and mortar. These stains were in the soil and stone pillars were identified supporting wood structures on the top and on the sides of the hill, particularly the small areas of terracing. It looks very much as if Silbury Hill had buildings on the plateau, at the top and in places these extended out from the plateau area on supporting stone pillars. The layout of the buildings should provide some indication as to what was going on. For those interested in ancient astronomy it is well to remember that most stone circles were in temples and those that were not were in covered ways or were part of a Dolmen. A manmade hill rising above ground mists and smoke from wood fires and with a good view of a distant horizon, would be a much more suitable site for astronomical observations. A final check was made on underground streams and it looked as if there were a number of water channels under the hill. The hill had been built on a sacred site defined by a multichannel stream, and a previous temple.

Bell Chambers

Having completed a pilot study of Silbury Hill and obtained some idea as to when it may have been constructed, its structure, why it is where it is, what it might have been used for and evidence that metallic lead was available in the Stone Age we left for a final look at the main circle. The task was to follow the underground river out of the circle and across the road to the Avenue. Then to follow the river up the Avenue. We were going to look for the oak bridges across the river as these would show us where to look for the tunnels leading to the Bell Chambers. If you are going to dig a Bell Chamber, then under a hill or a mountain is a good place. Waden Hill rising to the west of the West Kennet Avenue could be home to the subterranean halls of the Druids.

The first task was to pickup the underground stream as it left the Stone Circle near the southern entrance and confirm that it was in a subterranean cave. This was done and then it was followed as it went under the dyke. At the bottom of the dyke an oak bridge could be detected indicating that the Druids had access to the stream at this point. The stream was then followed until it entered the Avenue on the far side of the modern road going to the village of Avebury. Once in the Avenue the stream was easy to follow. We were now alongside Waden Hill and it was not long before an oak bridge was identified. To the east of the stream was a short tunnel leading to a room which we refer to as a shrine. The tunnel to the west was much longer and it was not until it was well under Waden Hill that it came to an end. As in all previous cases the tunnel from the stream entered a circular service tunnel which was surrounding a Bell Chamber. The chamber was approximately 15m in diameter. The Bell Chamber and surrounding tunnel was not looked at in detail. The aim was to show that the Avebury Stone Circle complex followed the normal pattern for a Druidic underground temple complex and this it did.

Marburgh Henge

Four days after returning home from Avebury I was on my way to Scotland and the Glamis music festival. Village Water, a small charity group based on amateur water diviners, had been given permission to make a collection at the festival in aid of the charity's activities in Africa. I, along with other volunteers, was going to be dressed in a smart yellow jacket and carry a bucket round the thousands of music lovers.

However, there was going to be one stop on the way at Penrith in the Lake District. Earlier in the summer when attending a meeting at Penrith I had visited a henge site called Marburgh Henge. The henge has a high bank of stones round it and a clearly defined eastern gate. The gate had been the site of a bloody battle between the Druids and Romans. There was blood everywhere in the gate area so I knew the site was in use up to Roman times. Inside the bank I had found one temple at the centre of which there was still a standing stone (Figure 22.8). I had picked up the traces of paint and gold round the base of the stone and the clay pool in which it stood. On this visit I wanted to have a closer look at the standing stone and gather evidence to show that it had been the central feature of the pool. Nigel and I had found the outline of plinths or of a central feature in every pool we had looked at. The standing stone in the Marburgh Henge was the first real central feature we had come across. We did not count the rocking stone at Pontyprith as we were not certain at this stage that the Druids had erected it. It may have been left by the Ice Age and the Druids then built their temple round it. On this second visit I was able to confirm the paint round the base of the standing stone (Figure 22.9) and that it had been in the centre of a clay or tiled pool. The standing stone appeared however to be set in a bed of boulders and there was no visible sign of clay. We had always looked upon the pool as being made of clay but it could have been constructed from clay tiles and bricks. If this were the case they would have been of value and soon removed to leave a clay stain in the ground. Round the standing stone, artwork had been drawn on the ground at sometime. The Thunderbird on the eastern side was plotted out on the ground. It was complete down to the eye, eyebrow and drops of blood. The Thunderbird appeared to go right up to the standing stone and may have been drawn on the ground long before the temple represented by the stone was built. The Marburgh Henge appeared to contain a typical temple in terms of what was above ground level and what was below the ground. The life of the Henge, like that of many other sacred sites, probably spanned thousands of years. However, the tombs which would prove this point were not looked for. I had only come to see if the standing stone was sentinel to the demise of its Druids. Before going further north I had to check the road running past the farm I was staying at in the Lake District. I soon found the diamonds of a Druidic road and they followed the modern day road from the north then went straight into the forecourt of the farmhouse while the modern road minus diamonds curved to the left to avoid the farmhouse and yard. Roads were now one of our main interests. Having discovered that the Druids had marked their roads so clearly we knew that in time it would be possible for dowsers to map out the topography of the Henge Age from its cities to its villages. They would identify its trunk roads and country lanes and the ones the Druids had bequeathed to the modern inhabitants of the British Isles.

Figure 22.8 Marburgh Henge

Part of the central megalith of the Henge can be seen in this photograph. One of the four Thunderbirds has been marked out using canes. The Thunderbird was painted on the ground using manganese dioxide, a black pigment. The eye and blood droplets are in red ochre and there is a manganese dioxide eyebrow over the eye. The canes on the right show where a complete circle of black pigment has been painted round the standing stone.

Figure 22.9 Marburgh Henge Central Megalith
The single standing stone of the Marburgh Henge was at the centre of a temple. Traces of the paint and gold used to decorate it can still be detected round its base. Beneath it will be the underground tomb complex. The stone is seen against the Eastern Gate. The ground outside the gate is covered with blood stains indicating a fierce battle with the Romans.

Central Scotland

On Thursday 10th August 2006 I left the Lake District and headed for Central Scotland and the countryside round Glamis. I had booked a room in a farmhouse near Lintrathen not far from Glamis and its castle. When on dowsing trips I always tried to find accommodation on farms as they very often had one or two fields providing a home for some interesting archaeology. The farm I was to stay at for the next five nights was no exception, although I did not know it when I arrived. My main aim and second reason for being in Scotland was to study some recumbent stone circles. On an earlier visit to Scotland I had looked at two such circles, the Easter Aquorthies and Loanhead of Daviot Stone Circles. Both are a little further north than Glamis and in Aberdeenshire. At the time of my first visit I had checked for energy rings round the stones and for the postholes of a woodhenge. At the time the stone circles did not look very convincing to me, partly because of the recumbent stone. When I first looked at recumbent stone circles in Ireland I thought the role of the recumbent stone was that of an altar or to provide a focal point for ceremonies. The size and shape of the stones I had seen in Scotland did not fit this idea. It was going to be sometime before I finally worked out what the purpose of the recumbent stone might be. It was probably the Drombeg Circle in Co. Cork that misled me. Its recumbent stone was dressed and looked as if it was an altar.

On the final stage of the journey north and as I was approaching the small town of Kirriemuir, I spotted a standing stone in a field. It was asking to be investigated so after finding a place to park I made my way back to the stone. The stone appeared to be on elevated ground and at first it was not clear what it might be. Lone standing stones are often part of a Dolmen and this proved to be the case with the one I was now looking at. The processional way from the Dolmen showed that it was an eastern Dolmen, the first stone belonging to an eastern Dolmen that I had found.

Having satisfied myself as to what the stone was I continued the journey. After getting lost in Kirriemuir I eventually found my way to the farm I was staying at for the Glamis music festival. The following morning I was up early and jogging round a reservoir not far from the farm when I spotted two Dells in the fields I was passing (Figure 22.10). Next morning I took my dowsing rods with me when I went for a jog. I made a diversion up to one of the Dells and started dowsing. It was as I had expected, the Dell indicated the remains of a Druidic Temple and it looked as if it followed the same pattern as Dells and Temples hundreds of miles to the south. If complex ceremonial buildings such as temples follow the same design in the south and north of the British Isles it indicates a very strong cultural link between them. It is often forgotten that the division of Britain into three parts, Scotland, Wales and England, came about as a result of the Roman occupation. For thousands of years prior to the arrival of the Romans the British Isles was almost certainly one cultural unit. At least we have not as yet discovered any evidence indicating otherwise. After recovering from the jog, a good breakfast and now armed with some evidence that the Druids were in Scotland I drove north up the A93, a road that runs from Blairgowrie in the south then north to Braemar and then east to Aberdeen. I soon started to recognise signs which said that Druidic engineers had been at work. Their earth works and barrows were so abundant that south of Braemar I decided to stop and see if the modern A93 road was following a Druidic road. Taking advantage of lulls in traffic I soon found the diamonds painted by the Druids on their roads. Two rows of the large diamonds with the smaller diamonds between the larger ones in each row. There were two chariot tracks with traces of leather, arsenic, tin, iron. The road, time wise, could span the Stone Age to the Iron Age to modern times. The A93 or at least one part of it was probably 4000 years old at least. The little bit of the A93 that I looked at appeared to be right on top of the ancient road. I checked the road further on and it was still Druidic. I then drew into a car parking area, the ground dropping away on the valley side down to the river. Having again found diamonds in the road running past the parking area I began to realise that there was something odd about the car park. The ground had been built up on the slope of the valley side and my first assumption had been that the parking area was modern and probably made by the Local Authority for the benefit of tourists. However, the flat area did not look as if it had been designed or built for cars. It looked more as if drivers had found a piece of land suitable for parking and were using it. There were other features that looked odd.

Figure 22.10 Dells in the Scottish Landscape

Dells are characteristic of temple sites. For some reason the land that has been used as a site for temples for possibly a thousand years or more becomes compacted, depressed into a depression or in some way made difficult to use for agriculture. Dells appear to be as common in parts of Scotland as in Southern England.

There was a drive way from the road along the outer edge of the parking area. Perhaps created when the spoil was brought in to level the area. That idea did not seem to fit as there were signs of the track going round the base of the elevated ground. The track then left the edge of the parking area and headed for the embankment the road (A93) was on (Figure 22.11).

Figure 22.11 Druidic Highway

The A93 road on the way north to Braemar lies on top of a Druidic road. There are two lines of diamonds indicating a two lane highway. The hard curb of the Druidic road is about where the present edge of the road is. The level car parking area is Druidic and the elevated terracing for the road could also be Druidic. The scale of their terraforming can be seen from the terracing which now supports the modern road.

Figure 22.12 The Tomnaverie Stone Circle

The large size of the recumbent stone and flanking stones appears to be deliberate. From them a blue line is projected across the valley. An analysis of the tomb and temple should enable the recumbent stone approach to energy engineering to be dated.

276

I checked the parking lot for postholes. There were plenty of them so something had been built there at one time. I then checked the track way for bronze chariot wheel marks and found them. The track was a Druidic road. The chariot wheels followed the track and went straight into the embankment, then under the road and I could follow the chariots by scrambling up the mountain side. The Druidic road round the car park was clearly ceremonial and was probably heading for a ceremonial chamber in the mountain. The similarity between Wales and Scotland was striking. After passing Braemar I checked the road again and found the chariot tracks and diamonds. I had a feeling that there had been a major Druidic road which was about the same width as the present one, running through central Scotland at the time the Egyptian Pyramids were being built in about 2600BC. As is the case with the modern Egyptians and their maintenance of the pyramids the Local Authority has done little to improve what their ancestors have left them. The road is still the same width it was 4000 years ago and the same parking areas are being used. After Braemar the A93 heads east and at the town of Aboyne a left turn onto the B9094 leads to the recumbent stone circle of Tomnaverie, (Figure 22.12) not far from Tarlan.

Figure 22.13 Energy Engineered Recumbent Stone Circles
Using bricks a recumbent stone circle can be modelled. This approach to producing a blue line is effective. It requires a stone (brick) circle, properly energy engineered, and a recumbent stone correctly orientated magnetically otherwise it will not work. The canes show the blue beam coming from the circle. The ancients knew how to do this three thousand or more years ago.

I had not intended to look at this circle and had hoped to make it to the Easter Aquorthies or the Loanhead at Daviot Circles. However, I had spent so much time looking at roads, dodging traffic and finding lunch in Braemar that time was now short and as Tomnaverie was close that was the most convenient circle to look at. On arriving at the stone circle it was clearly a rehabilitated one for the tourist trade. It was on high ground on the edge of a quarry so partly fenced around to prevent people falling over the edge. The fencing restricted movement round the site but the sites well kept grassy areas made the circle and its stones easy to access. The Tomnaverie stone circle is a genuine ancient recumbent stone circle. The encircling magnetic fields are there and there is a central tomb with a Prince and Princess. The textile fibres present were hemp and nettle. There was no linen or bronze but gold was present. The tomb contents indicated a Stone Age site which did not fit the presence of stones in a circle round a clay pool which, if they belonged to the tomb would be about 4000 years old. I had not come across stone circles that old but of course it is not impossible. I did not have time to study the detail of the site and see if there were a number of tombs in or around the circle. My main aim was to see if the circle was designed to produce a blue line or beam. Walking in front of and across the recumbent stone I picked up the blue line heading out across the valley. Much to my delight the real recumbent stone circle behaved as the model in the garden (Figure 22.13). It was now clear that the Druids had at least two methods of making powerful blue lines. The power of the line would be related to the size of the stones used. The bigger the stone the better and this was why there were recumbent stones so large that they could not be used as altars. The shape of the stone was also not important as it was its magnetic field that the ancients were after. The stones did not have to be dressed unless they were to be painted and smooth surfaces were required. The large rough misshapen stones now made sense.

Figure 22.14 A Sacred Stone
The large boulder on Brankam Hill looks as if it was left by an ice sheet. However, the hill is free of boulders and the question arises 'was it put there?' and if so how and why. The boulder would not be visible as it was in a building. The site of the altar is indicated by the white canes. Similar boulders have been found in S.Wales.

Brankam Hill

With a long way to go back to base I set off intending to return the following day to the site or to Easter Aquorthies. That evening I took advantage of the long mid summer evenings of the northern latitudes and started to look at the road past the farmhouse where I was staying. The diamonds and chariot tracks of a Druidic road were clear to 'see'. The wheel tracks contained copper, arsenic, tin and leather. There had been two sets of wattle walls with the inner pair indicating that the road may have been laid down in the Stone Age or early Bronze Age. Coming along the road from the east, the present day road follows the Druidic one until it is just past the farmhouse when it bends to the right and leaves the Druidic road. The road told me that I was in Henge Age Country. On talking to the farmer and his wife I discovered that their 800 acres were full of archaeological sites including a Bronze Age one on a near by hill called Brankam Hill. From the road outside the farmhouse it was possible to see a large lone stone on what appeared to be the brow of the hill. I say appeared as the true height of a hill is often hidden behind what appears to be the top and so what starts off as an easy climb turns into a mountain assault. However, as I could see it I decided to investigate the lone stone which had probably been left by the last ice sheet as it retreated north. It would make a good starting point for my own archaeological survey of the farm. The Druids would not have been able to resist a large stone in such a commanding position and there should be some traces of their activities round it. The following morning I started up Brankam Hill. It was not long before I realised that I should have looked for the path the Druids used as at times I was being forced to scramble and to take long looks at the countryside unfolding below me. The slope then eased and I could see the boulder a few yards ahead. Now that I was there facing an enigmatic boulder that travellers along the road below would have looked up at for thousands of years as they journeyed past, it suddenly dawned on me where the old army saying 'If it moves salute it if it doesn't paint it' came from. It must have come down from Druidic times and as the boulder did not move they would have painted it. Out came the witnesses and I started to investigate the area round the base of the stone (Figure 22.14). I found egg yolk, red ochre, gold, woad, henna and manganese dioxide. The next step was to see if the stone had been protected from the weather. I found that weather protection had been provided and there were traces of wattle walls round the stone in the form of a rectangular hut with a straw drip line a little way out from the walls. Outside the building that enclosed the stone there was a depression in the ground. An altar stone had stood in the depression and there were traces of terracotta tiles or clay and animal blood. There was no sign of human blood and the altar did not appear to have been enclosed. The next step was to pick up the processional ways. I identified a path from the stone and altar using a feather to identify the priests clothing. I could not identify linen but I did get a response with hemp and nettle. This may indicate that the stone and altar was in use during the Stone Age. However, as said many times, religious ceremony is by its nature conservative and the Druids using the site could be early Bronze Age. I followed the feather path round a contour of the hill to the west and came across another depression in the soil. This was the site of another altar for animal sacrifice. By now I could see a path coming up the hill from the west and the fields below. It was the route I should have taken. The gradient was shallow and suitable for priests in robes. The feather trail was just to one side of the path being used by present day walkers. On the other side of the path from where I was standing the ground rose to the true top of Brankam Hill. On climbing to the top it was clear that there was a wide expanse of almost level ground, the sort of place for a temple if you are going to build one. I started to look for signs of a temple and soon homed in on one and its central tomb. The rings of wood posts appeared to be the normal, 6 + 3 with a diameter of about 60m, that is provided I had not confused two temples. The reason for the caveat is that I could only identify one body in the tomb at the centre of the circle which raised the question as to whether I had found a tomb belonging to a pre-Roman temple or some other age. There was no bronze in the tomb and as luck would have it I had left my witnesses by the path below so could not investigate further. However, I now knew that the hill had temples on it and many signs of Druidic activity. The recumbent stone circles further north would now have to wait whilst I explored Brankam Hill. Later that day and after more discussions with the owners of the farm I discovered that there was a much easier route up the hill from the east. Nearly half a mile down the road from the farm there was a track leading through fields and then gently rising up the eastern side of the hill. On Sunday 14th August I left the car at the bottom and started along the path to Brankam Hill. After a few yards I knew I had to check the track. This I did and it was clear that at one time bronze tyred carts or chariots had used it. The painted diamonds confirmed that it was Druidic as was the modern road from which the path came. From where I was studying the track I was only about a hundred meters from the road and there in the field to the east of the track was the tell-tale imprint of a temple. The only problem was that lined up on the other side of the fence were the inquisitive faces of about twenty cattle watching my every move. I decided to give the temple a miss and instead to look for execution lines. I soon found them running along the track to the hill. I now knew that the Romans had made it to Central Scotland in their search for Druids.

The temple in the cattle field was probably in use at this time. Following the track up the hill it took a bend to the left to avoid a gate which was straight ahead and leading into a field. The imprint of one temple if not two could be seen in the field. The track I was following still had bronze wheel tracks with arsenic and tin and indications of Stone Age wheels. There was also iron and signs of deer, horses and oxen along with nettle, hemp and leather. The leather had been tanned with urine. The linen showed as a diffuse signal from the road and did not indicate a clear ceremonial path. All the signs were that the Druidic road had a long history. Looking at the landscape there were not many places for a roadway up the hill. The best path had been selected by the early settlers in the Neolithic and that path was still being used today four to five thousand years later. There were execution lines along the track indicating that the temples had been in use when the Romans arrived. I left the main track at this point and started heading further up the hill.

Ancient religious sites, sacred sites, did not cease to be sacred when the Romans and Christianity arrived in the British Isles. The burial practice changed dramatically first under the influence of the Romans then under Christianity when bodies were buried along an east west line instead of north south. Climbing the hill, stones marking graves could be seen. I stopped to look at an east west grave and found that it was a child dressed in linen. I had come across the use of Henge Age sacred sites for child burials in Ireland. My guide at the time had explained that if children died before being baptised they could not be buried in consecrated ground. For the parents the next best thing was a stone circle or known sacred site. The presence of child graves on Brankam Hill could indicate that it was at one time regarded as a sacred or holy place. Following the incline up I came to a flat level area. I was not sure if it had been levelled by engineers or was a naturally flat area. Whatever its origin it was an ideal place for a temple. I soon found a tomb more or less in the centre. It was about 2.7m (9ft) by 1.5m (4½ to 5ft) and it lay on a magnetic bearing of 325° - 330°. There was only one occupant and in that respect was similar to the tomb I had found the previous day at the top of the hill. There was no bronze or iron in the tomb but there was red ochre, henna, woad, nettle fibre clothing and bow string. There was also hemp clothing from the head to the knees. Possibly a cloak and hood with the nettle fabric worn under it. There was no linen but there was linseed oil on the body and on the yew bow. I could not find any ash wood to indicate a spear but there was a willow quiver. Oak seemed to be missing and there was no sign of a shield. The arrows had flint arrow heads at the bottom of the quiver. There were no feathers on the body. Feathers are used to identify Druids or priests so the body may not be that of a Druid. There appeared to be no ox hide leather but there was deer leather on the feet which had been tanned in urine. The body gave positive responses with foxglove, deadly nightshade, toadstool, honey and yeast. These indicate a ritual killing in line with Druidic practice. There was a ceramic pot by the feet which dowsed positive with wheat and honey. The tomb dowsed as if it was about 8ft or 2½m deep. The tomb contents indicated that it was not the burial place for a priest. There were no feathers and there was no female companion. The tomb looked quite different to the normal ones associated with temples. The temple had the usual 6 + 3 rings of postholes but I could not find a pool. The next step was to look for paintings on the ground. Using manganese dioxide I found centred on the tomb, four triangles connected to the centre. The triangles were the wrong way round with the base nearest the tomb. Outside the four triangles there was a circle painted with woad. Now it is of course not possible to say that the blue circle and black triangles belong to the same picture. The one indicator of connection is the symmetry, the cross being inside and centred on the circle. Using a gypsum witness another set of triangles was found, this time the base was away from the tomb and there was a thunderbird head in the woad circle. It is not possible to identify the dates of the different components of the paintings in relation to the tomb. My guess is that the temple is a very early one. It has the 6 + 3 configuration of postholes, there are portal stones at the south gate, and the thunderbird motif is present. The absence of a pool may be admitting that they could not get water to the site and painted floors were used in place of the pool. The importance of the find is that Brankam Hill may have temples and artefacts dating from the beginning of the Henge Age. If not the beginning then there may still be clues as to the origin of the Henge Age, its culture and Druidism. At the time the temple was built there were people who were aware of the oak's summer and winter auras. The temples in Scotland like those in England and Wales indicate that Druidism, or some core elements of it, was present possibly as early as four to five thousand years ago.

The second and last night of the Glamis music festival was over. The music had been beefed up with spectacular and noisy fireworks that were as tone deaf as myself. The following day, Monday, I returned to Bovingdon and prepared for a trip with Nigel to the West Country. On the Wednesday we drove down to a farmhouse which was conveniently situated on the edge of Dartmoor. The visit to Scotland had been aimed at determining the geographical extent of the Druids and Henge Culture and how far back in time they went. The visit to Dartmoor had these objectives but a few more were added.

Chapter 23

Some Surprises in Devon and Cornwall

The Henge Temples of Dartmoor

The reasons for visiting Dartmoor were many and complex. Having visited Dartmoor many times I knew that the moor was not good arable land. This state of the land has been ascribed by various authors to a combination of climate change something like 2000 or more years ago and over zealous farming by people who did not know how to farm marginal land. On previous visits I had often stood on the moor looking at the landscape and the stone circles wondering what might have happened to the people who had built them. When I first started visiting Dartmoor many years ago I went along with the conventional interpretation which was that the ancient people had been driven off the moor by climate change or poor soil management. They had of course deforested the moor hence the use of stone for stone circles. That was the conventional wisdom. When the climate change occurred depends on the source consulted but is said to have taken place between 1000BC and about 500BC. The current consensus relating to the origin of the moor points to the moors being deserted, apart from perhaps miners, when the Romans landed in 43AD. If this were the case there should have been no active temples on the moor when the Romans reached Devon and hence there should be no execution lines or battle blood round the temple sites and stone circles. If we were to find execution lines and battle blood on the moor it would indicate that Dartmoor was still supporting a large population and being successfully farmed in 43AD. The collapse of Dartmoor as an agricultural area must then have come after the Druids had been removed from the scene by the Romans. This would indicate that the maintenance of Dartmoor as a farming area had been dependent on the agricultural husbandry skills of the Druids. If new people moved onto the moor who did not know how to farm marginal land, for example Romans, the land would degrade and become the moorland we see today. Alternatives are that the Romans destroyed any hydro engineering that was controlling soil water levels on the moor or the actual climate change did not occur until much later. Another reason for the visit to the south west was to find out if its windy lanes were Druidic. The conventional wisdom relating to the origin of the deep lanes in Devon and Cornwall embraced such theories as pigs using them on their way to market or repeated use by carts and the soil being washed away. We thought that it was possible that many of the lanes would be Druidic complete with diamonds, chariot tracks, curbs, trails of linen and feathers. When required for either ceremonial purposes or to make the roads usable by chariots the Druids would have dug cuttings. We were also interested in the West Country because if the Henge Culture had come up from the south of Europe, the West Country might be the place to look for signs of its early arrival and the initial stages of its development in to late Henge Age Druidism. If we were right, many of the lanes and roads in the West Country could be Druidic ceremonial ones.

On our first evening, Wednesday 16th August 2006, we started looking at the lanes round the farmhouse where we were staying. These were in the main small, just wide enough for one vehicle and often between raised banks or with dry stone walling running along the side. It did not take long to find the diamonds, chariot tracks, cement curbs, feather trails. All the lanes round our B&B farmhouse were Druidic and in one case where two lanes joined at a junction there appeared to be a bollard? (Figure 23.1). Looking in the fields nearby and the garden of the farmhouse we found signs of temples. The immediate area round the farmhouse had received a lot of attention from the Druids and on walking down one of the lanes we found a tunnel entrance leading off the road and into the rise of a hill. The chariot tracks could be followed from the lanes into the tunnels. It looked as if the countryside on the northern boundary of

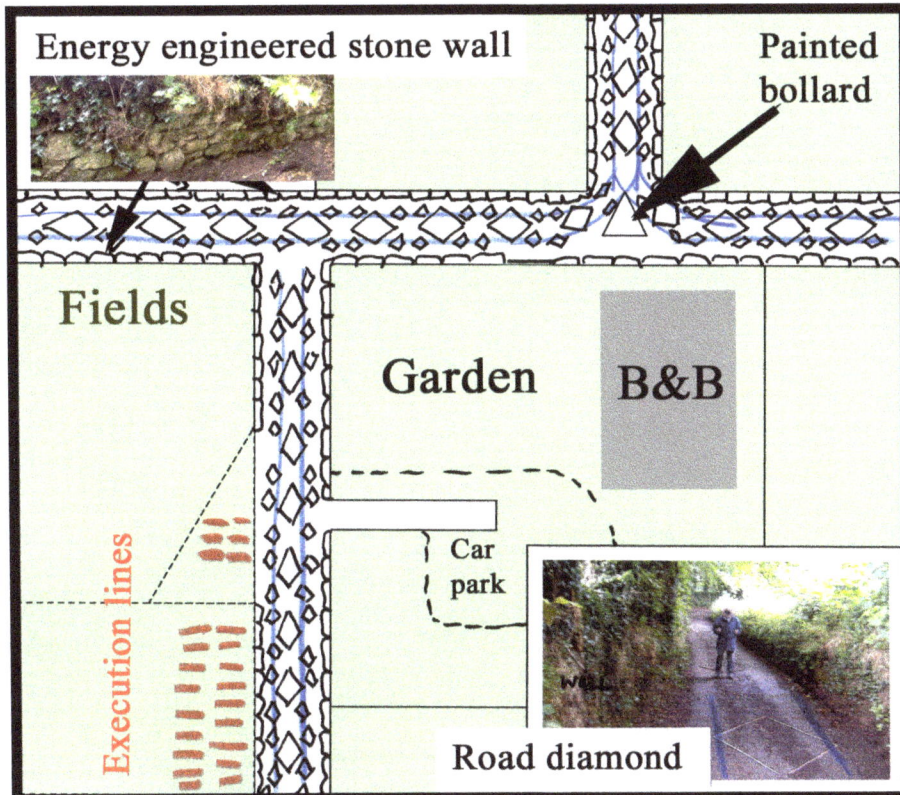

Figure 23.1 Druidic Roads

Following a lane outside the Farmhouse where we were staying we came across a road junction. The chariot tracks could be followed coming down to the junction then splitting to go either side of a painted bollard. The local field layout and the position of present day houses was determined by the road layout planned and executed possibly 4000 years ago.

Figure 23.2 The Merry Maidens, Cornwall

The stone circle in Cornwall, not far from Lands' End, known as the Merry Maidens is believed by archaeologists to date from about 2000 years BC, that is Bronze Age. The tomb contents indicate that a temple was built on the site round about that time. The stones of the circle may have been placed for load bearing or other purposes at the time or much later as sacred sites have a long history. The stones were clearly there when the Romans arrived and could have been part of a religious building at that time.

283

Dartmoor was as full of Druidic civil engineering as any other area we had studied. The next question was when did it come to an end. On an area of flat ground alongside a track running past the farmhouse we found the first of the execution lines. Then in the field alongside the track we found battle blood showing where Druidic warriors and Roman soldiers had fought. We had all the signs of an active Druidic centre which had been destroyed by the Romans and hundreds of people executed.

The Merry Maidens

The following day we decided to drive down to Land's End and visit a stone circle called The Merry Maidens (Figure 23.2). On the way to the circle we checked a number of modern A and B class roads. Everyone we checked was a former Druidic road. It was beginning to look as if a large part of the West Country's road infrastructure is based on a road system originally laid down by the Druids some three to four thousand years ago. When we arrived at the Merry Maiden Stone Circle we found a typical circle of stones which appeared to be placed on the 5th ring of postholes. All the stones were magnetically aligned and they had been painted on the radial and inside faces. Egg white, red ochre, manganese dioxide and woad were dowsed for. The radial faces had been painted with woad and manganese dioxide, the inside face with red ochre. Other colours were almost certainly used but our pigment witnesses were limited. The pudding stone cement witness gave a strong response round the stones indicating that Druidic cement may have been used to provide a footing for the stones. There was a clay pool in the centre approximately 14ft (4m) in diameter. Pottery piping was traced from the south side of the pool to a well in the south corner of the field. There were signs of a plinth in the centre of the pool. At the centre of the temple was a tomb with its Prince and Princess. The Princess wore a gold blouse or jacket, both had bronze crowns, and the prince had a bronze shield, the bronze being arsenical bronze. The bronze hinges on the gates of the temple were also arsenical bronze. The presence of arsenic could indicate the early Bronze Age, say 1500 years or more before the Romans arrived. However, bronze is durable and can last for thousands of years so all the arsenical bronze hinges may be saying is that a temple may have been built on the site or nearby in the early Bronze Age. The interpretation of the bronze in the tomb is also difficult as it could be recycled bronze. Iron and tin were checked for in the tomb and none could be found. The plumbing in the pool was also checked for lead and none found. I did not survey the whole site so it is still possible that a lead pipe was brought into the pool from another direction. To be reasonably certain of the age of the tomb a full analysis of its contents would be required. However, answering questions from tourists who were interested in what we were doing used up the time and so we had to leave the tomb and start to look for the cavities of the underground temple. We knew that the Druidic engineers could cut into hard rock in Wales and we wanted to know if they could cut into the granite of Devon and Cornwall. We identified a cavity where the tomb was at the centre of the temple. Then on the north side of the tomb first a tunnel and triangular antechamber was identified then a bone depository. We looked for the seven sided tunnel encircling the central complex and found a section of it on the north side. We also found what looked like the down shaft and access tunnel to the tomb complex. All of this was done in a bit of a hurry and between chats to tourists but the indications are that there is an underground tomb complex under the Merry Maidens Stone Circle similar in design to that found in other parts of the country. The implications of this is that the Druidic engineers in Cornwall could dig underground chambers and tunnels in hard rock but we did not obtain any indication of when they were dug or if bronze or iron tools were used.

The next step was to look for battle blood and execution lines. As usual the battle blood was round the outside of the temple and at a little distance from the temple. Near the road was a mass grave. The execution lines were along the road, now a modern metalled surfaced one but at the time of the executions a ceremonial road with its row of painted diamonds. Our last task was to find the underground river and evidence for a Bell Chamber. We soon found an underground river running through a cave system not far from the temple. I followed the river whilst Nigel tested it with an oak witness. Nigel soon found the place where there had been a bridge across the river. On the temple side was a shrine, on the other a tunnel leading off under the road and into a field which was not open to the public. We therefore missed the opportunity to find the first Bell Chamber in Cornwall. I should say, however, that in hard rock a 'Bell Chamber' is unlikely. The chamber will have a flat roof supported by columns. This design is also found in chalk.

Our time at the Merry Maidens Stone Circle was limited but within about two hours we had established that the stones had been painted and were inside a 60m diameter building. The stones were there for structural purposes on the fifth ring of posts. The potassium lines left by the burnt lintels or ring joists were between the stones and were easily dowsed. The building had the standard 6 + 3 pattern of rings and the stones clearly had no astronomical purpose that relied on visual contact with the sun, moon or stars. The presence of arsenic in the bronze and the Roman execution lines indicates that the site had a long history as a sacred centre for the local community. There is evidence of an underground temple complex indicating that hard rock was no barrier to the construction of tunnels and rooms below the temple.

The Michael Line

On the way back to Dartmoor we stopped to check some more side roads. They were all Druidic and the evidence was mounting that the main architects of the topography of the West Country were the Druids.

On the following day we were due to meet up with a famous West Country artist, John Christian, who was also an intuitive dowser. I had known him for many years and on previous visits he had acted as my guide to Dartmoor. His house, which was not far from where we were staying, had a field across which the Michael Line ran. The Michael Line is one of the better known Ley lines or energy lines running across the country from Land's End to the Norfolk coast in an almost straight line. Perhaps straight line is the wrong description as it has many bends but the bearing of the line is almost constant making it look reasonably straight when drawn on a map. A good description of the line is given in a book called 'The Sun and the Serpent' by Hamish Miller and Paul Broadhurst. I was interested in Ley lines mainly because they were never defined beyond the level of energy lines. I felt that they could probably be divided into three groups according to the origin of the energy that the dowser was responding to. Group 1 was where an ancient road was the source of the magnetic fields. The material from feet, shoes, clothes, wheels, lubricants, hooves, animal droppings and hair would be taken down into the ground over hundreds if not thousands of years of use. If the road was Druidic the chariot wheels, feathers, postholes, wattle walls and painted diamonds would be present. If Druidic there was also the possibility that stretches of the road would follow underground streams.

The second group of energies come from geological features such as long cracks in the underlying rock or changes in strata. Nowadays with long lengths of gas, water and oil pipes it could be possible to mistake them for a Ley line. The third group is where the energy line is travelling horizontally through the air from a source to some distant target. This happens with standing stones and with particularly man-made structures such as flint churches and geological features where rock is standing proud of the surrounding ground. The most common of these energy lines in my experience is the blue line.

In looking at the small stretch of the St. Michael line which crossed John Christian's field we had to determine where the energies were coming from and identify them. In our experience to date, when people say they have a Ley line it has always turned out to be a drover's road, a Druidic road or a more recent track or road. From John's house we walked across the field to where the St. Michael line was said to cross the field. We had no difficulty in locating it and had soon identified a 20ft wide line crossing the field. We normally go for chariot tracks first, which we did on this occasion, and had soon located two sets of wheel tracks separated by a gap in the centre. On the outside of the chariot tracks were postholes and a wattle fence. The bronze wheel tracks, deer and horse faeces, feathers and painted diamonds all indicated that this section of the St. Michael line was a Druidic ceremonial road. The question arose that if this part of the Ley line was a two lane highway was the St. Michael Line along its whole length a two lane Druidic highway, a motorway dating from 4000 years ago? The application of the latest dowsing techniques along the St. Michael line would be able to prove or disprove this conjecture but it was going to have to wait. However, the idea that a civilization, possibly 4000 years ago, constructed a road from the tip of Cornwall to the coast of Norfolk grabs the imagination. Discoveries we were making in other parts of the country were beginning to indicate that the Druids may have been master road builders long before the Romans and laid the foundation of our present road system. We could have spent hours on the St. Michael line unravelling its history but our objective was to get onto the moor and have a close look at one of the Tors and the stone ruins to be found there.

The Round Pound and Kestor

John Christian, our guide, took us to an interesting site called the Round Pound. The site is just beside a moorland road with ample parking not far from the ruins of the temple. On the other side of the road to the Round Pound and car park there is a Tor, Kestor. The starting point for the study was the car park, a nice flat area, and if you were a Roman soldier you would appreciate that it was ideal for executions. It did not take long to show that this is what it had been used for. There were execution lines running parallel with the road. Having established that the Romans had visited Dartmoor and that at the time there had been a sizable population on the moor for them to execute we moved on to the stone ruins. Like many ruins, the jumbled nature of the stones made it difficult to see what was there and what might have been there in the past. However, with some help from John we began to appreciate that the lower levels at least of the temple walls had probably been in stone. However, I could not make out if what remained had been the foundations of walls with wood flooring spanning the space between walls or if the stones were the remains of a low wall the task of which was to keep the wattle walls off the ground. Fortunately this was not too important to the task in hand as having identified what might be the centre of the temple Nigel and I started to search for the tomb and pool. The tomb was not difficult to find. There were two occupants with the woman being over five feet and the man over six feet. It looked as if they were adults. The bronze shield was copper and tin, that is late Bronze Age, they both had bronze crowns, the man had a dog or wolf hair garment, and the woman wore a gold blouse or jacket. There was linen and linseed oil on the bodies. The underground temple was comprised of annex rooms and bone depositories and followed the standard Thunderbird pattern of underground temples. The wall round the pool area was possibly where the second or third rings of posts would have been but this could not be confirmed. The pool itself was defined at one time by a wall, now gone, and tiles. There were two tile aprons on the outside of the pool and at the centre there had been a plinth (Figure 23.3).

Outside the wall which enclosed the central pool and tomb was what looked like a small room. When we first arrived at the site we could not place it because we had not come across anything like it in the design of other temples we had studied. We were at a loss as to what it might be.

Figure 23.3 Pool area of Kestor Temple
The temple at Kestor clearly differs from temples in Hertfordshire in having an extensive stone foundation. It is not clear how extensively stone was used in the building as standing stone structures would have been destroyed by the Romans. The Tor is to the right.

Figure 23.4 Kestor
The rocks of Kestor stand proud of the surrounding land. After 2000 years many engineered features will have disappeared. The main question is are the outlying granite stones placed deliberately to attract blue lines from the Tor or are they creating the blue lines just by chance?

However, after identifying the tomb complex we realised that the room was nicely situated for the access shaft to the tunnel going to the tomb complex. We soon found a rectangular area about 6 to 7 ft across. From it a tunnel appeared to lead to the tomb complex. Below the shaft was a cavity and the tunnel also showed up as a cavity. This was very interesting as it indicated that the shaft and tunnels had not been back filled. In chalk areas it is possible that many shafts and tunnels have collapsed in on themselves and the chances of getting into a tunnel and then walking round the tomb complex did not seem to be that good in Hertfordshire. Here on Dartmoor in a hard rock area there seemed to be no reason why the tunnels should not be clear and easily accessible. We therefore took a closer look at the access shaft. It appeared to be sealed by three blocks of sandstone which were only six to seven feet from the surface. The underground complex of the Round Pound Temple looked as if it could be the easiest one to access that we had come across. It would be an interesting temple to look at as the indications were that it was Bronze Age. If this were to be the case, it could have been as much as a thousand years old by the time the Romans arrived in Britain. As the temple was still being used when the Romans appeared on the scene it should be possible to obtain an estimate of the life time of the temple and the time during which the moors were being farmed. The mass graves from the executions were on the other side of the road to the temple, that is the Kestor side. This indicates that when the Romans arrived there was sufficient depth of soil to bury thousands of people. In one area we found 10 execution lines (3 rows in each line). The Tor must have been a very important religious site. The Tor itself and the land around it looks as if it has been landscaped. Walking up to the Tor from the road one is soon aware of what looks like lines of stones running into the distance from the Tor to the east (reeves). The stones are now scattered to a certain extent but it looks as if at one time there was a nice radiating linear array spread out over the gently sloping land to the east or on the walkers left as they head up to the Tor. The steeper slope from the Tor to the west seemed to lack the lines of stones and because of the more difficult terrain we did not investigate that side. Battle blood could be found from near the Kestor rocks down towards the road. A major battle had taken place round the Tor indicating that it was of considerable spiritual importance, possibly more so than the temple, and on a par with Glastonbury and Stonehenge in terms of religious and spiritual significance. If this were the case it is not inconceivable that the Romans deliberately destroyed the fragile soil and agricultural land maintained by the Druids and turned the area into the moorland as we know it today. However, so much for conjecture, we progressed up the Tor and eventually arrived at the top of Kestor. Kestor is a small Tor as far as Tors go and because of this is easier to study (Figure 23.4). Our visit was a quick look to see if there was any special religious or spiritual feature about a Tor. The first and obvious thing was why did the rock stick up proud of the landscape. Had the soil and debris been removed from around it to make it a prominent feature standing proud of the landscape? Was it quarried to provide stone? Was there an underground temple system under the Tor? Why the lines of stones to the east? As we had walked up to the Tor we had crossed terracing and chariot tracks and feather tracks so there were ceremonial routes round at least part of the Tor. Walking round the base of the rock pinnacle on the eastern side it was possible to pick up many blue lines radiating out from the rock of the Tor. A single block of granite placed on a table and used to simulate the Tor does not produce an array of blue lines but if small pieces of granite are placed in front of it, properly orientated magnetically, blue lines can be developed (Figure 23.5). It looked as if this is what the druids had done round the Tor. The blue lines had been produced deliberately possibly by placing lines of granite stones round the Tor and then sent off into the distance. To where and how far we do not know. When standing on top of the Tor some blue lines can be identified passing across it. There are also signs of ritual activity at the top such as incense burning and pigments. On the way back to the car park we looked for and found more battle blood and execution lines. The battle had been widespread and costly in terms of life confirming our earlier thoughts that the site was of considerable importance to the Druids. There was only time for a brief visit but the information gathered indicates that at least one Tor on Dartmoor, and I suspect it applies to most, is or was an important sacred site. The Tor and the land around it looks as if they have been modified for religious purposes. The Romans considered the sites on what we now call Dartmoor to be important, so important that, as already mentioned, it may have been the Romans who turned a productive agricultural countryside into moorland. They could have done this by destroying the civil engineering that controlled the soil and water.

Figure 23.5 Tor Energy Engineering
The granite block on the table does not appear to prouce blue lines if it is the only piece of granite on the table. If granite chips are placed round the stone as shown, a series of blue lines are produced coming from the granite block and going through the chip into the distance.

Scorhill Stone Circle

On Saturday the 19th August we drove to a car park near Scorhill Stone Circle (Figure 23.6). The circle is out in the moorland of Dartmoor so a walk into the moors is required. From the circle there is only moorland in all directions and if the archaeologists are to be believed the stones of the circle have been standing there since the Neolithic, over 4000 years ago. The people who built the moor's stone circles and stone rows would have been driven off the moor round about 500BC by climate change, at least so the theory goes. Looking round the moors from the circle it is difficult not to believe the standard view of the moor's history. Clearly something dramatic happened whether it was about 500BC, earlier or later.

Scorhill Stone Circle is well known and I suspect has been painted by many famous painters such as John Christian. It has an atmosphere all its own. Perhaps created by its position, over looked by nearby high ground but still being sufficiently elevated to command a marvellous view to the west with its valley and distant Tors. From previous visits I knew that the stones were laid correctly magnetically but I decided to recheck while Nigel had a look round and got his bearings. After checking the stones I looked for the central tomb. It was not long before I realised that there were at least four tombs inside the stone circle. We decided to do a quick dating of them by looking at the tomb contents. Table 23.1 lists the contents of two of the tombs.

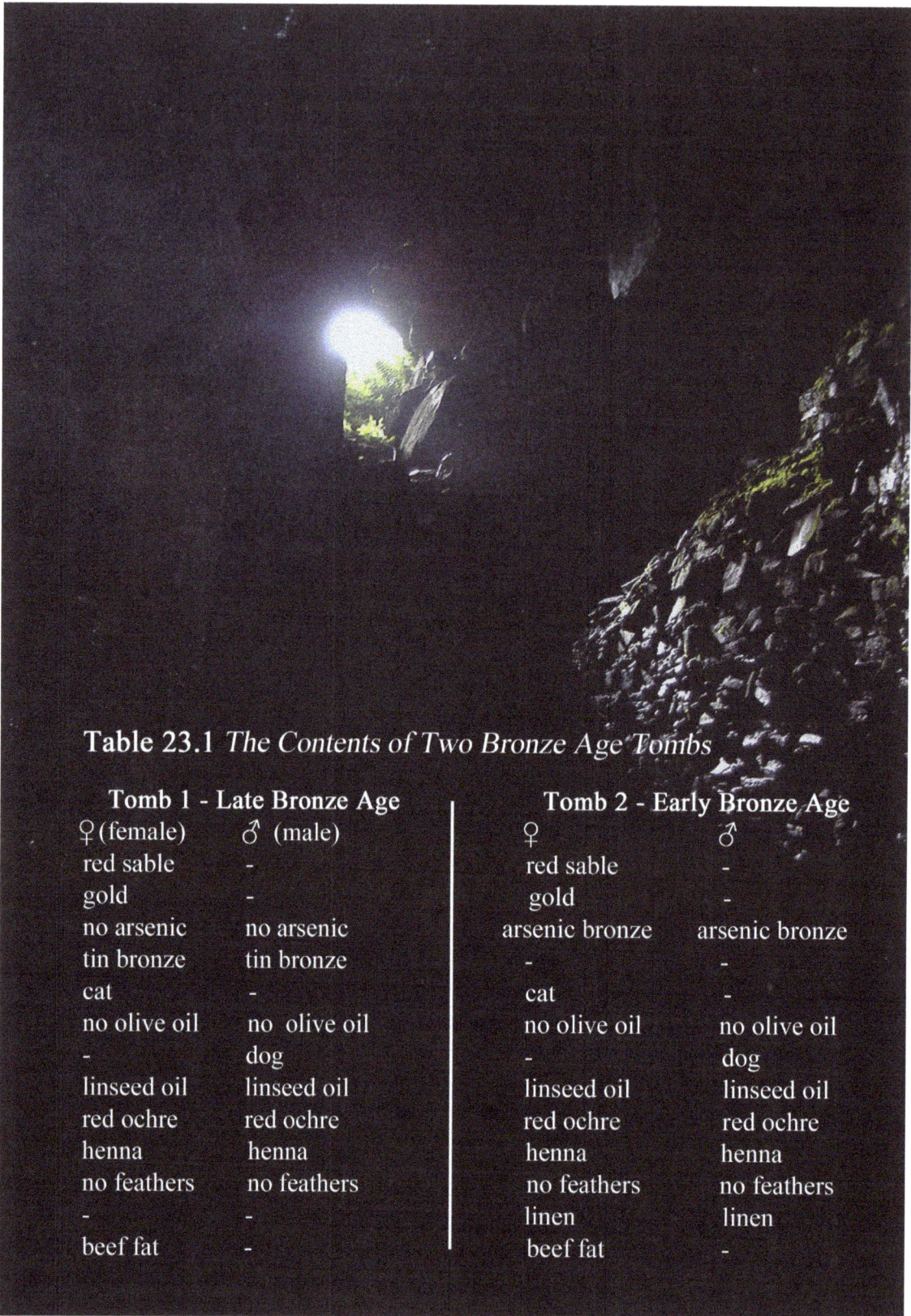

Table 23.1 *The Contents of Two Bronze Age Tombs*

Tomb 1 - Late Bronze Age		Tomb 2 - Early Bronze Age	
♀ (female)	♂ (male)	♀	♂
red sable	-	red sable	-
gold	-	gold	-
no arsenic	no arsenic	arsenic bronze	arsenic bronze
tin bronze	tin bronze	-	-
cat	-	cat	-
no olive oil	no olive oil	no olive oil	no olive oil
-	dog	-	dog
linseed oil	linseed oil	linseed oil	linseed oil
red ochre	red ochre	red ochre	red ochre
henna	henna	henna	henna
no feathers	no feathers	no feathers	no feathers
-	-	linen	linen
beef fat	-	beef fat	-

Figure 23.6 Dartmoor Stone Circles
Scorhill is a stone circle set well within the present day Dartmoor. Just past the circle is an artificial stream built by the Druids. At the time the stone circle was part of a religious building and just before the Romans arrived there is every reason to believe that Dartmoor was productive agricultural land supporting a large population.

Tomb 4 differed from the others in that it contained an iron sword and an iron tipped spear. The textiles included linen, cotton and on the woman something that dowsed as silk. The cotton covered the whole body of the male and female. The male had an oak shield with a bronze boss. There was no dog or wolf hair. The bronze crowns also had gold on them. From the size of both the tomb and bodies the occupants were children. The male had bird feathers on his body indicating that he was dressed as a Druid. The survey of tomb contents was a quick one as the aim was to identify which tomb, if any, matched the stone circle. Tombs 1 to 3 looked as if they could belong to the early and late Bronze Age (Table 23.1) and the Stone Age. Whichever age they were, they did not appear to be Iron Age and the stones of the circle almost certainly belonged to the Iron Age. The matching tomb was therefore tomb 4 which was not in the centre of the circle. The next step was to identify the pool and circles of postholes centred on tomb 4 and the outer wattle and daub wall of the building. When this was done the stone circle was clearly set to one side with the western stones on ring 8 of the postholes and the eastern stones on ring 3. The stone circle was offset within the temple for some reason. The building the stones were in had been burnt down leaving the potassium print of the roof structure so in theory it should be possible to identify the shape of the building and its roof.

Outside the north gate of the temple there was battle blood and two rows of execution lines. To the south we found more battle blood and three execution lines. When the Romans arrived this part of Dartmoor was supporting a large and prosperous community. That would only have been possible if the moor was rich agricultural land at the time. What we had found at Scorhill matched our findings at the Round Pound.

Although little time had been available for investigating the Henge Age on Dartmoor our findings provided some clear pointers for our next visit. Dartmoor was not moorland 2000 years ago. There was enough top soil to provide smooth surfaces for roads and paths and sufficient depth for burying people. Wood, in particular oak, was available for temples and population density was high indicating productive land. The Druidic engineers knew how to tunnel in hard rock and had the resources and technical know-how to terraform what we now know as Dartmoor.

Figure 23.7 Druidic Engineering

The Scorhill site was used as a religious site for temples from the Stone Age, say four to five thousand years ago, up to the arrival of the Romans, 43AD. The stream shown provided water for the temple and was constructed by the Druids.

Figure 23.8 St. Albans Abbey and Cathedral

The Abbey dates from the 7th century. The modern building is however much younger and much of it was built during the Victorian period. The Cathedral is built on the site of the execution of St. Alban in 303AD and the site of a major Druidic religious centre. The north side of the Cathedral was where the chapel over a well was found. On the floor of the chapel was a painting of a five pointed star.

Hell-Fire Caves

The large number of blue lines (narrow blue beams) that we had found on Silbury Hill and now at Kestor raised once again the question of whether any of them were naturally occurring. There was also the possibility that they were produced by accident when stone, particularly granite and flint were used in construction work.

An indication that the latter might be the case came during a visit to the Hell-Fire Caves at West Wycombe. The façade of the cave entrance is constructed from flint and looks like the interior of a church, that is an end wall flanked by two side walls. If one walks across the entrance but at a distance to it, it is possible to pick up a strong well defined blue line. The line almost certainly comes from what might be a resonant chamber created by the three flint walls. The evidence for this is that if a piece of flint is taken into the 'chamber' and placed on a table or just held by somebody the blue line disappears. It therefore looked as if some blue lines at least might be a spurious by-product of wall configurations. A good place to find a flint wall surface with two flanking flint walls is on churches and cathedrals. The buttresses provide the flanking walls. Using churches and one cathedral, blue lines were identified coming from the wall between two buttresses. Once produced the blue lines might well go on for miles. However, the ones we have found so far are not very strong. They are not in the same league as those detectable at the top of Silbury Hill. The realisation that blue lines could be a by-product of building design and materials caused me to set up a laboratory study as it now became important that we were able to distinguish between deliberately made blue lines and fortuitous or natural ones.

It was whilst studying the blue lines associated with the flint walls of St. Albans Cathedral that Nigel found an ancient structure or building close to the western wall of the north arm of the transept. Nigel, being somewhat easily distracted from the job he is suppose to be doing, started to look at the building in a little detail. It looked as if the building and what was in it could belong to or be part of something that was under the nave of the Cathedral at this point. The building had a wattle wall that seemed to head off into the Cathedral and it had a large clay or tiled rectangle occupying much of the floor space. In the rectangular area there appeared to be a painting, possibly an unsymmetrical five rayed star. It looked as if more Druidic painting had been discovered.

Summary

The visit to the West Country had been productive in that we had confirmed the widespread use of the Henge Temple design both above and below the ground. We had found that hard rock was no barrier to the Druidic engineers. The Tors looked as if they were more than natural outcrops of stone and the moor had been good agricultural land supporting a large population. The Romans had marched to Land's End in their search for Druids on Druidic roads that are still in use today. The blue lines that we had thought were man-made had been found to also have a natural origin and it became imperative that the natural and man-made blue lines could be distinguished one from the other.

Chapter 24

The Dawn of Christianity

The Lost Well of St. Albans Cathedral

Nigel had been making good progress studying ceremonial ways, blue lines and other pre-Roman artefacts round St. Albans Cathedral. Like many if not most Cathedrals St. Albans was built on an ancient Druidic sacred site which can be traced back to the early Bronze Age if not Stone Age (Figure 24.1). After his latest discovery Nigel called me and I agreed to go over and help check his findings on Monday 14th October 2006. The day selected turned out to be a pleasant autumn one and we made good progress. After a while we found and started to mark out the enigmatic building Nigel had found by the west wall of the North Transept. I confirmed the tiled floor which was Iron Age, the wattle and daub wall and signs of painted art work on the floor. It looked as if one picture was a star. As it was now lunch time we retired to the Cathedral restaurant and over a nice home cooked meal discussed what we had found. Just in case the building had either been or was at the time no more than a well house we decided to check the area for a well after lunch. On returning to the tiled area we soon found a well almost in the centre of the floor. The tracks of the oxen used to draw the water could be traced for at least 50ft, then they disappeared through a garden wall. This gave an oxen run of more than 50ft indicating that the well was at least that deep and possibly deeper. The fact that oxen were being used to draw water indicated that there was an underground stream running through a cave system. When we had first come across wells that looked as if they had been filled in we were a bit worried as we could not imagine the Druids blocking up an underground river. We had then discovered that wells which were going down on to streams were sealed and not filled in. We therefore checked the well we had just found and like previous ones it was sealed by three pieces of sandstone, at least that is what they dowsed as. The sandstone had been worked by iron tools. The mortar or cement used to bed and seal the stones was Druidic as it responded to pudding stone and urine witnesses. The seal appeared to be about eight to ten feet down. At this depth the well would not have been disturbed by grave diggers or by those building churches and eventually a Cathedral on the site. It was now late and time to pack up and go home but Nigel was still not satisfied that we had identified the main structure he had found. He felt that there was a major pre-Roman site of some sort extending under the Cathedral and that we were looking at a part of it. It was however difficult to fit a 'well house' into any Druidic building that we knew of. The sealed well, the tiled floor and signs of painted pictures on the floor, possibly a star, seemed to be saying something but even over a final cup of coffee in the Cathedral's restaurant we could not work out what it was. The sealing of wells and of a shaft access to an underground temple on Dartmoor was beginning to worry us because the sealing must have taken place before the Romans arrived. The sealing of the wells and access shafts was not done in a hurry. The capping stones were properly laid on cement and mortar and the gaps between the stones were sealed. We found another well near the Cathedral and that was also sealed by blocks of sandstone. It was possible that wells and shafts had always been sealed after centuries of use. However, we knew that not all had been sealed as access shafts to tunnels had been found that were still open. The fact that some shafts had not been sealed would have to be fitted in to any theory. Unfortunately we did not know if the open shafts we had found were sealed at a point below the surface as it had not occurred to us to check. Did that mean a trip to each one to find out? Perhaps not but a few had to be checked.

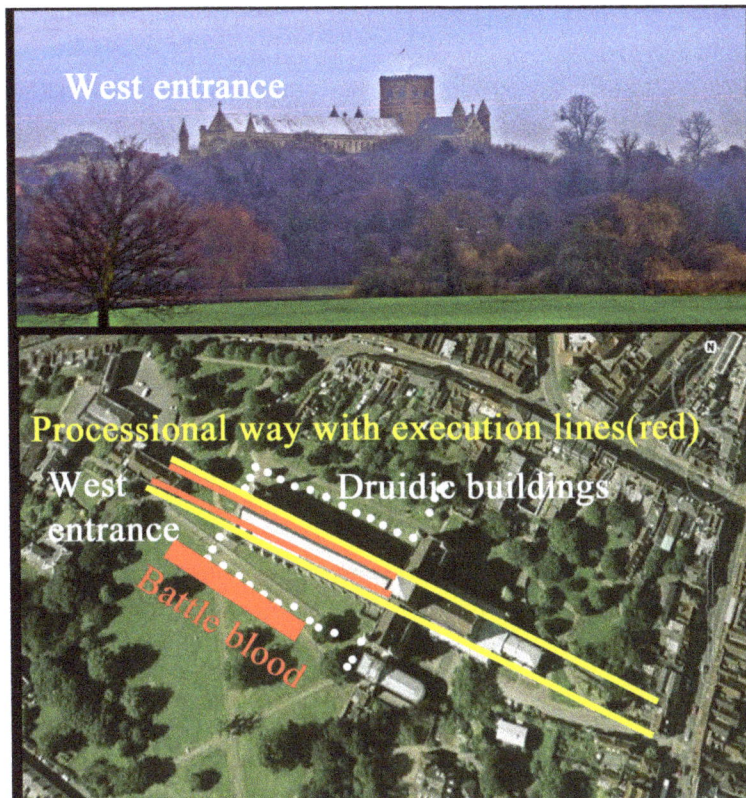

Figure 24.1 St. Albans Cathedral Druidic Alignments
The modern day Cathedral of St. Albans is built on a hill, (see Figure 23.8). The site is a rich archaeological one with the religious buildings and ceremonial ways of the Druids being much in evidence in and around its grounds. The site shows the transition from Druidism to Prot-Christianity and then Christianity before the Romans arrived. The Romans destroyed the early Christian buildings and executed everybody associated with them in execution lines some of which run through the modern parts of the Cathedral. The ground outside the West Gate and to its North side has the chemical memory of the major changes in Druidic buildings and processional ways that determined the alignment of the modern cathedral.

The Stoney Lane Temple Site

The first well to be visited was the one in Stoney Lane, a ten minute walk from my house. We knew a lot about this particular well and how it was used. It was probably first dug in the Stone Age or early Bronze Age, about four to five thousand years ago. When I arrived at the site on the 15th October 2006 it did not take long to locate the well. I found that the well was sealed with three blocks of sandstone. From the positive response to an iron oxide witness it looked as if the sandstone had been worked with iron tools. Henge Age cement had been used to bed the stones and make a good seal. Below the stones there was a cavity. The next step was to identify the tiled or clay floor area and then a wicker wall round the tiles. The tiled rectangle was about 32ft by 20ft with the well to the bottom of the rectangle. So far so good. The next step was to see if the tiles or ground had been painted and if it had been to work out what the artwork might have looked like.

However, to digress for a moment and consider the significance of sealing wells and other entrances to the subterranean world. Nigel and I had been involved in long discussions about the sealing of wells and shafts. For thousands of years the Henge culture and its priests had considered that access to pure water from mother earth was of considerable importance. Its importance was shown by its central role in ceremonies and in providing holy (uncontaminated) water for drinking. The evidence that water was transported along the road network and delivered to individual houses was growing. The illnesses associated with surface water would have been just as obvious four or five thousand years ago as they are today. The control of these waterborne diseases by using clean water from underground sources would also have been just as obvious then as now. The ability to identify accessible and copious supplies of underground water and control the distribution of clean water on a large scale would have given the Druids considerable power. It therefore looks as if the siting of Henge Temples along side underground streams was not done by accident. The role of the temple may have been to control the holy (clean) water. All the Henge Age Temples we have studies to date have a well sunk on to an underground river and a system employing oxen or deer to lift the water to a header tank. From the header tank there is always a plumbed supply to the temple and in some cases to other sites. This holds from the Stone Age Temples to the Iron Age ones. A span of at least two thousand years and possibly much longer. We know of some temples where the water is delivered by a water cart but they have a stream of clean water close by. For those in doubt about the importance to health of clean water, comparisons can be made between the health of a population with access to clean water and one having to use surface water. Such comparisons can be made using African villages with and without clean water. Part of the Druids power base could have come from its control of 'Holy' water. Another aspect of the Druids power over their communities lay in their role as priests helping the dead on their way to the 'after life'. At the moment it looks as if the funeral ceremony involved taking the deceased on a journey from the chapel of rest to the pool then to the Dolmen where the birds would strip the flesh from the bones. The bones were then 'dressed' in red ochre at the Dolmen before being taken to the temple and then down the access shaft to the underground temple and bone depositories. The bone depositories held the bones of the ancestors. Because of this very complex religious ceremony that shows as far as we can see few changes between the Stone Age and the Iron Age it was very perplexing to find that sometime in the Iron Age the wells and access shafts to the underground temples were sealed. This could only mean a very big change in religious philosophy. The question was 'what was the change to', 'when did it take place and over what period of time'. The wells and access shafts were not the only things to be sealed. Horizontal access tunnels to underground temples were also sealed and their sealing was precise and not hurried. The underground temples and what they represented were being sealed because they were no longer required for ceremonial purposes and they were being sealed with care and in a professional way. However, not all underground temples were sealed by the time the Romans arrived. The battle blood outside the Hell-Fire Caves in West Wycombe show that some were still being used for religious purposes. At this time we still thought that the religion was Druidic although we were conscious that religions change and evolve and after three thousand years there may have been many changes. The problem was spotting the changes. The technical ones were easy. They showed up in the tombs. The tombs are a succession of time capsules from the days when wool was not being used as a fabric to when cotton and silk were being imported. It was however quite another matter identifying when the altars and Dolmen ceased to be used. I was convinced that we could do it but it was going to involve a lot of work.

This is to digress from the story of the wells in order to mention our concern about sealed entrances to the underground temples of the Henge Age because we knew it signalled something big. What had been painted on the ground over the well in the Stoney Lane field would show just how big. I was now standing by a well which had been sealed, the well head removed and a building built over it. I knew that most of the Druidic artwork we had discovered so far carried a message and we had been able to trace some of it back to its origins in nature and to the world the ancient people could 'see' using the human magnetic sense. To the ancient people the magnetic sense revealed a very real spirit world. It was not a matter of belief or faith, the magnetic world was a reality like the worlds of the other senses. The subliminal magnetic sense was for many a conscious sense and it enabled them to 'see' the spiritual world as a real world sharing their own. Even today many people who are sensitive to the magnetic fields around them believe that they are part of a spiritual domain. Because of this I knew that the painting on the ground was going to be extremely important and it may well tell me why the Druids blocked and sealed their contacts with the world of the Earth Goddess and possibly what took the place of the Earth Goddess or if she remained, perhaps it was her place of abode or form that changed. At this moment I only had the impression of the misshapen gypsum star found by Nigel at St. Albans Cathedral to go on. I therefore selected a gypsum witness from the box of witnesses and started work. This time there were no people walking past wondering what we were doing amongst the grave stones and I had plenty of time. I found that there were indeed four rays and a short fifth one but they were not making a very convincing star. I started to place markers in the ground as I worked my way along the edge of the gypsum stain. The rays of the star were not rays they were the same width all the way along and had square ends. The use of a gypsum witness revealed the form of a Roman crucifix with a central rectangular area and a short stumpy fifth arm between two longer arms. I changed the witness to red ochre, then iron oxide and charcoal. At the end of the four long arms there was an area giving a positive response with red ochre and in the middle of the red area a small patch responding to iron oxide. The short stumpy fifth arm had black hair on its top and sides, the hair being indicated by charcoal. I started searching the field for holly and brambles. When I found them and used them as witnesses I obtained a positive response along the top of the head. The picture I was looking at on the ground was a bit of a shock to say the least. The figure appeared to be that of a crucified person on a Roman cross complete with nails and crown of thorns, in this case holly and brambles. The figure at this stage was hidden by the gypsum background of the cross which had colours round it – blue, a space, red, yellow and finally black charcoal. The hair was in charcoal, the eyes in manganese dioxide the mouth in red ochre. Part of the surprise at finding a picture of a person crucified on a Roman cross was that I knew it was not likely to have been painted after the Roman invasion. If it had, there was only a very small window between 43AD and 60AD. The Romans had destroyed temples and people during this period. However, the 'Druidic' painting had to be dated. The painting was on the top of a well, which had been sealed in the Iron Age. The pigments used and the painting of pictures on the ground were Druidic. The altar or table I had found with stains of wheat bread and beer yeast from the Eucharist ceremony clearly was Druidic. The ceremonies associated with the new religion would have to be identified to see if they were all Druidic or new imports. The structure and design of buildings, processional ways, altars, clothing used etc., would all have to be determined. Once this was done comparison could be made with the previous Druidic period and the later Christian period.

The picture of the crucifix now marked out on the ground differed in a number of ways from a modern Christian crucifix or picture of the crucifixion. The picture showed a Roman cross (Figure 24.2). The 'X' design of this cross enables the human body to be butchered and therefore its use may not be restricted to the Romans. The Druids at one time butchered human bodies and fed them to dogs and so may well have used such a cross. The cross could have come to represent the death of both the body and the soul. Above the head was the painting of a set of antlers entwined with oak, ivy and possibly other plants. To begin with, when the antlered head was first found we called it the 'Green Man' until we had time to look at it in detail. Later we realised that the antlers and vegetation represented Herne the Hunter (Box 24.1). Before leaving the Stoney Lane site I decided to check the main temple for signs of harvest festivals. The idea was that the harvest festival is a ceremony that should leave a clear imprint on the ground and as it had been carried forward into the Christian era it may be possible to identify it in the temple. I still thought of the site as a temple although it was obvious that if the well area had been converted into a small chapel or church that the main temple could now be quite a different shape and size to the 60m diameter temple of earlier days. I soon found rectangular areas giving positive responses to witnesses of oats, barley, mushrooms, apples and yew berries. It looked very much as if the locals did celebrate the harvest in the traditional way and that the festival was taken up by Christianity.

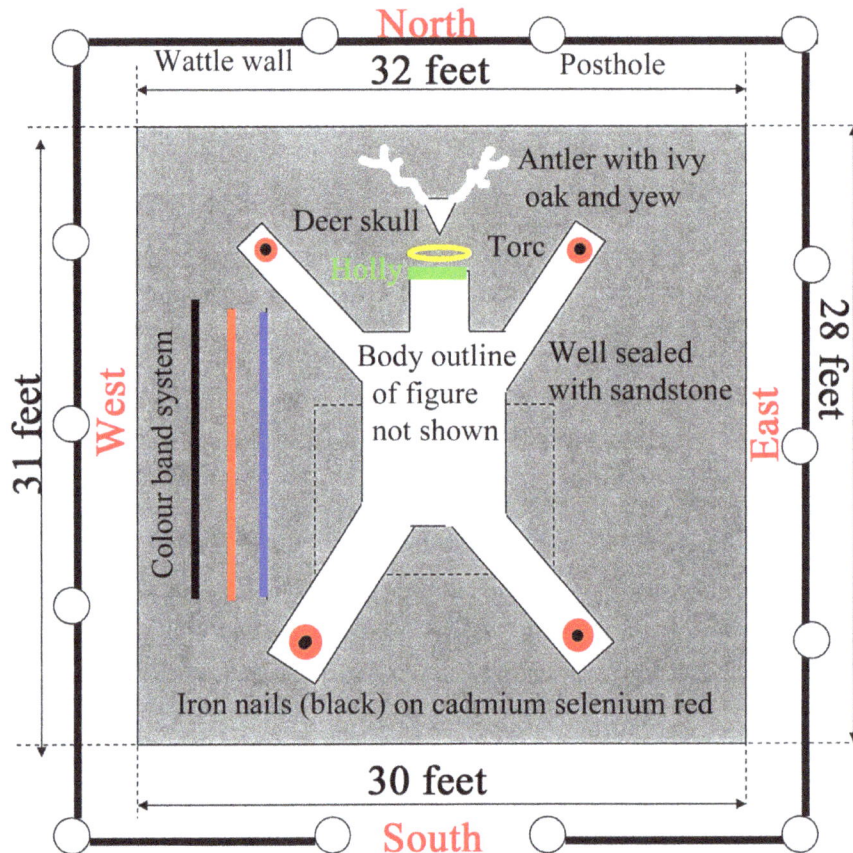

Figure 24.2 The Stoney Lane Well

The well has a floor of clay or tiles approximately 30ft by 30ft. Possibly underneath these tiles is a painting of a crucified person on a cross. The cross is in gypsum. There are bands of colour round the cross. At a distance of a few feet from the tiles there is a wattle and daub wall. When discovered no attempt was made to identify the body as separate to the gypsum cross. Red ochre was also found on the iron nails.

Return to the Lost Well

Having found one picture of a crucifixion it was necessary to return to St. Albans Cathedral and re check what we had found a few days earlier. On the first visit I had picked up what could be arms but at the time had thought in terms of a geometric figure. On the morning of the 18th October 2006 I arrived at the Cathedral and quickly found that there was a similar crucifix to the one I had found at the Stoney Lane temple. The figure of the crucified person came up well with all the detail. No disruption from the graves. The holly was there and the Stags horns that had been found by Nigel. There was a linen covered table. The picture was so good it could not have been disturbed by grave digging. This indicated that the picture was deep and there might be a wall round it creating a sort of crypt or sunken chapel. I found a sandstone wall with the urine mortar used by the Druids. The chapel was sunk in the ground and protected by a wattle and daub outer structure. I found traces of spilt wine, candles and potassium lines from burnt timbers. There were steps down to the floor of the chapel. The head and face of the crucified person had blood flowing down it and charcoal hair. The gypsum cross the body was on was outlined by blue, yellow, red and black (Mn02). This was all good evidence that there had been a small chapel or church on the site of St. Albans Cathedral. Dating the crucifixes with Herne above them could be difficult (Figure 24.3).

Box 24.1 Herne the Hunter and the Green Man

In the book by Miranda J Green entitled Exploring the World of the Druids, see page 24, reference is made to the religion of the Celts. The name 'Celt' is used in the book to describe not only the actual Celts on the European mainland but also the people inhabiting the British Isles, who were almost certainly not Celts. The pre-Roman people of the British Isles, and I suspect also people along the European Western seaboard, differed in culture from the Celts as their temples demonstrate. However, what Green says is very important and I quote "Our evidence for the nature of Celtic religion comes partly from ancient literature and partly from archaeological evidence……..". This limited knowledge base relating to the people of the British Isles has led to a reliance on a lot of conjecture. What we lack is written testimony from the people she calls the 'Celts' themselves. The lack of contemporary written records has led to the production of an effusive literature which I suspect relates more to the imagination of modern man than to the object of the ancients' prayers. Green gives some good advice on the interpretation and evaluation of literature and evidence relating to the ancient gods. However, with the discovery of the spirit paintings on the floors of temples and on roads there is now available a written record of sorts. One of the paintings relating to a possible god is of an antlered head with the antlers draped with the leaves of sacred plants and with a gold torc at the neck. The nearest description of a god matching this painting is that of Cernunnos or Herne the Hunter. The antlered god is described as the guardian of the underworld and the giver of life. When I first dowsed the greenery draped over the antlers it made me think of the Green Man. The face of the Green Man with vegetation round it and appearing to grow from it is very common in Britain and Europe. The face has been carved on many gothic churches. The face does not have antlers so we soon realised that the antlered head, even with its load of vegetation, was more likely to be Herne than the Green Man. As Herne appears on the floor of temples and mile after mile of Druidic road it is clear that he was quite an important deity. From what Green describes as a relatively small number of images of gods upon which to base our picture of them, we now have hundreds if not thousands of pictures of at least one of them.

The Access Shaft in the Horse Field

On the return to Bovingdon I decided to check the access shaft in the horse field near the house. Like the wells, I found that the shaft was sealed by three sandstone blocks which dowsed as if they had been worked using iron tools. There was a large rectangular area of tiles and an outer sandstone wall with urine mortar. There appeared to be steps leading down into the chapel from the south end of the building. The whole chapel was in a wattle and daub building and in this respect the structure was similar to the site at St. Albans Cathedral. On the floor of the building there was a painting of a crucified person with the antlered head of Herne, above the crucified body. The antlers were entwined with oak and ivy and possibly other vegetation which was not tested for. It looked very much as if the Druids had deliberately used the access shaft area as a site for a chapel. The chapel was much larger than the earlier chapel for the reception of bones. As the access shaft and its chapel is normally found within the temple this seemed to indicate that the main temple may no longer have been present or at least had been modified, when the new chapel was built.

I knew that the area of the access shaft chapel was one in which I had previously found execution lines so I checked to make sure that they were there. As expected the execution lines ran across the chapel. This could only mean one thing and that was that the chapel, with its picture of a crucifixion, predated the Roman attack on the temple or what we had thought was a Henge Temple. It looked as if the Henge Temple at the time of the Roman attack had been replaced by other buildings. But why should the Druids build a chapel over an access shaft and then fill it in? One possible answer was that the temple was modified and became the chapel or church. I started to look for a crucifix where the pool of the temple had been and soon found a crucifix on the floor of the main temple with a rectangular tiled floor area and seating round the central area. It looked very much as if the temple area had been used for a large sunken chapel. This central chapel, with a crucifix painted on the floor, had been built and was being used before the Romans arrived in Bovingdon.

The time sequence appeared to be: first the access shaft was carefully sealed and a sunken chapel built. Then the central pool of the temple was removed and a chapel built in its place. The chapel in the temple was also sunken but as the depression is still clearly visible two thousand years later it had not been filled in. There was no sign of a painted crucifix on the western Dolmen which could indicate that it was no longer in use. Later on crucifixes were found associated with some Dolmen.

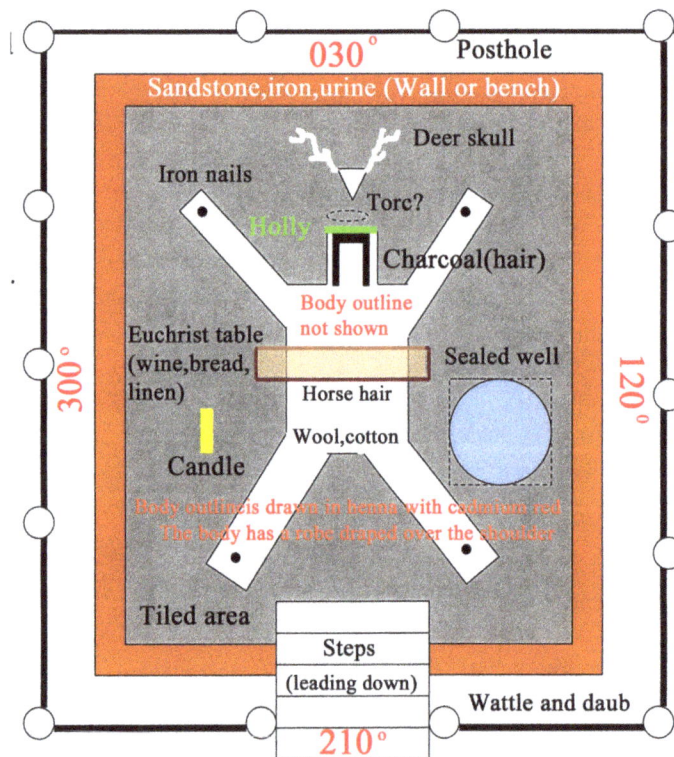

Figure 24.3 The Chapel On Top of a Sealed Well at St. Albans Cathedral.

The crucifix has a table (altar) across the body with stains from bread and wine, the Eucharist. There were candles and it is possible that the sandstone walls were in fact pews. There are steps leading down into the chapel. The head of Herne is above the cross and it is possible that there is a torc below Herne.

300

A Crucifix at Three More Sites

Whilst I had been working on sites at Bovingdon Nigel had found another sealed well with an irregular star like figure and rectangular tiled area in a wood at Chiswell Green on the outskirts of St. Albans. I had not told Nigel of my discoveries at the Stoney Lane or Horse Field sites as I thought it might be a good idea if he worked out the pattern for himself. So two days later on the 20th I met up with him and he showed me the site he had been working on. The gypsum witness he had been using for the study was not a very good one, it was plaster of some sort but not a good sample of gypsum. This made it difficult to follow gypsum lines and reveal the pattern. I gave Nigel one of my gypsum witnesses and the misshapen star soon started to look more like a cross. A rectangular area between two of the outstretched arms appeared and then it became clear that the two sets of arms were separated by a rectangular area. To my eye the crucifix was clear but it was not until red ochre and iron oxide revealed the blood and iron nails that Nigel tumbled to what was on the ground. Holly and bramble witnesses soon revealed the crown of thorns and then the antlers and greenery of Herne were found. Nigel was as amazed as I had been when he saw what an ancient people had painted on the ground. We knew that the implications of what we had found were profound in many ways. Religions do not change overnight and what we were looking at was almost certainly pre-Roman. The following days saw the gathering of more evidence that Druidic Temples had suddenly become possibly Christian chapels and churches. We visited Home Farm and found that the well had been sealed and a substantial chapel built on it. In addition to the clay or tiled floor and painting we could identify seats, entrance steps, an altar or table with a linen cloth, a separate entrance for the priest, candles, spilt wine and bread along the edge of the table. The congregation were wearing wool, the priest linen. After spending the morning of the 29th October 2006 on Home Farm the afternoon provided just sufficient daylight to check out a long shot. We knew that the arenas on the cursi were the site of ceremonial procedures. There were pictures painted on the ground and clear signs of the taking of drink. The question was, did the battles still go on but now with a Christianised system of ceremony and with a picture of the crucifixion on the ground. On returning home I headed for the Little Hay cursus. The arenas as always were easy to pick up using human blood and I soon had the four corners of the male and female arena marked out. I then checked for the gypsum cross. It was there, lying approximately east west along the length of the cursus and with the head of the crucified body to the west. Both arenas had a cross. In the Little Hay cursus the cross stretched to the corners of the arena but on a cursus that Nigel checked near St. Albans the cross appeared to be limited to a central area and was quite a bit smaller than the Little Hay one. Finding the crucifix in the arena did not mean that fighting was still taking place in the arena after Christianity arrived. It may be that the fighting had been replaced by something else.

From Druidry to Christianity

At this time there were a number of worrying aspects about our findings. We could not see how Druidry could change so quickly to Christianity. We were expecting to pick up signs of gradual change in Druidic procedures. Perhaps the battles to settle local disputes continued but in a Christian context. We knew we had a problem but did not know how to solve it. One of the possibilities that needed to be considered was that ancient Druidic sites had to be blessed or neutralized by painting a crucifix on them. To test this idea I checked Dolmen for crucifixes and found nothing apart from Thunderbirds. This could possibly be due to the Dolmen I was checking having fallen out of use long before and that I had not checked enough Dolmen. However, the initial evidence was that they were not being blessed by painting a crucifix on them.

Some evidence then turned up which made me doubt the sandstone wall theory of chapel construction. At Home Farm I had found wool on both sides of one of what we thought at this stage were walls. This was rather difficult to explain if it was a wall. Another possibility was that the 'walls' were seats or benches made of stone. I therefore carefully checked the main temple area in the horse field and found what could be two rows of seats on three sides with no seats on the western side. The centre of the chapel area was in a depression which is still visible to this day so it is possible that the chapel was sunk down in the ground. I did not look for a wall to the building as at that time I thought the Druidic Temple was still standing. Later it became obvious that at least part of the Henge Temple had been removed.

Crucifixes on Dartmoor

Two days later I was back on Dartmoor with my wife to visit an art exhibition. I was hoping to buy a painting of a Dartmoor scene and also persuade John Christian, the artist, to act as guide once again and to take us to some stone circles. This time the aim was to see if there was on the ground within the circle, a painting of the crucifix. We already knew that the circles we had visited were part of the structure of temples which were in use up to 43AD or possibly to just before 60AD.

On the 2nd November 2006 I set out for a circle I knew well, Scorhill, and was at work by 10am with a clear sky, almost ideal for dowsing. I found one of the tombs and confirmed the antechambers and bone depositories. The next step was to find the tunnel and access shaft. This did not take long. My notes indicate that I forgot to check if the access shaft had been sealed but it did have a crucifix over it. The next step was to find the central pool. I searched but could find no trace of it. There must have been a pool in the temple at one time but there was no trace of clay or of plumbing to give it away. The thought then occurred to me that as there was a stream nearby they may have carried water from the stream to the pool. All I had to do was look for where the water may have been emptied into the pool. The splashing of water on to the ground leaves a lime rich area. I soon found the splash zone, picked up the bronze wheel tracks of the water cart and traced them back to the stream. The Druids had brought the water up from the nearby stream using water carts. The stream was man made and followed a contour line from some distant source. Instead of digging a well, the Druids had diverted a stream to provide their temple with water. There was evidence that the stream had been maintained in post Druidic times indicating that just as with the roads, later people took advantage of the Druid's hydraulic engineering work. The area where the water cart was loaded was identified and a crucifix was found to have been painted on the ground at this point.

The present terrain round Scorhill is not suitable for wheeled transport such as a water cart and the ground has clearly degraded with boulders, tussocks and gullies appearing. The next step was to look inside the circle for a crucifix. It was not difficult to find and it had the head to the north. Scorhill is a typical Henge Temple site and like the temples in Hertfordshire it clearly shows the transition from Druidism to a religion which is using a crucifixion as its symbol. Importantly, this transition occurs before the Romans arrive on the scene. How far back in time before the Romans arrived it is not possible to say, at least at the moment. Later when discussing what I had found with John he picked up the point that had been worrying Nigel and myself. Is the painted crucifixion pre-Christ? At this time we were assuming that the crucifix with its nails, blood and crown of thorns was post crucifixion but pre 43AD. What was now being suggested was that the Druids may have evolved to a religion where they had death symbolised by a criminal's death and dismemberment which was then followed by resurrection due to the power of Herne, the God of life. Perhaps the Druids had, with their knowledge of drugs been able to resurrect people from apparent death. Further development of these ideas had to wait. In the afternoon my wife and I met up with John who took us to the Nine Stones Circle. A small circle in which the central pool showed up by using a red clay witness. Water was delivered by water cart to the pool. Round the central pool I found four rectangular tiled areas which I thought from a quick check might be associated with crucifixes. Identifying one crucifix might be a good test for John. John like many dowsers is an intuitive dowser. In other words the dowser tries to tune into what is there and identify it by using intuition. The 'sensitives' are the extreme end of the spectrum in that they can tune in, 'see' and describe what is there. The sensitives I work with are normally correct on the solid objects and major chemical stains. I was therefore interested in whether an intuitive dowser like John could identify what was only a faint chemical stain from a painting on the ground now 2000 years old. I left John to dowse one of the tiled rectangles using his intuitive methods whilst I went off to confirm that the Nine Stones Circle had been a working temple when the Romans arrived. Walking round the site I found battle blood and execution lines which identified to within a few years when the temple complex was destroyed. On finishing my survey of the site I returned to see how John's intuition was dealing with a possible crucifix site. Not very well as it turned out. I then dowsed the site and found that the picture of the crucifix was not easy to pick up. The reason was that at one time the head had been to the north and on another to the south. Such a target where one picture is placed in reverse on a previous one is very difficult to deal with using any type of dowsing. Perhaps this was not a good target for an intuitive dowser. However, the three day trip to Dartmoor resulted in the acquisition of a superb painting of a Dartmoor scene by a master and confirmation that the temples were using a crucifix and Herne as an icon. At Scorhill the crucifix had been in a church or chapel with three rows of benches on three sides.

Was the Crucifix a pre-Christ Logo?

On returning home I met up with Nigel and discussed the possibility that the Druids crucifix could predate Christ's crucifixion. We could not see how it could because it was correct down to the detail of the crown of thorns and in one the spear wound had been painted on the side of the body. There was only one thing for it another visit to Stonehenge so Sunday 12th November 2006 found us looking for the well that had supplied water to the pool in the Stonehenge Temple. We did this by picking up the plumbing to the henge from the well. The pipes lead us to the well and we then confirmed that it was on an underground river. Satisfied that it was the well we had found on a previous visit we checked to see if it was sealed. The well was sealed and below the stone seal there was a cavity. On top of the well we found a tile or clay floor and the picture of the crucifix with the painting of Herne above it. We also found pews or seats round the chapel and the wall of the chapel. The site must have been filled in after being used for a period as Roman execution lines went across it. This indicated that the building must have gone by the time the Romans arrived on the scene. The next step was to find the tunnel leading from the shrine on the bank of the underground river to the Bell Chamber. At this stage we did not know that they were there but on the basis of what we had found in Hertfordshire and Avebury felt confident that we would find them at Stonehenge. In fact the shrine chamber was not far from the well (Figure 24.4) and we had no difficulty in finding it. As the shrine is always on the far side of the stream to the Chamber all we had to do now was follow the tunnel to the Chamber. The tunnel seemed to go on for ever but we eventually arrived at the circular service tunnel that goes round the Bell Chamber (Figure 24.4). Inside the service tunnel was the Bell Chamber. Once the service tunnel is identified it is not too difficult to locate the tunnel leading to the access shaft. We followed this tunnel for about a hundred feet and came to the shaft. It was sealed and like the well it was almost certainly sealed before the Romans destroyed Stonehenge. Once the access shaft was found we spent some time studying it.

We had always assumed that the Druids entered the access shaft down some steps to a lower level and had then climbed down a ladder or steps set in the wall of the shaft to the bottom.

It looked a tight fit but it could be done, provided you were not too old or too important. However, Nigel had the bright idea of looking for the traces of oxen tracks between the shaft and some distant point. After all, if the Druids used oxen to lift water from a well why not use them to lift spoil from the underground workings during construction and important people once the chamber was in use. We found the oxen tracks on one side of the shaft. On the other side were the tracks of Druids going to and from the shaft. It looked very much as if a lift had been installed to enable priest to move easily between the surface and the underground temple. We picked up signs that the lift had been in use in the early Bronze Age. A lift three to four thousand years old! If confirmed, that was going to take some beating and it was going to be yet more evidence that the Henge culture knew a thing or two about engineering. The track of the oxen was forty feet long so we knew how deep the shaft was. The tunnel from the access shaft to the service tunnel of the Bell chamber is nearly one hundred feet long. The tunnel is sealed by a wall as it opens into the access shaft.

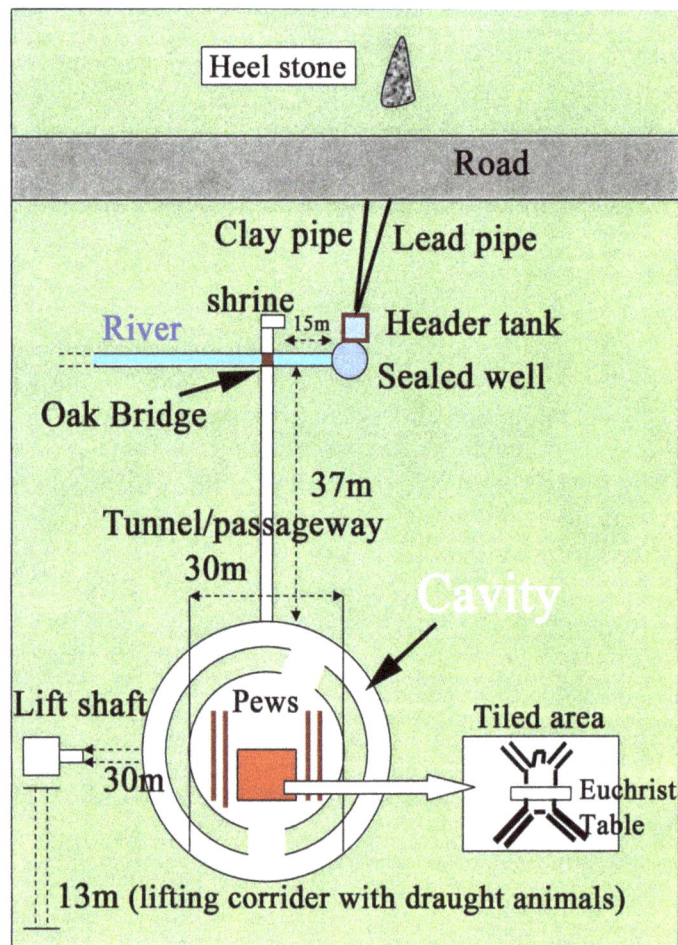

Figure 24.4 An Underground Chapel at Stonehenge
At Stonehenge a chamber was found by identifying the underground stream feeding the well, then finding the oak bridge crossing the stream from the shrine to the tunnel. The tunnel was followed to the service tunnel and chamber. The drawing is not to scale but shows the approximate position of the underground system in relation to Stonehenge.

If anybody was to break the shaft seal and reached the bottom it would appear as if they were in a blind shaft. The opening or entrance from the tunnel into the service tunnel was not sealed but the nearby connecting link between the service tunnel and Bell Chamber was. There was a two to three foot thick wall blocking the entrance to the chamber. The chamber dowsed as if it was about 30m in diameter. This is a very large chamber so it may be that the dowsing is wrong. To put things into perspective, a Bell Chamber 30m in diameter would probably need a vaulted ceiling about 15m high in order to stabilize the chalk and have the right curvature to stop it collapsing. If this was the case it would mean that the chamber would have a greater height than the depth of the access shaft. The vaulted ceiling of the chamber also requires a minimal thickness of rock above it to provide the necessary strength to prevent collapse. This minimum thickness of chalk or rock could be say fifteen to twenty feet (5 to 6m). Adding this thickness of rock to the height of the chamber places the floor of the chamber about 50ft (20m) below the surface. That is the surface above the Bell Chamber. Looking from the access shaft to the Bell Chamber it was easy to see that the ground above the chamber is lower than that where the entrance shaft is situated possibly by 3m. This means that the bottom of the shaft at 13m is about 10m higher than the chamber floor i.e. about 33 feet. To reach the chamber floor the tunnel from the access shaft would have to drop by 33 feet over the 96 foot length of the tunnel, say one foot in three. This is a steep slope and not the sort of mistake the Druid engineers are likely to make. Having recognised that we had a problem we had to leave its solution to another day. It was not just the steepness of the slope for access by the priests but also for the removal of spoil. Another access shaft for spoil removal started to look a distinct possibility. However, in the limited time left the important thing was to determine if there was a chapel on the access shaft, which there was, and if the Bell chamber had a crucified person on its floor.

When dowsing from the surface for the contents of a Bell Chamber or for paintings on its floor there is the problem of knowing if the dowsed target is a few feet below the surface or on the floor of the chamber about sixty feet down. It was decided that one approach was to mark out the chamber so that we knew where its walls were. If anything went through them it must be on the surface. If all the artefacts stayed within the confines of the wall then they might be on the floor of the chamber. We soon found that we were dealing with a complete chapel that stayed within the confines of the chamber wall. We checked the pews and found that they only went as far as the walls. The only clothing fabric associated with the benches was linen. We could not find any trace of wool. This indicated that the chapel was a select one for priests only. The tile or clay rectangle was twenty by fourteen feet (6m by 4m). The crucified figure had its head to the south. There was a rectangle of linen on the figure with a line of bread (wheat) and red wine along one edge indicating that the Eucharist had been part of the ceremony held in the chamber. Having established that the Bell Chamber had been used as a chapel for what appeared to be Christian like ceremonies we did not study the painted figure in any more detail. To be sure that the paintings were not on the surface we looked for signs of a surface wall and building round the area of the chapel but could not find one. This indicated that the chapel must be in the chamber and not on the surface.

Review of Findings

With three major Druidic sites come Christian structures located and plotted out in the field on the opposite side of the road to Stonehenge and just to the east of the Heel Stone, it was clear that a dramatic change had taken place in the pre-Roman history of Britain. The elaborately engineered Druidic well with its plumbed water supply to the Stonehenge Temple was sealed and decommissioned. Its elaborate header tank and highly engineered water drawing mechanism removed to be replaced by a building that has all the hallmarks of a Christian Church or Chapel. A large clay or tiled rectangular floor bears a painting of a crucifixion with a

Christ like figure nailed to a cross. The painting bears remarkable detail, appropriately placed blood, nails and a crown of thorns. The clothing is draped and worn in a characteristic manner over the body. The spear wound on the body is also there. The congregation sits on top of the crucifix on benches with a table/altar suitably positioned in the aisle for the congregation to take the wine and bread of the Eucharist.

A vast Bell Chamber 30m across excavated deep down in the chalk has a similar structure with a central crucifixion painting, and leading out to the east from the chamber there is a tunnel linked to a shaft that housed an elevator for the Druidic priests and then later Christian priests. The well and biolocation techniques identifies and dates these as Iron Age Christian Churches in use in a pre-Roman Britain. What we had found at Stonehenge confirmed our findings in the St. Albans area but it still left us with problems. One of these problems was the chronology using the accepted historical dating of these times.

For example:

- The Romans invaded Britain in force in 43AD

- Jesus Christ was said to have been crucified between 29AD and 33AD

- The religion in the Stone, Bronze and early Iron Age was Druidic

- By 43AD the country had changed its religion relatively suddenly and dramatically to embraced Christianity on a massive scale all over the south of the country at least

- Within 10 to 14 years of Christ's death the Romans had arrived in Britain to find the Druidic system being built over with Churches bearing painted depictions of the crucifixion and the Eucharist as part of religious ceremony

Religions do not change quickly so this takes some explaining.

For example:

- Evidence from biolocation is erroneous and flawed leading to a complete misinterpretation of the situation

- The evidence is fine but the structures and paintings are another aspect or development of the Druidic religion and its by chance that they resemble Christian worship

- They are Christian Churches but are wrongly dated

- We are looking at a direct link between Druidism and Christianity

The first explanation that the biolocation is wrong could be partly correct. That is depths, witnesses or something is wrong and leading to false interpretations. However, biolocation relies on a convergence of evidence and although some of the detail may be in error the overall picture is likely to hold.

The second explanation has to be taken seriously and the dating of artefacts is going to be very important. From looking at changes over centuries and millennia as the Henge Age developed it is now necessary to consider decades as the story moves from pre-Roman Britain to the Roman period.

The third explanation is a very real possibility and good evidence will be required to identify the dates of the Churches or Chapels. It is not necessary to date all Church remains found on Druidic wells and shafts as churches were subsequently built on such sites for about 2000 years. It will be necessary however to find some that clearly predate the Roman occupation. At the moment the evidence is based on the sealing of wells and shafts as being pre-Roman and the presence of execution lines going across church sites indicating that they had been filled in before the Romans arrived. The early churches always appear to be below ground level.

The fourth possible explanation ties in with mythological tales linking Jesus and his family to the British Isles but any such conjectural linkage will require a clear demonstration that Druidism evolved, prior to the birth of Jesus, into a religion with much in common with the teachings of Jesus and his family and Christianity. This would enable Druidism to slip straight into Christianity.

If the integrity of the biolocation studies is to be believed, although very much conjecture at the moment, 4 brings up some extremely important points to consider in the development of religion in the west. Not the least of these is that if the painting on the ground represents Christ's crucifixion it is the first historical documented evidence of the person and the event from the time at which it took place. Jesus with his uncle, Joseph of Arimathea (Box 24.2) and possibly other members of his family, is said to have visited Britain when aged 12/13. If this is the case then what would be the purpose of his visit? Certainly he would be coming to a deeply spiritual country with a society ensconced in the Druidic practice and ritual. It is perhaps the nature of this spirituality that he came to experience and learn about. He would have found a society based on a powerful religion (Box 24.3) with care in the community, few if any beggars and little if any poverty. An ancient socialism or communism. He would have seen the Henge people as a healthy and intellectual people with a devout faith and adherence to worship, ritual and a firm belief in reincarnation and life after death. What a wealth of experience learning and teachings could Jesus and other members of his group have taken with them on their return to Palestine. An alternative is that Jesus himself impacted on the Druidic religion in a major way. However, it seems unlikely that a religion thousands of years old, even if it was gradually evolving, could change suddenly and reorganise itself to embrace Christianity.

Box 24.2 The Grail Legend

The Grail legend is one of the most widespread legends and can be traced back at least to the Middle Ages. According to the legend, the Grail was the cup or bowl used at the last supper. Joseph of Arimathea, an uncle of Jesus, is said to have brought it to Britain in 37AD or 63AD. From what we now know 37AD is the most likely date. The story of a Grail may go back to a much earlier date, see pg. 157 'Celtic' Myth and Legend by Mike Dixon-Kennedy. One story is that the myth evolved from the magical cauldron of 'Celtic' tradition. However, from what we know about the Druids the story could be based on the temple pool that appeared to enter into many ritual practices. Another source for the myth is the font that may have appeared in temples before the pool. The pool with its water supply, its central position in the temple with the font - would only have been created if it was believed to have magical or mystic powers.

According to mythology, Joseph of Arimathea's connections with Britain go back to the first years of the first millennium. He is said to have been a tin or metal trader and sailed between the Eastern Mediterranean and the West of Britain. If this is the case he could easily have brought his twelve year old nephew Jesus to Britain for education by the Druids. The Druids at that time were well into the Proto-Christian era with what we thought at the time were crucifixes, also Herne, a torc and the fish as their logos. Later and after the crucifixion, there would have been no problem in linking the pool to the concept of the Grail and its mystical powers.

It is also unlikely that the death of Jesus would invoke such a wide spread response involving the building of churches and the painting of their floors on the scale observed. The only plausible explanation for this very rapid change from Druidry to Christianity appeared to us to be that Druidry had already evolved into a Proto-Christianity long before Jesus arrived on the scene.

With the end of the drive home from Stonehenge the discussion also ended. We were both now convinced more than ever that we had stumbled on something of great importance. There was the very real possibility that in those years when Jesus disappeared from the records, that is between being a boy visiting the temple and appearing as a mature preacher and meeting up with his brothers and relatives, he was in Britain learning from the Druids. But how to show that the Druids had evolved a religion that possessed many elements of Christianity. A religion that could side step into Christianity with the major icons and ceremony of the new religion already in place.

Needless to say this presented us with some tough problems. Then one evening Nigel came round to the house and we started yet again going through the data we had. We knew that the Druids sealed their wells, access shafts, the openings into tunnels but did not know when. We knew that they used the crucifix and Herne as icons and painted them on the ground. Were they using any other icons? At this stage we did not know. They used the crucifix very widely and by now we had found it on the cursi, in the arena and on the ground of the Maypole also in Dyke Temples. Then Nigel came up with the idea that it might be on Druidic roads. Now we knew that roads fall out of use and new ones are constructed. In other words we should be able to find Stone Age, early and late Bronze Age and Iron Age roads and if the crucifix was on them see how far back its use went. That realisation made us aware that the answer to the problem was probably in the detail of the crucifix. If it predated the death of Jesus then it may well show an evolution of design, the last stage of which was the Christian one, with a crown of thorns and a spear wound.

The Search for the pre-Christian Icons

The following day, the 12th January 2007 we found the crucifix on Druidic roads. Every road looked at had a line of crucifixes running along them. I marked one out on the Iron Age road running through the garden. It was a Roman cross with the head on the south side. Above the head were the antlers of Herne complete with yew, ivy, oak, holly and possibly other greenery that I did not test for. The main difference was the absence of a crown of thorns and blood from the nails. The body at this stage looked as if it was a dead body that did not bleed when nailed to the cross. As a result of this discovery we now knew that the crucifix took at least two forms and the task was to determine if the one without the blood and crown of thorns predated the one with. Our starting point was to be Bell Chambers and we regretted not having studied the Stonehenge one in more detail and with greater care. At the time we had assumed crucifixes were crucifixes and so had cut corners. Later on the 12th I visited one Bell Chamber in a nearby field. It had a crucifix with no crown of thorns or blood. A few days later I visited a chamber on the south coast of Devon at Hope Cove it also had a crucifix with no crown of thorns or blood. I could find no battle blood round the entrance to the Hope Cove chamber which I took to indicate that it was sealed before the Romans arrived. The chariot tracks leading to the tunnel entrance indicated that the chamber had been dug in the late Bronze Age and was used into the Iron Age. Knowing that the chamber was sealed before the Romans arrived still did not tell us when it was sealed. The Romans landed in 43AD but their extermination of the Druids was still going on in 60AD. The Hope Cove Bell Chamber crucifix had not been altered so news of the crucifixion may not have reached them when it was sealed. This indicates a date prior to 33AD for when it was sealed. But was it 10 years or a 100 year prior to the crucifixion?

When marking out crucifixes on roads we had noted that unlike the diamonds there was a big space between them. This indicated that there might be another painting between the crosses. Nigel had found some data on the internet relating to the fish as a religious symbol and I had found a reference saying that the fish symbol could be traced back to 300BC. On the 19th of January 2007 I started to look for a fish on the Iron Age road in the garden. The sort of fish I had in mind was the Christian fish symbol and started by looking for the curved body outline. I soon found something but as the dowsing progressed, the shape of a real fish emerged. The body and tail was outlined in gypsum, fins in manganese dioxide, eye in a cadmium selenium red, a mouth in gypsum and manganese dioxide and an eyebrow. The fish was distinctive and complex, the sort of icon that could show both regional variation and variation over time

309

(Figure 24.5). We had possibly another symbol that might enable us to fit a time scale to the Proto-Christian era. The evidence indicating that Druidism had developed into a religion very similar to Christianity was beginning to build up. Showing the evolutionary development of Proto-Christianity from Druidism and dating it was going to be a major challenge and would have to wait for the summer   At the end of January 2007 it was beginning to look to us as if Christianity developed from Druidism, the religion of the Henge culture. With the good communications and freedom of travel within the Roman Empire, Proto-Christianity could have become widely known and used as a basis of religious philosophical discussion if not teaching and belief in many parts of the Empire.

Figure 24.5 Fish and Crucifix
A fish was found on the Iron Age road passing through the garden. The fish lies between two crucifixes. The painting is a spirit painting and contains chemicals from fish. Such paintings are easy to identify on roads using a trout witness. The detail on the fish is remarkable with fins, eye and mouth well defined.

Chapter 25

The Cathedrals of the Druids

To South Devon

The year 2006 came to an end and it still seemed impossible to draw the project to a close. The year had finished with a realisation that Druidry had evolved during the Iron Age into Christianity. There were, however, many loose ends and tantalising questions that required answers.

It was by chance that I and my wife had decided to spend some of January and February 2007 in a little cove on the South Devon coast. The apartment we rented provided a splendid view overlooking the sea and nearby cliffs. The idea was to see the winter from a different perspective. The cove, called Hope Cove, was tucked in behind the headland of Bolt Tail. Within a few days of arriving I knew that the area round Hope Cove and Hope Cove itself was a landscape carved by the Druids. Every road appeared to have the diamonds indicating a Druidic road. The dry stone walling was in the main Druidic and magnetically active. Where the Local Authority had made a passing place for cars in the narrow lanes and had removed the original dry stone walling the new walling was 'dead' magnetically. The stones had not been laid according to their magnetic fields by the modern dry stone wall craftsmen. Ancient and modern dry stone walling could easily be identified by the presence or absence of magnetic fields. Looking across Inner Hope to the cliffs of Bolt Tail I could see a Long Barrow above the cliffs, in fact more than one (Figure 25.1). I soon realised that I had a very busy two months ahead of me. However, most of what I was to find is another story but part of it does belong to this one. From the final days of the Henge Age described in the previous chapter it is fast back in time to the early days of the Henge Age. Possibly three thousand or more years before the Romans destroyed the Henge civilization. At this time the sea was lower, the shore line of Bolt Tail and Hope Cove many hundreds of meters further out and the climate quite different to today's climate. Possibly more like that of Scandinavia today. On Sunday morning 11th February Nigel arrived in Hope Cove. He decided to come down once he heard that I had found a Long Barrow. These are considered by archaeologists to date from 3000 to 4000BC and to be burial mounds. That is date wise from the early Bronze Age back to the Neolithic, late Stone Age. One of Nigel's theories was that barrows, round or long, were in buildings. This theory had found support at Stonehenge where there are three nice round barrows lying alongside the cursus. All three appeared to be inside a building as did the one at the western end of the cursus. The barrows were often providing the platform for raised alters but could also be there for other reasons. The chance of studying a Long Barrow to see if it had been within a timber building was too good to miss.

The Hurlers, Bodmin Moor

On Sunday afternoon I took Nigel round the Hope Cove area showing him the Bell Chamber I had found, the Henge Temples, roads, dry stone walling and the Long Barrow. A quick look at the area before starting work on the following day. During the night the weather changed and on Monday 12th February Hope Cove, like the rest of the West Country, had a gale blowing through it carrying rain drops that felt as hard as ice. Dowsing on Bolt Tail was out of the question. After about an hour of watching spectacular waves crashing over the breakwater I suggested that we could go to the Hurlers on Bodmin Moor. My reasoning was that the wind and rain might not be quite so severe inland and we would only be on the moor for about thirty minutes. The reason for going was quite simple. Hamish Miller and Paul Broadhurst in their book "The Sun and the Serpent", Page 61, reported finding a misshapen pentagram in the centre of the middle stone circle. When Nigel had first found the painting near the transept of St. Albans Cathedral it had at first looked like a misshapen star or pentagram. It was only when I had studied the figure closely and marked it out as I went and then checked the pigments and colours that we realised that it was a crucifix. Identifying a crucifix was therefore not easy and it could be missed. However, the chances that the Druids had painted a pentagram at the centre of a temple on Bodmin Moor and not a crucifix were small to say the least. We were therefore reasonably confident about what we were going to find.

Figure 25.1 Bolt Tail
The view across Inner Hope shows the ridge of the Long Barrow coming down to the cliff. When the Barrow was built the cliff face was much further out and the Cove was dry land with at one time a bronze foundry on the present day beach. The temple is in the left hand valley.

Collecting our wet weather clothing and dowsing gear together we set off from Hope Cove for Bodmin Moor and the Hurlers. Just over an hour later we arrived at the small village of Minions. We found the car park for The Hurlers and the other ancient sites on the nearby part of the moor. After parking the car it was clear that the rain and the wind were just as severe as at Hope Cove if not worse. After dressing up we started out for The Hurlers. Once out on the moor we felt the full force of the wind and it was a real gale, making walking difficult. The rain had eased a little but the moor had not drained. After walking into the wind and driving rain for what seemed ages we arrived at The Hurlers and looking around, there was not a sheep to be seen on the moor (Figure 25.2). The birds were however enjoying the wind, practicing speed runs and stationary flight. The circle to our left looked as if it was in a moonscape, pock marked as a result of past mining activity. The middle circle looked in a reasonable state with the ground flat. The circle to our right also looked as if we could do something with it. I had never dowsed in a gale before but I suppose there is a first time for everything. With heavy iron rods it is possible to dowse in surprisingly high winds. We started work on the central circle and the first step was to check the magnetic alignment of the stones. They all seemed to be set up correctly magnetically and it appeared as if they were the outer wall of the temple. At least there had been wattle walls between the stones at one time. Two of the temple gates were marked by larger stones and some, possibly all, stones had been painted. This is a good indicator that they had been protected from the weather and under a roof. At the centre of the circle was the tomb of the Prince and Princess with the associated underground temple complex. A search outside the temple led to the execution lines being found so we knew that the temple had been in use up to about 50AD. We then started to look for signs of a crucifix. I took the sealed access shaft to the underground temple and Nigel took the centre of the temple. It was not easy under the prevailing weather conditions but both sites had a crucifix.

It looked very much as if Hamish Millar on the 3rd April 1988 had discovered a crucifix at the centre of The Hurlers, not a pentagram or star as he had thought. There was also a crucifix in the centre of the right hand circle. Although the Hurlers appear to be standard Henge Age Temples the question arises as to why three in a row. The answer to this question will have to wait for a second visit.

Figure 25.2 The Hurlers, Bodmin Moor

The middle circle of the Hurlers on Bodmin Moor. The central pool area and the top of the access shaft had crucifixes on them. The stones of the circle were painted and formed part of the outside wall of a building. The paintings were sheltered from the weather by the roof of the building.

The Long Barrow on Bolt Tail

We retreated from the moor to the car and then made our way back to Hope Cove to don dry clothing and sit the storm out. The following morning the wind had dropped and it was dry. Almost ideal for a visit to the Long Barrow on Bolt Tail. When I had first arrived at Hope Cove early in January 2007 I had spotted the Long Barrow from the apartment I was renting with my wife. Archaeologists refer to it as a 'fort'. The term 'fort' is used by archaeologists to cover a whole host of earthworks about which little is known. I was soon investigating it and felt sure that it was a Stone Age Long Barrow, Stone Age as I could find no sign of metal. From the barrow it was possible to trace chariot wheel tracks to a garage tucked in behind the barrow and to a temple about 200m from the barrow. There was no sign of metal at the Long Barrow or at the temple apart from the gold in the tomb. Also I could not pick up any sign of linen or wool. I therefore felt sure that I had stumbled across a Henge Age Temple directly connected by a ceremonial route to a Long Barrow. The significance of this was that Long Barrows are considered to be the oldest structures in the landscape of the British Isles. The Long Barrow on Bolt Tail could be 4000 years old or perhaps 5000 years or even older. The West Kennet Long Barrow, possibly the best known one in Britain, is said to have been built between 5000 and 6000 years ago. Six thousand years ago the climate in Britain was quite different to what it is today. The tree cover was dominated by pine, birch, alder, and hazel. The oak, beach and elm were only just coming up from the south as the ice sheet retreated north. I had tried to identify the wood of the temple's postholes but had found no response when using oak. I had therefore asked Nigel to bring a selection of wood witnesses down with him.

Bolt Tail is only a twenty to thirty minute walk from where we were staying, close enough to carry our dowsing survey equipment and for us to set up 'camp' next to the Long Barrow (Figure 25.3). Whilst I confirmed the chariot track Nigel started looking for postholes marked them out and identified them as pine posts. The chariot wheel lubricant was deer fat and the wheels rotated on the axle. The draft animal was a deer. There were signs of the use of birch brakes on the wheels. The reason for saying this is that as the chariot approached the unloading point two areas of birch wood were found on each wheel track where brakes would have been applied. The bones were carried in a wicker basket which was unloaded onto a wood stand. This procedure produced a very sharp edged rectangular red ochre stain on the ground. From the unloading point (Figure 25.4) the basket was taken to the entrance of the Long Barrow. Here there was another red ochre stain from the basket as it was handed to priests in the Long Barrow. The interior of the barrow can be dowsed by walking along the top. Not an easy task but the central tunnel, small side chambers and the position of lamps, which were burning deer fat, can be identified (Figure 25.4). A second chariot track with a separate unloading point was found running parallel to the first one. It seemed to serve a separate burial chamber. Nigel looked at the Long Barrow in some detail and found that it had at one time steps going up onto the top. The steps lead to a terrace that ran along the eastern side of the barrow. The terrace, due to thousands of years of weathering, is not now readily visible and it is easy to see how archaeologists miss the detailed structure of barrows. The postholes were marked out, the roof timbers, the wall panels and drip line identified. Nigel was right, the Long Barrow was in a large building which rose possibly two meters above the top of the barrow (Figure 25.5). On one side there was an entrance for the chariot which drove down to the barrow entrance under cover. On the other outward facing side the barrow had a dry stone wall face and above it a terrace along which priests walked and possibly spoke to people in the large covered area below them. The Long Barrow was the centre piece in a large ceremonial building. The stones from which it was constructed were arranged magnetically. This resulted in the dry stone wall on the outer face generating magnetic fields which could be identified on walking in towards the wall. The areas of wall which have been repaired in modern times did not have the stones laid magnetically and they appeared to be 'dead' i.e. no dowsing response on walking in towards the wall. There were signs of paint along the wall so it is possible that there was at one time a line of flat painted stones in front of it.

Figure 25.3 The Long Barrow on Bolt Tail
On the outside of the Barrow is a dry stone wall which appears to hold the Barrow back and defines its edge. The path in the foreground is going in through the gates between two Barrows.

314

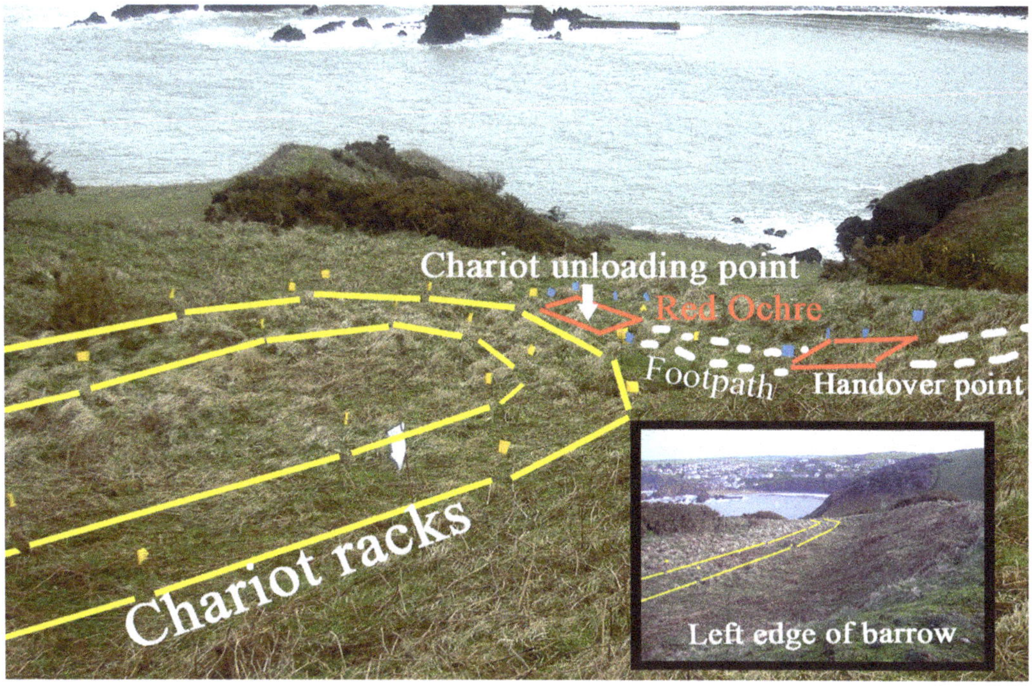

Figure 25.4 The Long Barrow Bone Unloading Points

The chariot comes in from the front of the building and runs along the left edge of the Barrow. The chariot turns left and stops at a point where it unloads the bones in a basket, onto a stand. The bones are then taken to priests at the entrance to the Barrow and handed to them. The cart then swings in a circle and returns to the entrance of the building.

Figure 25.5 Concept Sketch of a Long Barrow Building

The building containing the Long Barrow is based on the postholes (yellow flags) found and the potassium lines (blue flags) showing the position of beams and rafters. The terrace running along the right hand side of the barrow was used by priests as indicated by feathers from their clothing. In the open area next to the wall there were benches for people to sit on.

The Henge Temple

If the chariot tracks are followed out of the gate they bend north and go down into a small valley to arrive at the north entrance of a Henge Temple. The tracks enter the temple and come to a rectangular area of red ochre which is the loading or unloading area for bones. The significance of this as already mentioned is that by chance we had found a direct physical link between a Long Barrow, which are amongst the earliest known burial structures, and a Henge Temple, the design of which we knew was still in use along with associated ceremonies between two and three thousand years later. Druidry as a priesthood had therefore been in existence for at least 3000 years. Whatever age the temple was, it was to show us that Druidry was already advanced by the time the temple and the Long Barrow were built. The temple was not a standard circular building with 6 + 3 rings and 60m across. The building was 45m north to south and 34m west to east. It had a central tomb with two children as the Prince and Princess. At each of the four gates there were eight sacrificed Druids with ash staffs. At an estimated distance of about 50m from each gate there was a Dolmen. There appeared to be no pool at the centre of the temple but on the floor was a painting of the four thunderbirds (Figure 25.6). Round them were possibly four circles. Along the radius studied there was first white then a gap then a yellow, a gap, a blue area another gap and then a red area. It looked as if the gaps had been painted with something but I could not identify what it was. The presumed circles were not complete as they bent in at the north gate. About four feet to the south and level with the tombs western side (Figure 25.7) is a Eucharist table about 2ft by 6ft. On the northern side of the table is an area dowsing positive on a bread witness (wheat) and a yeast witness. The table cloth was made from a nettle fabric. We had already found Eucharist tables in Iron Age Temples but this was the first evidence that the ceremony may date from the late Stone Age. Searches for the crucifix and fish drew a blank in the temple and nearby roads.

The HGV Roads of Stonehenge

It was whilst enjoying a glass of wine in the Hope and Anchor one night that it occurred to me that the Druids might paint temporary roads with logos or emblems, particularly if they were used for transporting religious loads. For some time we had been wondering how to tie down time wise, the transport of the large sarsen stones at Stonehenge. We knew that they had been moved in the Iron Age and that we could follow the wheel tracks of the HGV over the Salisbury Plain. We also knew that the access shaft to the underground temple at Stonehenge had been filled in or sealed before the stones were moved onto the site. We had found the iron tyre marks of a HGV going across it. At the time this did not mean a great deal to us as it seemed obvious that if you are moving large heavy loads you fill up all the cavities in the ground and possibly back fill all the underground chambers. That was in 2005. In 2007 we now knew that there was considerable religious significance to the sealing of wells and access shafts. The sealing indicated what we were calling the Proto-Christian phase of Druidism with its change from the diamonds and spirals indicating underground streams to the new Herne's head, crucifix and fish. The reasoning taking place, with the help of a glass of wine, was that if the sarsen stones had been moved in the Proto-Christian period the routes taken by the HGVs would not be marked by diamonds but by crucifixes and fish. This would not give us the year in which the sarsens were erected but it would tie them to the old Druidic period or the new Proto-Christian period. If the stones were moved in the Proto-Christian period it would indicate that this period started early in the Iron Age and certainly a long time, possibly hundreds of years, before the Romans arrived in 43AD. As we had already worked out the energy engineering side of Stonehenge it was possible that Stonehenge was telling us how the old mystic side of Druidism was linked to the new Proto-Christian form of Druidism.

We were due to return home to Bovingdon for a few days on Wednesday 14th February at

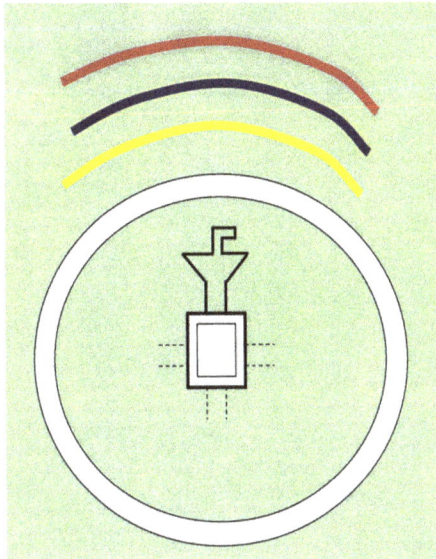

Figure 25.6 The Thunderbirds of the Tomb
The tomb in the temple had a white (gypsum) rectangle then a black rectangle. There was a Thunderbird on each side then a series of bands of colours. They were not traced all the way round so may only represent a sector of a floor painting.

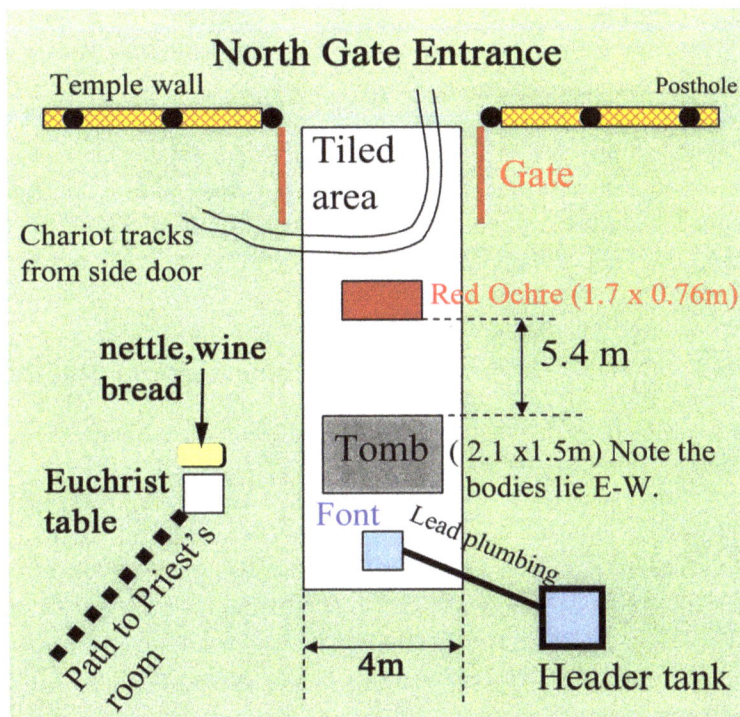

North Gate Entrance

Temple wall

Posthole

Tiled area

Gate

Chariot tracks from side door

Red Ochre (1.7 x 0.76m)

5.4 m

nettle,wine bread

Tomb (2.1 x1.5m) Note the bodies lie E-W.

Euchrist table

Font   Lead plumbing

Path to Priest's room

4m

Header tank

Figure 25.7 The General Layout of the North Gate Area of the Temple
There is an apron of tiles from the North Gate to just the other side of the font. In the centre of the temple it covers the tomb and towards the North Gate there is an area where the bones were handled and transferred to or from the chariot. The chariot entered or departed via a door to the west of the gate.

the end of Nigel's visit. We therefore agreed to return via Stonehenge and take a close look at the routes taken by the HGVs as they headed for the henge. Before looking for the HGV tracks we found the processional way coming west from the Heel Stone and following the split beams. We knew that there were diamonds along it and it did not take long to confirm that they were there. The next step was to look for a pre-Christian crucifix and one was soon found. As on previous roads we had studied there was a big gap between crucifixes which was filled by a fish. Having confirmed that the processional way was marked with Proto-Christian symbols we moved onto an HGV track. The vehicles using it had six wheels on an axle. We took a section of the track and found that the Druids had painted pre-Christian crucifixes and the fish on the track. The Druids had marked the route taken by the very large vehicles bringing

317

stone to the henge site. The crucifix was a pre-Christian one. The stones were therefore moved after the use of iron was well established and before Christ was crucified. We could not find any diamonds connected with the HGV route. If this can be confirmed, then the Sarsen stones at Stonehenge were almost certainly moved during the Iron Age when Proto-Christianity had been established as the latest development of Druidism. The fields set up by the resonant chamber and the Heel Stone were important to the new Proto-Christian form of Druidism.

A picture was gradually emerging in which Proto-Christianity had developed perhaps early in the Iron Age and had adopted as its icons the crucifix, Herne and the fish. It was not long before I realised that yet another visit to the Stones was required. We had to find out if there was a crucifix in the pool area and on top of the access shaft to the underground temple.

Is Stonehenge a Christian Cathedral?

On the 16th March 2007 Nigel and I headed for Stonehenge once again. We had permission to visit the henge between 6am and 8am on the17th. I had drawn up a list of objectives for our visit, Table 25.1, which was intended to help us work fast and efficiently. The significance of Stonehenge proving to be possibly the first Christian Cathedral had not been lost on us and we knew that if it was, it was going to be an archaeological find of major international significance let alone national significance. If Stonehenge followed the pattern of previous temple sites there should be a crucifix and chapel on the access shaft to the underground temple and a crucifix more or less where the pool had been in the main temple. We arrived early at 05.30am whilst it was still dark and made ourselves known to the security people. By 05.45 we were walking through the tunnel under the road behind a security guard and on our way to the stones. As we passed the wall paintings of men hauling on ropes and trying to erect stones Nigel reminded me of a comment I had made on an earlier visit along the lines that 'There is going to be a lot of repainting required". The guard opened a gap in the rope fence and we walked from the path to the Stones. There was now reasonable daylight for working round the stones. We were the only visitors so had the place to ourselves. Nigel was first to start looking for signs of a crucifix and by the time I had sorted my witnesses out and was ready to join him he had identified what could be an arm or leg. We now knew a crucifix of some sort was there. From then on it was a process of mapping out the figure that had been painted on the ground about nineteen hundred and seventy four years ago. As we mapped the figure out by placing white and red chopsticks on the ground I was aware that the dowsing was crisp and clear. When we looked for signs of other crucifixes we could not find them. What we were dowsing appeared to be the one and only crucifix present. Soon the crucifix was marked out with its nails, blood, crown of thorns, spear wound, eyes, hair, and mouth. It was a Christian crucifix and no sign of a Proto-Christian one (Figure 25.8). The next stage was to find the gold torc, Herne and the fish. The torc was there but much of Herne's head was under a fallen stone. The fish was a real fish with eye, eyebrow and mouth but some of it was under fallen stones but this time and very importantly part of the tail had an upright standing stone on it. This was clear evidence that some of the stones had been erected after the picture had been painted on the ground. There is the possibility that we have missed a Proto-Christian crucifix and the stone was erected during the Proto-Christian period. In the absence of such evidence it looks very much as if some of the stones were being erected after Jesus was crucified. That means that a Christian place of worship was being erected at Stonehenge between 33AD and 43AD. Of course, construction could have started well before 33AD and it was the final stages of construction that took place after the crucifixion. Our first objective had been achieved and we now knew that Stonehenge is what remains of possibly the first Christian Cathedral to be built and it was built in the British Isles because that is where the roots of Christianity were. We later found the pews and identified the position of the Eucharist table. The users of the pews appeared to be dressed in wool and so presumably were not priests. The next step was

318

to find the access shaft which at Stonehenge lies just outside the wall of the temple. The shaft was sealed, there was a Proto-Christian crucifix and a sunken chapel. The chapel had been filled in and the six iron tyre marks of a HGV went straight across the shaft and chapel site. This indicated that the large stones were being moved in the Proto-Christian period and as they went straight through where some of the present standing stones are still standing it is further evidence that at least part of the present day Stonehenge was built during what is called the Iron Age and its Proto-Christian period. Crucifixes were looked for in the dyke but we could not find any. We found where an oak bridge, taking iron wheeled vehicles, had spanned the dyke and confirmed that iron wheeled vehicles delivered stones to where they stand today. We looked for signs of lintel stones on the blue stones. If the blue stones were used for energy engineering, as we believe they were, they would almost certainly have required lintel stones. The lintel stones help to generate the field produced by the horse shoe of blue stones but they are also required on the blue stone ring if the field is to penetrate the ring. The only way we could think of for identifying the use of lintels was to look for traces of paint where the lintels would have been. We identified paint between the stones which could indicate that there was something spanning the gaps between them. The paint does not confirm the use of stone lintels but it is supporting evidence. We also found signs of wood seating in front of the circle of linteled sarsen stones, perhaps additional pews for the congregation. The two hours amongst the stones soon went and it was time to return to our accommodation for breakfast. It is after two hours on site in the cold of a March dawn that one appreciates the virtues of a full English breakfast.

After recovering from the early morning survey, still hardly able to believe what we had found we returned to Stonehenge, at least the fields to the north of the A344.

Figure 25.8 The Crucifix at Stonehenge
At Stonehenge there is a gypsum outline of a crucifix in the pool area. The blood and iron nails are represented by red ochre and iron oxide. The crown of thorns is on top of the head giving rise to the blood running down the side of the head. The body is painted on the gypsum cross.

319

The Processional Way Following the Split Beam

The processional way following the split blue line from the Heel stone could be two processional ways. One based on the beam from the blue stones when they were being used, if they were used in such a way, the other from the sarsens. The blue stone beam generator could be Bronze Age as that is when the stones are said to have arrived. The sarsen stone beam generator should be Iron Age. We realised that one processional way could be on top of the other with possibly no way of separating them. From previous visits we knew that there were two processional ways, one marked with diamonds the other with crucifixes and fish. At the time we did the survey it appeared as if the tracks were separate, one being slightly to one side of the other. If this were the case there might be a way of separating the two processional ways. This time we were going to study them with greater care. On arriving on site we first marked out the iron and bronze tracks. They were separate. The bronze track was both arsenic and tin bronze so spanned the early Bronze Age when presumably the blue stones were set up. The iron tracks only had the crucifix, Herne and the fish associated with them. We could find no sign of diamonds. We then looked for the covered ways. There was one for the bronze tracks. The posts used were pine. There was also two rows of oak postholes associated with the iron tracks. The inner row of oak postholes were set up on a bronze wheel track. This was because the Iron Age ceremonial way was to one side of the Bronze Age one. The Bronze Age and Iron Age processional ways were therefore quite different and built at different times, early Bronze Age and possibly mid Iron Age, say 1000 to 1500 years apart (Figure 25.9). We could find no sign of Stone Age tracks. The identification of a Proto-Christian ceremonial avenue clearly in the Iron Age indicates that Proto-Christianity may not have appeared until the Iron Age. It also indicates that Stonehenge is likely to have had a beam splitter for a sarsen energy beam at the time the processional way was being used. If this is the case, much of Stonehenge may have been operational, particularly the five trilithons, long before Jesus was crucified. It is therefore perhaps the final stages only that were completed after the crucifixion. We confirmed that the crucifixes were Proto-Christian and then moved on to find the Bell Chamber.

The Bell Chamber

We had some engineering that did not add up if what we thought was a Bell Chamber was one and we also had to check the crucifix. After finding the well and stream we quickly identified the shrine and the tunnel leading to the Bell Chamber. The service tunnel was found and then inside it the chamber. The wall of the chamber was marked out with flags and we could see that it was not quite a circle. We then identified the crucifix, which was a Proto-Christian one, then the torc, Herne's head and the fish. The chamber had been used during the Iron Age but then closed and sealed. This was possibly quite early in the Proto-Christian era as there had been time to build a chapel on the access shaft.

We next set about solving the engineering problem of holding up the ceiling of a large chamber hewn out of chalk. One obvious answer was to leave supporting columns of chalk. It did not take long to find rectangular supports holding the roof up *(Figure 25.10)*. The chalk surfaces were protected by ceramic tiles or bricks which were painted. The wall of the chamber was also lined by painted tiles. If everything was left in place when the chamber was sealed this chamber is going to be quite a sight when it is opened up. It is the most complex one found to date.

Having found answers to all the questions we had started out with it was time to retire for lunch and then make our way home. At this stage there were only two people on the planet to know that Stonehenge was what remained of the first Cathedral to the then new Christian religion.

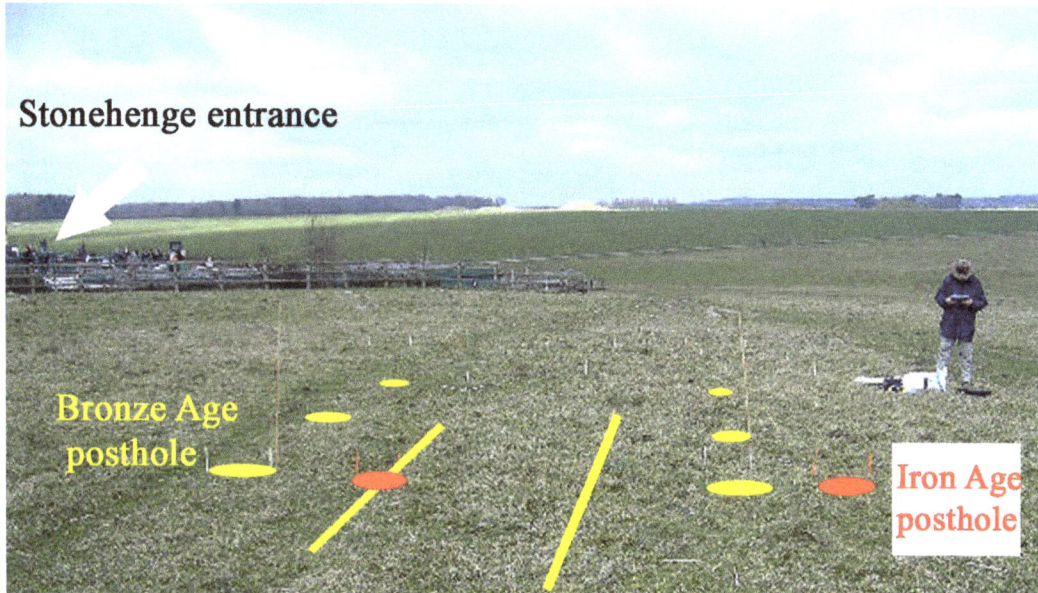

Stonehenge entrance

Bronze Age posthole

Iron Age posthole

Figure 25.9 The Processional Way Under the Western Split Beam
The ceremonial ways are indicated by the bronze (yellow) and iron (red) wheel tracks. The roads were enclosed. The Bronze Age covered way postholes are indicated in yellow. One of the iron wheel tracks is almost on top of these posts. It therefore looks as if the blue line which marks out the processional way moved to the right as seen from the Heel Stone in the Iron Age.

Figure 25.10 Chamber Wall and Roof Supports
The wall of the underground chamber is indicated by the yellow flags. The blue flags show where three rectangular supporting columns are. There were rows of these supporting columns going across the chamber.

321

Bolt Tail: A Stone Age Temple

On Monday the 2nd April 2007 Nigel and I arrived at the Sun Bay Hotel, Hope Cove for a follow up study of a temple. The temple was a rather special one as the contents of its central tomb indicated that it was from the Stone Age. It was also linked physically by a ceremonial path to a nearby Long Barrow. As Long Barrows are considered to belong to the Stone Age and early Bronze Age it appeared to confirm the date indicated by the tomb (Box 25.1). Back in February my study of the temple had been brought to an end by gales and cold. However, not before I had detected evidence of paintings on the floor of the temple. The purpose of our visit was therefore to confirm that the temple was Stone Age and to map out the paintings on the ground. The Henge Age has left us with little in the way of written history and it is possible that their ceremonial art may be their only 'written' record.

We were familiar with at least some of the Druid's paintings. The diamonds, Thunderbirds, a range of small paintings on some processional ways and more recently the crucifix, the face of Herne and the fish. To this could be added some geometric shapes that we had discovered on hillsides. When I first picked up signs of a painting on the Bolt Tail Temple floor it looked as if it filled most of the floor area and could be geometric in shape. It was also painted in a number of colours (Figure 25.6).

After recovering from the five hour drive to Hope Cove with a cup of tea in the hotel lounge we found the bright spring sunshine impossible to resist and started out for Bolt Tail and the temple. A matter of a twenty minute walk. The first stage of the study was to peg out the wall of the temple, mark the four main gates, two small doors and the tomb (Figure 25.11). The idea was to obtain a picture of the temple upon which the painting could be superimposed. The route of the chariot coming in from the Long Barrow and then returning to it was confirmed. The north gate had been used for this purpose along with a small gate a few yards to its west. Which way the chariot was travelling was not determined. Next we identified the tomb and realised that we had a problem. The tomb was not in the centre of the temple. This normally means that we have picked up the wrong tomb. However, we could not find a second one and there was only one wall to the temple. Not the multiple walls associated with multiple temples. We therefore had to conclude that the tomb was set to one side for a reason and left it at that as we wanted to make progress on the painting. We started by picking up the outside of the painting and followed it round the inside of the temple wall.

The pigment used was henna which is a dark yellow. Although I was not sure what we were going to find, about three weeks previously I had been in mid Wales studying a temple site. It was an early Bronze Age Temple and I looked for a painting on the floor. I had found one and was quite surprised at what animal it appeared to be. However, I did not have time to study it in detail although the ears appeared to be very characteristic. It was now late afternoon on Bolt Tail and as we went round trying to identify the outline of the painting up came the ears. To me they said 'cat' but Nigel was not too sure. However, if the ears were those of a cat then the rest of the face should be present. The next to be identified were the eyes, then the nose followed by the mouth. By now it was approaching supper time and we were sure that we had a picture of a cat. It was just a question of the detail and we decided to leave that to the following day although a quick look for the whiskers showed that they were there. That evening we were pretty excited about what we had found. Was it the Cait-Sith, the fairy cat of 'Celtic' legend. Whatever it was it could possibly predate the domestication date of the cat, which was given as 4000 years ago in one reference I had. It was clear that the Druids were familiar with the cat as the Princess in the tombs always had one and that applied to Stone Age tombs which were well over 4000 years old. The cat was also an important component of the Druids religion. The High Priestess had a cat on her Dolmen, the human sacrifice in the tomb

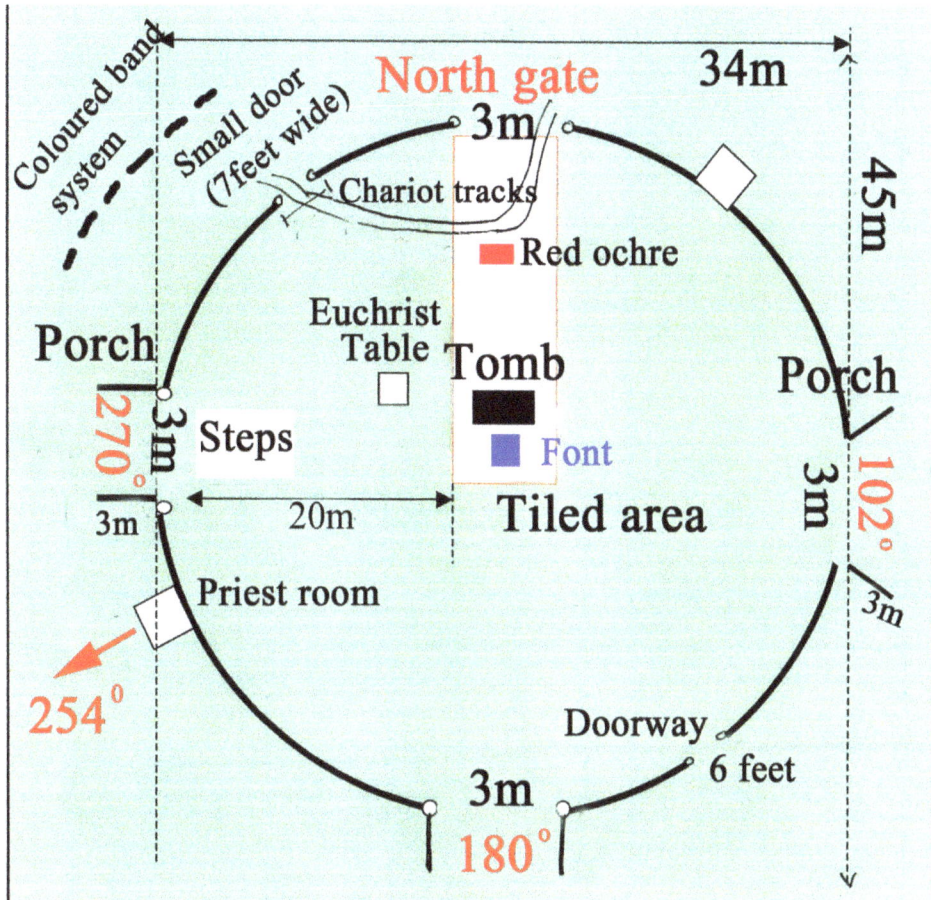

Figure 25.11 The Plan of the Bolt Tail Temple

The Bolt Tail Temple was more oval than circular being about 45m long (N to S) and 24m wide (E to W). The postholes indicated pine posts with hazel wattle walls. The main entrance was approximately north and faced down the valley to the sea. The floor had been painted in succession with a number of designs. Possibly the first being the Thunderbirds and then the Cait -Sith. At the West Gate there were bars or strips of colour.

Figure 25.12 The Temple Wall

The photograph shows the temple wall marked out with yellow flags. The temple is on a slope in a valley. This may have been dealt with by terracing within the temple to create flat areas but no evidence of it was found.

323

appeared to have a cat and now the temple floor had a cat painted on it. Why the cat was so important we did not know. It was however clear that cats had been domesticated long before the dates given in the archaeology books I had read. The morning of Tuesday 3rd April was bright but windy. After breakfast we set out for the temple on Bolt Tail. On arriving we marked out the gates and tomb as reference points. The next step was to mark out the face of the cat but we could not find any trace of the face. It soon became clear that there was a magnetic storm of some sort and it was only with great difficulty that we started to pick up some henna paint lines. To make things easier I decided to find the nose of the cat and Nigel the eyes. The nose I found using a gypsum witness was somewhat bigger and more angular than the one we had found the previous afternoon. It was also not in the right place. Once Nigel had marked the eyes out it was possible to see where the nose should be and it was soon found. That left the problem of what it was that I had found. The rectangular area was more or less where I had found the Eucharist table during my last visit but it was bigger and the stain was gypsum and not from a nettle fabric. Now it happened that one of the things we were on the lookout for was a font. We felt that in the absence of a temple pool there must be something that was taking its place. It could be a stone construction, wood or possibly lead. I checked for lead and found a lead stain indicating a lead vessel about two foot square at least. On the south side there was a blood stain possibly indicating circumcision and a footpath from the font to a single door in the wall of the temple which lead into a small room or cubicle. The cubicle had a tile or clay floor. Whilst I was doing this Nigel traced a lead pipe from the font to just outside the wall of the temple and what was possible a water tank. The Druids had a plumbed water supply for the font in a temple that serviced a Long Barrow. See figure 25.13 for details of the font. The chemical stains from wheat, oats, yeast and red wine ran in front of the font. There was no sign of barley so it looked as if wine and not beer was used for the ceremony. The font occupied a nearly central position in the temple.

Having sorted out the font, noted the use of what could be wine and lead in the Stone Age we returned to tracing the cat. The magnetic storm was easing and it was becoming easier to trace the henna outlines. Round the sides of the face and towards the ears there appeared to be more than one outline. This presented us with problems. There could be one painting on another and we did not know which outline belonged to the eyes and nose that we had identified or if they were all part of the same picture. However, we pressed on and gradually the face of a cat emerged (Figure 25.14). The cat had an open mouth with red teeth and its paws, with four claws on each paw, were clearly marked. It looked as if the cat could be holding something in its paws. Once we could stand back and see the outline marked out on the ground it was possible to identify some of the detail in the picture. The tail of the cat came round on its left side, there were facial marks and the nose was clearly drawn. It was later that night that Nigel realised that the outer lines we had identified were indicating the body of the cat and not its face. The painting of the cat was sophisticated with a great deal of detail. At the moment we cannot identify shades of a colour so there may be more detail that is not dowsable at the moment.

The previous evening we had been discussing the practical aspects of painting floors. We realised that with sacred paintings it is often the doing that is important. Once the painting has been completed it can be walked over and destroyed. However, it is not easy to paint on grass, soil or sand and some hard surface would have to be prepared. For our last session on Wednesday morning we drew up a list of targets the first of which was to look for the material the floor was constructed from. Using a terracotta witness for tiles I checked all the gates and doors for signs of tiles. Only two came up positive. The priests' small room to the south west had been tiled and then there was a large apron of tile coming up from the porch on the north gate. The tiles formed a rectangular apron the width of the gate and extending to just past the font (Figure 25.7). the area for the water supply to the font was also tiled. The general floor

Tiles extending to North Gate

Euchrist table

5.5 feet

Puddingstone/gypsum base

1 foot

FONT

Lead silver

Human blood

4.5 feet

Header tank

Lead plumbing

Tiles

4 feet

Southern edge of tiles

31 feet to southern wall

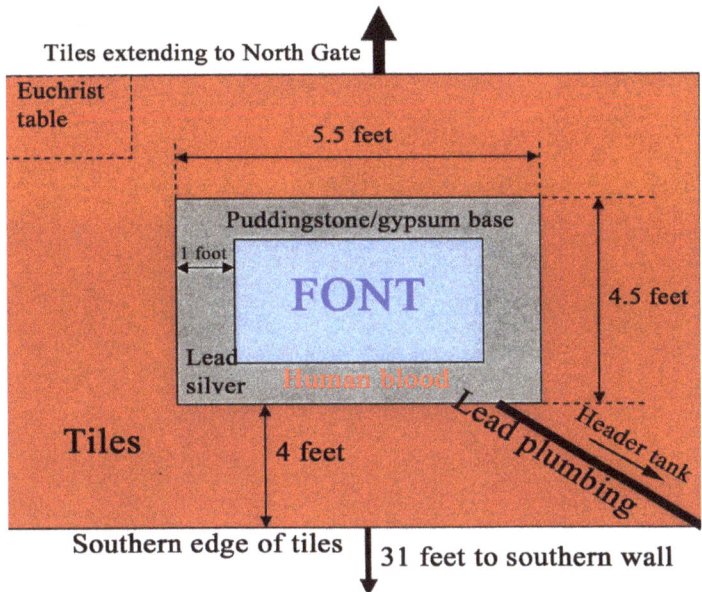

Figure 25.13 The Font Area
The Font was either made of lead or lined with it. At about 2ft square and with its own plumbing it looks as if it was designed for more than the present day christening ceremony. The Eucharist table is next to the Font. The footpath of the priest to the Eucharist table from a robing room is clear but no footpath was found leading to the Font. This means that a more careful search is required or that the priest moves from the table (altar) to the Font.

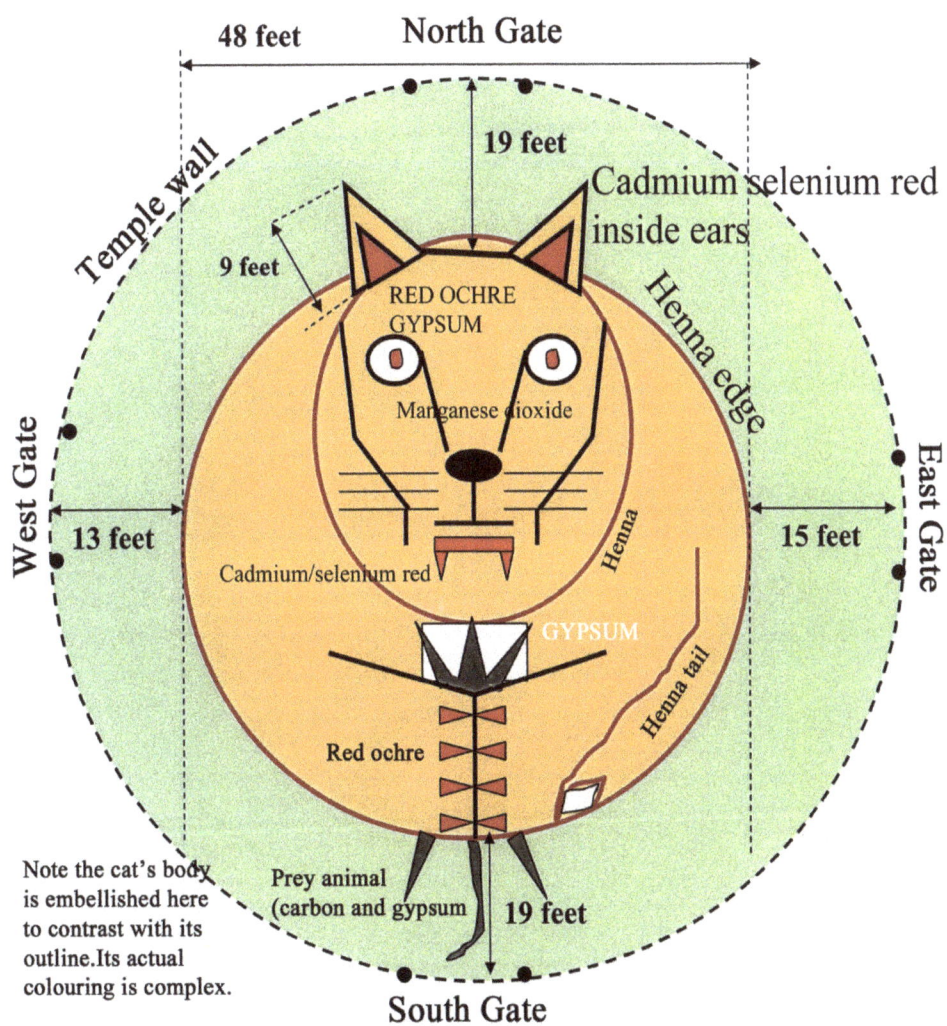

48 feet

North Gate

19 feet

Cadmium selenium red inside ears

Temple wall

9 feet

RED OCHRE GYPSUM

Manganese dioxide

Henna edge

West Gate

13 feet

Cadmium/selenium red

Henna

East Gate

15 feet

GYPSUM

Henna tail

Red ochre

Note the cat's body is embellished here to contrast with its outline. Its actual colouring is complex.

Prey animal (carbon and gypsum)

19 feet

South Gate

Figure 25.14a Artists Impression of the Cait-Sith
The size and detail of the Cait-Sith indicates that the Druids had skilled painters and probably used some of the methodologies that are in use today for doing large paintings of the theatre back drop type.

325

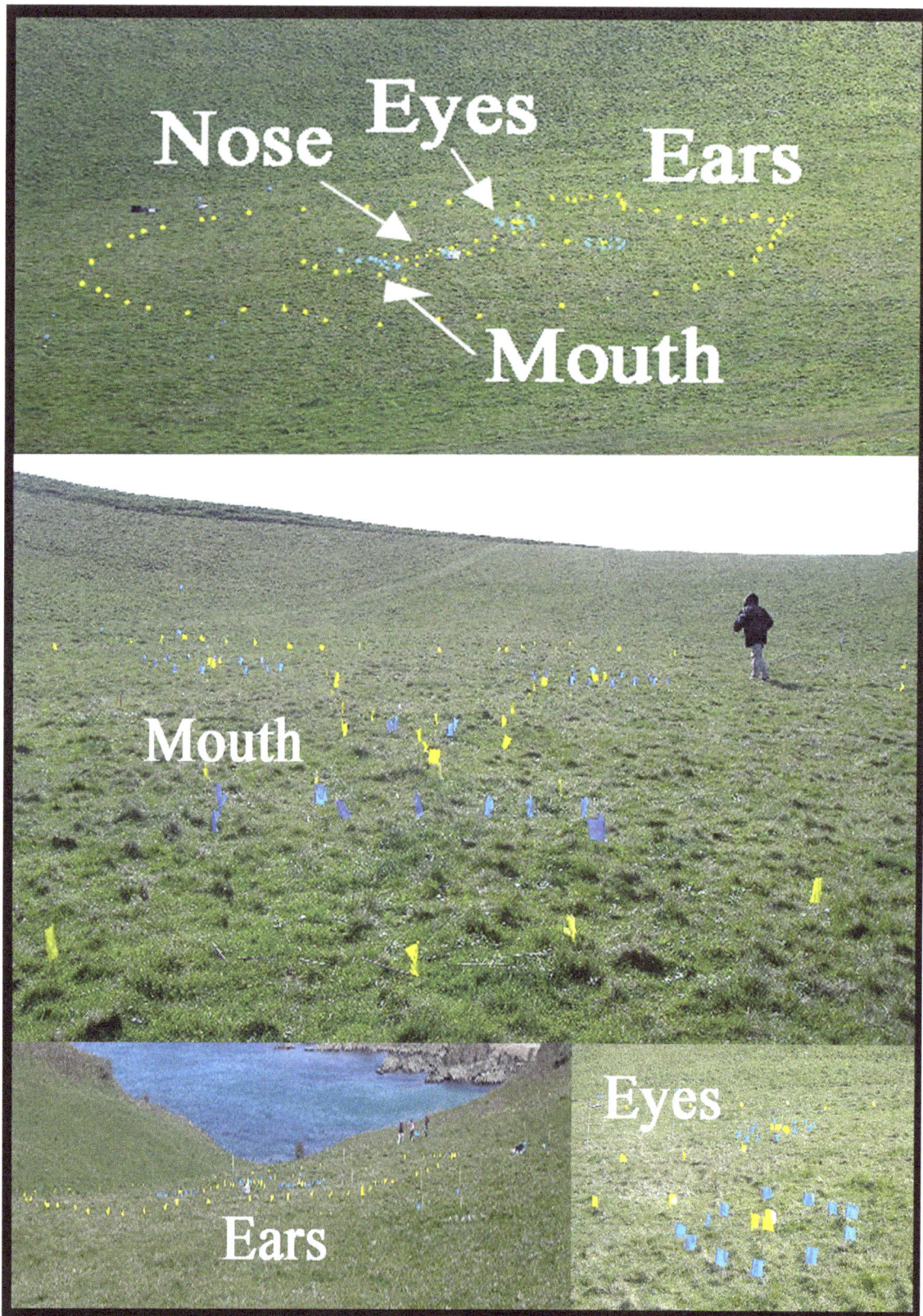

Figure 25.14b The Cait-Sith

The Cait-Sith is outlined with flags on the floor of the temple. The ears, eyes, nose and mouth can be clearly seen. This is a painting done in the early part of the Henge Age (The Neolithic). The skills, paint technology and trade to bring in the pigments were present at this time.

area had been prepared in some way as chalk (lime) and urine could be detected. The font area was looked at in some detail, see (Figure 25.13).

The final task was to look for what the cat was holding. This was not easy but finally we had an object which we felt could fit the legend of the Cait-Sith.

On returning to Bovingdon I started to look for the painting of cats on one of the local cursi. I started the search by looking at an arena on the cursus. The male arena had the cat occupying most of the space with a border in red and white round it. I looked for the white patch on the chest of the cat which is said to be diagnostic of the Cait-Sith and found it. It therefore looks as if the Druid's cat is the fairy cat of Scottish and Celtic legend. Having satisfied myself that the male arena had a cat painted on its floor I moved onto the female arena. I was half expecting that the women warriors would have a different emblem. I soon realised however that a cat was involved but this time it was a kitten, complete with a bowl of milk. I could not believe it either but it is there.

Whilst I had been looking at sites in Bovingdon Nigel had been doing the same round St. Albans and in South Wales. He had found the cat in all the temples he had checked. The cat appeared to have been painted on the floor of all temples at some stage in their lives. The question is, did the men insist on a kitten for the female warriors or did the women? However a word of caution, we must not look at the mouse or the kitten and bowl of milk with modern 21st Century eyes. They may appear humorous today but over 2000 years ago they contained a lot of religious symbolism. The problem is how to interpret it and determine what they meant to the Henge people.

Review

The eight weeks spent in South Devon had been very rewarding. We now knew that Druidry and its ceremonies had a very ancient history indeed. It had been possible to reach back at least five thousand years and possibly even further. The use of what dowsed as wine and the use of pigments in large quantity by a Stone Age people indicated well established trade with Europe and the rest of Britain. This in turn may perhaps indicate a period of fine weather and because of lower sea levels, a closer geographical proximity of Britain and Europe than today. The Temple and Long Barrow show that many aspects of the Henge culture were already well established 5000 years or more into the past. It is not inconceivable that the Henge culture goes back to the retreat of the ice sheet some 10,000 years ago. When the people of the Henge culture reached the British Isles they did the same as the more modern Europeans did when they reached the Americas. They set up their cultural institutions. The equivalent of their churches, schools, farms, roads, trade, craft institutions, governance and planning. The ability to construct large buildings and burial chambers came with them. Their use of draft animals was well developed. They were used for raising water, moving soil and rock, building monuments as well as the normal ploughing and agricultural tasks. With time the culture developed and became a complex and advanced one in the technical and engineering areas. Their religion and philosophy did not stand still and the evidence indicates that it developed first into a Proto-Christianity, which took them away from the old form of their gods, and then developed into a more Christian form. The subterranean realm was deserted and the tunnels and chambers left as one of their monuments for the 21st century to marvel at. The old gods were still there but perhaps now they were wiser and possibly more human. From Proto-Christianity it was a small step into Christianity itself which happened possibly when news of the crucifixion reached the priests of the Henge culture. But how much of the Henge culture was going to move into the era of Christianity? The concluding phase of the project was aimed at providing some answers. Excarnation and the use of energy engineering were soon dropped but when? As for other elements of the Henge Age culture, it looks as if they moved into Christianity en mass and now play a role in defining our culture.

Box 25.1 The Stone Age Tomb

The bodies in the tomb lay in a nearly east west direction with the heads to the east. The tomb was properly constructed with large stones sealing it. The bodies or clothing had been treated with linseed oil and wood resin. The hemp dress on both extended from the shoulders to the feet. The nettle fibre garment covered the trunk only so may be the comfort layer. Nettle fibre fabric also appeared to be used for the horse hair pillows. The girl had a gold garment covering the body to the waist. There appeared to be no sign of a spear, bow or shield. No feathers, wool, linen, copper or iron were detected but lead was. The lead appeared to be a torc round the neck of the Prince and gave a positive response to a silver witness. Silver is a contaminant of lead. There was a cat alongside the Princess and a rectangular terracotta object between the two bodies. There was also a terracotta object in the bottom right hand corner of the tomb. Red ochre was used on the bodies. No tunnels, anterooms or cavities could be detected on the outside of the tomb. On the surface the tomb had been marked by a black manganese dioxide rectangle and from it four Thunderbirds could be traced.

Box 25.2 The Henge Age

In the context of the British Isles the traditional way of referring to periods of prehistory, that is the period prior to the Roman invasion is to call them Late Stone Age (Neolithic) Chalcolithic (Copper Age, Lead Age) Bronze Age and Iron Age. The period covered is from when the Romans arrived in 43AD back in time to 4000BC when the Neolithic started. The Mesolithic or Middle Stone Age which predates the Neolithic extends from 4000BC back to 8500BC. This period could also be relevant to the Henge Age. Using some aspect of the technology of the time for classifying a period has advantages. The first appearance of a metal in a geographical area can often be given a date. However, as already discussed, the technology of a people or of an Age cannot be used to describe its culture and it is the development of culture that is of most interest. In the present context it is the development of the Henge Culture and of its priesthood, the Druids, that is of greatest interest and of greatest relevance to post Roman Britain. The use of biolocation for archaeological studies enables a very large amount of information to be gathered. Importantly one piece of information can often be related to other pieces in time so that temporal sequences can be identified. The speed at which information can be obtained by biolocation enables ideas to change and be checked within hours whilst still on site. An example will illustrate this. On the 4th July 2007 there was an announcement on the radio that a 60m long image of a serpent had been found on a construction site near Hereford. The stones used to form it on the ground were known and the remains of posts supporting the walls and roof of the building that contained it had been found. The announcement gave no further information. By chance Nigel had brought me some snake skin on the 2nd July which he had been using to identify paintings of snakes. Using it we had found the painting of a snake belonging to the temple I have in the garden. The paintings of animals appeared to be on the quarters. On the evening of the 3rd over a period of two to three hours I had marked out the temple snake on the ground, identified the foundations of the painting, the type of paint used, the pigments, the colour band along its back and the paint brushes they were using (Figure B25.2.1).

Figure B25.2.1 The Snake God. The snake is lined up towardsthe north gate of the early Bronze Age temple. The tail is near the pool and the body is swollen near the mid point with a meal. The head disappears into a planter but the head markings and eyes could be identified.

The head presented a problem as it was under some planters. As biolocation does not destroy the artefact the snake is there for others to check. It is about 3500 year old so may well survive another few thousand years. In response to the news of a Snake Temple I went up to the Little Hay area near the cursus on the 6th July and soon found a Snake Temple. This wealth of information revealed by biolocation enables many technical aspects of a culture to be followed. For example the use of flax appears to start with oil from the seeds (linseed oil) and its use as thatch. The flax then appears as threads or cordage then finally as linen. Sheep appear before wool, lead before bronze, and arsenical bronze before tin bronze. The temple paintings and the contents of tombs on the same site change with time. It is too early to structure the Henge Age into time periods but as more information becomes available it will be possible.The need is for the Henge Age to be divided into sections which relate to the culture of the time and cultural changes rather than to its technology. For example the Iron Age we can now divide into three – the early or Druidic Iron Age, the Proto-Christian and the Christian phases. The Christian phase only lasted a few years but it may have been responsible for carrying the Henge Culture through the period of the Roman occupation and into the post Roman and early Christian era. At the moment the Henge Culture first appears with the Long Barrows dated as 4000 years ago but possibly as far back as 6000 years. If the latter age is correct then the Henge Age can trace its origin to the Middle Stone Age. If this is the case the Henge Age is likely to have entered the British Isles as a well developed culture from the south and its origins lie outside the British Isles.

Chapter 26

Has History Got it Wrong?

The Hell-Fire Caves

The visit to Stonehenge on the 17th March enabled Nigel and myself to confirm that like all the other Henge Temples we had looked at, after 3000 years or more of religious evolution Stonehenge had become a Christian ceremonial centre. Stonehenge is not the best place for a detailed biolocation study as many other people have paid their money to hug stones, take photographs and just stare and wonder. We also did not have time to look for and compare Proto-Christian crucifixes with Christian ones. Such detailed work had to be done elsewhere. However, before starting on the crucifixes there was one thing we wanted to clear up. The Hell-Fire Caves at West Wycombe presented a bit of an enigma. The caves had not been sealed and as a result there had been a big battle between the Romans and Druids at the entrance. What we wanted to know was, did the fact that the underground temple complex was not sealed when the Romans arrived mean that Proto-Christianity had continued to be practiced alongside the new Christianity. To find out we had to return to the caves and the Bell Chamber. On the 12th April 2007 Nigel and I arrived at the caves with about half an hour to wait before they opened. This enabled us to walk up the hill above the temple complex. Once on the hill it did not take long to identify an underground stream and then by working our way along it we found the shrine chamber and the tunnel to the Bell Chamber. We were then able to pick up the detail of the tunnel network. Feeling rather pleased with our efforts we returned to the entrance to the caves, inserted our tokens in the turnstile and made our way rapidly to the Bell Chamber. We waited a short while for our eyes to adjust to the low lighting level and then set about finding the crucifix. The first one we found was a Christian one complete with the gold torc as a halo, the crown of thorns, the long hair, nails into the hands and feet, blood and spear wound. The head was only about five feet from the wall (Figure 26.1) so there was no room for Herne and the fish and I could not identify them. The crucifix was at 305°. The second crucifix was a Proto-Christian one with its head below the Christian one. This provided sufficient room for the gold torc, Herne's head and a fish. It was now clear that the Bell Chamber had been used for Proto-Christianity and then Christianity. The latter period came to an end when the Romans arrived and systematically cleared the country of Druids. For a long time we had realised that we could identify Herne's antlers in the floor paintings by using an antler witness and the greenery by using ivy, oak and other plants. Also the crown of thorns could be identified by using holly and brambles. We had always assumed, without devoting too much thought to it, that plant foliage had been placed on the painting as a ritual and had either rotted or had left a chemical trace in the ground. Finding Herne's antlers so well marked in the Bell Chamber where there was no rain and no rotting antlers, started us thinking. We were aware of spirit painting and one possibility was that the paints used for the pictures contained the 'spirit' of the object being painted. The spirit was imparted to the paint by mixing in something from the object. In this case powdered antler, ivy, holly and hair for example. If the ancient artists had done this it should be possible to check all the main features of their paintings using witnesses made from the subject matter of the painting. To test the theory I decided to try it out on what the Cait-Sith was holding in its paws. That evening our black long haired cat 'Shadow' was put out in the garden on mouse patrol. The following morning the beheaded body of a mouse was laid out on the backdoor mat. I used its blood and tissue to make some witnesses and later that day walked over to the arena on the Little Hay cursus. I had no difficulty in identifying the outline of the mouse being held by the Cait-Sith. Nigel tried the technique on the face of the cat in the temple near his house. Using cat's hair, whiskers and claws he was able to identify the painting and some of its detail. As all

330

the witnesses worked, we now had good evidence that the Druids were using what is called 'spirit painting'. Painting in which part of the object is captured and with it the spirit of the object.

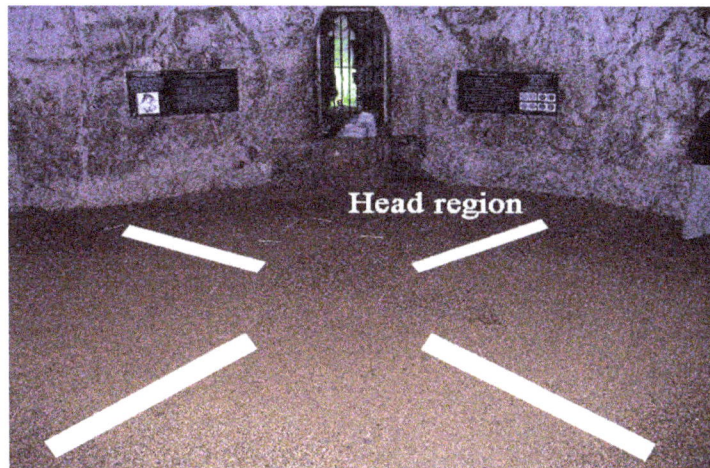

Figure 26.1 The Hell-Fire Cave Crucifix
The position of the Christian Crucifix in the Banqueting Hall of the Hell-Fire Caves. The chamber was being used as a chapel up to the time when the Romans attacked the site.

The Crucifix Presents a Problem

On the 14th April, two days after visiting the Hell-Fire Caves I was back on the Little Hay cursus and the arena. My task was to determine if the body on the Proto-Christian cross was male or female. To determine the sex I used male and female urine. Our assumption had always been that the body on the cross was male but as we had thought that we had no means of identifying the sex of a painting there it had to rest. Now there was a possibility that the paint used would have been mixed with urine, blood or something from the human body and that we would be able to identify the sex of the crucified person. Using the urine witnesses I soon identified the figure on the cross as female. But why a dead one? Perhaps she was not dead and I had made an incorrect assumption. The nails were looked at more closely and there was something else there. Nigel had reported that he thought there were straps holding the body. It did not take long to identify leather straps. There was also a leather strap round the head. For some reason the idea of a woman (Goddess) in childbirth came to mind and so I started looking for a child between her legs. What I found was rather big for a newborn infant but there it was, clear and unmistakable. The Proto-Christian crucifix was a birth table with Herne and the fish both symbols of life, above the cross. I phoned Nigel to let him know what I had found and it was not long before he was able to confirm that the figure was female and to add additional detail.

Although the use of male and female urine could identify the sex of a body in a grave and now it appeared in a Druidic painting we had always been on the lookout for female hormone pills. However, the women we knew guarded them as if they were gold dust and so we had to make do with urine witnesses. Then one day Nigel managed to obtain some hormone pills. He was very quiet about the sacrifices he had to make to obtain them but we now had them and that was the main thing. I soon made some witnesses and decided to test them out on the Proto-Christian and Christian crucifixes. The first should be female and the second male, shouldn't it? On the 15th April I walked to a temple site where I had been studying the crucifixes. It had a Christian one with its head to the west and a Proto-Christian one with its head to the east. I started on the Proto-Christian crucifix using a gypsum witness to mark the cross out. We had never picked up the detail of the body apart from the hair, eyes and mouth as we did not have a witness for the paint pigments used to paint the body. With the hormone pill it was now possible to pick up the outline of the body very clearly. The hormone witness worked a treat. We still did not know the pigments being used to simulate the skin but the outline of the body came up clearly.

I then marked out the gypsum Christian crucifix. Once done it was a simple matter to go across it with a hormone witness. It was female, I checked the outline of the pubic hair, it was female. The hair from the head was down to the shoulders. Again a sign of it being female. There was no doubting that the crucified Christ complete with crown of thorns, spear wound and nails was female. The picture on the ground must therefore have been painted between the actual crucifixion round about 33AD and when the Romans reached Bovingdon which would have been soon after 43AD. The Druids would probably have known Jesus on first name terms so as to speak. They would certainly have known whether Jesus was a man or woman. We are therefore left with two possibilities. Either the Druids changed the sex of who was crucified by the Romans to fit their religious and cultural beliefs or the Romans and Jews, who made up the early Christian community, changed the sex of the person crucified to fit their cultural beliefs. The paintings of the cross are contemporaneous with the crucifixion. The written accounts are not. The truth will be hard to determine but out of all the religions of the world is it not Christianity that sees the world as a woman sees it! However, one possible approach is to find as many crucifixes as possible and determine the sex of the crucified persons. The sex represented by most paintings might provide the true answer. We are working on it.

The Rollright Stone Circle

After Stonehenge I thought things would be quiet for a while with no more discoveries. However, the Rollrights had been on my mind since I had recognised what they might be earlier in the year. At last I was able to find a day for a visit and on the 27th May 2007 Nigel picked me up and we were on our way to the stones. It was raining and during the day it just got worse with the temperature dropping lower and lower.

For years I had found the stones to be an enigma, not fitting into the overall picture of stone circles. A complete circle of flat stones was difficult to make sense of. Were they there just to be painted? Were they a wall round something? They clearly post-dated the temple(s) that had been on the site. The stones are not Late Neolithic, around 3000BC, as mentioned in Aubrey Burl's book on the Rollright Stones and their projected use into the Bronze Age is pure conjecture. The site on which the stones stand may have been used since the Late Neolithic to build temples on, the postholes and tombs of which are easily identified. If there is one thing that can be said with certainty about the stones it is that they were erected on the top of the postholes of earlier temples. The stones must therefore postdate the temples. The ring of stones is complete apart from entrances and exits. It was while I was editing a draft section on the Rollrights and contemplating what had been written that the answer to the enigma suddenly occurred to me, almost as if my guardian angel Galano had spoken to me. I rang up Nigel to try the idea out on him and he agreed that it made sense. From then on I knew that we had yet one more piece of tidying up to do before the final draft of the Book was ready. Being early on a Sunday morning the journey from Bovingdon to the Rollrights only took about an hour so by 9am I had passed through the gate to the site and was walking along a hedge hiding the Stones form view. Reaching the end of the hedge and turning to face the Stones my heart sank. The circle must surely be far too big for what I thought it might be. I returned to the car where Nigel was unloading and dressing up for the weather. I said that my idea as to what the stone circle was may be wrong but that we should still test the idea out before checking the stones for paintings. Having prepared ourselves for the weather we returned to the stones. I thought the best approach was to first test for the red ochre patches which are to be found under the biers of a Dolmen. This I did and almost at once found one and very quickly had identified a row of ten patches. Then another row and another. There were six rows of ten in all. There was then a gap across the centre of the stones between the two main gates of the circle before another four rows of ten were found. The Rollrights were a massive Dolmen designed to deal with a hundred bodies at a time. The stones had been set up to produce an animal proof area. The scale of the operation was immense. If the site dealt with a hundred bodies at a time and it took say one month for the birds to strip the bones the Dolmen could deal with about a thousand bodies a year. Say that the average life span at that time was fifty years then the Rollright Dolmen could serve a population of about 50,000. It was the equivalent of the modern crematoria but with some important differences. The Dolmen was a recycling unit in that the biological mass of the bodies was used to feed birds. The droppings of the birds would be distributed over the countryside but mostly round the Dolmen and nesting sites. The "Sacred Groves" of the Druids may well have been woods of tall trees set aside for the nesting of crows, rooks and other birds.

Having identified what could be a Dolmen we have to ask ourselves that very important question 'if we are right what follows?' The 'what follows' involves the whole of the engineering and architecture of an important structure. Was the floor tiled, where were the drains, how were the bodies delivered and the bones taken away, did they paint the stones? In fact there are so many questions that it would have taken days to study the place. We therefore limited ourselves to asking a few questions. Figure 26.2 presents most of the information gathered.

The first step was to make sure that the birds were in attendance. We found the bird perches on either side of the bodies and the design of the perches indicates that there were a lot of birds present. Using bird faeces we found that the stone wall had a line of bird droppings a little way from the stones. The line was a broken one in that there was a line of bird droppings about 4m long then it disappeared for a similar distance before reappearing. Looking at the inside of the stones there was a similar picture. Where the bird droppings disappeared on the outside they appeared on the inside. The wall of stones had a capping or protective roof of some sort to provide it with protection but it was built to a design. Working out what the design was will have to wait. The main gates were interesting. The Eastern Gate near to the information hut received the bodies on a small iron wheeled cart , possibly a handcart. There was no sign of bronze wheels but the bodies had been washed nearby as indicated by a trail of lime produced by water dripping from bodies. The cart entered the stone circle to unload and to turn round and exit the way it had come in. The bones, after being dressed with red ochre, left by the West Gate. Due to the vegetation it was not possible to determine in detail what happened at the West Gate. However, the bones were brought to the gate and possibly placed on a stand outside the gate, under the stand the red ochre produced a stain. A cart came up to the stand to pick the bones up. Inside the circle the staff wore wool clothing. No sign of linen was found within the circle. Outside the West Gate priests dressed in linen received the bones. There is a narrow gate on the south side of the stones. This was not investigated. No attempt was made to find the drainage system and soakaway or cesspit that would have dealt with effluent from the Dolmen. There is also the possibility of a plumbed water supply and certainly a washroom and toilet facility for the staff.

The supporting evidence for the Rollright Stone Circle being a huge Dolmen, a house of death, was now convincing. What we did not know was its date. The Rollrights are certainly an Iron Age structure but the Iron Age is a period of about 800 years. The next step was to look at the associated artwork or pictograms. Inside the circle there is the artwork, such as

Thunderbirds, which is associated with the temples. That could make it difficult to identify specific artwork with the Dolmen. It was therefore decided to look outside the stone circle. On the assumption that the Dolmen started life early in the iron Age I looked for Thunderbirds. I found a manganese dioxide ring about 18" wide round the stones and Thunderbirds on the two quarters that were accessible, that is east and south. The Thunderbird was about 16 meters to the top of the head and complete in its detail. The presence of Thunderbirds indicates that the Rollrights date back to possibly early in the Iron Age. I then looked for the Proto-Christian and Christian crucifixes. The only one I could find was the Christian crucifix but with Herne and the fish above it. Unfortunately the Proto-Christian crucifix can easily be changed to a Christian one and it takes a lot of careful dowsing to sort the two crucifixes out when one is on top of the other. The important point is that the Rollrights could possibly have been used as a Dolmen into the Christian era. There is however another possibility and that is that when the Christian era arrived soon after 33AD the Rollrights had ceased to be a Dolmen but as an ancient sacred site it had to be blessed with the new icons such as a crucifix. There is a Christian crucifix within the circle so it will be possible to determine if the Rollrights did become a church after a period as a Dolmen. At the moment all that can be said is that the Rollrights could be archaeologically very important. Using the Rollrights it may be possible to unravel the time at which excarnation ceased and determine if the site was used for Proto-Christianity. If it was, did a Christian church follow. Answering these questions will require a lot of careful work.

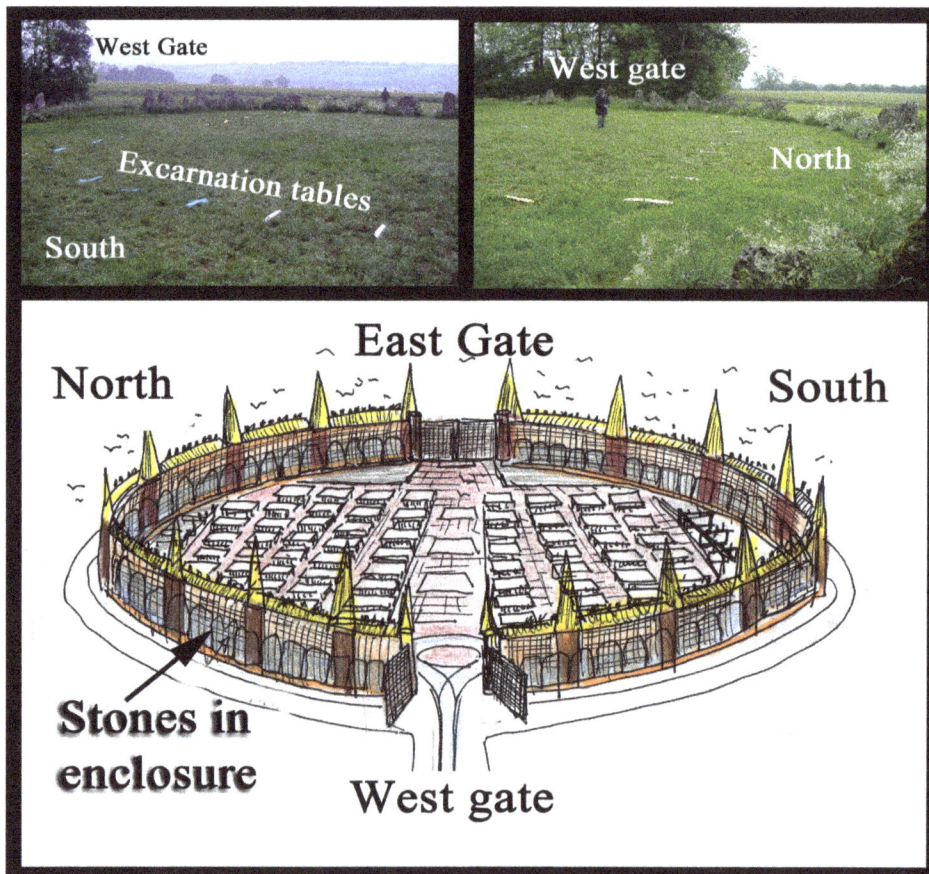

Figure 26.2 The Rollright Stones Give Up their Secret
A sketch of the Rollright Stones when they were being used as a Dolmen. There were excarnation tables for a hundred bodies. With so many bodies passing through records (probably written) would have to be kept of each body and table.

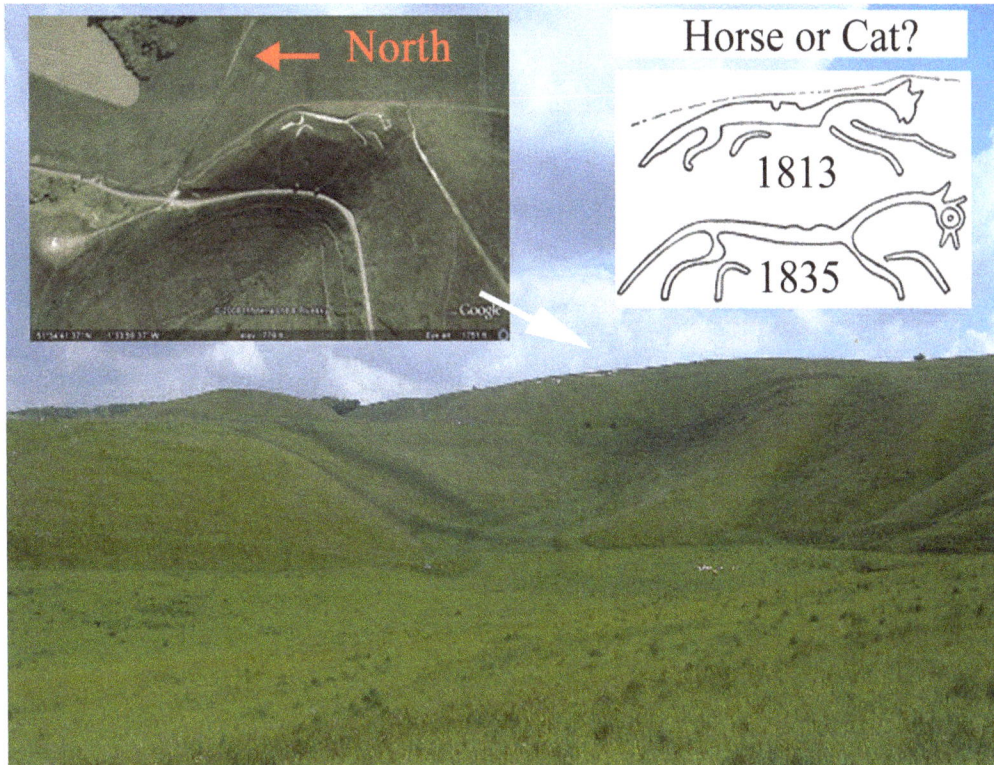

Figure 26.3 The 'White Horse' at Uffington
The 'White Horse' is shown in its landscape setting.

Figure 26.4 Dragon Hill
Dragon Hill when viewed from the 'White Horse' is an impressive man-made structure associated with many myths and legends.

The White Horse at Uffington

Having made it to the Rollright Stones and feeling rather pleased at having sorted out the enigma of the stones, the weather did not stop us from considering another mystery of the British Landscape.

The aerial photographs of the Uffington White Horse indicate that Picasso had an earlier life as a Druidic artist (Figure 26.3). He must have been out of place amongst the artists of the day as most if not all of the pre-Roman Henge Age artwork I had seen appeared to represent the reality. If an animal was painted it looked like the animal so if an animal was carved out of a hillside it would be expected to look like the animal it was representing. The Picassoin horse was not convincing. It was either new, that is post Roman, or over the millennia it had been modified and evolved into its present shape. If it was ancient there was the possibility that it had been in a building or that it served the valley leading up to it and was related to local features such as Dragon Hill. When we arrived at the car park the rain was torrential. There was a lone ice cream van, water cascading down its sides and windows. The blurred form of a very unhappy driver just visible through the condensation and cascading water. We thought that our task was simple and could be done rain or no rain. All we had to do was determine if the 'horse' had been painted at one stage of its life and to pick up the original form of the horse if it differed from the present one. There was also a third objective and that was to check for other painted figures which were now grassed over. The reasoning was that if the ancients had cut and painted one figure into the landscape then they may have painted many more onto the chalk downland. Our standard method of checking for painting activity had developed to the use of the bristles the ancients used for their brushes. It was therefore very easy to identify, using the brush witness, the outline of a painting. As a result, soon after reaching the Uffington Horse we new that it had been painted. Following the painted outline we could see that the original did not fit the modern version, so some changes in form had taken place. Because of the weather we could not mark out the original to see just how great the changes had been. We also discovered another problem. There appeared to be mounds of soil over parts of the original. This was going to make dowsing the outline difficult in fine weather let alone at a time when the month's rainfall was arriving in a matter of hours. We then had a quick look around for signs of other painted areas and found them. The Uffington Horse area was going to be worth looking at in detail – on a fine day. We had achieved our objective in that we now knew that the 'horse' had been painted at one stage in its life. This almost certainly meant that all other Henge Culture hill carvings had been painted and could now be identified and studied in their original form. We also felt that the hill carving was not just a pretty picture. It would have played a role in ceremonies and was where it was for a reason. On the way back to the car we realised that we should have checked to see if the carving was inside a building. Something to do on the next visit. At the time we did not realise just how soon the next visit was going to be.

It was whilst writing up my notes on the White Horse that the worry about the shape of the carving surfaced again. I had an aerial photograph in front of me, pg. 121 of Richard Muir's book Riddles in the British Landscape. The outline of the carving just did not make sense as a horse. I started to think of it as other Henge Culture icons to see if they would fit. I looked at the shape of the green (grass) pieces enclosed by white chalk. Then, as is so often the case, it became obvious what it was. I gave Nigel a ring and asked him to check the idea out. Nigel had also been worrying about the shape of the so called horse and said that the only way to settle the matter was to go back to Uffington. Five days later we were on our way back wondering if we were about to crack another of the mysteries of the British Landscape. This time fine weather was forecast and the forecasters were true to their word. When we arrived at the small car park near the horse the sun was shinning, the visibility over the Vale of White Horse was excellent, the wind was almost non existent and it was warming up to be one of those days when everybody feels that they have to visit the Uffington White Horse with as many children as they can muster. Whilst Nigel laboured to make some coffee I had a look round the car park and at the single track road running past it. The road was clearly Druidic with the painting of diamonds and the later icons of the Henge Culture along it. Round the car park, acting as edging, were some big stones. The stones were not native to the chalk downlands and were from the temples and Druidic civil engineering. There were some very large pieces of Druidic cement that is the puddingstone without the ballast or aggregate. Later in the day we found other areas where stones from the temples and buildings were being used by the present owners of the site. After a coffee break we walked towards the 'Horse'. Leaving the road we went along a path through a small cutting. It was a Druidic feature with diamonds on the path and the potassium lines from burning showing where the rafters had been. Soon after leaving the mini dyke we came across a depression in the ground to our right. It was just before the path divides with the left fork going to the 'Horse'. The depression was clearly the entrance to a tunnel. Chariot wheel tracks went up to the entrance and into the hillside where they could be followed. There were two branches from the tunnel entrance running into underground chambers. The tunnel had the usual entrance design and was sealed. The engineering, the ceremonial structures were the same on the Berkshire Downs as in Hertfordshire. The next stop was the 'Horse' itself. We had a number of tasks. The first was to determine if the horse was indeed a horse. We were going to be able to do this because the Druids, like some other ancient peoples, used 'spiritual' painting. When they painted an animal they put something of that animal into the paint to represent its 'spirit'. The Uffington 'Horse' should therefore have horse hair, horse blood, skin and probably a mixture of biological materials from a horse. If these did not show up then you try something else until you find out what the painters intended the painting to be. We were already certain that we knew what the 'Horse' was and would go straight for it. We arrived at the horse, selected the witnessed required and started work. There was no sign of horse hair or horse blood. This meant that the carving in the downs was not a horse. The next set of witnesses gave a resounding positive and identified the true outline of the animal which was quite a bit different to the present one. It was the Cait-Sith, the Fairy Cat of legend complete with the white spot on its chest (gypsum) and a mouse between its paws. The eyes had been blue with red slit pupils and the body possibly yellow, at least henna was one of the pigments used. I marked out the ears and a bit of the head and it did not take some of the passers by long to spot that the Horse was in fact a cat. We checked the 'Horse' and found that its front legs were modern, the body of the cat had been in gypsum the new areas were not. The use of gypsum made it look as if a surface had been prepared to take some serious painting. If that were the case the Horse, now Cat, would have to be in a building. We soon found the postholes, wattle walls and drip lines associated with a building or weather protection. The outline of the building confirmed which parts of the modern 'Horse' can trace their ancestry back to pre-Roman times and which parts are modern. For example, the present tail extends beyond the building so the end part of the tail is modern. We checked selected parts of the cat and found that it was far more extensive than the present. In other words there is a 'real' body and the legs are the right size complete with paws.

From the building housing the Cait-Sith there was a tiled apron descending (possibly stepped) to a flat terrace overlooking the Manger Valley. This area was later used by the Romans for executions. Above the Cait-Sith on the flat ground there is battle blood and execution lines. I counted the execution lines. They were four deep above the Cait-Sith so with 60 at least in each there were 240 people killed in quite a small area. The execution lines were also along the roads and at Dragon Hill so in the area of the White Horse and Uffington Castle the Romans probably executed thousands and more or less cleared the area of its population.

Returning to the White Horse. How is it possible to mistake a cat for a horse? The cat is likely to have faded from view as religious icons changed. After the cat three crucifixes were painted on the hillside where the cat had been. They are paintings of Christian crucifixes. It is therefore reasonable to suppose that for a while the cat faded from view. Once the Romans had passed through the Uffington area the cat might have reappeared during dry weather because of its gypsum base. On one of these occasions the locals may have decided to dig out what they thought it was. They clearly did not think it was a horse. If they had they would have made it look like a horse. However, if your neighbours have horses and giants on their hills you may not want to own up to having a cat and a mouse on yours so why not call it a horse. You then do your best to modify it so although it does not look much like a horse at least the cat is not obvious and the mouse is blotted out. A few hundred years pass maybe a thousand and everybody is convinced that the odd creature above the Manger Valley is a horse. By now of course so many people have become attached to the White Horse that many places have been named after it and that is now going to present a problem. A lot of renaming is going to be required as the Vale of White Horse should be renamed the Vale of the Yellow Cat and Brown Mouse. However, not to worry, mankind has been presented with bigger problems and survived.

Having sorted out the enigma of the 'White Horse' we headed for Dragon Hill. Seen from high on the Downs above it, Dragon Hill looks like a typical man-made construction (Figure 26.4). It has a symmetry not normally seen in the landscape and looks a bit like Silbury Hill.

Figure 26.5 The Manger Valley
The topography of the Manger Valley looks as if it has been engineered. An Ice Age origin has been suggested but is unlikely.

Not withstanding its artificial appearance, some authorities consider that Dragon Hill is a natural hill with its top levelled and others that it is a man-made hill or natural hill that has been reengineered to make a suitable ceremonial site. As we walked down to the road which ran between Dragon Hill and the Downs we discussed how we were going to approach the problem. We knew that if it was man-made we would find traces of the engineering used – the hoists and lifts. However, that would not be convincing as the same engineering would be used for modifying a natural hill. The answer probably lay in determining why the ancients wanted to build a hill in the first place, if indeed it was a built hill.

We knew why Silbury Hill was where it was. It was over a network of underground streams and a temple which had been built on the site during the Stone Age. There was something about a multiple channel underground stream, or at least its magnetic fields, which attracted the ancients. It was a sufficiently important phenomena for temples and hills to be built on them. Was Dragon Hill built on a multiple channel stream and a Stone Age Henge Temple. Approaching Dragon Hill I thought that it was quite a bit smaller than Silbury but once on it and walking round the plateau I was not so sure and it could easily be of comparable size. It did not take long to find the tomb of the Prince and Princess. There was no iron, bronze or linen in the tomb so it could be Stone Age. The rings of postholes and the tomb were not centred on the plateau. Counting the rings of postholes out to the east the 9th posthole was well away from the edge of the plateau. On the other side, the nine postholes did not fit in and the western edge of the temple was down the slope. A ceremonial path also went straight over the edge of the plateau. It looked very much as if Dragon Hill was on the site of a Stone Age Temple. There was an underground stream going under the hill and round the middle of the plateau it split up into a number of channels before joining up again. The next step was to find the lifts being used to raise the spoil. Two were found and they were being operated by deer which would indicate a Stone Age origin for Dragon Hill. Looking down the steep eastern side of the hill the area round the bottom still looked as if there are mounds of spoil waiting to be lifted to the top of the hill.

The visit was a quick one aimed at obtaining some basic information. The important information is that Dragon Hill is a man-made structure and is likely to have been built in the late Stone Age, say 2500BC. The hill postdates the tomb but at the moment we have not identified bronze being associated with the construction of the hill. It is of course possible that the tomb and hill are earlier than 2500BC. The use of deer as draft animals and nettle and hemp as fibres for cordage and rope indicates that an earlier construction date is possible.

Finding that Dragon Hill is a clone of Silbury Hill raises the question of how many more there are around. It is their size that will make them invisible but it is unlikely that only two were built.

Making our way back to the car we stopped to look down the length of the Manger Valley (Figure 26.5). The hillsides surrounding it formed a horseshoe and the question arose as to whether the hills generated a blue line. Did they form a natural trilithon amplifier? If it did we should be able to identify the blue line at the entrance to the valley. Fortunately the road we were to take home went past the end of the valley so we stopped to check for a blue line. The Manger Valley is almost an out of this world landscape. Standing by the road and looking towards the 'horse', now a cat, high above the end of the valley, the terraced ceremonial roads on either side, a succession of what looks like spoil slopes cascading down on the right hand side, Dragon Hill to the right of the valley. Although the spoil cascades may be actual spoil from local tunnelling the other possibility is that the gap between has been mined. This could be to engineer the landscape or it could be to take spoil for building Dragon Hill.

Unfortunately we did not have time to investigate and had to settle for a blue beam. There was one present coming out of the valley, possibly thirty feet wide. It cannot be assumed that the valley generates the beam. That still has to be proved. However, the fact that it is there indicates why the valley may have been important enough to have a cat carved into the hillside above it. The Manger Valley looks as if it holds a lot of secrets about the Henge Society.

Chapter 27

A View Through the Stone Portal

Energy Engineering

It is now about eight years since late one afternoon I sat on a low stone wall looking out over Clew Bay thinking of the task ahead of me. The task was to use a nearby ancient stone circle and its outlying stones to unravel the mechanism by which the stones set up circles of magnetic fields around themselves. I also had to look at the mechanism behind the beams of energy going off to other stones or just into the distance. I referred to the mechanism involved as 'The Secret of the Stones'. The secret, as I saw it, was simple. How did the ancients build the stone circles so that, as I thought at the time, some 5000 years later people would still be able to 'feel' the magnetic or energy fields as they are called circling round them. Not only that, the fields would induce today's generation of stone circle visitors to weave all sorts of imaginations into and around the stones. The energy fields as I had discovered very early on, are magnetic fields and it is salutary to remember that that is all that is there. There are no spirits whispering the secrets of the Druids or conveying messages from another world. There are only magnetic fields and from the simple piece of information 'yes' 'a field is there'. 'No' 'there is not one' the biolocator has to work out what the field or fields might be telling them. What information they contain.

By the year 2000 I was well advanced with the 'Diamagnetic and Paramagnetic Theory of Biolocation' (Appendix 3) so I was in a position to identify fields as diamagnetic or paramagnetic and in the latter case to say what the source was, for example that it was silica, calcium or carbon etc. I had also developed techniques which enabled me to study the fields particularly the direction in which they were travelling. My understanding and knowledge of the underlying physics has developed greatly in the intervening years. As the project progressed I had to ensure that the theory kept pace with the observations that Nigel and I were making and the problems we were meeting. I also had to study the behaviour of magnetic fields and what influenced them. In short the laboratory work had to keep up with the field work.

To help with the task of identifying artefacts hidden in the ground biolocators construct models in their minds such as a Roman Villa. These mental models then guide the search for more information. In June 2000 the models being used were very simple. They were based on stones being magnets and having magnetic fields. The ancients would have seen the fields as 'spirits' or some such mental image. The mental models I was using for the stone circle had one big advantage over those of the ancients, that is they could be turned into solid physical models by identifying the right stones and building a mini stone circle to experiment with. If the mini stone circle constructed on the lawn turns out to have the same fields as the real one said to be four to five thousand years old I would feel fairly confident that I had identified the way the ancients built their stone circles. I may not know exactly how they did it but at least the method I was using was producing the same end result and I could then say that the theory had been confirmed and the sequence of reasoning had been validated. The circle of reasoning had been closed. This is a very common approach in science and it has been used many times in this study. For example with the aura of the oak. A theory of how the magnetic aura is generated by the oak tree in winter was first constructed based on the materials known to be present in the oak during the winter. A physical model was then made using the identified materials which produced the four Thunderbirds. The reasoning was therefore vindicated and the circle closed. This circle of reasoning was applied to stone circles and it is now possible, after a break of about 2000 years, to build stone circles to the energy engineering principles used by the Druids.

The Killadangan Circle was a small simple circle and almost certainly marked out the pool in the centre of a temple. At the other end of the range of stone circle complexity is Stonehenge. Stonehenge did present a problem with its two sets of linteled stones that is the five trilithons in which there is a lintel stone spanning two uprights and a circle of linteled stones. (Chapter 19 and 20). My view of Stonehenge had been for sometime that the ancients may have sacrificed energy engineering for mechanical and visual performance. This was found not to be the case. In fact the Stonehenge Druids and their counterparts at Avebury took energy engineering to new heights, and it is going to be a while before we know what they were doing with the awesome power from Stonehenge, or indeed from the very large recumbent stone circles found in the northern parts of the British Isles.

The use of beam splitters by the Druids takes some explaining. Just what were they doing with what appears to be quantum entangled magnetic fields. They knew the split fields were there as they constructed processional ways along them confirmed by the painted pictures on the road surface. We still do not know why they considered the ordinary blue beam, that is before they are split, to be important. There is plenty of conjecture but little hard observation. The blue beams were still important to the Druids when first Proto-Christianity then Christianity arrived on the scene. Stonehenge was still being used in the Christian period before the Romans arrived. The blue beams therefore appear to have fitted into the new religion but what their role was remains a mystery. Perhaps science, that is a rational approach to the observed world and religion were not separate domains in the Henge Age. We may be looking at the latest in both scientific achievement and religious development. Could we be looking at a religion that embraced the science of the day and to understand the Henge Age we may have to accept that this is what happened.

The Civil Engineering Legacy of the Druids

Underground Temples

It is not possible to make physical models of everything that is biolocated and certainly not of tombs tens of feet below the ground. We therefore still do not know if we are picking up actual tombs or just the chemical remanence of what stood on the ground or just below the ground or in man-made cavities tens of feet below the surface at sometime in the past. All that can be said is that there is a convergence of evidence which indicates that there are tombs and underground tomb and temple complexes. This is quite an acceptable approach as theories such as Darwin's Theory of Evolution are based on convergence of evidence. The presence of structures that were above ground, such as the Woodhenge Temples, is not in doubt as a physical mark is left on the landscape. With some temples the access shaft to the underground system is still visible, the debris fields contain artefacts from the temple and the clay pools can still be identified. There is also the evidence of what remains of the stone circles that once formed part of the temple structure. The collapse of ground into underground temples and shafts testify to their presence and some underground temples are even maintained and open to public view as follies.

The Sacred Landscape

The archaeological literature in dealing with the pre-Roman period often refers to a sacred landscape without defining the features of such a landscape. However, see Chapter 8 in Nigel Pennick's book "The Sacred World of the Celts" for some detail on landscape. After eight years it is now possible to start on the process of identifying what the 'giants' did to the landscape to make it sacred in addition to moving 20 ton blocks of stone around. The monuments that they have left to say that they considered the land to be sacred and to be the body of the Earth

Goddess can now be described. The giants were the Druids and I should say that we have no evidence at the moment that very large numbers of people were involved in the massive terraforming projects visible all over Britain. What evidence there is, points to the Druids being skilled civil engineers whether it was in the building of a plumbing system, moving a 20 ton block of stone or digging a 20ft deep dyke. First the deer then the oxen provided the muscle, the Druids provided the skill and machines. From what the authors have seen in travelling round Britain, the scale of Druidic civil engineering projects over a period of 2000 years or more may well dwarf what went on in Egypt over a similar period of time. Not in the scale of individual buildings or monuments or the masonry and artistic skills but in the quantity of earth and rock moved, the underground work and the timber and stone temples built. Just as the ancient Egyptians worshiped a Sun God so the ancient Britons appear to have worshiped an Earth Goddess. In Egypt the secular and religious command structure was combined, in Britain they appear to have been two separate departments, each possibly with its own resource base with the Druids concentrating on certain areas and the secular people possibly on others. The Druids hold on the people was however powerful and lasted possibly for 3000 years or more. This is not unexpected where the Priests, who teach a continuity of life and rebirth are prepared to sacrifice themselves and so demonstrate their belief and confidence in the Earth Goddess and a realm the other side of the grave. The priests held ceremonies in grand structures for the beginning and the end of life that is the newborn and the dead, traceable from the Neolithic. Most people do not have much control over these events and just go along with the ceremonies involved. The fertility bit in between is more exciting and I suspect that most people would not want to miss out on any of the ceremonies associated with it. The Druids therefore had a firm hand on the general population through out their life. As a result they could direct the resources of the ancient inhabitants of the British Isles to the construction and maintenance of great earthworks, tunnel systems, roads and temples all over the land. The magnitude of the maintenance task associated with the constructions once built must not be overlooked. Wood, stone and thatch buildings 60m across with all the outbuildings and services. Dyke systems miles long, roofed and enclosed. Every local area with miles of ceremonial roads to maintain.

On one occasion, when leading a dowsing course on Henge Temples members had marked out the wall of the temple. I was standing on one side of the temple looking at the sweep of yellow flags to the far side from a slightly elevated position. The far side was 60m away and for some reason, perhaps it was the presence of an apple orchard within it, it looked bigger than previous temples I had seen marked out. A feeling of unease began to creep over me as the thought that this sweep of field could not be the size of a completely roofed in building. The roof of such a building would have to be about 30m (approximately 100ft) high as thatch has to be at an angle of about 45°. When we got to the stage of investigating the roof it was with some relief that the class found the stone columns supporting the roof were about 1.5m square and that the posts towards the centre were about 50cm or more in diameter. The rafters also reached from the temple wall to nearly the centre of the building. Whatever the slope on the thatched or tiled roof, the building was completely roofed over with an appropriately robust supporting structure.

The tunnels and underground temples left by the Druids still impinge on the 21st century as collapsing 'chalk mines', subsidence under buildings, dents in hillsides and deep dells round which farmers still plough 2000 years after they were last used. It is what might be called in civil engineering parlance the 'ground works' of the Druids that define the sacred landscape. The ceremonial roads were taken across hillsides and to ensure that they were safe for the chariots the hillside was terraced to ensure a level road. Edges were hardened using stone and cement. Over a period of 3000 years a great deal of terracing can be done but the scale on which it was done is strange. Having built one ceremonial road round a hill or up a hill why build more? Having built one dyke round a "Hill Fort" (Hill Temple) why build three? Roads do not make good agricultural land and yet on a hillside terraces can often be seen one above another. Could the terracing have been a way of disposing of the spoil from underground workings? The roads must have remained in place for a significant period of time because the chemical stains from the wheels, horses, deer, stones, mortar, cement and colour pigment are still there. In the lounge of my house, a diamond painted on the ceremonial road running through the house is clearly dowsable after possibly 3000 years. Also the tracks of wood wheeled vehicles and the deer pulling them. The road was not laid one year to bless the land and then dug up to plant corn the following year. The ceremonial roads were well defined and in some cases they could and did become commercial roads. Some of today's country roads and 'A' class roads round St. Albans follow the puddingstone verges, bronze tracks and diamonds of ceremonial roads laid across the landscape thousands of years ago. These ceremonial roads ran into and out of mini dykes. Many minor roads or lanes in Hertfordshire still follow the ceremonial routes. When they do, they may appear to run through cuttings with high banks on either side. These cuttings are often mini dykes and indicate the antiquity of the road, possibly 3000 years or more old. In Devon and Cornwall the minor roads and lanes may still have dry stone walling from the Henge Age. Wider modern roads may run through dykes which have been enlarged. The original banks have gone but there is still raised woodland on either side. The original Druidic road paintings are under the modern road and it is still possible to dowse them. Where the Romans have followed the routes of the Druidic roads the different archaeological levels may be Druidic, Roman or Modern. The cuttings are likely to have been dug by the Druids not by the Romans or Modern engineers.

In many parts of Britain you cannot go very far without seeing round shallow depressions in fields and sometimes in the larger gardens. The depressions vary in size from 15m to about 60m across. The most common depressions round the Bovingdon area of Hertfordshire are the smaller ones. Once you have an eye for them they can be spotted easily in the landscape. Many depressions have been flattened out by farming so there are many more Henge Temples than visible depressions. With visible and hidden temple sites plus the sites where temples have been built one on top of another there are so many of them scattered over the Hertfordshire landscape that at one time I thought temples must have had a limited life. But as the size and complexity of the Druidic Temple began to emerge I was not sure that this could be the explanation. The answer almost certainly lies with the 3000 years or so that the Druids had at their disposal. Three thousand years is about twice as long as the modern Christian era in Britain. Some temples have left marked depressions of various sizes which in some cases appear to have degraded the agricultural land. This resulted in the formation of what we now call Dells. The Dells appear in different forms. In one form the trees, shrubs and undergrowth centre on a depression which was the pool of the temple or a sunken Christian chapel. Sometimes this may be full of water and used as a pond. There are several examples of this type of 'Dell' around Bovingdon and its neighbouring villages. Another type of Dell is produced where the original excavation went down to a tunnel system. These have collapsed in but often still have steep sides. The trees in the local ones tend to be large as it is not possible to do much with it apart from filling it with waste and junk. A third type of Dell seen on hillsides is where the entrance to the tunnel system is dug into the hillside and a flat area created in front of it for the temple buildings (Figure 27.1).

The Druids also built upwards and created Long Barrows and Round Barrows which were in buildings. These barrows can be a distinctive part of the landscape in some parts of the country. All the ones we have looked at appear to have been inside buildings. The Druids were not the only people to build Round Barrows so the contribution of Round Barrows to the landscape cannot be claimed entirely by the Druids. The Long Barrows, which according to the archaeologists are amongst the earliest structures in Britain, also contribute to the landscape

344

in some parts of the country. Their date of origin, possibly as early as 4000BC, indicates that they may be from the very earliest days of Druidism and the Henge culture. The best evidence we have found for a link between Long Barrows and Druidism or the Henge culture is in South Devon. On Bolt Tail there is a Druidic temple which is connected by a ceremonial road to a Stone Age Long Barrow. In the park at St. Albans there is another Long Barrow linked to a temple. At the moment we have only looked at two Long Barrow sites and found no sign of metal associated with them.. We have however picked up bone depositories, the use of small carts to move bones around ceremonial tracks, the patches of red ochre and altars. Long Barrows are not the only long earth structures. For example in Yorkshire at Ulshaw near Bedale there is a colossal earth work running alongside the River Ure. It is so large that on first sight it is difficult to believe that it is not a natural feature of the landscape. Yet another type of Druidic structure and perhaps the best known are what are referred to as Hill Forts. The ones we have looked at so far have turned out to be temples. Either simple temple complexes as at Glastonbury Tor, or going to the other end of the spectrum, they are complex Dyke Temples as at Maiden Castle. It is when it comes to interpreting what the Druidic structures on hills

Figure 27.1 The Entrance to a Tunnel System
The entrances to tunnels and underground temples have been sealed and filled in as is the case in the example shown. The depression in the ground still remains with chariot tracks going up to and into it from a levelled area where the temple once stood. There are execution lines in front of this temple indicating that the site was being used when the Romans arrived.

are that the weaknesses of the conventional archaeological approach and mind set reveal themselves most clearly. Archaeologists of the past hundred years or more have been brought up in a land of castles and forts. To them a hill looks a good place to build a castle or fort so that is where pre-Roman society would have built them. The assumption is that pre-Roman societies were as warlike as post Roman ones. The ramparts may look as if they were built by amateurs but then people thousands of years ago may not have known much about building forts. Then of course all societies build forts, castles, defensive positions so we need to find some! This is how the modern mind reasons. Once you have convinced yourself that you have a fort, conjecture moves in to flesh it out and the conjectures of the last century becomes this century's fact. The actual military and social context in which forts and castles work and the ones in which they do not, seldom if ever seems to be addressed. As previously mentioned, fortifications are built on the principle that those inside can be relieved or that they have the endurance to outlast that of the enemy. A close examination of a hill fort such as Maiden Castle or an actual castle such as Berkhamsted Castle, which is built on a Dyke Temple and Round Barrow complex, shows that the builders could not have had defence high on their list of priorities.

Box 27.1   The Proto-Christian Crucifix Becomes a Birth Table

It was during the final stages of compiling the book that we came across some very large paintings of the Proto-Christian Crucifix. We had already discovered some of the detail of the crucifix such as it being a woman strapped to a crucifix with a baby between her legs. The thought had occurred to us that the crucifix was in fact a birth table and might be telling us how the Henge people gave birth. If the woman was strapped to the table it might be for a reason. One possible reason could be to stop the women sliding off when the table is tipped to aid the birth process. We looked for signs of a tipping axle but had not found one. Then we found a very large painting where every aspect was accessible and easily dowsed. What we had been calling a crucifix was indeed a birth table, made from oak and with an axle to rotate the table from the horizontal. There is not only the birth table but a place for the midwife to stand, vessels for water and the placenta, moss is used to deal with body fluids and to place the infant on. What we had assumed was a crucifix is something quite different. We do not yet know if it is the earth Goddess on the table, she has a gold torc or halo above her head but this does not say that she is a Goddess. However, we are working on it. We now know that the transition was from a birth table to a real crucifix, or do we. Perhaps we had better wait until we have a large picture of a crucifix to work on.

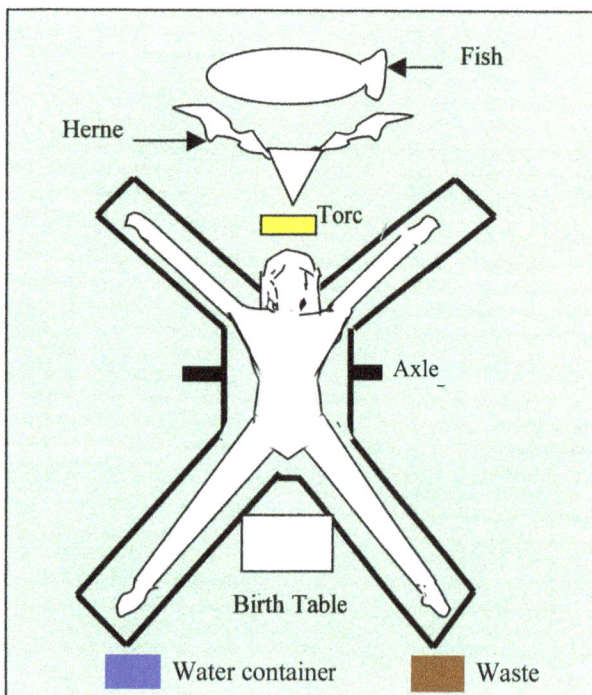

Figure 27.2 The Proto-Christian Crucifix

A diagram of the Proto-Christian Crucifix with Herne, a Torc and the Fish. The person on the crucifix is a woman and there is an infant below the woman. Later discoveries showed that the crucifix was a birth table.

346

The possibility that pre-Roman Britain was a reasonably peaceful society with methods of resolving disputes has not been considered by archaeologists and historians and it needs to be before inventing forts. No tribal group may have required forts, castles or defensive positions. The Druids had power and they may well have used it to ensure things ran smoothly. To understand a place like Maiden Castle you have to step back and look at what is there not at what you think is there. Science based biolocation does of course provide a large body of data not available to conventional archaeology. This data helps to strip away conjecture and enables the investigator to get that little bit closer to the truth. Without looking at all Hill Forts it is not possible to say that they are all Druidic Temples and that none are defensive. Some could be Roman or post Roman conversions to defensive positions and some may have been built as forts in much more recent times. However, the presence of terracing and dykes does not make a fortress out of a temple and the name 'Hill Temple' is probably a better descriptor than 'Hill fort'. The reference to the Iron Age is also likely to be wrong as in most cases sacred sites can normally trace their ancestry back well before the Iron Age, in fact is most cases to the Neolithic . All the sacred sites we have looked at to date go back to long before the Iron Age, often to the Stone Age and all belong culturally to the 'Henge Civilization'.

In the St. Albans area the major legacy from the Henge Age is the dyke system and the Dyke Temples, that is a temple dug into the ground with a roof over it. In walking into these temples you descend into the ground. There will often have been a pool somewhere on the lower level. However, pools are not restricted to lower levels and have been found at the top of a Round Barrow. The basic elements of the Druidic Temple are still there in a Dyke Temple but it is now a roofed over cavity in the ground which comes in different designs. The amount of spoil removed from the Dyke Temples and from the dykes is very large, as would be expected from excavations that are maybe 5m deep, 20m across and are hundreds of metres long. Such gashes in the ground can be seen round St. Albans and all over the country. Other examples of Dyke Temples are Maiden Castle, Avebury and Berkhamsted Castle.

At one time I thought that the temples the Druids constructed were perhaps following an evolutionary series with the simple temple buildings being the start and the complex Dyke Temples as found in Bury Woods being the last of the series. This does not appear to be the case. When the Romans wiped out Druidism the whole range of temple types appear to have been in use. Maiden Castle, Glastonbury Tor, a temple at a tunnel entrance where the temple acted as the entrance hall to a complex of underground chambers, the stand alone temples, the temple complexes associated with running water, Dyke Temples such as Durrington Walls, they all appear to have been in use. Druidism may have imposed some design features on temples such as a pool of water, seven altars, an underground section, the use of ceremonial routes, and roofing over the dykes but provided the temple complied with core requirements there was a lot of scope for design changes. The changes are often adaptations to the local geography or the designers taking advantage of geographical features such as hillsides and streams. The Druids were engineers and craftsmen. They had cement, concrete, bricks, tiles, lead and later bronze and iron at their disposal. A host of materials such as oil based and water based paints, lubricants, ropes and leather. There was a trade in pigments and it was possibly paint pigments that started and formed the early basis of trade long before metals, pottery and food. Pigments have the virtue of being high value and low bulk and weight. As far back as the Stone Age there is evidence for a trade in pigments.

By July 2007 it was beginning to emerge that there was possibly yet another form of temple. First the Uffington Cat (White Horse) was found to be in a building that could be a temple. Then the snake, about 60m long, was discovered by archaeologists. We soon found them round St. Albans and they were in their own buildings or temple. This was followed by the discovery of a fish temple.

The aim of this very brief review of the civil engineering legacy of the Druids is to emphasise the scale of their operations and that in many places where people walk and drive their cars today the roads and paths were laid down by the Druids 2000 to 6000 years ago. The countryside in many areas shows the remains of terracing and temples. I suspect that most Roman roads follow Druidic roads but this has yet to be confirmed.

Woodhenges and Stone Circles

Since finding the first tomb of a Prince and Princess in the Welsh heartland the total number of tombs found must be about fifty and if I was to be asked to find more could find another 50 without difficulty. Of these only about 10 have been looked at in any detail. It took us a little while to realise that the tombs may provide a method of dating the temple to which they belonged. We recognised that we were dealing with an elaborate form of human sacrifice in which the body is preserved and protected by a stone structure and possibly mummified. The purpose of the sacrifice is not known but we have worked on the premises that the two people were seen as envoys to the spirit world or to the Earth Goddess. This human sacrifice, with the sets of eight Druids at each of the four temple gates, can be traced back from clearly Iron Age Temples to what we believe are Stone Age Temples possibly dated about 2000BC, or in the case of Bolt Tail it could be as early as 4000BC. A Stone Age tomb may only have flint weapons, a wood shield, nettle as a fibre, urine tanned leather, no wool garments, no tiles, no tomb wall plaster and only dog hair and bird feathers as ceremonial garments. The tomb contents say that it is Stone Age' but because religious societies are conservative the actual time may be early Bronze Age. However there is the alternative possibility that the Druids were one of the few groups of people who could afford bronze when it first became available and that a Stone Age tomb does mean Stone Age. Particularly when they are linked with a Long Barrow. The construction of Long Barrows, the enclosing building, wheeled transport and temples shows that Neolithic people had advanced engineering and construction skills, trade skills and surveying, management and planning skills. By the time humans reached the Neolithic, society was organised and disciplined. The use of wheeled vehicles when Long Barrows were being built and used should be noted as wheels are said by some authorities to date from about 3500BC, see Renfrew and Bahn. If Long Barrows in the British Isles can be dated as from 3000BC, then wheeled transport was being used by the local people from about 3000BC.

We base the time line of the Druids on the design of the tomb complex and temple. After the Stone Age and over a period of possibly 2000 years, maybe more, the design of the tomb complex, the use of envoys to the other world, the design of the Woodhenge Temple appears to remain constant. The quality of materials used and the range of materials used increases with time so that eventually the Princess wears a gold blouse, semi precious stones, bronze headdress, pig skin leather. In the clothing worn there is the change from nettle and hemp to wool to linen and with metals the first to appear are gold and lead then arsenical bronze followed by tin bronze and then iron.

The temple building first increases in size during the Stone Age and then in the Bronze Age remains constant at about 60m across. The design and size of the clay pool changes over time. In the Stone Age it is absent or small and nearly circular but some later temples also have small pools. The actual shape of the pool and the stone circles derived from them, seems to be dictated by the need to enter and exit the pool and to include a baptismal font. However, the pool gets larger in the Bronze and Iron Ages with some appearing to be much larger than required for their purpose. They also take on non circular shapes to accommodate more elaborate entrances. In the centre of the pool is a plinth to support something or possibly an obelisk. The pool and plinth are a common feature from the Stone to the Iron Age. The actual energy engineering aspects of pools and obelisks has not been worked out but we do believe they may contribute to the overall energy plan. Other features running through time are the Dolmen and the ceremonial vehicles, first small carts then chariots appear on the scene. Some things change such as gate hinges. They are first fibrous ones – nettle and hemp then bronze, then iron. There are changes in dress. The priests in the Stone Age wear nettle, hemp and feathers then wool when it becomes available and feathers then linen comes in and the feathers are reduced in number. Belts are first made from fibres then leather. This overview is based on a small sample of temples but graves do tend to have a common design in any age.

Bronze pins or nails are first introduced into the structure of the temple roof and then iron pins replace them. Stone columns to support roof rafters at about the pivotal point of the rafter can be identified in the Stone Age and they remain a design feature up to the end of the Henge Age. One of the more interesting aspects of temple design is the use of materials. The Druids used stone, bricks, tiles, mortar, gypsum plaster, cement, concrete and lead, at least in the Hertfordshire area and a number of other sites. The use of cement and concrete may go back to the early Bronze Age. Its main use appears to have been strengthening the edges of roads, stairs and the making of concrete for various tasks. The cement was also used for making knives and possibly axes. The central pool in the temple was often enclosed by a stone wall with a tiled area inside to stand on. The central plinth may have been made of stone or concrete and either it or what was on it was painted and possibly gilded with gold leaf. The stone used for making the wall of the pool has in some cases survived to the present time as the small diameter stone circles. Sometimes the surviving stones made up the wall of the pool, as in the North Yorkshire stone circle we studied, or large stones are spaced out round the pool wall and in some cases have survived so that they now appear as small circles with a limited number of scattered stones round the circle. The wall between the stones has gone. The next group of stone circles are the remains of the structural stone columns, normally supporting ring joists. These stone columns are normally found on rings 3 to 6. The stone columns have generally been made from small stones set in mortar or cement. As a result the stones were easily removed from the site for other construction work. The last of the three standard circle types is the large diameter circle where the stones were used as part of the outer wall. So far we have not obtained much information on this circle, for example was it structural, decorative or was it mainly contributing to the energy engineering. All three circles appear to have one thing in common which is that the stones were erected and positioned according to their magnetic fields.

The discovery that our national heritage of stone circles were at one time but part of Woodhenge Temples may sound as if they have been demoted. Certainly a lot of people who have woven great conjectures about their astronomical role or how people from Atlantis or other planets came and built them will now need a plan B. The reality is the stones were used as part of the temple building. The stones had a functional role such as the wall of a pool, defining portals and processional routes, a structural role holding up the roof but even so they hold a great mystery. For example, why go to so much trouble to erect and position the stones in relation to their magnetic axes.

By the Iron Age, if not much earlier the Druids had realised that the field strength of a stone was proportional to its mass. Either that or they just got hung up on large stones or they knew what the Egyptians had done with stone and were aiming to do the same. It may be that Stonehenge was possibly the cutting edge of Druidic energy engineering. If it was, we have to understand it. There was a pool of water with a wall of large stones in a horseshoe (blue stones) and let us assume that they were all correctly aligned magnetically and the horseshoe produces an energy beam. Behind them there were five trilithons and again let us assume they have the magnetic field of each magnified by the lintel stone. Next comes a circle of blue stones and then a circle of Sarsens with their fields possibly magnified by lintel stones. At first it looked to me as if the fields of the inner horseshoes would not be able to break through the outer circles. As many water diviners know a pipe within a pipe of the same material is not visible to the dowser. The magnetic fields short out. The same should happen with the blue stone and sarsen horseshoes. Their fields should short out to the blue stone and sarsen circles round them. But as the laboratory experiments showed they do not. So what were the Druids up to? The chances of working out what was happening by studying Stonehenge are not very good as there are so many fallen stones at different angles. The best way forward is to build a model but even then we will not know if some of the missing stones had special alignments to ensure fields penetrated the circles. When building a mini stone circle the stones that pull the blue field or rays out of the circle have to be set at a certain angle to the circle. It must also be remembered that the Druids may have wanted to contain the fields from the two stone horseshoes. A possible argument against this is that the magnetic sensory system can be saturated and the dowser (Druid) no longer 'sees' the field, they are blinded by the intensity of the field. This happens when dowsing at least some types of modern cell phone. If you get too close to the phone there is no dowsing response to its fields. The structural role of the circle of linteled stones is easy to see and other stones can be assigned tasks such as portal stones. It was the energy engineering aspects that presented a problem, at least for a while. However, as is described in Chapter 18 the problem was solved – up to a point and that point is that at the time of writing we still do not know what the Druids did with the energy beams. As dowsers they were way ahead of us and there is still a lot of catching up to do. The Druids in terms of their knowledge of and use of their sixth sense, the magnetic sense, must have been awesome figures. Just as today there are people, we call then sensitives, who pick up the magnetic fields and then convert them into visual images so in Druidic times such people must have existed.

Perhaps one of the biggest conflicts between our findings and the views of conventional archaeology is in the dating of sites such as Stonehenge, Avebury Stone Circles and Stanton Drew. The fact that Stonehenge was inside a building is not likely to be too much of a problem as some archaeologists and engineers have already arrived at that conclusion. They got in right. The main problem is likely to be the dating. The historians know when Stonehenge was closed down and destroyed by the Romans so it was in active use up to the time of the Roman invasion. But when was it built? The tracks of the vehicles carrying the stones can be traced right up to individual stones and there is no doubt that these vehicles had iron tyred wheels. In the case of Stonehenge these tracks have not been traced back to where the stones came from because of the distance involved. However, at Avebury the iron vehicle tracks have been traced back to the sites where the stones were dressed and loaded onto the vehicles. In the case of Stonehenge the well and the change in materials used in its construction and plumbing show that it was supplying water to the Stonehenge site from possibly late in the Stone Age up to the Iron Age. Like the Stoney Lane site in Bovingdon the Stonehenge site spans possibly 2000 to 3000 years or more. The Stonehenge site was next to a city and over the ages the temples are likely to have become more and more grand. The final temple was the greatest and grandest of them all and was clearly put up by Druids who had a good command of mechanical engineering, materials and energy engineering. When in the Iron Age it was constructed maybe difficult to determine but at the time of construction the Iron Age was sufficiently advanced in Britain for iron tyred wheels to be available. Iron may also have been used in the HGVs which were carrying 50 ton stone blocks for about 30 miles. With the additional evidence of the tomb complex, which may now be back filled to take the load of the stones, the one thing that can be said with certainty is that Stonehenge, as we see it today, is not Neolithic or early Bronze Age but Iron Age and belongs to the Proto-Christian era of the Henge Age. We know also that it was in use in the Christian period and stones were being positioned on top of previous ground paintings of the Proto-Christian period.

Returning to the people who built and used the temple. The use of linen clad priests with staffs, ceremonial chariot roads, the 8 by 4 sacrificed Druids at the four quarters, the tomb complex with its Prince and Princess all point to Stonehenge being Druidic. In fact the Druids were the only people who could build and run such a temple. It is also clear that the Romans looked on the people running the temple as Druids as they met the same fate as Druids all over the country. It can therefore be said with some degree of confidence that Stonehenge was a Druidic Temple. Designed, built and run by the Druids. The same can be said for the Avebury Stone Circles and Stanton Drew. The evolution of Stonehenge continued into the Proto-Christian era and then into the brief Christian phase. It is almost certainly one of the first Christian Cathedrals.

Another area in which the present study is perhaps in conflict with conventional thinking is in the relationship between stone circles and woodhenges. I have heard archaeologists say that the ancients moved to stone when they ran out of wood that is woodhenges predate stonehenges. From what has been said earlier the stone circles and standing stones are all that has survived from temples and Dolmen. The wood building was the main structure from the late Neolithic through to the Iron Age and the Druids never ran out of wood. Their use of woodland plantations i.e. Groves, ensured a steady supply just as their farming of deer ensured a supply of antlers. The mass graves in Anglesey and Hertfordshire were dug using antlers.

The Development of Temples

Back in 2000 my view of the six rings of postholes was that they were a very simple religious site possibly with no roof. My explanation of what appeared to be a very large posthole at the centre was that a tree trunk had been used. A view reinforced by the discovery of Seahenge on the north Norfolk coast. Seahenge had an oak tree with the trunk in the ground and the roots in the air. It was only later that I began to realise that Seahenge was almost certainly a Dolmen. The body could be placed on the roots for excarnation by the birds. The tight circle of posts kept animals out, hid the body from view and a door closed the narrow entrance. Over the next four years as Nigel and I discovered more and more detail the temple became increasingly complex, it became larger and it acquired a basement. The temple complex just kept on growing in size and complexity as we discovered more detail but it still remained a basic circular structure with first six rings as we thought then 6 + 3 rings of postholes. Then the large excavation in some woods near Bovingdon and St. Albans were found to have the underground tomb at their centre, then the eight sacrificed Druids at the entrance to the Dyke Temples and then signs of the presence of pools were found in the excavations. It slowly dawned on Nigel and myself that we had another type of Henge Temple and it was roofed over so creating very large buildings with internal terraces, altars, alcoves and tunnels running out from them. These Dyke Temples as we called them were complex. We then realised that they came in different sizes and shapes. One in particular I thought was designed for chariots to enter the temple, spiral down to the bottom and then back out. Later study indicated that the chariot tracks were underground. We still have this one to sort out. Next we realized that the well known earthworks referred to as dykes could be connected with them. The temples and associated structures just kept on getting bigger and bigger. At first it looked like an evolutionary progression but this is not supported by sites such as Durrington Walls and Maiden Castle. Large temples and earthworks appear to have been part of the Druidic scene from very early on in the Bronze Age if not the Neolithic. The latest developments are the discovery of what might be called water temples and large temples for the snake, cat, deer and fish. There are also complex road systems with possible religious significance.

Then towards the end of the project we found the extensive underground tunnel and Bell

Chamber systems. They are clearly a separate development to the tomb and its system of tunnels and chambers under the temple building. With this discovery our concept of what lay below the ground started to change. It was clear that a few metres or tens of metres of tunnel were not involved. There were in fact some very extensive tunnel systems and some very large chambers deep within the ground. Tunnel runs of fifty metres and more were not uncommon. However, what is perhaps the most impressive aspect of the engineering was the ability the Druids had to navigate in 3D underground and find a target 50 to 100m away. The paintings on the floors of chambers also show that the Druids when tens of metres underground still knew where the compass bearings were. Having gained some insight into the scale of the Druids underground engineering we then found that they started to seal the entrances to their underground world. In place of Henge Temples and Bell Chambers sunken chapels and churches appeared. This more than anything made us aware that with the evolving engineering skills there had also been an evolving concept of the spiritual or religious world. The last phase of the evolution, the Proto-Christian phase, resulted in real design changes to the temples. The underground system was being sealed off and sunken chapels or churches constructed. With no underground system for the deposit of bones a new system of dealing with the dead must have evolved. The sacred sites of the temples were used for the new buildings and many ceremonies such as the Eucharist were carried over from the Druidic period to become part of Proto-Christianity. We have not been able to determine with any precision when the chambers and the wells were sealed and the religion moved to Proto-Christianity. It took place in what is called the Iron Age, that is sometime after 800BC. Iron tools and iron tyred wheels were in use so the middle of the Iron Age say 500 to 400 BC is probably as good a guess as any. We also do not know how Proto-Christianity developed over this period of possibly 500 years before it became Christianity after the death of Jesus. Even after the death of Jesus the emblems of Proto-Christianity remained. The crucified person is now a live person not a woman giving birth. There is a gold torc, Herne and the fish were all in use up to Roman times, there were snakes on either side of the crucifix. (Figure 27.2). Later on Christianity modified the cross to the upright one, retained the torc as a halo, removed Herne but for a while retained the fish. The snake disappeared but became part of Christian mythology. The form of an actual fish complete with fins was not used. The fish became stylised to the curved lines of two circles overlapping. The shape of the final temples,, the churches or cathedrals of the Christian era, have not been determined. They appear to be depressions in the ground with the floor painted, rows of pews, a small entrance for the priest and a large one for the congregation. There is a fabric table cloth on a table from which the Eucharist is served. The chapels or churches, just before the Romans arrived could well have looked somewhat similar to a modern chapel but with more paintings and colour.

Druidic Roads

Just as buildings tell you a lot about a people and their culture, so the roads tell you about the development of social and commercial organisation and of course the technology available for road building. An advanced society is required to plan, build and maintain roads. Advanced in the sense of having planning skills, engineering skills, control of resources, adequate resources, chains of command for maintenance, the need for a road system which comes from trade and communications over large distances and the desire to meet these needs. When the Romans arrived they appear to have been able to move around quickly and the indication is that they were able to take advantage of an advanced road network. There was a ceremonial road system but as that was designed to roam around the landscape, it was not suitable for commercial and military purposes. In the St. Albans area there are signs of Druidic commercial roads which show some signs of having been engineered.

To build roads certain technical skills have to be available and it is the technical skills of the

Druids that present a puzzle. While the Romans were advancing over the land the Druids were still knapping flint to make tools and weapons. They were still using deer antlers and bones and yet they had bronze and iron door hinges, iron spits, used iron utensils in their washrooms, used iron tyred wheels. In one part of the country deer antlers were used for tunnelling whilst in another bronze tools were being used. The Druids could make a cement and concrete so hard that it can be polished for decorative purposes. They used a chalk urine mortar. The Druids could make massive carts to carry 50 ton loads, had command of the logistics to build massive temples and yet could not adapt and learn new combat skills to deal with the Romans. The basic technical skills associated with textiles, leather, painting, wood preservation, making beer were all there. Their bricks and tiles survive to this day and the shards of their pottery and cement structures are everywhere on debris fields and round temples (Figure 27.3). This society was advanced in some ways but backward in others. The question is was the use of flint tools, antlers and bones part of the religious system. Was their use required by their religion? Was the use of flint weapons required for formalised battles fought to a set of rules? Perhaps one day we will be able to work it out. However, this is moving away from roads. In the Hope Cove area for example there are foundations for roads, constructed using a urine mortar and these foundations now form a footpath for today's travellers. The diamonds once painted on the road can still be identified. It is the paintings on the roads that have the potential for not only dating roads but throwing some light on the development of the Druidic religion from the Neolithic to the arrival of the Romans. The lack of metalled pre-Roman road surfaces is almost certainly due to the removal of the surface which may have had considerable value as building material. This would be the case when tiles had been used. However the cuttings made into both soft ground and hard rock when constructing a road have not been destroyed and in fact they have been used as part of the road infra structure of the British Isles for the past 2000 years. These cuttings remain as perhaps one of the most impressive monuments to the civil engineers of the Henge Age (Figure 27.4). The extent of the Druidic roads is so great that in some parts of the country if you want to find a non Druidic road it is necessary to go onto a modern housing estate.

The People of pre-Roman Britain

One thing that has not been discussed so far is the origin of the pre-Roman people of the British Isles. The story to date has been following one group of people who left a time trail in the buildings they build and the terraforming of the landscape. The trail left by that group of people appears to run consistently through at least two thousand years prior to the arrival of the Romans. Our latest finds indicate an origin possibly 2000 years earlier still. So far we have not picked up any sign of another group of people coming in to the British Isles and causing things to change. If Celts did arrive in Britain they must have adapted and blended in with the society causing little if any change. However, it is much more likely that Celts did not make it to Britain. If they had, iron making skills would have improved and iron weapons and armour would perhaps have been available to the Britons. Iron arrow heads would have been able to penetrate Roman armour and so enabled the 'Celtic' warrior to stand back and not mix it with the Romans in hand to hand combat. The Welsh, Scots and Irish may feel a bit upset at being told that their 'Celtic' credentials no longer hold up. The reality is, I suspect, that the so called 'Celtic' people of the British Isles can trace their ancestry and culture back to a far more ancient people. A people who moulded the landscape and created its legends and music when ancient Egypt and China were in their infancy. How has the idea of 'Celtic' peoples in the British Isles gained such a hold on our imagination? There is a tendency for archaeologists to identify a people by the artefacts they leave behind. Quite a reasonable supposition. But does that mean archaeologists a few thousand years from now are going to classify the Japanese as Europeans because they find traces of railways and cars? At least one hopes not. The people of these islands, like the Japanese today, would have picked up from overseas ideas, technology and art; they would have imported goods and materials. Also just as today, skilled people and artefacts would have moved around the ancient world, perhaps more freely than we like to think. The closest neighbours to the Britons of the day were a continental people who from about 800BC had a Celtic culture. It might therefore be expected that the dominant influence in the immediate pre-Roman period would come from a Celtic culture on the continent. The chances of the Celts being able to successfully land and invade the British Isles in pre-Roman times would not have been very great. The indigenous population was well organised and almost certainly could have mustered warriors and war dogs in their thousands at short notice.

Figure 27.3 Pottery Shards
Selection of pottery shards found on the debris fields of temples.

Figure 27.4 Druidic Road Building
A road cutting dug by the Druids in use 3000 years later.

It is possible to trace the ancestry of the Druids back to the Stone Age using their temples, tombs, ceremonies and art. So what sort of people were they. They appear to have been a religious, possibly monastic group. At the moment we have no evidence that says they were a separate group to the secular society, although the possibility of a cast system has to be considered. The Druids had families and at one stage it looks as if the wives were sacrificed when the husband died and possibly the husbands when their wife died. The matching number of male and female biers in the East Dolmen makes it look as if it was a rule with few exceptions. The size of the East Dolmen compared with the South and North Dolmen may indicate the relative sizes of the religious and secular sections of society. If it does, Druids were about 20% of the population. However, temples and their Dolmen will have to be checked all over the country to arrive at a reasonable estimate. The temple is only one aspect of the Druids lives and it will be necessary to study the 'housing estates' round the temples to learn more about them. The Thornborough Henge in Yorkshire would be an ideal place for such a study. The size of the henge indicates a possible city close by and certainly the presence of large areas of housing and manufacture. Like Stonehenge, the Thornborough Henges will have a halo of Druidic and secular settlements around them. There is only one problem. Much of the area round the Thornborough Henges has been used for quarrying and a lot more is due to be destroyed as the quarry extends. The halo of buildings, roads, secular and religious activities round a major structure such as a henge helps to develop an understanding of that structure. Fortunately, every temple has its own halo containing the information of its support staff and so offers an opportunity to study the Druids in a little detail. If the Druids did comprise 20% of the population it is possible that they had the resources to do most of the engineering work. That is building temples, roads and whatever else had to be done. An alternative is that the Druids used contractors who were part of the secular society and the contractors would move around as required. The consequence of this is that money would have been required and although it may well have been in use sometime before the Romans arrived I know of no evidence that its use can be traced back to the Neolithic, the late Bronze Age perhaps. If you are going to buy copper from North Wales and lead from the North of Britain you need something to do the purchasing with. Prayers and a word with the local God, then as now, are not hard currency. Now this is speculation and it is going to be a while before we get a lead on the economic and financial system of pre-Roman Britain. The archaeologists at the moment have not found any clear evidence of the wide spread use of money and pre-Roman Britain may have relied to some extent on barter. But, the scale of engineering operations, trade, religious activity, industrial activity and its coordination over the British Isles required a sophisticated economic and trade accounting system and this would have been very difficult without a written language of some sort. Records, accounts, information transfer, postal system, all would have been required. It could have been managed by a dedicated group of people such as the Druids. Right up to the Roman invasion many aspects of the Druid's ceremonial life appears to have been constant over a period of about 2000 to 3000 years, possibly more. As a group of both male and female priests they were responsible for a number of religious services to the general population. They baptised infants and may have circumcised them. One ceremony involved the ceremonial taking of bread and drink. We have not yet identified their role in marriage but their role in dealing with the dead is very clear. The way the dead are dealt with may differ round the country in some details but the core procedures are likely to be followed everywhere. The dead are brought to a 'chapel' of rest, then taken to the temple pool for immersion or washing in the 'water of life'. The body is then purified by excarnation in a sacred Dolmen by birds. The birds may be considered as something special, possibly holy as their feathers are used as a symbol of priesthood. The Raven also appears in the ground art (geoglyphs) as the Thunderbird. In the Dolmen, red ochre is applied to the bones and this enables the pathway of the bones to their final resting place to be followed. The pigment leaves a stain wherever the bones are handled. The resting place appears to be underground first in Long Barrows then in underground tunnels and then in a sub temple tunnel complex. There is always a shaft and tunnel in the temple to the underground tomb complex. This also applied to the one temple looked at in Ireland. Underground there are the triangular antechambers with small altars so the priests conducted some sort of ceremony below ground. Above ground the tracks of the priests, carts and chariots reveal quite complex procedures and ceremony associated with the burial of bones. The Dolmen indicate that the temples were servicing populous communities which would have produced very large numbers of bones, particularly on sites such as Stoney Lane which span thousands of years. Bones have a strong phosphate signal. For example the graves round a church can be picked up at well over a hundred metres. There does not appear to be such a signal from a temple site. One possibility considered was that bones were only allowed to 'rest' in the temple complex for a limited time and then they were moved onto another burial site or were burnt.

The absence of bones began to worry me. I rang up Nigel one evening and said 'we have a problem'. 'We are missing 2000 years of bones'. I drove over to St. Albans to see him and to discuss what the options were. If the bones were burnt, the phosphate would not disappear so we should have picked up phosphate areas when scanning for Dolmen and mass graves. We have not come across any signs of burning. This could mean that we might have to look further a field for the signs of bone burning. Another possibility was that the bones were in bone depositories but for some reason were invisible to the dowser. If this were the case why did the bones in the four bone depositories show up so well. Nevertheless, the possibility of the phosphate fields of bones shorting out to each other had to be faced. If they did there was a simple experiment which could prove it. Put a few bones into a pile and see what happened. Having reached this point I left Nigel to work his way through a tray of chicken drum sticks and returned home. Later that night Nigel rang 'Geoff have you ever dowsed a bone on the Dowson Bench?' the answer was 'no'. I knew of the crystalline structure of bone and that the body had a diamagnetic axis running through it from head to toe but I had never studied a bone. What Nigel had found was that the bones do behave magnetically as a crystal. There is a strong diamagnetic field with its axis running along the length of a long bone. At right angles to it there are two strong paramagnetic axes. It turned out that the long bone of an animal is ideal for demonstrating that the human diamagnetic sensor is in the head. Because the diamagnetic axis is along the length of a bone, which is highly visible to an audience, the bone can be pointed like a torch at different parts of the body and the dowsing response noted. The bones have a strong magnetic field associated with them just like a tree and a stone. The ancients would have known this. To them the magnetic fields were part of the spiritual world and they would have seen the spirit of a person as residing in the bones. Nigel had discovered why the ancients revered the bones and why they cleaned the body of flesh before burying the bones. When the bones went through the ceremonies they were holding the spirit of the deceased. Excarnation was a deliberate act aimed at cleaning the spirit or soul of a person of its earthly trappings before it went to reside with the Earth Goddess and ancestors. However, that still did not tell us where 2000 years of bones were.

The next step was to place bones alongside each other in a pile. The dowser, using a phosphate witness could still 'see' them. Then the bones were jumbled up so that they were crossing each other and facing in different directions, the pile of bones became 'invisible'. They could not be seen by the dowser. We now had the evidence that the bones could be in the underground temple complex. It was just that we could not dowse them. So, if the bones were there, how were we going to identify them, make them 'visible' so as to speak. Bones are not just phosphate. There would be calcium, tendon, blood, red ochre, contamination from the birds and insects, perhaps shelving for the bones. If the bones were in a chamber there would be antler pick residue left on the walls. If these other materials did not short themselves out as phosphate did they might be detectable and reveal the bone depository. We knew that the access tunnel leading to the tomb complex had rooms leading off it at intervals (Figure 27.5).

356

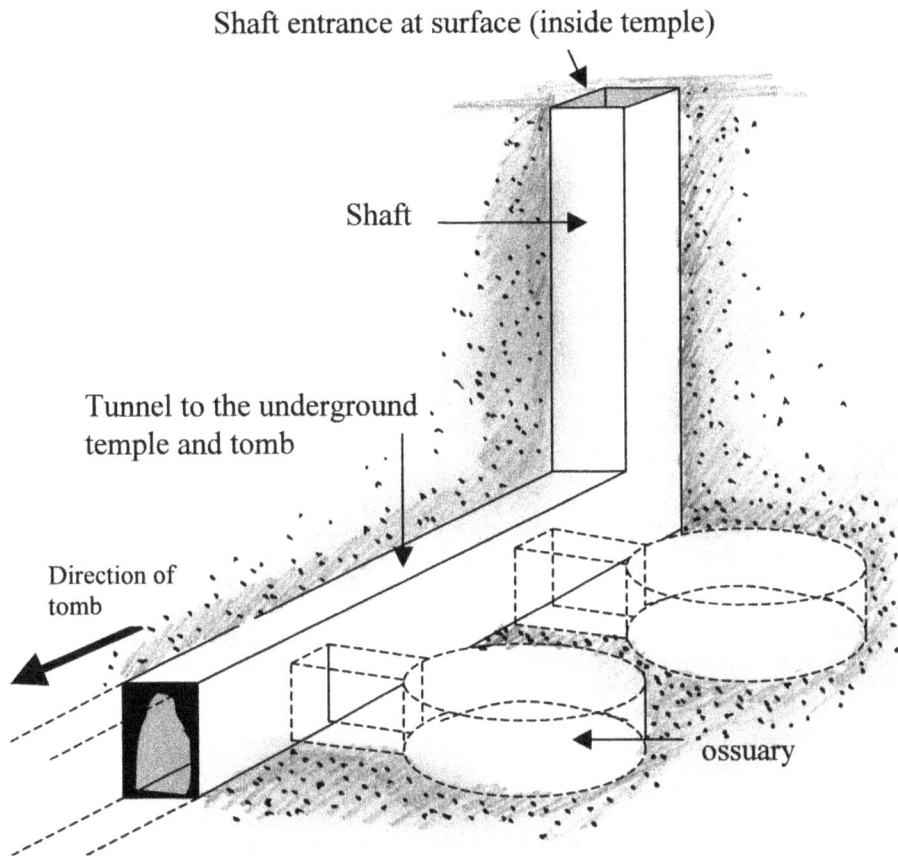

Figure 27.5 A Burial Place for the Bones
Diagram of the access shaft and the ossuary's along the tunnel to the tomb.

There always appeared to be a pair of rooms not far from the bottom of the shaft and then a variable number along the tunnel. There was also some evidence for chambers close to the tomb complex.

Nigel with his chicken drumsticks and witnesses started work on the temple in the near by recreation ground. He soon found some rooms off a tunnel and that the rooms could be identified with red ochre, cedar wood oil, tendon, hair and teeth. Back in my garden I found that the rooms containing the bones were off the tunnel to the tomb complex under the flowering cherry. We were convinced that we had solved the problem of the missing bones but just in case we still look for signs of bone burning when on site in case both methods were used.

Apart from dealing with the needs of humans the Druids were involved with dogs which they used in ceremonies, and also in the slaughter and butchery of animals. The path taken by the animal to the altar and then to the station where it was cut up can be traced. One of the seven altars was for humans. Traces of foxglove, deadly nightshade and toadstool can be found at this altar which may indicate ceremonial sacrifice. The Druids sacrificed their own as the eight guards at each of the four gates show. Four of them are female and four male at least on those gates where we have checked the sex. Children were included also cats and dogs. The clothing worn indicates that Druids of different rank would be sacrificed. The Prince and Princess in the tomb also clearly show that human sacrifice was a very important part of Druidic ritual when a temple was built. The presence of a butchery station outside the temple for human bodies and the feeding of dogs on human remains indicate that criminals also met their end on an altar. Garlic has also been found in the bloodstains at the human altar of some temples which were in use until the Romans destroyed them. This is taken to indicate that Roman prisoners may have met their end on the altar. However, garlic alone is not sufficient evidence to say that Romans were sacrificed. In fact with the development of Proto-Christianity and then Christianity it may be unlikely.

The Artwork of the Britons – a written record of pre-history

One of the mysteries of the 'Celts' as the Henge people tend to be called is in their artwork (Figure 27.6). But first a word of warning. Because the actual Celts on the European mainland have been combined with the Henge people of the British Isles their art has been considered as a whole. It is therefore likely that there are two art streams. One due to the Henge culture and one the true Celtic art. There could be a lot of overlap but two artistic traditions should be discernable. However, that is for the artists to muse over and what follows is in relation to the Henge art as we have found it.

Art is often based on the world about the ancient artists or the mythology of the time. One of the big enigmas of Henge art is the spiral. When we discovered the three underground spiral chariot ways we thought that we had the answer to the spirals seen on the New Grange Kerbstone 67 and 52 (see Lost Science of the Stone Age, page 148) also seen as a triple spiral County Meath, Ireland (page147). The simple interpretation was that the three spirals represented the three breasts of the Mother Goddess. This was reinforced when we picked up prayer stones and human milk at the centre of each spiral. Then Nigel identified spirals along the bank of a river and we were not so sure. Running water generates at least two dowsable patterns, the spirals and the rectangles which we refer to as diamonds. The diamonds are very characteristic of Henge art and can easily be generated. The only problem is that there are two sources for the diamonds – Water, the bringer of life and the winter aura of the oak, death. The diamonds on the roads may be the water diamonds whilst those in the anterooms to the bone depository could be based on the oak. But that is only an idea. What is important is that the spiral and diamond, so characteristic of Henge Age artwork, is probably a symbol for life, the spirit of the Earth Goddess. The zigzag can be dowsed anywhere on the ground and could easily represent the earth. The carved design on an upright in the passage at New Grange could easily represent life by combining the symbols of water and earth.

Perhaps the clearest examples of the Druids dowsing abilities and knowledge of auras are the Thunderbird (Figure 27.7) and the starburst. The Thunderbird is clearly associated with death. The pattern shows up in the winter aura of the oak and four Thunderbirds in the form of a cross are found round the Dolmen of the Priestesses. Larger Thunderbirds are also on the site of the temple. At the Stoney Lane site there were many sets of Thunderbirds. If the Thunderbirds represent death then it is possible that they were drawn on the ground after a temple was destroyed or burnt down. Perhaps temples were looked on as living entity's and had a limited life, after which they were destroyed or died. At the moment it is not possible to say. The period of time spanned by Henge Temple building could be 3000 years. If each one was burnt down by a lightning strike or accident, ten temples on a site gives each one a life of 300 years. At the moment it is not possible to say what factors determined the life of a temple. Lightning, accident, deliberate destruction, so far there is no indicator of what happened to them. It is also possible that the four Thunderbirds are looked upon as a 'door' to the realm of the Goddess. If the Prince and Princess were to visit the Earth Goddess they would need a door to go through. The ancients may have thought that the life of the oak went through the door in the autumn and when the oak came to life again in the spring the door closed and disappeared. This is conjecture as at the moment we do not know the link between the

358

Figure 27.6 The Origin of Henge Art Work
Some of the elements of Henge art work are taken from dowsed energy fields.
The zigzag, spiral and diamond are perhaps the best known ones.
When brought together as in this carving they could symbolise life.

Figure 27.7 The Thunderbird Motif
The four Thunderbirds based on a central rectangle or circle are a common theme.
There is normally an outer circle in black or white and sometimes coloured circles.
When the Thunderbirds are painted round a tomb it looks very much as if they represent the door to the other world.
Round the Dolmen and in the arenas they are later replaced by the birth table and crucifix.

Thunderbirds and Henge Age ceremony. We have not found the Thunderbirds in the Proto-Christian chapels when the chapels are not in the centre of the temple and on top of previous paintings.

The starburst has only been found on a few occasions and then with the temple. The starburst appears near the outer limit of the temple and because of its size and complexity we have only ever followed it for a few metres. It will require much more study before its purpose is identified. The starburst could come from the oak, possibly other trees as well, and as with the Thunderbird, it can easily be produced with a model. There are other pictures not associated with tree auras, such as the zigzag. The zigzag can be identified during the day on open ground.

The Druids painted the stone columns in the temples and wood work. The pigments can easily be identified. There is also some evidence that the Druids painted pictures on the walls of houses and tunnels but they are only likely to survive in the driest of tomb complexes and on hard rock. The pictures that do survive are the ones that were painted on the ground. As more comes to be known about these pictures we may find that there is a written record of the Henge Age spanning thousands of years. The pictures of the cat, fish, Herne and snake are detailed and done by competent artists.

There is plenty of evidence for decorative metal ware particularly candle sticks and lamp stands. The pool and font appear to have had gold, silver and bronze lavished on them. Probably nothing survived the pillaging by the Romans apart from what might still be hidden in Anglesey. Even the lead plumbing, bricks, tiles, worked or dressed stone were taken from the temples or perhaps removed when Proto-Christianity developed. The concrete slabs and artefacts (puddingstone) were broken up although some large pieces do remain. The Druids appear to have used their cement technology for both construction purposes (roads, steps) and for making sacred objects such as altars, plinths, knifes and vessels such as poison chalices and bowls. All of this in a Stone Age society. Whilst they were knapping flint and using stone tools for making bows and arrows and sacrificing humans on altars they were making bricks and tiles which are still usable 2000 years later. They were building carts capable of carrying 50 ton stones and making door hinges from bronze and iron, plumbing systems and highly crafted metal work. Their chariots were highly crafted and decorated with paint and gold leaf. Druidic society was a mixture of ancient procedures and materials with the latest materials and engineering practice of the day. Because the Henge Temple has been the focus of our studies the secular section of society has been neglected. It has only been involved when it becomes entwined with the religious activities of the Druids or in some ceremonies such as animal dipping, May Pole dancing and battles in the arena.

The Druidic Religion

The temples the Druids built remained essentially the same in design for 2000 years and possibly 3000 years. Whatever the religious beliefs of the Druids they were certainly durable lasting until at least the Iron Age. It might be expected that a religious society that required its members to offer themselves as human sacrifice and possibly live to a monastic code with the accent on disciplined work might fade with time. Druidism did not fade with time. The reasons are possibly many but amongst them will be that they appear to have brought up their children in the Druidic system. Once born into a religious system and educated in it from childhood it is very difficult for a person to break out of it. If that person has secret knowledge and skills they might even be hunted down and killed if they try to leave. The few Druidic houses so far investigated indicate that Druids had families but so far there is no indication of how the children were brought up and educated.

The large number of temples and ceremonial sites on the tops of hills, particularly in Wales presents a problem as it might be thought that the top of a hill could not have been the best real estate available. A hill is a difficult place to deliver water to and other commodities such as food and fuel. The hill top sites are uneconomic and would have been costly to maintain. However, something induced the Druids to build on hill tops and ridges. One possibility is that the sites were deliberately set apart from the community for religious and educational purposes. This may be an idea worth investigating but it must be remembered that the climate of the time was quite different to todays and the communities may have been on the hill tops because they provided a good place to live in the climate of the time. Maybe they were away from flies and mosquitoes.

According to the literature the Celts/Druids had a Mother Goddess or Earth Goddess. Everything about the temples the tunnels and the terraforming that went on indicates that the Druids did in fact worship a Mother Goddess or some deity close to it. The Goddess was identified with the earth and water and access to the Goddess was via tunnels in the ground. The four bone depositories, the division of the skeletons into four sections, the four anterooms, the four gates and the four Dolmen indicate that there was a spiritual element associated with the Goddess or that the Goddess was seen as comprising four components. The four components or spirits are probably derived from the aura of the oak with its 'Celtic' Cross or Thunderbird design. The linkage between the designs of the underground tomb complex and of the above ground wood temple with that of the aura of the oak is very strong indeed and very significant. The Druids were not dowsers in the modern sense of the word. Modern Dowsers in their thousands have dowsed tree auras, hugged trees, dowsed stone circles, dowsed earth energies, claim to be able to set up stone circles in the manner of the ancients but none to my knowledge have recognised what the Druids saw as the oak's aura. They also failed to see what was and still is under the Stone Circles, and the energy engineering involved in the stone component of the temples. By comparison with modern dowsers the Druids appear to have been very accomplished and skilled dowsers with a disciplined rational approach. They could identify the magnetic fields associated with the objects around them with precision and accuracy. They could set up piles of stones and build walls so that their magnetic fields were in harmony. They could read the aura of the oak and reproduce it below ground and above ground in the structures they built. They were aware of the energy fields associated with running water and to be found on the land and reproduced them in their artwork. Where this skill and knowledge came from is not known but it looks as if it has incredibly early origins. As this knowledge is there in the late Stone Age, the Neolithic people must have obtained it from an earlier source. The Druids may have researched it and developed it over the years so that they learnt how to work with stones weighing tens of tons, how to put stone lintels across stones, how to split energy beams and send beams in specific directions. There is evidence that they used colour in their dowsing. But somewhere, way back in time people perhaps the Shaman, were aware of the 6 + 3 ring pattern of at least the oak. They were also aware of the four Thunderbirds in the oaks aura. The Druids were aware of the fields associated with running water and that a pool of 'clean' water stabilized the fields of stone circles and the obelisk and possibly of people and other objects. Dirty water from a pond does not protect the fields. The Druids may have associated the magnetic fields of objects with another world, the spiritual world. They may have used them as proof of another world that was present in the ground and in the stones and trees around them. A world that they, the Druids, could access using hallucinogenic plants and fungi. Now this is all conjecture but the dowsing skills of the Druids, that is their ability to 'see' and touch another world when combined with an hallucinogen that takes their minds into that world is a pretty strong mix. So strong in fact that unlike present day religions faith is not required. They were in contact with another real world in their minds and they could touch and feel that world with their dowsing. The Druidism of the pre-Roman period was a very strong, possibly mainly a non faith religion. Talk about a God to a Druid and I suspect they would put you in direct contact so that you could touch and feel the God. That is if your magnetic sense is well developed. Their Gods and spirits, the world they inhabited were there with them. When Druidism developed into Christianity it could have taken many of its beliefs with it into the Christian era. Three hundred years later when Roman Christianity came to Britain the priests would have good reason to know that if they wanted to make headway they would have to accept many of Druidic Christianities ceremonies and rituals, which they did.

It is worth noting that the Druids did not look on the magnetic fields or 'spirits' as a god. People do not question their gods, they do not investigate them, measure or assess them. The Druids treated fields as something to be investigated and measured and then painted pictures of them on the ground and on surfaces. Their god or gods must therefore have been much further back in the system. Hidden deep underground or in an alternative world. They were not in the world inhabited by people and fields. The world and its fields or spirits could be questioned measured and challenged. It should also be noted that at least by the iron Age the religion was evolving and changing so questioning and assessing was going on. There must have been a theology which enabled the development of the Druidic religion into Proto-Christianity and then into Christianity. The Druidic religion of the later Iron Age was therefore quite unlike most modern religions. It was a religion involving questioning and change.

Druidism did not operate in a vacuum. It was part of a real physical world. The Druids were a significant part of the population. Even as I write we are finding more and more execution lines, the latest being found along the Watford Road leading out of St. Albans. Over a mile of them so far. The scale of killing by the Romans is beyond belief. Every Druid, every child of the Druids appears to have been put to death. They were not killed by soldiers on the rampage. They were lined up in rows in front of temples (Churches), along roads. Places where their execution could be seen. They were then properly buried. The Druids were also executed outside their houses and in small groups as if there was a final clear up. This massive killing of Druids and then the secular tribal people needs to be remembered by geneticists and people plotting population movements. If an area such as a large part of central England has its population removed. People are going to move in to fill the vacuum. The date of the movement can be identified along with the source of the people. The source is unlikely to be other parts of Britain as they would be too easily identified by the Romans.

The Druids being hunted and put to death by the Romans were not the Druids of say 1000BC in beliefs and culture. It was a new breed of Druids who had probably left human sacrifice behind apart from criminals and captured Romans. The method of dealing with the dead had changed and excarnation may no longer have been practiced. Society, according to the accounts of the time, was still well run, well fed and healthy. According to historical accounts the society looked after its sick and elderly members. The houses were well built, toilets and sewage system, washrooms and drainage, water supply, roads and communications. Trade was well developed and although, according to the historians, money was not widely used, bartering is said by the historians to have provided the basis of much trade. If the historical accounts of the Britons in 100BC to 60AD and the biolocation findings are reasonably accurate they paint a picture of a well developed civilization A civilization with its towns, cities, monumental structures, buildings, temples, art and religion. The civilization stretched across the British Isles and from the records of megaliths probably covered the western seaboard of Europe including the Iberian peninsula. The civilization in terms of its religious beliefs was evolving and by the time the Iron Age was well developed it had adopted what we have called Proto-Christianity. The name is derived from the use of the God of life, Herne, the birth table that we at first thought was a crucifix with a body strapped to it so that there is no bleeding from the use of nails, and the fish as its symbols. There is no conjecture about these symbols. Any dowser of reasonable ability is going to be able to find them and determine the pigments that were used 2000 years ago or more to paint them on the ground. There is no conjecture about the buildings used for religious ceremony. These were initially the standard Druidic shrines and parts of the temple complex both above and below ground. Then individual chapels were built on and around the Henge Temple site. There is one problem with the picture of a well run land without at least gross inter group conflict. As John Bligh puts it in his book called 'The Fatal Inheritance', a must for a well-read person's library, there has to be a mechanism for dealing with population growth. On the continent inter group conflict that is war with the help of disease and natural disasters played the main role. On a large landmass a social group could not limit their population voluntarily as it would make them vulnerable. On an island with nowhere to go the population had to be limited or it would bounce up and down with conflict being the driver. It is therefore conceivable that the Druids had a method of dealing with the fatal inheritance. In fact if they did not they would not have been able to maintain social stability and organisation for 3000 years and more within the confines of an island. At the moment the arenas found on every cursus appear to possibly be the Druids method of dealing with the fatal inheritance of population pressure. Moving from fact to conjecture. It is suggested that Druidic society had evolved to a point where it had many elements of a society based on socialism. It was a centrally planned system at the local level following a common pattern over the whole of the British Isles. The administrative class, the Druids, ensured the use of good agricultural practice so that food and raw materials such as wood were never in short supply. There was clean water delivered to the house, washroom and toilets and local entertainment on the cursus. In fact, to a visitor from the Middle East or the Roman Empire with its high cereal low protein diet, poverty, slavery, castles and walled cities Britain must have appeared to be a Shangri-La. Shangri-La, a heaven on earth. The society was made possible by a strict religious code, respect for people and a strong belief in a God of life who's land they were guardians of and the continuance of life after death. Coupled to this was the education of an administrative and engineering elite. It is into this world that the young Jesus was possibly brought by his uncle Joseph of Arimathea. The twenty missing years of Jesus's life could have been spent in Britain, in effect learning from the Druids and perhaps becoming a Druid. Importantly he would have learnt about the socialistic or communist approach to social organisation and the concepts upon which it is based. It may be worth noting that in an economy based on bartering and not money it is going to be more difficult for individuals to accumulate wealth. Many of the bartered materials are perishable and have to be passed on. By contrast, with an economy based on money where it is possible to accumulate wealth. This accumulation by individuals can go to the point where enough coins can be removed from the economy to produce deflation and reduced economic activity. An economic system based on bartering might therefore be more stable in the long run than one based on money. The influence of money on an economic system is summed up in Oliver Goldsmith's poem 'The Deserted Village'

'Ill fares the land, to hastening ills
a prey
where wealth accumulates and
men decay'.

When Jesus returned to Palestine his Druidic culture and ideas would have been in direct conflict with those of a society based on capitalism and on an ancient religion that embraced capitalism. He would have been considered to be very dangerous. His ministry only lasted three years so it did not take the local community long to deal with him. It should be noted that some early Christians tried living in communes possibly following the Druidic system. The above is conjecture but what is not conjecture is that there is a great deal of mythology relating to members of Jesus's family visiting Britain. Also the Druidic crucifix is suddenly modified to incorporate details of Christ's crucifixion at about the time of the crucifixion. The body is now a living one which bleeds. The dates of the Christian crucifixes painted on the ground can be defined within a few years, they can be drawn on the ground by any reasonably competent dowser and they are the first known historical record of the crucifixion made at the time of the crucifixion. They predate the Gospels and I suspect all texts relating to the crucifixion. The speed of the Druidic response to the crucifixion means that they must have been in contact with Jesus. He was one of their own and he eventually provided them with the ultimate proof of a life after death. Proto-Christianity became Christianity overnight and when the Romans arrived they possibly found two societies. A secular society, perhaps keen on entering the capitalistic fold of the empire, and a separate religious society based on social ideals incompatible with those upon which the empire was based. The religious people, the Druids, would have pulled some of the secular society into their field of influence and perhaps this lead to what little resistance there was to the Romans. It would therefore make sense for the Romans to set up execution lines and publicly execute the Druids and those associated with them. The Roman fear of the Druids may have been due to the Druids philosophy as much as anything else. The large Druidic buildings, now devoted to Christianity, had to be destroyed including Stonehenge. Stonehenge, because of its size and solid structure could be called a Cathedral. Two thousand years after its destruction it remains as the ruins of possibly the first and certainly the oldest Christian Cathedral. It and the land around it holds the secrets of the development of religion over thousands of years, of man's endeavours to answer fundamental questions, from the Neolithic to the dawn of Christianity. At the moment the only way of reading the story locked up in the ground is to use the methods the Druids would have used. Using their methods it looks to me as if the birthplace of Christian concepts was the British Isles. These concepts are still with us today two thousand years after the Henge Civilization and the Druids were destroyed. The Druids not only built lasting monuments on the land but lasting monuments in the minds of men. They may have needed a little help from a Palestinian but some of the ideas which they developed and upon which they may well have founded their own society almost certainly underpin ours today.

The Druidic and Secular Societies

Using the size of the Dolmen associated with the Yorkshire Temple the secular society may have comprised about 80% of the whole. If this is so there appears to have been a sufficient schism between the Druids and the secular society for the main group in Britain and the one with most soldiers and modern weapons, to, in the main, stand aside and watch the Romans kill every Druid. The Druids do not appear to have removed their linen robes and melted into the general population. The children, as far as we can tell at the moment, were not sent away to safety or hidden in the general population. The bodies are there, at least in the mass graves we have looked at. If these conjectures are near the truth why did the bulk of the secular society not see the Druids as one of their own? Did the Druids keep themselves to themselves, as many religious groups do, afraid of the wind of reason? The consequence of this could have been that they were not seen as part of the whole society. Or perhaps the piety and conservatism of the Druids and their power over the more religious members of society posed a threat to the secular authorities. The ease with which the Romans landed on the South Coast of Britain and moved around the country seems to have been quite different to Caesars reception about a hundred years earlier. Caesar had to retreat because of the difficulty of maintaining his forces. A few victories, negotiated peace settlements and Caesar departed. On the second occasion nearly a hundred years later the speed of movement and the Romans use of small forts and outposts indicates a mainly passive if not friendly society with only minor resistance during the first few years. What fighting there was appears to have been limited to small groups of Druids round their temples and some chiefs who were not too sympathetic to the Romans. This again indicates a mainly passive society, apart from Coracatus and the tribes in modern Wales and Scotland. This passive approach of the natives held at least until the Romans made a mistake in handling a secular leader, Boudicca, Queen of the Iceni. But until 60AD the Romans appear to have been able to roam the country murdering all the Druids they found without the secular part of society appearing to be too worried. They were in fact involved in eliminating the Druids when Boudicca decided to revolt. One possible reason for this is that the Romans had been invited in to remove the Druids and free society from what might have been perceived as a religious tyranny. After two to three thousand years of conservative religious dominance by the Druids, society may have thought it was time to catch up with their neighbours in Europe. The Druids, in order to run their temples and sacred landscape schemes, required control of resources that could be used in other ways. The Kings and Princes were not strong enough to deal with the Druids and may not have been able to rely on the support of their own people in such a confrontation. Inviting the Romans in to remove the Druids was perhaps a risk worth taking. As the Druids were the arch enemy of the Romans and the possessors of great wealth, it was perhaps too good an invitation for the Romans to turn down.

The Romans

The Romans are part of written history and so are well known down to the buckles on their uniforms. However, little is known of why they hated the Druids to the point of paranoia. The conventional view is that the Druids were always behind revolts and were nationalistic. Perhaps it was because they controlled a lot of wealth and would not pay taxes. Whether they were behind revolts or not is going to be difficult to determine. The presence of priests in an army does not mean they are behind the fighting and the reason for it, no matter what the Romans thought at the time. For example the presence of priests with the British Army does not mean they are responsible for the fighting. The reasons for sending tens of thousands of troops to Britain to systematically clear the place of every Druid adult and child may lie elsewhere. For example it may lie in the stark contrast between Druidic society and Roman society. Druidism was almost certainly a monastic and socialistic system. Roman society a capitalist system based on slavery. Wealth in a Druidic system was used for the benefit of their God and the people. It did not go to the 'Captains' of industry and commerce or the State as taxes. Women appear to have held high positions in Druidic society and may have been considered equal to men. The Boudicca incident involved the Romans demonstrating to Britain's in a graphic way that they had quite a different view on the role of women in society. The primitiveness of Druidic society in Roman eyes may have been another factor. The Britons were in many respects still in the Stone Age. The writers of the day believed that they indulged in human sacrifice, animal sacrifice and that their God(s) were quite different to the Roman Gods. A Roman could probably have drawn up a long list of reasons why Britons and in particular Druids were primitive savages. The fact that Druids were instrumental in running a mainly peaceful and prosperous society would not have carried much weight. There may have been, however, one aspect of Druidism that the Romans were fearful of. The Romans were no strangers to what today we call dowsing or biolocation. If a dowser is working on a Roman Villa and wants to know where the well is they locate the nearest underground stream and then follow it along until the well is found. The Romans used water diviners to find water for their villas. It is also believed that they used energy engineering. The dowsable 'blue line' up either side of an amphitheatre, the seven or so blue rays from Roman shrines. Although dowsing mythology believes that these blue lines have been created deliberately we now have evidence that they, or at least some of them, may be a product of the materials and design. If the Romans deliberately created blue lines they would have been well aware of the skills of the Druids and their powers in the spiritual world. The Romans general attitude to religion was that the more gods there were the better provided people could pick and choose and move around. That is people could move between gods according to their needs. A single powerful religion with power in the secular sector of society could be a threat to the Empire, particularly if the priests could enter the 'other world' and converse with the spirits in it. Not only that. The Druids had such a belief in the reality of the 'other world' they would sacrifice themselves and be fearless in battle, even if they relied mainly on Stone Age weapons, easily broken on Roman armour and shields. By the time of Gaius (Caligula), Claudius and Nero 37AD to 68AD the socially relevant ideas of the Druids as well as their ideas relating to a single or restricted number of deities may have started to worry a secular empire based on capitalism and slavery. Because beliefs are difficult to destroy there was only one answer, the total destruction of the Druids, which the Romans achieved. By contrast the engineering works of the Druids were not so easily destroyed and many remain to this day to bear witness to their skills. Also the stains of the Druid's blood in its neat ranks still bear testimony to what happened nearly 2000 years ago. The scale of the atrocity and the possible ignominy of being associated with it meant that the events do not appear to have been recorded by Roman writers and there is scant record of what might have been the world's first great Witch hunt and genocidal act.

The Roots of Native Britons

This brief study of what the Druids left behind is fairly clear on one thing. The pre-Roman ancestors of the people inhabiting the British Isles were not Celts. They were a people who could trace their ancestry back at least 2000 years before the Romans arrived and possibly 3000 years and more. The question therefore arises as to where did the ancestors of the 'Celtic' fringe come from. The repopulation of Britain occurred as the ice retreated north, a retreat that was happening over Europe with the mountains and high ground hanging onto glaciers for much longer than the low ground. This would indicate that repopulation was from the south of Europe, modern day France and Spain and happened thousands of years before the Celts reached that part of Europe. Whether this is the case remains to be seen. Experts in culture will have to examine what the Druids left behind them and compare it with early cultures in France and Spain and perhaps even North America where the Thunderbird is part of an ancient culture. Possibly even Asia to compare Shamanistic beliefs, which still exist to this day, with Druidic practice. At the moment archaeological and genetic studies, see 'The origins of the British by Stephen Oppenheimer, seem to favour the Western Mediterranean or North Africa as the area from which the Henge people came. Once in Britain the Henge people were protected from the tribal conflicts in Europe, at least for a few thousand years. This enabled the Henge people to develop a civilization that embraced the Earth Goddess in a big way and made use of the human magnetic sense. The magnetic sense was used to identify fields associated with underground water, stones and geological formations. The Earth Goddess could speak to them and tell them where water and minerals were. Using stones the Henge people could engineer the fields to achieve objectives and they had skills and knowledge which modern dowsing or biolocation cannot yet match. The Henge people were possibly unique in developing a civilization that depended so much on the human magnetic sense and the magnetic fields of their environment and the objects around them.

Chapter 28

A Never Ending Journey

The Knowledge of the Ancients

The journey and adventures reviewed in the preceding pages started with a desire to discover some of the knowledge of the ancients. The questions to start with were quite simple 'How did stone circles work?' 'How were they built?' 'What were they used for?' To answer these questions the phenomena of dowsing had to be put on a scientific footing. This was done with the development of the Diamagnetic and Paramagnetic Theory of Biolocation. (Appendix 3) Its discovery is a story in it own right. The theory led to the development of dowsing techniques along scientific lines. To distinguish the form of dowsing upon which this book is based from conventional dowsing I have used the term 'science based dowsing' or biolocation. It must be remembered that the story as it unfolds is still based on a dowsing response to a magnetic field with all its weaknesses and opportunity for error. The use of witnesses must always be treated with caution. When silk is found in a tomb all that is found is something that dowses as if it is silk and not necessarily silk. Fortunately the biolocation is confirmed in many areas by visible artefacts such as clay pools, stone tools, the bricks and tiles used by Druids in construction, the access shafts to the underground tomb complexes, the shape and design of the temples and dykes. The large imprints of Druidic art, in some cases 20 to 30m across, left on the ground 2000 years and more ago are a 'written' record of the Henge Age. These pictures can still be drawn out today with precision by a dowser just looking at the ground in front of them with no reference points or lines to guide them. The tunnel entrances, subsidence into the Druids underground workings, the massive earthworks through which the chariots ran provide a physical backdrop to the biolocation results.

Perhaps just as important as the physical evidence is the way the biolocation results fit into an interlocking picture. There is a convergence of dowsing evidence. Time and again the question 'if this observation is correct what follows?" has been asked and it has unlocked another step forward. The sheer amount of dowsing that has gone into the study is enormous. Nigel, a dowsing high drive, out twice a day seven days a week dowsing with his dogs. Picking up details, challenging ideas, trying to find what should be there. Trying witness after witness to see if they reveal something new. The laboratory work to develop and test ideas, to close the circle on observations such as tree auras, solar interference, to analyse stone and ceramic artefacts to find out what they were used for. Biolocation has revealed a whole new world. Practically everything we look at reveals new information and takes the story of the Henge civilization forward by another step so that it has not been possible to draw the story to a close. In fact it is not going to be possible to bring the story of pre-Roman Britain to a close. There was a civilization over the whole of the British Isles, protected from Europe by the sea for thousands of years. The civilization, like animals and plants on an isolated island developed in its own way and like those civilizations in the Americas became very vulnerable to the more technically advanced societies developing in Europe. Many things are still under investigation. However, it is clear that the Stones did not hold one secret alone; they are the portal to a whole civilization that we have traced back for 3000 years possibly more. The civilization is likely to have come from Southern Europe as the ice sheet moved north. This is in agreement with Stephen Oppenheimer's genetical studies which show that the ancestors of today's British and Irish people arrived from Spain as the ice sheet retreated north. The people will have maintained trade links with the south as their frontier moved north, at least until rising sea levels or deteriorating climate and weather presented problems. The Henge civilization was a civilization in which the human magnetic sense played a large and possibly more crucial role than in any other civilization. It may therefore be a civilization quite unlike any other and possibly unique. The magnetic sense is often referred to as the God sense. People have the feeling that something is there but they cannot see it, hear it, smell it, touch it. This subliminal sense is almost certainly the basis of most God orientated religions and consequently, through religion, it is the basis of at least one of the foundations of many civilizations. Because of that the magnetic sense, unbeknown to us, may still be influencing our lives and behaviour and the way we see the world around us.

The silent Stones, the only visible remains of large temples, caused all manner of people to struggle to give meaning to them. The books and records of their efforts will now be a monument to the limitations of the human intellect in the 20th century. The multitude of books that are witness to this struggle will pose their own questions. How come so many people misinterpreted the Stones? Why did so few people, for example, realise that they were but part of a much bigger enclosing structure based on wood and thatch. How did so many people come to weave fantasies about the Stones? Why relate them to astronomy instead of the ergonomics of daylight and prevailing winds? Because of the written word and records these people have created, the Stones will now show future generations of scholars some aspects of how the human mind works. There is an irresistible appeal of magic, phantoms and of worlds where the laws of thermodynamics do not apply, whether it be in 2000BC or 2000AD. The professional mind also likes short cuts and simplifications and it is easy to overlook the fact that technical and cultural evolution or development are two quite different time lines. The appearance of bronze or iron does not mean that the culture changes, they 'eventually' interact but 'eventually' may be a very long time. For this reason it is not a good idea to define cultural development by reference to technical development. A technical development can be used as a time marker but that is all it is. It is not an indicator of a change in culture or of the way people look at the world around them.

Conjecture

For a scientist coming into archaeology from the outside perhaps one of the main problems to be found with archaeology as a field of study is its recourse to speculation and conjecture. Yesterdays conjecture can often come to be today's fact. A stone tool or pottery shard is related to an age and then to a people, a society and culture. It is used to date artefacts that cannot be dated. The errors accumulate and eventually stones which are clearly structural members of a building and constructed towards the end of the Henge Age are attributed to a date 2000 years earlier and deprived of their enclosing building. As a result the Stones must be given some purpose so what could be better than astronomy. There is a failure to appreciate that a piece of real estate can be built on for thousands of years. As a result there will be pieces of carbon and shards of something from every age but they will date little apart from themselves.. The ancient world that then builds up moves further and further away from the reality and people hesitate to question it, as Albert Einstein said 'blind respect for authority is the greatest enemy of truth'.

It is easy to see why archaeologists develop stories. If after days of digging there are only a few flint tools, some shards of pottery and a few postholes to show for it, it is very tempting to develop stories that the artefacts just cannot support. Conjecture has to be limited, based on facts and used as a tool to guide the next step. Above all ancient peoples must be credited with some common sense and intelligence. It must not be assumed that just because an ancient people were locked into a phantom world created by their attempts to make sense of the real world about them that they lacked common sense and intelligence. Their phantom worlds of spirits and Gods and their sacrifices to them are no more bizarre than the phantom worlds of the 21st Century and the 'sacrifices' people today make to their 'Gods'. It must also be accepted

Box 28.1 The Interpretation of the Beltane Fire Pits

Traditionally books such as Miranda Green's (page 35) shows the mythological impression of Beltane. A dramatic painting specially commissioned for the book is based on the knowledge that was available to the artist, the bit that had survived 2000 years. Figure 28.1.1a is a typical illustration of this mythology.

Figure 28.1.1a Beltane - Mythology
The mythology of the Beltane Fires has passed down the generations for 2000 years. The agricultural or animal husbandry purpose of the event became lost but the event entered the 'race' memory, possibly because it was also a social event.

New information obtained by biolocation reveals that the two fires used are only half correct, they should in fact be fire pits with embers for heating branding tools. Animals passing through the system included sheep, cattle and goats as well as deer. From what might be called the original box of knowledge relating to the event, little survived and what did survive has had to be padded out with conjecture. What is now known about the 'Beltane Fires' is illustrated in Figure 28.1.1b.

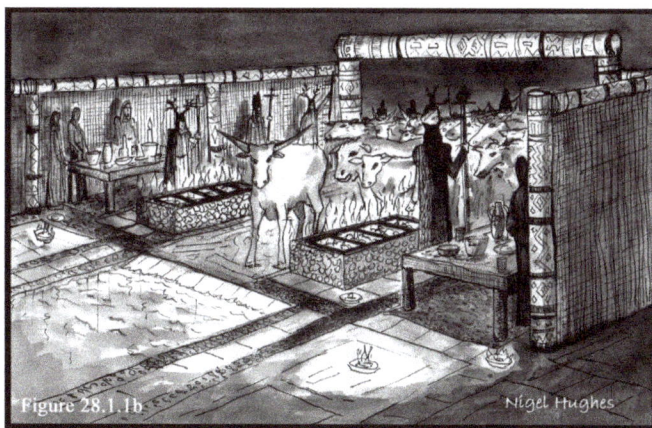

Figure 28.1.1.b Biolocation - Beltane
With the latest information the event can be reconstructed to show what was likely to be happening at Beltane. After the work of blessing, recording, branding and dipping the animals had been done they had a big barbecue using the fire pits. The traces of the BBQ are easily identified round the fire pits.

The Beltane Fires were two fire pits with branding stations on each side. The animals came down a fenced way into a pen where the Druids sprinkled holy water on to them. The Druids arrived at their station by walking down a covered way to the branding station. The animals were passed to the branders and then into a cattle dip then out along a pathway. This picture clearly shows an ordered and planned process with a clear animal husbandry objective. The presence of the Druid priests and the construction of a covered way for their procession to the site shows how closely secular activities and religion were linked. The fire pits were used after the branding for a barbecue. The iron spits go across the fire pit and each one is dedicated to one animal – lamb, deer, ox, pig. Each animal had its place and it was held for hundreds of years so cross contamination between spits is rare.

that the people of the Henge Age looked at the world and interpreted it in quite a different way to people brought up in the 20th century. Very early on in the work on the Stones it became apparent that conventional archaeological interpretations very often just did not fit the facts as we were finding them. The mystical interpretation placed on the stones by intuitive dowsers and mystics also did not fit. Because of this the conjecture in this book is based on what we have found not on what conventional archaeology says. The lack of reference to archaeological sources is deliberate. In some areas there is agreement between our findings and current archaeological ideas. For example, the site of Stonehenge does appear to be an ancient sacred site possibly going back to the Neolithic but the present stone circle is Iron Age. This does not mean that stones have not been carried forward from an earlier age. There was a stone circle there in the Bronze Age. The smaller stones may have been brought on site during the Bronze Age or earlier but if they were, the tracks of the vehicles that brought them to the site will be there and can be looked for.

Perhaps one of the big mistakes of archaeologists is to look at earthworks as defence systems when their design and construction clearly does not fit a defence role. It may be that the size of many dykes is such that having them roofed over seemed preposterous. It does, but the imprint of the posts, rafters and lamps are there in the ground. The imprint of the ceremonies is also there in the ground.

Returning to the world of mythology. The new information on the Henge Age and its people provides a base for assessing some of the mythology we have inherited from the past. The basis of the Cait-Sith (Fairy Cat) legend comes down from the Henge people, the May Pole, Beltane Fires, human sacrifice, women warriors, angels (white linen cloaks and feathers) halo (Herne's torc), the Eucharist, Christening, all from the Henge Age and many thousands of years old. Festivals such as Harvest Festival and Christmas with it holly and ivy can be traced back to the Henge people.

It also looks as if the new archaeological methods will show that Christianity can be traced back to the Henge people. At the global level the new methods will start to unravel the history of mankind. Caves, hopefully, will not be dug out and the time capsule destroyed. They will have their magnetic record read page by page with no page destroyed. Camp sites, occupied spaces will hopefully be 'read' not dug.

When talking to people about our discoveries I am often asked "What are the physicists and chemists doing? "Why do they not measure these things?" The answer is that they try but they just do not have the right equipment. I also find that they are often locked into the idea that a magnetic field is a magnetic field. That is they believe a ferromagnetic field is the only type of magnetic field there is. The ones I have talked to find it very difficult to take on board that there are many different forms of magnetic field. A bit like a colour blind person having difficulty appreciating what colour is. After all light is light. Telling such people that one of Faradays big intellectual leaps was to realise that magnetic fields were an entity in their own right and independent of the source seldom seems to help. They insist that paramagnetic and diamagnetism are the names of materials not fields. If this were so, dowsing and biolocation would not be possible.

Returning to the Druids. They appear from written records and their physical achievements to have been intellectually powerful. They almost certainly had a philosophy of life as is indicated by the fact that there was no race to build the biggest temple or biggest hill, no attempt to outdo others by turning wood and straw into stone structures. Towards the end they appear to have been socially and in terms of their religion a people from the future. When time lines cross, technical or social, there is conflict. This happened when the Henge people

370

and the Romans came into contact. A technically advanced people came into contact with a possibly socially advanced people. The result was that a whole civilization was destroyed and consigned to oblivion for 2000 years. The civilization was demoted to prehistory. It is now clear that the Henge Civilization is very much a part of our history and the pages of the history book can now be read.

A Three Element Approach

As the story unfolded over the years it became apparent that the approach we were using had three important elements. The first was that we were using the fields, the magnetic fields that played such a vital part in the development of the Henge culture and civilization. The fields were part of the world they were trying to make sense of. We were using their methods and finding their world. The second was the development of biolocation to a stage where we could identify the chemical stains in the soil and use it for forensic archaeology. I use the term forensic to indicate obtaining information which can be used for rational argument and discussion. Also information that can be used in building up the weight of evidence for or against an argument or idea. Archaeologists are no strangers to the chemical analysis of soil and artefacts but they have lagged behind the soil and environmental scientists in not appreciating just what modern chemistry is capable of and its application in field surveys. The spade and JCB appears to be the preferred tool and skeletons, flint tools and shards of pottery the preferred output. Keyhole digs and chemical analysis just does not have the same appeal. Perhaps a third element in our approach is the use of laboratory studies of magnetic fields and the modelling of artefacts to close the circle on field observations.

It must also be remembered that as Dr Betz showed with water divining some years ago, the human magnetometer out perform the physical instruments that measure magnetic fields. This will not always be so but at the moment the only thing preventing the human magnetic sense from realizing its full potential is a lack of understanding of the magnetic environment within which it operates such as the moving fields from the sun and the earth. As the understanding of the magnetic environment and the magnetic sensory system develops the human magnetic sense will become more and more accurate and reliable and I suspect that it is going to be a long time before the physicists and chemists catch up.

When I sit in the back garden of my house I can 'see' the Druidic house that once stood there, its work stations, toilets, washrooms. The fireplace, spit, oven. The beds, dog kennel, brewing pot and many more details such as the food they ate. Just in front of the Druids house and just before the planters and steps to the present conservatory and house stand the man, his wife and two daughters. They are there everyday telling more and more of the story of their lives The pools of blood left from their execution by Roman soldiers remain and I can 'see' the drag lines to the mass grave nearby. The human magnetic sense is indeed a powerful and awesome sense.

End

Epilogue

The Secret of the Stones

So what is the secret of the stones? Is it the magnetic alignment, the blue lines, the energy engineering, do they just tell us where a temple once stood or perhaps they are the memorial to a people? I have often wondered what the true secret of the stones is. Perhaps we have discovered it but I do not think so and I believe that it is still waiting to be discovered. However, I would like to think that we have discovered if not the secret, possibly one facet of it. The secret of the stones could be that they are a portal, the key to which has been lost for two thousand years. Discover the key and the door opens onto the Henge World and the Druids. Their world is still there, invisible, hidden, only entered by those who have the key and the wisdom to know how to tread the land they find.

Rocking Stone Rainbow, Pontypridd

In the following poem, if I may call it such, it is a dowser or biolocator, who is extending the invitation to visit the land of the Druids.

Walk with me

When the gates of the evening gently close,
Do not hide.
Pass through, and welcome the night.
Walk with me
and share the secret memories of the land.
Walk with me,
Through the shadows of an ancient world.

The memories of men are fleeting, fade,
and soon lost.
The memories of the land endure
and await our bidding.
They wait for us so they may live again.
So walk with me,
if you will, through the memories of the past.

Walk with me when memories
no longer hide from the day.
When the nights grey light will guide our way
through memories
of fear, hatred, reverence and love that once
walked this land.
Memories, now held safe, hiding from the day.

Walk with me through the memories of the past.
See the houses.
See a long gone people at work and play.
Now just shadows.
The chariots, ceremonies, wheeling black birds.
Memories.
Waking as we walk the night.

See the great temple of wood, stone and water.
Come enter.
See the seven altars, the pool of life,
the path from life to memory.
See the moving lines of those,
who for their moment of time, believed.
They are now just a memory from the past.

The great gates of the temple are guarded still.
Eight by four the spirits lie,
held fast where the bodies died.
They wait to rise
and walk again the sacred halls
when beckoned
by those who do not fear to walk the night.

See the great gates open to the quarters.
The gold chariots.
The lines of robed priests, now silent.
The four black clouds.
The cleansed souls, dressed in red
to meet their Lords
in halls carved from living rock
To last the life time of their Gods.

Beyond the shadowy houses, against the sky,
A dark grove.
Sacred oaks, a gift from Nemetona to the priests.
A place to dance,
To worship the Goddess of the grove
by a silvery light.
To hold time still for her duties to complete.

See the dells, as the sky lightens across the land.
Still sacred shrines.
For millennia honoured by men, ox, horse
And now machine.
Witness to man's folly, arrogance and hopes
The search for that eternal truth.

See the lands sculptured for a Goddess
with love and fear.
The roads for racing gold chariots
to caress her body and pay reverence.
Labour in due season to earn her bounty.
Memories now
Not to be forgotten
When we walk through the gates of the morning
as they so slowly open

GWC 12/05

Druidic Road, Devon
Nigel Hughes

Appendix 1

Biolocation (Science Based Dowsing)

Before discussing what science based dowsing (SBD) is a brief review of present day traditional dowsing will help to provide a starting point. Most people when they are introduced to dowsing start by first learning to obtain a response from a pendulum or 'L' rods as they cross a known dowsable object, the target. The target may be water, or electrical services, or the buried foundations of a building that is no longer there. The idea is for the trainee to develop a dowsing response such as the movement of the 'L' rods to known dowsable objects. Once responses are easily obtained the next step is to find unknown objects and then try to identify what they are. The technique the trainee is taught for identifying the target or some feature of the target such as its age is to ask questions. Is it a water main, a sewage system, a buried Roman road, the foundations of a Saxon church or whatever. This is where the skill of asking questions starts to come in and trainee dowsers spend a great deal of time on this aspect of dowsing. Learning to ask the right question is considered to be very important. Examples of the questioning technique can be found in some of the papers published in the Journal of the British Society of Dowsers 'Dowsing Today'. Hamish Miller's book on dowsing also explains the technique. The questions take the form of 'where did the inhabitants go when they wanted water'? The dowser then does a 360° sweep search and the rods cross when they are pointing in the direction of the source of water used by the inhabitants.

Conventional dowsing is based on asking questions about what the dowser wants to know. This form of dowsing is often referred to as 'Intuitive Dowsing'. This is partly because the dowser often finds it difficult to explain the stimulus they are receiving from the target. Many dowsers therefore resort to saying that they have an intuitive feeling that they are identifying a target.

The question and answer technique does not however allow a link to be made to the scientific and rational methods of archaeology or any other science. Also many people find the method difficult to come to terms with due to the lack of a rational basis for the method.

The use of questions is one of the aspects of dowsing which many people have difficulty in accepting. However, there is a basis for it in that the magnetic sense in man is used in a similar way to the other senses. We can focus on a particular magnetic target just as we do with sight, hearing and smell. If I am looking for say a drain, I will walk over many things without 'seeing' them because I am focusing on drains. How effective focusing can be when dowsing is not known or even if it truly does exist. It may be that the dowser knows the direction of what they are looking for and so picks it up in preference to other targets which cross the ground at different angles. In practice, whatever the explanation, the ability to focus on a target is a very useful ability. The asking of questions helps with the focusing process. If we are looking at a lake we may know that there are some water birds on the water or round the lake edge. The chances are that we may have to focus on finding kingfishers or other less obvious birds by asking ourselves if there are kingfishers round the lake. In practice the questioning is a flash of intent, we do not say out loud 'how many kingfishers are there round this lake'. It is the same for the more experienced dowsers, no more than a flash of intent which leads to focusing on the target. The difference between the conventional dowser and the bird watcher in this scenario is that the bird watcher will see and count the Kingfishers, the intuitive dowser expects to be given the answer from some ethereal source.

Science based dowsing (SBD) has been developed to deal with at least some of the problems of traditional intuitive dowsing and has been discussed in detail in a paper by Crockford and

Hughes, see Dowsing Today Vol. 48, No. 292, pages 10-15. It is also covered in a book on the human magnetic sense by Crockford and Hughes. The clear and objective approach of biolocation or SBD can be seen using the road as an example. The dowser first develops the 'form' or shape of the object, normally by placing canes or pegs in the ground along the edge of the object being dowsed. In the case of the road the outside edges of the suspected road would be marked out. Inner edges would then be looked for and marked as they were found. The length of edge being assessed would have to be sufficient to rule out the foundations of a building or other artefacts. When dealing with supposed roads or paths in the vicinity of buildings, side roads or paths can often be found and marked. The dowser, by using 'form' is building up evidence which can be used to determine if a road or path might be present. The dowser's or bio locator's next task is to gather more evidence and challenge the evidence already gained. For example many Roman roads have a particular structure evident in a cross section of the road. First a ditch, then a metalled surface, then a central ditch, another metalled surface and finally another ditch. Is this structure present? The actual dimensions of ditches and road surfaces are obviously important evidence. Witnesses can be used to show that a particular type of stone, cement or mortar was used in construction and the ditches may come up on a carbon or pond sludge witness indicating organic deposits, see figure B1.3.1. The metalled surface in my experience does not normally respond to an iron oxide witness which is an indicator of iron rimmed wheels being used on the road. The above pattern can still be obtained even if the road has long gone and all the road making materials are now scattered over the field as a result of ploughing. The original road after a few hundred years of use will have left a stain in the ground. The depth of the stain in the soil can normally be determined by taking a core sample.

With the scientific approach the dowser is building up the probability for a particular interpretation of the results. There are no rods crossing to say it is 'x' i.e. a yes or no, just a probability based on evidence. Importantly, the dowsing can be checked by soil analysis, very small scale digs and by other dowsers following a set standardised procedure and methodology.

With biolocation (SBD) because of the dowser's high sensitivity to traces of material in the ground, it has to be accepted that 'chemical stains' in the ground, possibly at some depth, will be identified. The dowser is identifying what was there at sometime in the past. Inorganic materials may have an indefinite life in the soil. Organic materials also have a very long life in the soil and as contamination on artefacts. The duration is measured in thousands of years so it is quite feasible to study 4000 year old fire places and identify some of the things cooked and eaten. At the moment it is not possible to distinguish between stains and real objects by dowsing. Other methods have to be used.

Dowsing can be used in a disciplined way. The relationship between biolocation (SBD) and the sciences is illustrated in Figure A1.1. The dowser is responding to magnetic fields or the gradients in them. All objects have a magnetic field so in theory the dowser can identify the presence of the magnetic anomalies created by concentration gradients of any chemical or substance in the ground. It does not appear to matter if these magnetic gradients are in the ground, in people's bodies or produced by solar activity. The dowser will respond to them. In Table A1.1 the relationship between particular aspects of dowsing and archaeology are indicated. The basic theory is that provided the biolocation is done correctly the archaeologist can check it. The bio locator is able to say what is there and where it is. The spatial resolving power of dowsing is high and the position of the magnetic field can be measured to ± 2 to 5 centimetres. The definition of what is there is sufficiently high for the archaeologist to take samples and do mini digs. Perhaps a simple example is the analysis of a target using witnesses. If the dowser reports high phosphate levels or the presence of clay or limestone, these can

Geophysics

Solar physics

Physics

Geology

Archaeology

Geography

History

Chemistry ———————— **Biolocation** ————

Biology

Isomagnetics

Building Services

Medicine

Health

Industry

Agriculture

Food

Mathematics

Statistics   Epidemiology

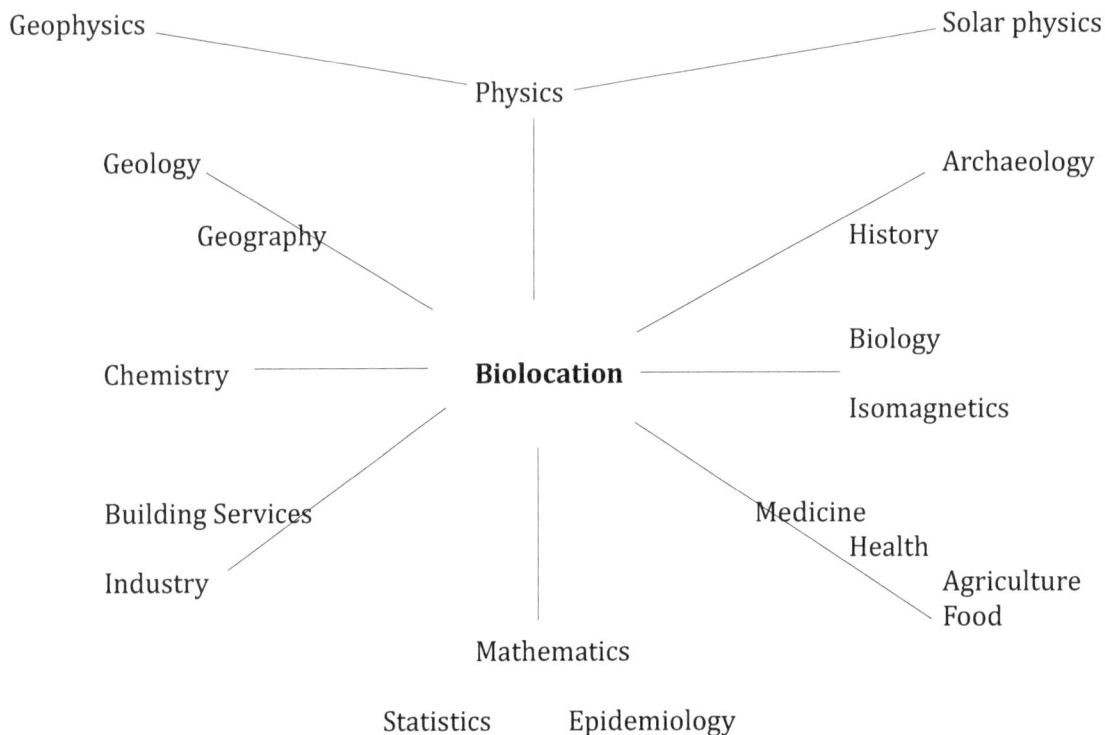

Figure A1.1 Biolocation Links to the Sciences

Because biolocation is based on a scientific approach it has a direct link to the sciences. The diagram indicates these links. The significance of these links is that a science based on a conditioned reflex initiated by the human magnetic sensory system can be assessed and evaluated using the conventional sciences.

be checked for by taking a core or auger sample and analysing it for an elevated level of the material. For example round a Dolmen the soil can be analysed for phosphate, red ochre (red iron oxide) and the organic carbon of the footpaths.

To distinguish science based dowsing from intuitive dowsing and to try and prevent confusion the term 'biolocation' is used. Biolocation is characterised by its use of procedures such as form, colour and witnesses and a rational science based form of enquiry and investigation.

Science

In this appendix and in the preceding pages the term science is used as a defining term, as if everybody is agreed on its meaning. Unfortunately this is not the case and the term has been used to justify many areas of intellectual enquiry and views of the world which just do not measure up to the real demands of scientific enquiry.

Using a dictionary definition as a starting point, science is defined as the acquisition of knowledge which is ascertained by observation and experiment then critically tested and challenged. The knowledge is then systematized and brought under general principles in relation to the physical world and its laws.

In planning observations and designing experiments a scientist would develop a model of the system being studied, a paradigm. The paradigm would be based on known facts and the

Table A1.1

The link between biolocation and archaeology. Dowsing is broken down into a number of techniques which are related to an aspect of archaeological science.

Biolocation Techniques	Link to	Archaeological Methods
• Determining the depth of a target		Archaeological dig, ground penetrating radar
• Determining the 'colour' of a signal (its chemical composition)		Core analysis
• Determining the series of a target		---
• Determining if a target is an image or a real target		The archaeologist should only deal with real targets not images
• Determining magnetic field direction and the magnetic dip of targets		Geophysics and magnetic fields
• The form and dimensions of targets e.g. foundation of a building		Available archaeological data
• Analysis of fields using witnesses for chemical composition		Chemical analysis of cores
• The magnetic axes of target or artefacts		Geophysics
• The interpretation of biolocation data		Archaeological data and current knowledge of the site

laws of nature. The experiments developed must be repeated and be repeatable. The Experimenter must try and identify all the variables that might affect the results and produce errors. From the knowledge gained and observations made it should be possible to make predictions about the results of new experiments and new observations. Now in some areas of scientific endeavour it may not be possible to do experiments. For example astronomy is an observational not an experimental science. Astronomy still makes models upon which predictions are based. These predictions can be checked by observation. In the present context intuitive dowsing is an observational activity whilst biolocation includes a theoretical basis, modelling both mind and physical models and experimentation. All experiments and their results must be repeatable by other equally skilled scientists. This normally requires that all the variables are known and are controllable.

Appendix 2

The Concept of Earth Energies (EE) and their Possible Role in the Culture and Religion of the Henge Age

According to dowsing tradition the earth has bands or lines of 'energy' (EE) crossing its surface. These energy lines are considered to be signs of a spirituality or of a deity connected with the 'living' earth. They are the manifestation of a possible Earth Goddess or some form of spirituality associated with the land and countryside, Gaia.

The British Society of Dowsers have published an Encyclopaedia of Terms in which the term 'Earth Energy' is distinguished from 'Earth Energies'. Earth Energy relates to 'energy' emanating from the earth. That is somewhere in the ground there is a source for an 'Earth Energy'. The term 'Earth Energies' differ in that it encompasses all 'energies' that can be dowsed on the surface of the earth. These include all sources such as the earth, atmosphere and sun.

Some dowsers consider that there is a cosmic connection between EE and the cosmos so EE may be cosmic as well as terrestrial in nature. One form of the theory is that early man or Stone Age man had a heightened awareness of EE and so was much closer to the deity or spirits residing in the landscape and planet. Dowsable 'energy' is also associated with trees and living things so there is a direct connection between visibly living things and the earth. As both share the same type of energy then both must share life. Trees for example have a direct connection to the earth and possibly it's life force through a root system. The ancient people developed methods of reverence, worship and communication with the 'earth spirits' and in some cases with the living world around them for example, the earth-centred spirituality of shamanism.

The ancient peoples are believed to have selected areas for their ceremonies where the contact with the EE was particularly good. This is believed by many dowsers to be over underground water streams, underground steams being the blood vessels of the earth along which energy flows. The EE is said to be particularly strong where two underground streams cross each other and where water rises or falls vertically in what is called a blind spring, a spring that does not reach the surface. Dowsers believe that ancient man sought out these areas of special 'energy' as their ceremonial structures appear to be associated with them. Dowsers therefore use what are believed to be Neolithic stone monuments as indicators of where EE are particularly strong. There is however a problem. The activities of people create EE and the human activity derived energies can easily be confused with natural EE and mistaken for them. For example, people and animals using a path create a linear band of energy, two paths in parallel and the fields interact to produce more complicated bands of energy which can be analysed with a colour wheel. Ceremonial paths round a central point or structure create energy patterns. Any construction of ditches, banks, stone structures, all now long gone, leave energy patterns for dowser's millennia later to find and ponder on.

The reason why energies are associated with footpaths is that the ground is compressed and leather and feet leave chemicals in the soil. The modern dowsers therefore have quite a problem if they are to sort out the energies that are associated with an ancient sacred site and identify their sources. The only way to do it is to use the dowsing techniques of 'form' that is the shape of the energy source, the 'colour' of the energy, colour reveals some patterns and helps to identify the nature of the energy source, and witnesses which help to identify the source i.e. iron oxide, chalk, flint. Most if not all sacred sites do have a very clear pattern of EE lines associated with them. Most of these lines are due to foot and wheeled traffic. The wheeled vehicles are associated with draft animals, that is horses, deer and cattle. It is also possible that the paths and roads fall into three groups time-wise and this must be remembered when studying a site. Those paths or roads present before the site became a religious or cultural centre, those associated with the active period of the sacred site and those produced by human activity after the sacred site had ceased to be used as such.

In dowsing a sacred site the dowser has to use biolocation techniques and methods which enable the sources of the energy being dowsed to be identified with some degree of reliability.

A true blind spring has to be distinguished from the circular 'black' energy field of a Druidic Pool at the centre of a woodhenge. This can be done by using carbon or pond sludge as a witness. These witnesses pick up the remains of the organic matter that settles at the bottom of a pool of water. A clay witness will pick up the clay used to make the pool. Both the clay and carbon should reveal the shape of the structure. A blind spring is not likely to be circular if it is making its way up or down a fissure and will not contain organic matter and clay. A true blind spring will also have streams coming into it and going out of it. The water in the different underground streams will have their own 'signature' in terms of what is dissolved in the water. For example calcium, magnesium, sulphate, iron salts, aluminium salts, possible surface contaminants such as nitrate and phosphate.

It is therefore in theory possible to separate the true EE that may have caused ancient people to select the site in the first place from the man-made EE associated with the use of the site.

The main EE source that the ancients would have focused on is that from underground water. Our latest work indicates that just any underground water was not suitable. It had to be an underground stream running in a cave system or open channel in the rock. The temple was often placed over a point where the stream broke up into a number of more or less parallel streams (Figure A2.1) Access to a large quantity of water was required for filling up the temple pool and possibly other uses. Although the temple(s) may be placed on top of the stream or near to it the well accessing the underground stream is dug outside the temple complex.

Stones

Clay

Carbon

Chariot tracks

Footpath

Diamonds from the
underground river.

Energies from the
stones

Small multiple diamonds
from underground riverlets

Underground
temple complex

Underground Riverlet
system

Underground River

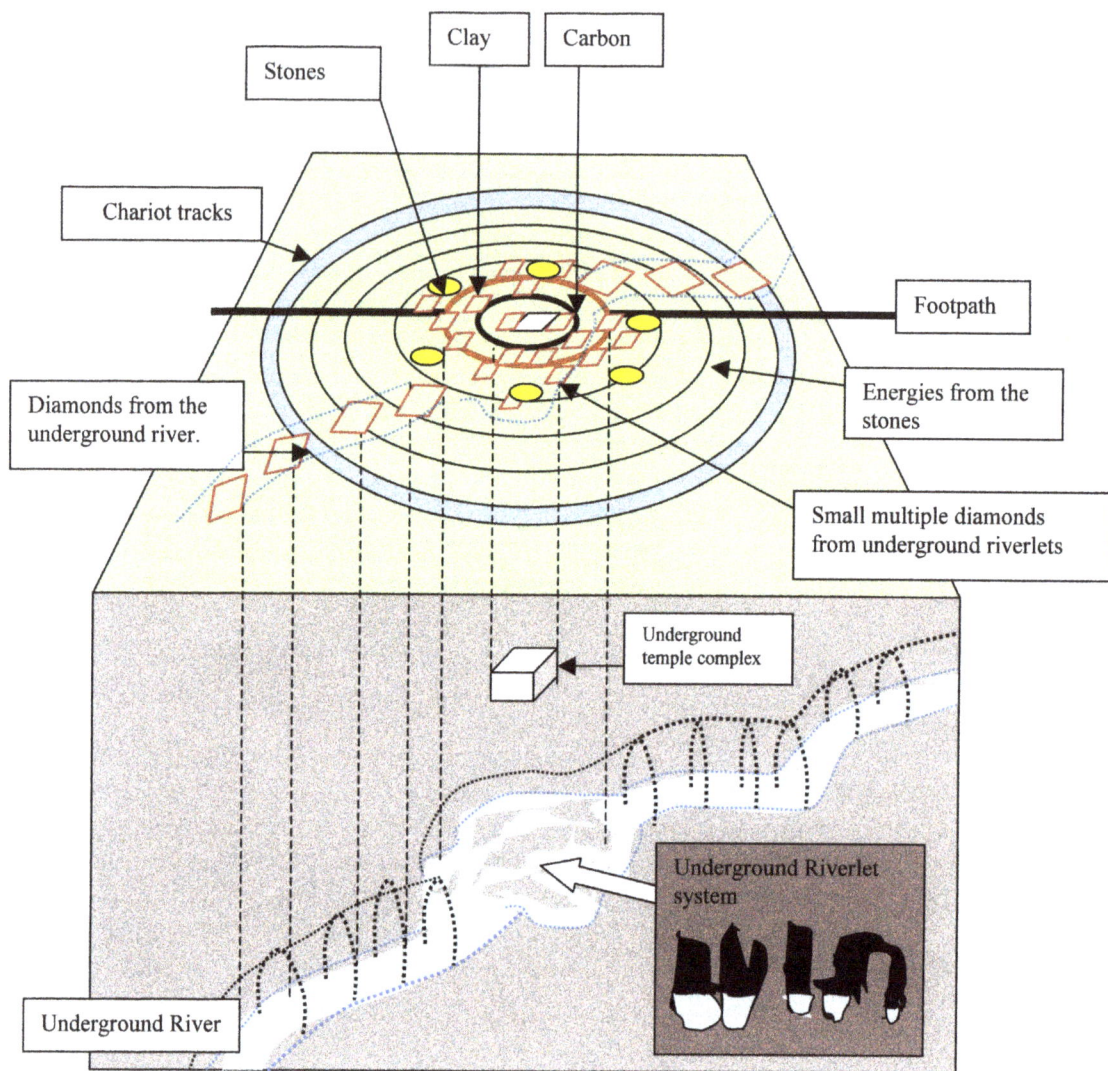

Figure A2.1 Earth Energies Associated with a Temple Site
With such complex magnetic patterns on the site considerable care
must be taken to plot out the form of the patterns correctly. It would
be quite easy to find any pattern you wish within this complexity.

A Synopsis of
The Diamagnetic and Paramagnetic Theory of Biolocation

G W Crockford

Appendix 3 is for those readers with some knowledge of science who may be interested in the scientific basis of biolocation. What is described here is not required for understanding the story of the stones, but it does support the validity of biolocation. The theory is developed in detail in a book by the authors 'The Human Magnetic Sense'.

When I first entered the field of dowsing in 1995 there was no theoretical basis for the dowsing phenomena, either the energy fields the dowser was responding to or the sensory system that was responding to them. This was not because people had not tried to identify the forces and sensory system involved.

However, in 1995 there appeared to be general but by no means complete acceptance that the field to which the dowser was responding was magnetic in nature. This view was reviewed by Bird in his book 'The Divining Hand'. In Chapter 14 Bird considers the evidence for magnetic fields being involved. He refers to the work of Harvalik in the 1960s and 70s who by experimental work came to the conclusion that there is a sixth bodily sense in humans for detecting changes in magnetic fields and that dowsers may be the world's most sensitive magnetometers. The sensor Harvalik identified was in the head.

It seems odd that Harvalik's work was not developed. However, the fact that dowsers would on the one hand not respond to quite strong magnetic fields but on the other would respond to fields that were so minute that they were immeasurable by normal physical methods at the time, probably created confusion and doubt. It was also not appreciated that if some vertebrates have a magnetic sense, particularly some mammals, then it was possible that humans still had a magnetic sense even if only at the vestigial and subliminal level.

Perhaps it was also not appreciated that if a sensory system is to be developed by an organism it would be aimed at those energies that contained most relevant information for the organism. The ferromagnetic field of the planet, although very strong compared with other magnetic fields, contains little information for an animal. The sensory system might therefore be blind to the strong magnetic fields but show the high sensitivity characteristic of the other five senses to the very low energy levels associated with diamagnetism and paramagnetism. Paramagnetic fields contain a considerable amount of information. Another factor that might have held back research is that it appears to be a common conception that magnetism is a homogeneous force. This is not the case as reference to text books and physical data on magnetic properties of materials refer to a number of forms of magnetism. Texts often refer to the material producing the magnetic field as being diamagnetic, paramagnetic or ferromagnetic etc. and this has possibly led people to associate the form of magnetism with the material and not with the field. In other words it is considered that it is not the field that is for example diamagnetic but the material. This is in spite of the fact that one of Faraday's big intellectual achievements was to realise that the magnetic field and its source were two different entities. In other words in dealing with the magnetic field you could forget about its source. This means for example that it is possible to study the field and then work back to its source as is done with electromagnetic radiation from stars. Once it is realised that magnetic fields from different points of origin within the atom and molecule may differ the possibility of biological sensors arise. The magnetic field environment like the acoustic, electromagnetic

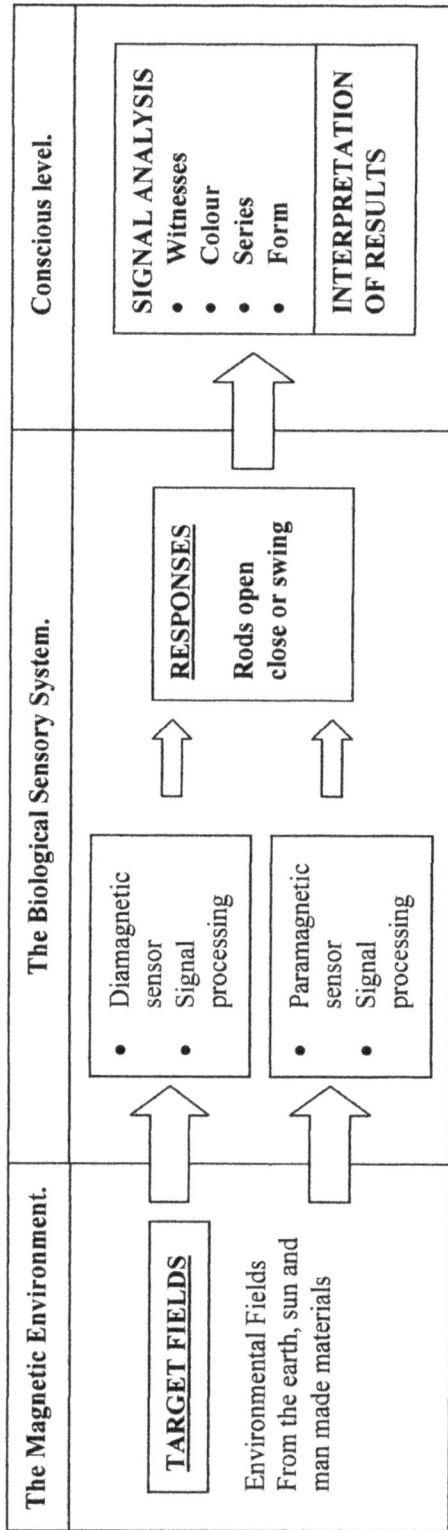

The Magnetic Environment.	The Biological Sensory System.	Conscious level.
TARGET FIELDS Environmental Fields From the earth, sun and man made materials	• Diamagnetic sensor • Signal processing • Paramagnetic sensor • Signal processing **RESPONSES** Rods open close or swing	**SIGNAL ANALYSIS** • Witnesses • Colour • Series • Form **INTERPRETATION OF RESULTS**

Figure A3.1 A Biolocation Model

The paradigm developed for the study of biolocation consists of three parts. The magnetic environment of the dowser with its multitude of components. This environment is dynamic and changing with many of its fields produced as a result of variable applied incoming fields. The magnetic sensing system is in most people subliminal but a small proportion of the population are conscious of the fields and can construct images from them. Such people are referred to as sensitives. The sensors for the diamagnetic fields are in the head. The sensors for the paramagnetic fields are in the skin. The third component of the model involves the dowsers experience and knowledge.

382

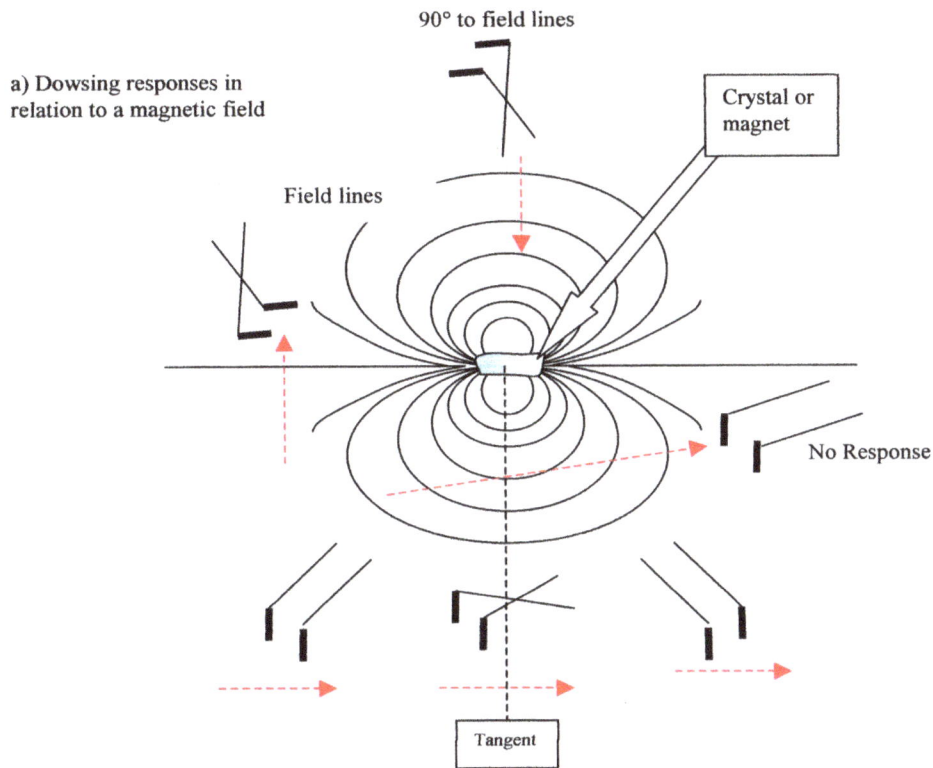

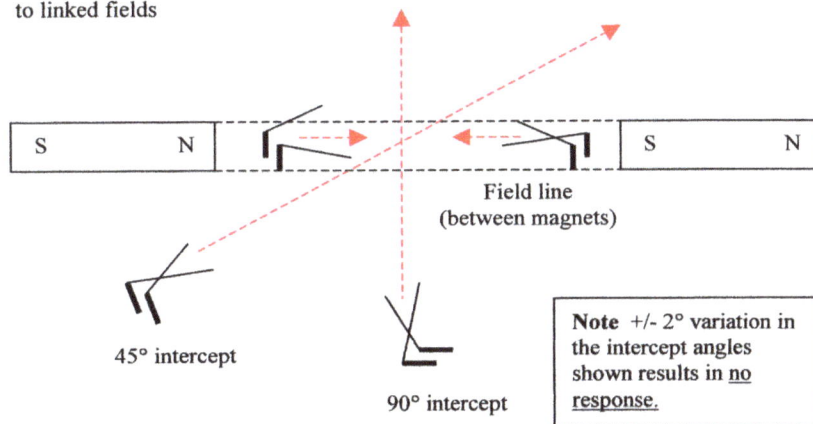

Figure A3.2 Magnetic Field Lines and Dowsing Response
The diagram shows the field lines round a magnet and some of the responses obtained with 'L' rods.

a) *In walking past the target the rods may turn towards the target to indicate where it is. When crossing field lines the rods cross and if walking past the target along a curved field line the rods will cross when at right angles i.e. at a tangent to the field of the target.*

b) *The field lines between the north and south poles of quartz crystals enable a dowser to identify the angles of approach at which clear responses are obtained.*

and chemical environments will contain a lot of information which may be of value to biological organisms.

Dowsing is a bit like learning to play a musical instrument. It is very easy to obtain a noise from the instrument or get the rods to move but a lot of practice is required before you can produce much in the way of a tune, or sensible and reliable dowsing results. It was during this phase of 'practicing my scales' so as to speak that with my dowsing mentors I began to appreciate more of what dowsing was and the dowsing phenomena that any theory of dowsing would have to explain. The model of the dowsing system that I developed is given in figure A3.1. The system has to be able to take on board in one way or another a number of dowsing phenomena. If a dowsing phenomena does not fit into the model then another sensory system may be involved and this should be considered. One of the first of such phenomena that had to be dealt with was remanence. Remanence occurs when a dowsable object has been placed on the ground for a while and successfully dowsed. If it is then removed the object can still be dowsed. The object has left a memory of itself on the ground to which the dowser responds. This memory effect has to be explained. Fortunately it is by magnetic theory. When dowsing, the rods show a range of responses which are used by dowsers to identify the direction of a target and its polarity (see Box 1.1. and Figure A3.2). Another phenomena is the search mode. Dowsers would often use a search mode in which they rotate through 360° to see if they could identify the direction of an object such as an underground stream. The rods cross when the dowser is facing towards the target which may be a 100m or more away. Dowsers also use colour wheels (Box 1.4) to identify particular energies, for example the bands of energy surrounding a stone circle or those associated with an ancient road or footpath. Yet another is the phenomena called series in which the dowser counts the number of times they obtain a response from a target such as an iron water pipe or a coin. The dowser repeatedly crosses the field from the target by stepping forward and back. The number of times the rods cross before failing to respond is the series number of the target which may range from 1 to 8 but is commonly 3 or 4. In this case the dowsing sense looks as if it is behaving like the sense of smell and touch both of which also rapidly fatigue. One of the very important tools the dowser has is that of the witness which has already been described (Box 8.1). Finally a method of dowsing called map dowsing should also fit into the scheme. In this form of dowsing a map of the area the dowser is interested in is used to identify underground water, graves and other objects under the ground. The dowser should ideally be in the vicinity (say within 1 Km) of the area being investigated. Some dowsers claim to be able to map dowse at considerable distances, for example maps of Australia from England, I know of no evidence that this can be done. If it could, it would clearly fall outside a model based on the Diamagnetic and Paramagnetic Theory of Biolocation. The idea that map dowsing is possible if the dowser is close to the site is based on the fact that you can do the equivalent of map dowsing with hearing, sight and smell. You do not need to see the road to know where it is likely to be or in which direction and how far away the pigsty is. If you have a rough map in front of you it will be possible to draw in the road and pigsty. The magnetic fields like noise and smell may well indicate to a dowser where a restricted range of certain things are to be found particularly those important to survival such as water. The dowser can then indicate on a map where they are without actually seeing the fields or areas involved.

The dowsing sense shows similar properties to the other senses in that it can focus on particular targets. For example drains, if that is what the dowser is looking for or Druidic tombs. When looking for specified targets the dowser will often miss much of what else is there and only respond to drains or something that is similar. The dowser is extremely sensitive to the angle of approach to a field. In some cases being 2 or 3 degrees off and the dowser will miss the field (figure A3.2). For further comparison of the magnetic sense with the other senses see the article on 'Dowsing as Perception' in which Ian Pegler develops the possible similarities

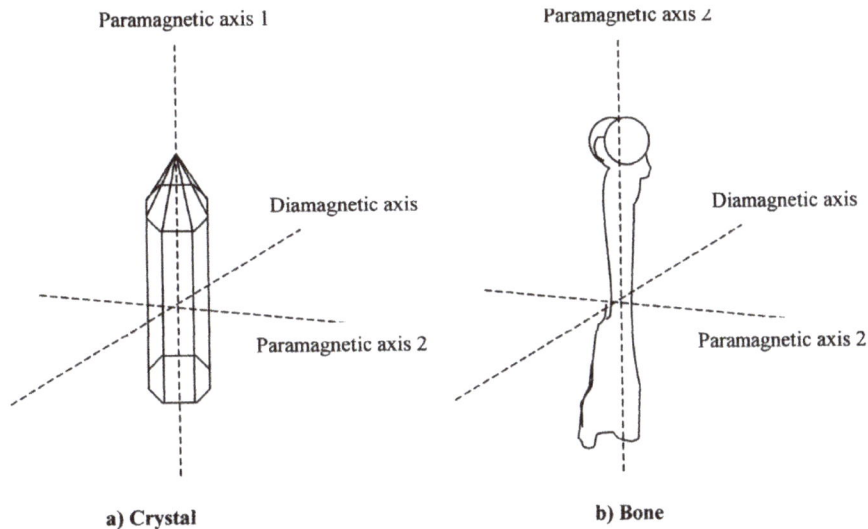

Figure A3.3 Magnetic Axes

The diagram of the quartz crystal shows the three magnetic axes of the crystal. In a long bone the diamagnetic axis is along the length of the bone and the paramagnetic axes at right angles to it.

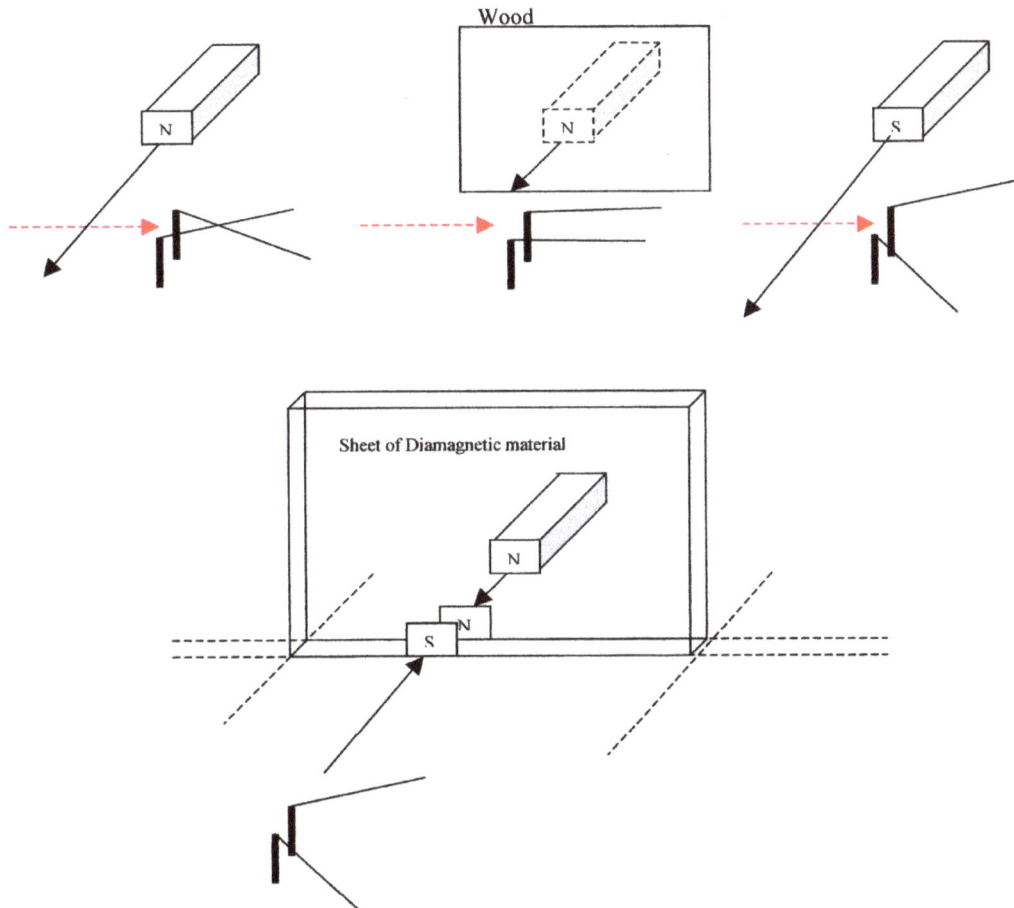

Figure A3.4 Dowsing Response to a Diamagnetic Field

The diamagnetic field of a magnet has the same axis as the ferromagnetic field. In walking past a series of magnets arranged as shown the dowser will obtain rods crossing indicating a north pole, no response if the field is blocked, rods opening when crossing the south pole and rods opening when crossing the north pole with a sheet of diamagnetic material in front of it.

385

between the magnetic sense and the other senses.

Returning to magnetic fields. Every object, every material has a magnetic field associated with it, in fact normally more than one. For example a brick may have a ferromagnetic field due to any iron or ferromagnetic materials it contains. There will also be the diamagnetic fields associated with all the atoms and molecules in the brick and then the paramagnetic fields associated with most of the materials from which the brick is made. The axis of the ferromagnetic field and the diamagnetic field tend to be common i.e., run together as they do in magnets and the planet itself. There are normally two paramagnetic axes at right angles to the diamagnetic axis and in some complex materials additional ones at various angles to the diamagnetic field (Figure A3.3).

When I started dowsing I had a general knowledge of magnetism and like many people, thought magnetism was magnetism. You could have more or less of it but that was about all.

The initial and very important observation that this was not so was made when I was dowsing the north and south poles of a magnet which was at shoulder height and in a glass fronted kitchen cupboard. The north pole of the magnet was being identified when the dowsing response suddenly disappeared. I could not understand this and I thought that it might be something to do with me so I just kept walking back and forth and two or three minutes later the dowsing response returned. On opening the cupboard door to change the magnetic pole facing me I found that the magnet was no longer behind the glass but that it had slipped down behind the wood frame of the door. Knowing that wood did not block ferromagnetic fields it did not take me long to realise that another form of magnetism was involved. This was soon proved by placing a breadboard over the magnet. A compass still responded to the magnet but I could not detect the magnet through the board.

All the work done by the author to date on many dowsers points very clearly to the dowser not being directly sensitive to ferromagnetic fields, hence the dowsers 'blindness' to strong fields. It is possible that this blindness to ferromagnetic fields also applies to animals and birds. This 'blindness' can be demonstrated by asking a dowser to identify which pole of a magnet, placed at shoulder or head height at a distance of about 2m is facing them. Most dowsers will identify a field as they pass the magnet, and identify if it is a north or south pole facing them by the rods closing or opening (Figure A3.4).

If a wood breadboard or cork mat is now placed in front of the magnet the dowser can no longer identify the field. It is no longer there. After a period of time the field from the magnet will penetrate the cork or wood and the dowser can again identify its presence (Figure A3.4). The ferromagnetic field on the other hand is able to immediately penetrate the barrier and can be identified using a compass. The dowser however cannot identify it. In other words the dowser is sensitive to diamagnetic fields – the weakest of the magnetic forces but not to ferromagnetic fields. The presence of the diamagnetic field can be demonstrated by placing a sheet of diamagnetic material between the magnets say north pole and the dowser. A diamagnetic material moves away from the applied diamagnetic field because of a physical law called Lenz's Law. What this law says is that a north pole develops to face the north pole of the applied field. This means that the other side of the material facing the dowser is now a south pole. As all materials are diamagnetic a sheet of plastic, aluminium foil, or a sheet of steel can be used to demonstrate the apparent change in polarity of the magnet.

Some caution is required here because the plastic or metal sheet or whatever material is used is subjected to two magnetic fields, the strong ferromagnetic is one and the weak diamagnetic field is the other. The ferromagnetic field is not responsible for changing the polarity of the

386

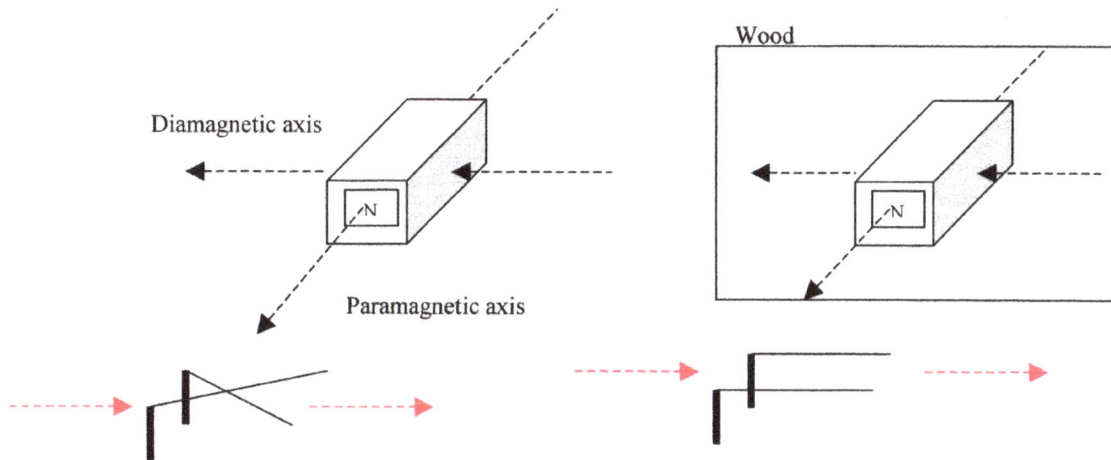

Figure A3.6 Collimated Fields
A collimated magnetic field either diamagnetic or paramagnetic can be used to identify which part of the body contains magnetic sensors.

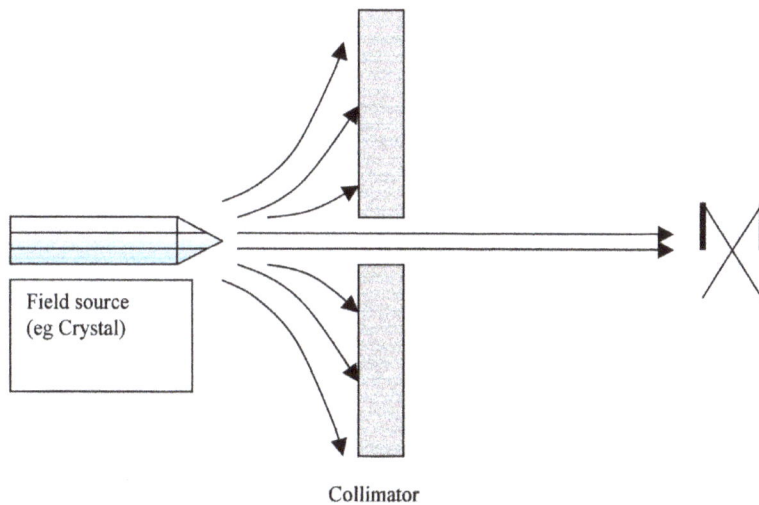

Figure A3.5 Dowsing Response to a Paramagnetic Field
The paramagnetic axis of a magnet is at right angles to the diamagnetic axis. If the dowser walks across the field the rods either cross (north pole) or open (south pole). A wood screen will block the field, a metal or plastic screen will not appear to reverse the polarity of the field.

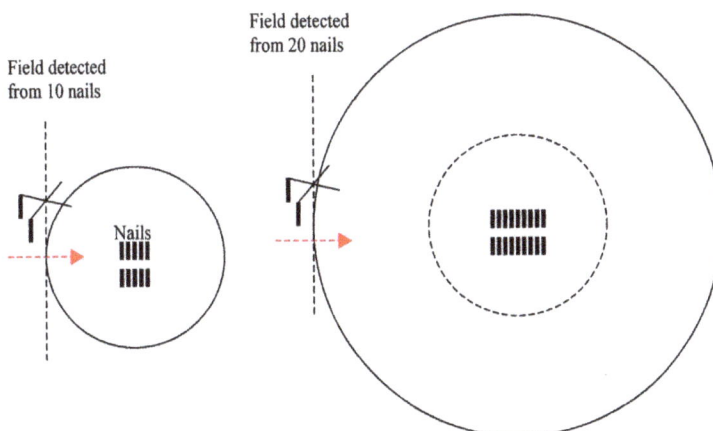

Figure A3.7 Magnetic Field andTarget Mass
If the mass of the target is varied the detection limit of its field will increase or decrease with its mass.

387

magnetic field of the plastic or metal. This can easily be demonstrated by placing a sheet of cork between the magnet and plastic sheet. When this is done the plastic looses its field in spite of the fact that it is still subjected to a ferromagnetic field. A pure diamagnetic magnet such as a salt crystal which has no ferromagnetic field provides supporting evidence as it induces a diamagnetic field of opposite polarity in the plastic. The Lenz's Law effect therefore appears to be due to an applied diamagnetic field which in magnets seem to have the same magnetic axis as the ferromagnetic one. It is therefore easy to confuse the effects of the two fields.

If the magnet is moved so that the ferromagnetic north and south poles are parallel to the dowser's paths the dowser will cross a paramagnetic axis. The magnet may have to be lowered so that it is level with the shoulders or chest. The paramagnetic field of the magnet, which is at right angles to the diamagnetic field, does not have its polarity reversed by a sheet of plastic. It does however penetrate wood and cork more readily than the diamagnetic field (Figure 3.5).

The use of a magnet for this type of experiment enables the dowser's blindness to ferromagnetic fields to be demonstrated. A dowser will respond to non ferromagnetic materials such as a large pure salt crystal. Salt, sodium chloride, is a pure diamagnetic material, ignoring for the moment any magnetic fields from the nucleus of the atoms. During Nigel's study of the fields from bones (Chapter 27) he discovered that a long bone is also an excellent device for demonstrating both diamagnetic and paramagnetic fields. A long bone is also more readily available than pure salt crystals. The diamagnetic field runs along the length of the bone so that lining the bone up also lines the diamagnetic field up with a specific part of a person, for example their head or trunk.

The next question was whether there were one or two magnetic sensors. Because diamagnetic fields are easily blocked by mats of chaotically arranged organic fibres such as cork mats, and by wood with the grain at right angles to the field, it is possible to collimate the field from a diamagnetic magnet such as a pure sodium chloride crystal. If the collimated beam is set up at different heights and the dowser walks through the beam a dowsing response is only obtained when the beam is at head height (Figure A3.6). Thrusting the hands into the diamagnetic beam does not produce a response. Gawn's work on muscle responses to diamagnetic fields confirms this observation. Gawn has shown that the deleterious (diamagnetic) earth energy does not act on the muscles directly but on a sensor in the dowsers head.

The paramagnetic axis of an object such as a crystal seems to be determined by the crystal structure. For a quartz crystal this axis is along the length of the crystal so it is possible to use the crystal to point a paramagnetic field at the trunk or head. To be certain that you are dealing with a beam the beam can be collimated using cork or blocks of wood. With a paramagnetic field a dowsing response is obtained whichever part of the body enters the beam apart from the head. For example a dowsing response is obtained if the hands are thrust into the beam. The dowsing response does not confirm that a sensory system is involved. However, if one hand is cooled below 10°C at which temperature the sensory nerves no longer function, test with a pin for anaesthesia, and the hand is then inserted into the paramagnetic beam no dowsing response is obtained. The response returns much later as the hand slowly rewarms. The same experiment can be done with an arterial occlusion at the wrist, but do not try it unless you are also a human physiologist. As the sensory nerves in the hand fail so the dowsing response disappears. Recovery of the dowsing response on removal of the cuff is rapid.

Based on the use of diamagnetic and paramagnetic fields the evidence so far indicates that there are at least two sensory systems. More information can be obtained on each of these sensory systems. For example, it is possible to block or prevent the target's paramagnetic field from reaching the skin by using a cork screen. If one side of the body is screened from a paramagnetic field by cork the rods swing out towards the unprotected side instead of crossing. This and other methods of exposing only one side of the body to a field indicates that the paramagnetic sensory system consists of two halves, one on each side, at least on the trunk. A witness for example does not work if on the mid line of the body but becomes more effective as it is moved to the side.

The strength of a targets paramagnetic field can be determined by walking towards the target, say an iron nail. At a certain point on the approach the rods will cross indicating that the dowser has reached the boundary of the nails paramagnetic field or aura as it is often called (Figure 3.7). The boundary is defined by the dowser's sensitivity. The field goes to infinity. If the number of nails is increased the point at which the rods cross moves further and further away from the nails. The strength of the field appears to be proportional to the mass of the target, in this case the number of nails. The experiment can now be repeated with aluminium rods, copper wire, and granite. In all cases the detection distance increases with the mass of the target. It looks therefore as if the paramagnetic field is related to the mass in the same way as ferromagnetic fields. Now what happens if the paramagnetic fields of different materials are added? If say the nails give a detection distance of 3m and the aluminium gives a detection distance of 3m it might be expected that by adding them together the detection distance of the combined targets would increase. In fact it does not. The paramagnetic sensory system is 'seeing' different fields from different materials and it does not add them together. It only adds like fields or so the present studies indicate. Do the same experiment with diamagnetic fields and the diamagnetic sensor and it is found that dissimilar diamagnetic fields are added. That is the detection distance of the diamagnetic field increases as dissimilar fields are added together. The two sensory systems appear to be different in the way they work. This work on the magnetic sense is at a very early stage but it is indicating that the two systems have different biological functions. It also shows that in carrying out dowsing research or biomagnetic studies the researcher has to be clear as to the fields and the sensory system which is relevant and involved.

It should be noted that the sensitivity of the paramagnetic sensory system can be influenced by how far apart the hands are held with the sensitivity to fields increasing as the hands are held further and further apart. This does not apply when responding to diamagnetic fields. Response distance is independent of how far apart the hands are.

Holding witnesses to blind the dowser to fields other than those of the witness they hold enables a dowser to extract considerable information about a target. Witnesses are used with the paramagnetic system. They can be used with diamagnetic fields but they cause the sensory system of the dowser to block out the target fields so they are not seen. That is the dowser cannot detect the target but will pick up everything else. Not a useful system. A witness placed between the target and the dowser will block diamagnetic fields just as it does paramagnetic fields (Figure 3.8).

Elevation

Shadow area

Iron target   Iron witness

Plan view

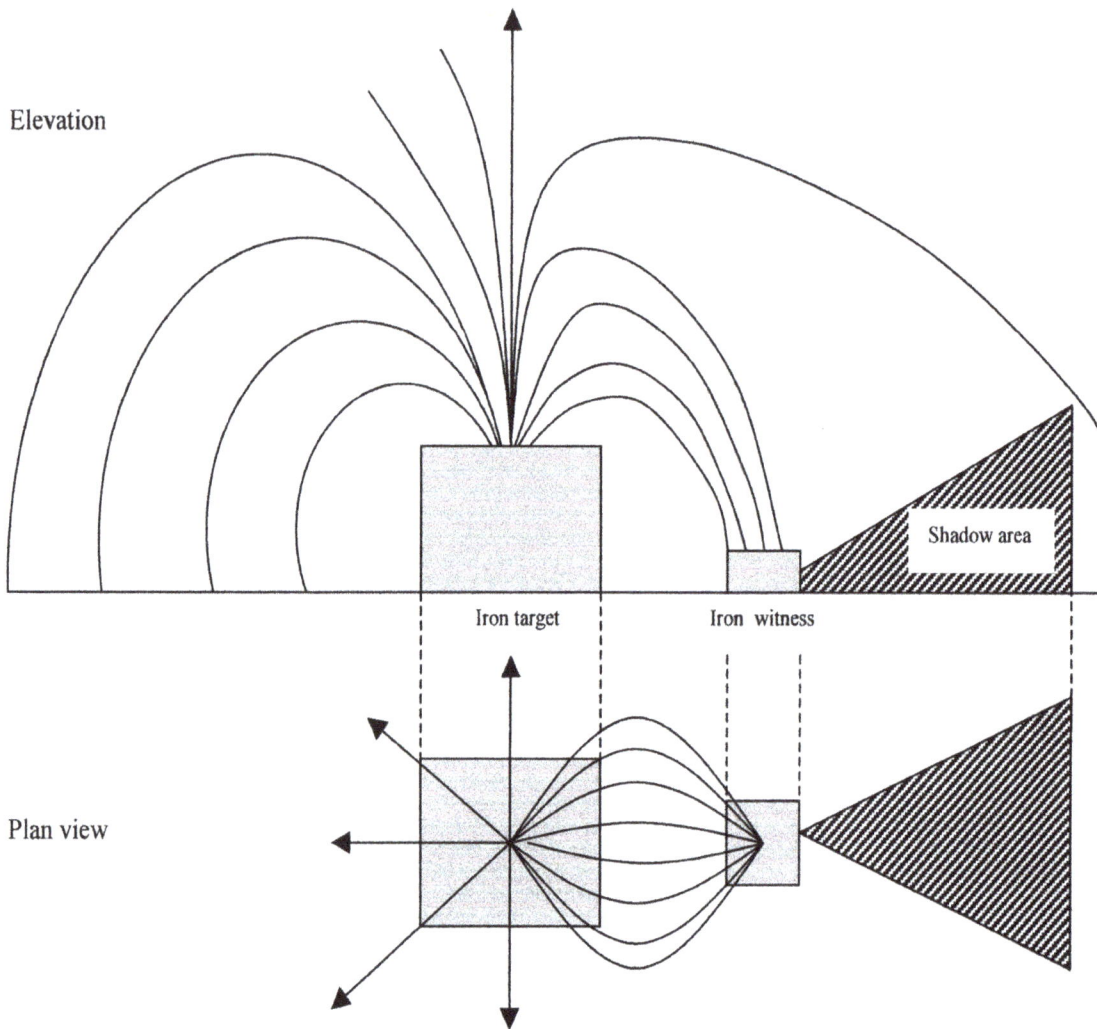

Figure A3.8 Field Shadow From a Witness

The magnetic field lines from a target are attracted to a witness of the same material. The witness creates a field shadow behind it.

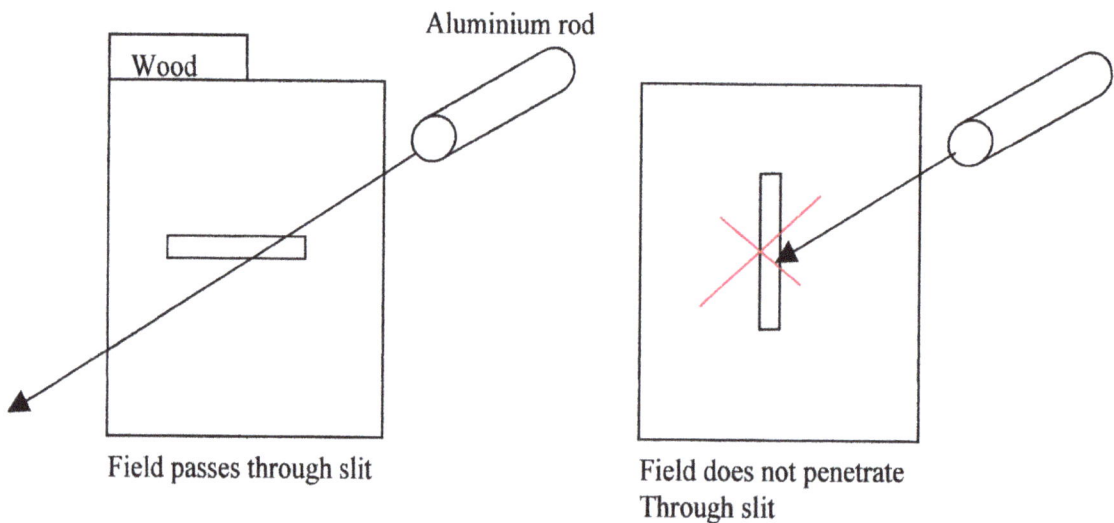

Aluminium rod

Wood

Field passes through slit

Field does not penetrate
Through slit

Figure A3.9 Polarized Fields

The paramagnetic field from a target such as an aluminium rod appears to be polarized. The field may not penetrate a narrow vertical slit but will pass through when the slit is horizontal.

390

Summary

To summarise, the dowser appears to respond when cutting magnetic field lines at particular angles. There appears to be a sensor in the dowsers head responding to diamagnetic fields, whilst the rest of the body is sensitive to paramagnetic fields but divided into two halves. The dowser is able to sense what is called the polarity of a field that is whether the field is from the north or south pole of the source. Because the dowser responds when cutting field lines at certain angles e.g. at 90°, 45° and at a tangent, the magnetic sense is remarkably directionally sensitive. Not as good as sight but better than hearing and smell. Just as with the other senses there is a considerable skill element in obtaining information about the environment using the magnetic sense. It also appears to be possible to focus the magnetic attention in a way similar to the other senses. The sensitivity of the dowsing sense in terms of energy levels to which it responds is almost certainly at the same level as that of hearing and sight. There only has to be a slight change in magnetic gradient and the dowser responds to it.

The underlying theory for the dowsing research reported in this book is that:

a) All dowsing phenomena are due to the dowser interacting with diamagnetic and paramagnetic fields.

b) Paramagnetic fields are characteristic of their source. That is with the aid of witnesses the source of a paramagnetic field can normally be identified. This does not rule out different sources having the same paramagnetic field however.

c) All materials are magnetic generating diamagnetic fields and most generate paramagnetic fields as well.

d) The diamagnetic and paramagnetic fields developed by a material tend to oppose each other and are normally at right angles to each other. Diamagnetic fields will tend to oppose and redirect a paramagnetic field. The relative strengths of the fields will also determine what happens.

e) The paramagnetic fields do not appear to interfere with one another and they retain their identity. The dowser will however 'see' some combination of fields as something else and not be able to distinguish them as separate fields. This is likely to be a property of the sensory system.

f) There are commonly two paramagnetic axis at right angles to each other but some rocks and man-made devices have more than two.

g) The dowsing sensory system is directionally sensitive and has a sensitivity to energy of the same order as other sense organs. The dowser has to move through the magnetic field in order to detect it.

h) The detection distance of a specific unfocused field is determined by the mass of material giving rise to the field, the sensory system for paramagnetic fields does not summate different fields only similar fields. The diamagnetic sensor summates different diamagnetic fields.

i) There appear to be separate sensory systems for the diamagnetic and paramagnetic fields. The head contains the sensor for diamagnetic fields the skin the paramagnetic sensor. The paramagnetic sensor has two halves one on each side of the body.

j) Both the diamagnetic sensor and the paramagnetic sensor become saturated if the applied field is too strong. The dowser becomes blind to the field and does not respond.

It has been noted that the paramagnetic sensory system does not add different fields together. If this were to be completely true it would indicate an extremely complex sensory system able to distinguish thousands of different fields. This is not likely to be the case and some fields are going to be seen as the same and together increase the detection distance.

The term paramagnetic field is used here to include the fields from protons and the atomic nucleus.

This synopsis does not cover the focusing, refection and amplification of fields. The Druids knew a few things about these aspects of magnetic field engineering

A more detailed knowledge of the Diamagnetic and Paramagnetic Theory than that given here is probably not required for understanding the physical basis of the dowsing used in the study of the Henge Age. However, there are a number of features of magnetic fields that do raise problems when dowsing. One is that the fields behave as if they have a plane of polarisation. Another is that incoming fields from the earth, sun and possibly upper atmosphere act both on each other and also on the source of the target's field i.e. the nucleus, electron shells, electron clouds of the molecules and conducting electrons in metals. Of these perhaps the easiest feature to demonstrate is the plane of polarization. Many plastics such as PVC, acetate and polystyrene are crystalline and have the molecules lined up as a result of the manufacturing and moulding process. A sheet of acetate used for overhead projection, preferably mounted so that it can be stood on edge is one of the easiest to use. It the sheet is placed in front of a paramagnetic source such as a quartz crystal it will either block the field or allow it to pass through. If the field is blocked rotate the crystal or the plastic sheet by 90° when the plastic should allow the field to pass through. Rotate the plastic a further 90° and the field will be blocked again. The plane of polarization can also be identified by using a slit (Figure A3.9). With the slit in one direction, say vertical, the field passes but when the slit is rotated 90° the field is blocked.

Another feature and the bane of dowsing are images of targets. Magnetic field lines have to go somewhere. On occasions this results in the field lines from a target making a loop back to the ground and creating both a magnetic loop and a phantom target or image. When dowsing the image, the dowser will think they have a real target but unfortunately it is only an image. This can be demonstrated by placing a large piece of cork or thick cardboard over what appears to be the target. If it is a real target it will disappear as the field lines coming up from the ground are blocked. If it is an image it will not be affected as the field lines are coming down onto it. If the position of the real target and image are known the corkboard or a suitable witness to block the fields can be used to plot the magnetic loop from the target to the image (Figure A3.10).

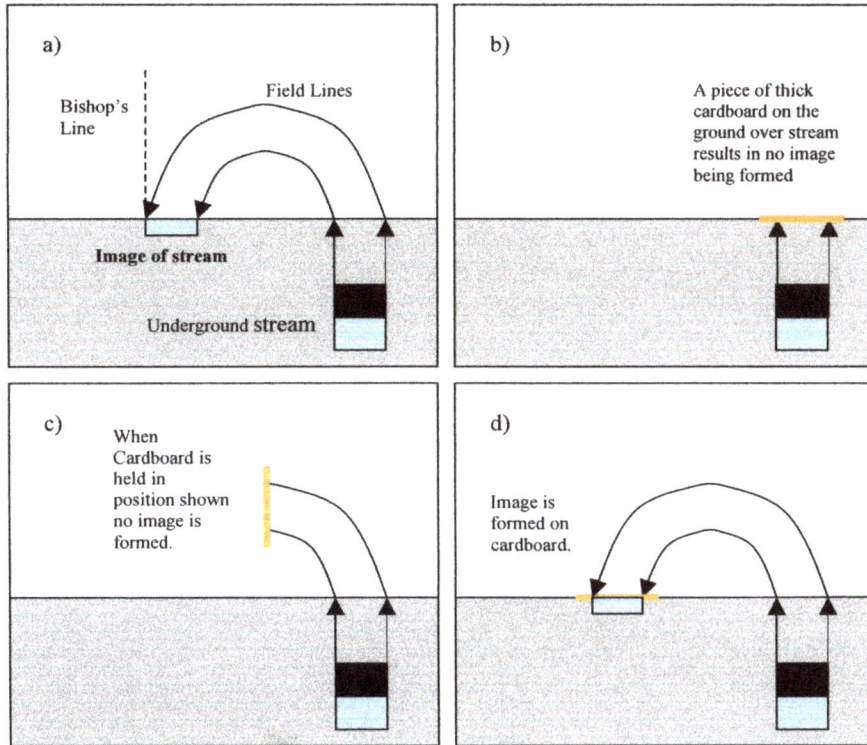

Figure A3.10 Magnetic Images

The magnetic image from a source such as an underground water pipe can be demonstrated. First by blocking the field at source and then by blocking the magnetic loop on its way to the image.

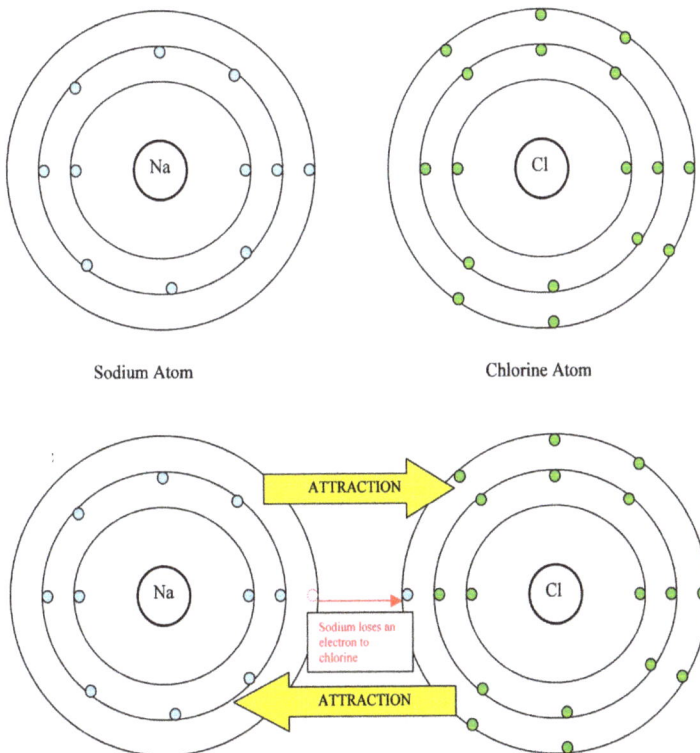

Sodium Atom

Chlorine Atom

Figure A3.11
A Diamagnetic Magnet

The electron structure of sodium chloride (NaCl) showing that the two ions have a complete outer ring of electrons and so consequently lack a paramagnetic field.

A sodium chloride crystal is a diamagnetic magnet.

Positively charged Sodium ion with a noble gas structure (Neon). The full electron shell makes it Diamagnetic.

Negatively charged Chlorine ion with a noble gas structure (Argon). The full electron shell makes it Diamagnetic.

Box A 3.1 Experimental Proof of Dowsing

Although many research workers have over the years tried to identify the mechanism and reality of dowsing none have produced an accepted theory of biolocation. And yet the results and proposed mechanisms in some cases appear reasonably plausible. Certainly some findings and proposed mechanisms appear to have been worthy of further study. One of the reasons for this lack of follow through may be that simple experiments were not developed by the workers which could be done by any dowser interested in exploring the mechanisms and the reality of dowsing. Also logical explanations have not been provided as to why dowsers so often get it wrong and find it so difficult to consistently obtain the right answer when subjected to a blind test. The reasons why something does not work on some occasions are closely knit to the reasons why it does work on others. The two go hand-in-hand and must be understood. At the moment it is not possible to do a blind test successfully and consistently (Appendix 4) for very good reasons. If in doubt cut out 12 pieces of thick cardboard about 15cm square, each piece being identical. Then tape a 2p piece under one, ask somebody to distribute the cards on the lawn about 0.50 to 1.0m apart when you are not looking, then try and find the card with the 2p piece. A simple and easy dowsing task but see what happens when you repeat it 10 times. The most reliable experiments are those in which the dowser has a visual input to confirm that extraneous fields are not interfering or to enable them to be compensated for the fields. If the right answer is not obtained the dowser can pause for a minute or two or check to see what might be interfering with the dowsing. Experiments of this type include cutting field lines at different angles (Figure A3.2). For this experiment I use two quartz crystals, north facing south, to set up field lines. Another is using collimated diamagnetic and paramagnetic fields to demonstrate the sensors in the head and skin (Figure A3.4, A.3.5).

Images of a target can be easily demonstrated by using a water main. The pipe is about 1m or a little more deep. What is called the Bishop's Line which relates to the depth of the pipe is therefore about 1m from the water main. A piece of cork or thick cardboard will block the field from the pipe if placed over it. If the card is now moved and placed over the Bishop's Line the line remains and can be dowsed. To block the Bishop's Line the card has to be placed over the water pipe or held somewhere in the magnetic loop, see figure A3.10.

Using a witness it is possible to determine a person's blood glucose level by approaching the person with a glucose witness in one hand. There comes a point where the rods cross at about three feet from the person. Now ask the subject to eat a Mars Bar and repeat. Using the appropriate witnesses it is possible to determine if a person smokes and the nicotine level in the body. If not nicotine then a sweetener in a glass of drink can be used. A sweetener tablet is used as the witness. The blood levels of substances such as glucose can be changed by a person eating something containing the material e.g. sugar, or in the case of nicotine by smoking. By using witnesses the body burden before and after can be demonstrated. In fact by taking readings at intervals and recording the distance from the body the biological half-life of the nicotine and sweetener can be demonstrated and also the glucose clearance rate determined. These are all simple experiments that can be used to demonstrate some aspects of the mechanism and reality of dowsing.

Box A3.2 Magnetic Fields

Most people are familiar with the magnets that cling to the sides of refrigerators and may even remember playing with them at school. If you approach an iron nail with a magnet sooner or later the nail will jump to the magnet. A classic school experiment is sprinkling iron filings on a card and holding a magnet under the card. The filings then show the actual magnetic field lines by forming lines running from the north to south poles of the magnet. The magnetism the filings are responding to is called ferromagnetism and it is the strongest of the five forms of magnetism. The human magnetic sensory system used by dowsers (the 6th sense) is not sensitive to ferromagnetism but it is incredibly sensitive to two other forms of magnetism. They were discovered by Michael Faraday who called them Diamagnetism when the material moved away from the applied field and Paramagnetism when a material moved towards the applied field. The dowser is using these two magnetic fields when looking for objects. The origin of these fields within the atom and molecule differs from that of ferromagnetism. Traditionally only iron, nickel, cobalt and their alloys are ferromagnetic. However, all materials are diamagnetic and most are also paramagnetic (See the Concise Science Dictionary, Oxford University Press). An example of a diamagnetic magnet is a pure salt (NaCl) crystal. This is because both the sodium and chlorine irons have an outer orbiting ring of electrons which are balanced and have a noble gas structure (Figure A3.11).

The diamagnetic field comes from balanced rings of electrons. If one electron was missing the ring would be unbalanced and a paramagnetic field then develops.

A quartz crystal (SiO_2) is both diamagnetic and paramagnetic, in fact it has two paramagnetic axis, one for the field derived from the oxygen and one for the field from silicon. Two or more paramagnetic axis are quite common in materials and most stones have at least two. Many dowsers can detect the stronger of the paramagnetic fields coming from a stone or a piece of pottery using their hands. Stone Age people may have been even more sensitive to these fields than modern day dowsers and Feng Shui practitioners. Arranging stones so that their fields augmented each other may not have been a problem for them.

Appendix 4

Is the Blind Experiment to Prove Dowsing is a Real Phenomena Possible

I have repeatedly been asked to do a blind experiment that will show that dowsing is a real phenomena. Many people have the idea that if dowsers can find underground streams, leaking pipes and other artefacts in the ground then there should be no problem in demonstrating the reality of peoples dowsing ability. When I first started to dowse I thought, like many others, that all I had to do was design a blind experiment, such as finding coins hidden under pieces of cardboard, and I would have the proof in the bag. After repeatedly failing at such experiments and then trying to identify the reason for each failure I began to realize that the magnetic world in which the dowser operates is a very complex world indeed. So complex that it is not possible to design a blind experiment. The reason why is quite simple. It is that we do not yet know the science of the subject. To be more specific we do not know the science of the magnetic fields or of the sensory systems involved in dowsing. Because of this, when setting up an experiment we do not know the questions to ask, the variables to be controlled and we are blind to the environmental factors that are involved and which influence results. After years of work trying to find the experiment that clinches it I now know some of the factors which stand in the way and can control them but there is a long way to go before I know enough to design an experiment in which all the extraneous factors that can influence the dowsers response are controlled. To illustrate why a controlled blind experiment is not possible at the moment I have developed a thought experiment.

On the planet Zaiton all the people are deaf (you can also use the sense of smell for this thought experiment). One day a Zaitonian claims that he can hear things such as birds. He knows if there is a bird in a tree without seeing it. Now the Zaitonion scientists are expert in many disciplines and are familiar with vibration and the transmission of forces in fluids and solids. If a bomb goes off they know the physics of the pressure and shock waves, the energy levels involved. However, they know that Zaitonions do not have a sensory system capable of picking up the infinitesimal amounts of energy that might radiate as pressure waves from birds or any other creature. A number of Zaitonions however claim they have the power to hear birds and other creatures. They are often women or eccentric males. The simple experiments that the Zaitonian scientists have done all demonstrate that Zaitonions cannot hear and those who say they can have well developed imaginations. Swans have been used, parrots, all sorts of animals but when the cages had a cover on them so that the subject could not see if there was a bird in the cage or not then the subjects could not say which cage was occupied by a bird. When the results indicated the subject could hear, a stray feather on the floor or smell was responsible.

The thought experiment is to design a blind experiment that will prove or disprove the thesis that Zaitonions have a sense of hearing. You may not draw on any knowledge you have of the hearing sense or the science of sound or acoustics. Remember, because Zaitonions are deaf there is no control of noise. No silencers on cars and lorries, shoes last a long time because everybody uses steel studs, clothing fabrics are not designed or treated to reduce noise, central heating, air conditioning, kitchen appliances, aircraft, machinery, buildings, nothing is designed or treated to reduce noise. Acoustic reverberation, standing waves, resonance, transfer of noise in structures, all not known. The transmission values of materials for different frequencies are not known. There was, at the time of writing, a million dollars on offer (now withdrawn) for those who could solve the problem for the human magnetic sense. For those who solve the Zaitonian hearing sense problem there is only fame. My bet is that the presence of a Zaitonian hearing sense or human magnetic sense can only be done using epidemiological techniques. Statistical techniques alone are too blunt, and ineffective. The task is a bit like trying to measure a person's response time to light and noise signals in a disco or canning factory.

References and Further Reading

Barclay, A et al   Lines in the Landscape
                        Oxford Archaeology Unit (2003)
                        ISBN 0-947816 798

Betz, H. D.   Unconventional water detection
                        Published by G.T.Z. Germany (1993)
                        ISBN 3-88085-488-2

Bird, C.   The divining hand
                        Whitford Press (1993)
                        ISBN 0-924608-16-1

Bligh, J.   The Fatal Inheritance
                        Athena Press (2004)
                        ISBN 1-84401-336-7

Burl, A.   The Rollright Stones
                        Tattooed Kitten Publications (2000)

Burl, A.   The Stone Circles of Britain, Ireland and Brittany
                        Yale University Press, (2000)
                        ISBN 0-300-08347-5

Carroll, R.T.   The Skeptic's Dictionary
                        John Wiley & Sons Inc. (2003)
                        ISBN 0-471-27242-6

Corlett, C.   Journal of the Galway Archaeological and Historical Society 49 (1997)
                        pp 135 – 150

Corliss, W.R.   Ancient Infrastructure
                        The Sourcebook Project (1999)
                        ISBN 0-915554-33-X

Cotterell, A.   A Dictionary of World Mythology
                        Oxford University Press (1986)
                        ISBN 0-19-217747-8

Crockford & Hughes   Chemical remanence and convergence of evidence
                        Dowsing Today, Vol. 48, No. 292, pg. 10-15 (2006)

Dixon-Kennedy, M.   Celtic Myth & Legend
                        Blandford (1996)
                        ISBN 0-7137-2571-0

Dodd, R.J. et al   Towards a physics of dowsing : inverse effects in the northern and
                        southern hemispheres
                        Transactions of the Royal Society of Edinburgh:
                        Earth Sciences, 93, 95-99, (2002)

Dubrov, A.P.   Modern Achievements of Dowsing – Part 1
Dowsing Today Vol. 39, No. 275, pg. 8-9 (2002)

Edmonds, D.T.   Electricity and Magnetism in Biological Systems
Oxford University Press (2001)
ISBN 0-19-850679-1

Fricke, B.   The use of polarised stones as a method of clearing houses and communities of detrimental radiation (mid. 1980s)
Translated by Nicolas Fick

Gawn, W.A.   The behaviour of lines of earth radiation and their action on the neurological system (2001)
E-mail *wgawn@utvinternet.com*

Green, M.J.   Exploring the World of the Druids
Thames & Hudson (1997)
ISBN 0-500-05083-X

Hadingham, E.   Circles and Standing Stones
Book Club Associates, London (1975)
William Heinemann Ltd

Horte, J.   The Green Man
The Pitkin Guide

Jones, T. & Ereira, A.   Barbarians
BBC Books (2006)
ISBN 0-563-49318-6

Lee, E.W.   Magnetism: an introductory survey
Dover Publications Inc. (1970)
ISBN 0-486-24689-2

Maby & Franklin.   The Physics of the divining rod, (1939)
(Quoted in Bird)

Mann, N.R.   Glastonbury Tor - A guide to the history and legends
Triskele (1993)

MacManaway, P.   Energy Dowsing for Everyone.
Southwater, (2004)
ISBN 1-84476-001-4

Miller, A.   Dowsing: a review
Network No. 66, pg. 3-8 April 1998

Miller, H.   Dowsing
Penwith Press (2002)
ISBN 0-9533316-1-X

Muir, R.   Riddles in the British Landscape
                        Thames & Hudson (1981)
                        ISBN 0-500-24108-2

Niblett, R.   Verulamium
                        Tempus Publishing Ltd (2001)
                        ISBN 0-7524-1915-3

Oppenheimer, S.   Medical geneticist, University of Oxford.
                        The New York Times, 6 March 2007

Oppenheimer, S.   The Origins of the British
                        Robinson (2007)
                        ISBN 978-1-84529-482-3

Peddie, J.   Conquest The Roman Invasion of Britain
                        Sutton Publishing (2005)
                        ISBN 0-7509 3798 X

Pennick, N.   The Sacred World of the Celts
                        Godsfield Press (1997)
                        ISBN 1-84181-135-1

Pegler, I.   Dowsing as perception
                        Dowsing Today Vol. 40, No. 282, pg. 11-13 (2003)

Pitts, M.   Hengeworld
                        Arrow Books (2001)
                        ISBN 0-09-927875-8

Powell, D.   Eccentric, the life of Dr. William Price
                        Dean Powell (2005)
                        ISBN 0-9550854-0-3

Poynder, M.   Lost Science of the Stone Age
                        Green Magic (2004)
                        ISBN 0-9542-9639-7

Reddish, V.C.   The D-Force
                        The Jane Street Print Co. (1993)
                        ISBN 0-9522525-0-3

Reddish, V.C.   Dowsing Interferometry
                        Transactions of the Royal Society of Edinburgh:
                        Earth Sciences, 89, 1-9, 1998

Renfrew C.   Before Civilization
                        Jonathan Cape Ltd (1973)
                        ISBN 0-224-00790-4

Renfrew, C & Bahn, P.   Archaeology: Theories, Methods and Practice 4th Ed.
Thames & Hudson, (2004)
ISBN 0-500-2844-5

Roberts, J.   The Stone Circles of Cork and Kerry
Bandia Publishing, Drumfin, County Sligo, Ireland
ISBN 1-9010-8351-9

Shalatonin, V.   Paper presentation to the Congress of the BSD 2003

Timpson, J.   Timpson's Ley lines
Cassell & Co.
ISBN 0-304-35402-3

Tromp, S.W.   Psychical Physics
Elsevier, (1949)

Verschuur, G.L.   Hidden attractions. The mystery and history of magnetism
Oxford University Press (1993)
ISBN   0-19-510655-5

Watkins, A.   The Old Straight Track (first published 1925)
Abacus (1974)

Watkins, A.   The Ley Hunters' Manual (first published 1927)
Turnstone Press, (1983)

Welfare, S &   Arthur C Clarke's Mysterious World
Fairley, J.   Book Club Associates, London (1970)

Wilcock, J.   The use of dowsing for the location of caves, with some results from
the first Royal Forest of Dean Caving Symposium, June 1994
http://www.sop.inria.fr/agos-sophia/sis/dowsdean.html
4 January 2006

Zirker, J.B.   Journey from the centre of the Sun
Princeton University Press (2002)
ISBN 0-691-05781-8

Analysis

The process of analysing something for some or all of its constituent parts. Chemical analysis is when chemical procedures are applied to determining the chemical composition of a material. In biolocation the magnetic fields characteristic of a substance are used to identify if that substance is present.

Angle of Approach

In dowsing the angle at which a dowser approaches a field line is referred to as the angle of approach. Clear dowsing responses are only obtained at certain angles of approach e.g. normal (90°), 45° or at a tangent.

Aura

An invisible subtle emanation from living things which some people claim to be able to see as a luminous or coloured glow. In the context of biolocation the aura is the complex of magnetic fields emanating from an object. All the components of the object are considered to contribute magnetic fields to the aura. The aura therefore contains a great deal of information about the object.

Beam Splitter

If a magnetic field impinges on a stone set at a certain angle the incident beam appears to split into two beams going in opposite directions and at right angles to the incident beam. The stone is referred to as the beam splitter but it is the magnetic fields of the stone that do the splitting.

Bell Chambers

The first underground Druidic chamber to be found and entered was 'bell' shaped. Not all chambers are 'bell' shaped and some have flat roofs which are supported by many columns. The term 'Bell Chamber' was used by the authors for a while to cover all the chambers being found which were linked to the underground river associated with a temple.

Biolocation   See Appendix 1 and 3

Blind Spring

A vertical movement of water that does not break the surface. The spring is fed through one or more horizontal fissures and the water may leave through a single or a number of fissures. It is believed by dowsers that blind springs are associated with stone circles and churches. To date we have not found a blind spring associated with any of the sacred sites we have looked at.

Blue Energy Line

Magnetic fields which give a positive response when dowsing with blue on the Mager Wheel are referred to as blue. See Box 20.1.

Bone Depository

An underground chamber designed for the storage of bones. A Henge Temple has four bone depositories, each at one of the four quarters. There are also chambers for bones along the access tunnel to the central tomb.

Celt

The people of Gaul who moved there from Eastern Europe in about 800BC. The modern usage of the term Celt dates from 1707 when an antiquarian, Edward Lhayd, used it to describe the Irish, Welsh, Cornish and Breton languages as a distinct group. (see Barbarians by Terry Jones and Alan Ereira). Before 1700 the peoples of the Atlantic seaboard were not known as Celts and they almost certainly belong to a group that migrated up the Atlantic coast from the south. The Henge people might be a better term for the inhabitants of the British Isles prior to 43AD as it would indicate a different origin and culture to the true Celts from Eastern Europe.

Civilization
The presence of towns, monumental buildings and writing. These indicate the presence of princes and priests, fulltime professional craftsmen, art and technical achievement. *(See Renfrew 'Before Civilization' page 193).*

Closing the Circle
When pursuing a line of reasoning the idea is to make deductions based on the original idea or hypothesis and test them. If the deductions are correct they support the original hypothesis. For example if the magnetic fields (aura) of a tree are due to the chemicals from which it is made being acted on by an incident field it should be possible to generate the tree's aura from the chemicals. If when this is done the same aura is generated as that of the tree the hypothesis has been vindicated. If the incident magnetic field is identified and the aura disappears when it is blocked by a physical object such as a board then this again supports the hypothesis. If the incident field is also blocked by a witness of the field this again supports the hypothesis and the circle of reasoning is closed.

Colour
The visual sensation produced by different wavelengths in the visible range of electro-magnetic radiation. In dowsing, colour refers to the colours of the Mager Rosette or colour wheel which are black, white, purple, blue, green, yellow, red and grey. These colours are used to characterise magnetic fields.

Conjecture
An opinion or a guess formed without supporting proof or on slight or defective evidence.

Convergence of Evidence
Where direct proof of an idea is not possible then a number of lines of evidence may point to the idea as being the most likely interpretation. For example, it is not possible to say that the Eucharist was practiced in Stone Age Temples. However, if on finding a rectangular stone or wood table in the middle of a temple or room which is covered by a fabric cloth and it is also found that on one side of the table there is a footpath to a small room and the footpath is used by somebody with feathers as part of their clothing, also that there is a line of wheat bread and beer or wine on the other side of the table and that the people consuming it are dressed in hessian, nettle or wool fabrics then there is a convergence of evidence pointing to a Eucharistic type ceremony. None of the evidence can prove that the Eucharist took place but together they provide strong support for it.

Cromlech
The dictionary defines a cromlech as a prehistoric stone circle. The term was also used to describe Dolmen and megalithic tombs.

Crucifix
A cross is a gibbet used by the Romans to display criminals. It is made from two pieces of timber placed transversely to each other in the form of an 'X' or '†'. The crucifix is a cross with a picture or a model of Christ mounted on it. In this text the term 'crucifix' has been used

to indicate a Roman cross 'X' with a body painted on it. The first 'crucifixes' to be identified appeared to be of Christ but then others were found with no blood or crown of thorns. These were then found to be birth-tables and not crucifixes. On the birth-table the woman has her arms and legs splayed as if on a Roman cross and held by leather straps.

Culture
The term is used to identify the characteristics of a group of people who have developed cities and towns, roads, trade, the arts, engineering techniques, administrative systems and cultural traditions which link people over a large area.

Diamagnetic
When a material is subjected to an applied magnetic field, say from a magnet, it develops its own magnetic field. If the field that is developed is in the opposite direction to the applied field i.e. a north pole faces a north pole, the material is referred to as diamagnetic and the field developed is a diamagnetic field. All biological and organic materials are diamagnetic. Because the diamagnetic field is produced by complete shells of orbiting electrons in the atom or molecule the magnetic fields of all materials have a diamagnetic component. Some materials such as salt crystals ($NaC\ell$) appear to be pure diamagnetic magnets. The human magnetic sensory system has a sensor in the head for diamagnetic fields.

Dolmen
The word *(see W R Corliss)* is derived from the 'Celtic' dual and means table with men. The 'tables' that have survived are made of stone. Archaeologists consider dolmen to be burial sites but there is little evidence for this. There are a number of different structures with different functions referred to as dolmen and it is possible that some may have been associated with burial chambers. All the stone dolmen checked by the authors in the British Isles have been used for excarnation. Because of this both the wood and stone excarnation enclosures have been called dolmen. At the moment the stone dolmen appear to be mainly associated with the western dolmen which was used for the High Priestess. This may mean that her rank warranted a special dolmen or possibly as there was only a call for one once in a while stone provided the durability to ensure that one did not have to be built in a hurry. The other three dolmen of the temples studied were built from wood and brick but it is possible that stone was used where rock and stones were available. The wooden dolmen were elaborate and designed to keep animals out and invite birds in. They were designed to be easily cleaned, and could handle many bodies at a time. The coating of bones with red ochre was done in the dolmen.

Dowsing
The art of identifying the presence of the magnetic fields associated with objects usually in the ground and hidden. A device is often used such as a pendulum or 'L' rods. The device amplifies the small changes in the tone of muscles in the arm which are caused by passing through a magnetic field or crossing a magnetic gradient. The word dowsing is said to be derived from medieval German (da sein) which means 'it is there'. The first recorded use of the word is ascribed to John Lock in 1692. By using 'L' rods most people can identify the change in muscle tone as they walk over a buried object. To try and identify the origin of the field and to determine what the object is two systems are used. One is intuitive dowsing based on asking questions the other is science based and rational and referred to as biolocation.

Dowsers North
The earth has a diamagnetic field with its north pole at the same point as magnetic north. A dowser can identify the direction of north by doing a sweep search. The rods cross when the dowser is facing north. The diamagnetic north is subjected to solar interference and can drift to the east or disappear.

Dyke
A large ditch with a bank on one side and a road or path at the bottom used for ceremonial purposes.

Earth Energy
Objects or chemical stains buried in the ground create dowsable energy patterns. Flowing underground water and geological features also create magnetic fields coming from the earth.

Earth Goddess
The Greek name for mother earth is Gaia but other people would have had their own names. The Goddess is said to date from man's earliest days going back to the Neolithic and possibly earlier. The Goddess is celebrated in spring festivals around the world. The literature is full of stories about Earth Goddesses for example see 'The Celts' by Juliette Wood. However, we now know that the Henge people were not Celts and the separate beliefs of the Henge people and the Celts are going to take some disentangling. At the moment we have not discovered any conclusive evidence of the Henge people having an Earth Goddess. It is just that their engineering activities make it look as if an Earth Goddess was part of their belief system. Later on and in the Iron Age there is the picture in their temples of a Goddess giving birth.

Energy
The strict meaning of the word is as a measure of a system's ability to do work and is measured in joules. In the dowsing world the term 'energy' is often used to describe the magnetic fields that can be sensed by people.

Energy Engineering
When it was discovered that the stones in stone circles were arranged so as to create circular magnetic fields and that other stones attracted specific parts of the field out of the circle the term 'energy engineering' was coined. To be correct the term should be magnetic field engineering but the term energy to describe magnetic fields is very widely used. Energy engineering is used to describe any arrangement of stones intended to create and direct magnetic fields.

Energy (field) Shadow
If a witness for a field is placed in that field, for example a piece of iron in an iron paramagnetic field, there is an energy shadow behind the witness. Because a shadow is produced the technique is used to provide protection from undesirable magnetic fields such as those from phosphate and carbon.

Energy Line
When magnetic fields travel in a line from a source to a distant object or destination the dowser detects what appears to be a line of energy or a beam of emitted energy.

Evidence
Evidence is what can be seen and is clear to the mind. Evidence is what makes something clear as a fact. In the present context much of the presented data or facts depends on evidence obtained by dowsing. For this reason it is important to have more than one line of evidence supporting data or what is claimed as a fact. Evidence and observations are also considered to be stronger or more reliable if there is a system of standardisation and quality control involved. The courts in the US use the Daubert Criteria which includes that the theory or technique is testable.

Excarnation

The stripping of the flesh from the bones of a dead person before burial of the bones has been practiced by a number of religions round the world. The body is normally placed on an elevated rock or timber structure which enables birds to feed on the body and prevents animals accessing it. The Henge society used excarnation for a least 3000 years. There is evidence that it was used until shortly before the Romans arrived in Britain (see page 31 Verulamium by R. Niblett). The reason for using excarnation appears to be that the 'spirit' was believed to be connected with the bones and not the flesh.

Ferromagnetic

Materials such as iron, nickel, cobalt and their alloys are ferromagnetic. They are used for making permanent magnets. Modern high strength magnets often use rare earth elements or neodymium.

Fields

The dictionary defines a field as a region in which a body experiences a force as a result of the presence of another body in the vicinity. In the present context the term field is applied to magnetic fields. The force of these fields can vary in direction and appears to follow field lines. All material produces magnetic fields. The theory underlying the dowsing phenomena is that when the dowser moves through a field the magnetic field sensors are activated if they cross the field lines at certain angles. In dowsing, a field is an area of energy or magnetic fields round an object. It is often referred to as the aura of the body.

Flax

A slender stemmed plant with blue flowers. Today there are two varieties grown, one for its seeds from which linseed oil is made and one for its fibres which is used for making linen. Linen was being woven in ancient Egypt about 4000 years ago.

Focus

In dowsing it is the ability to concentrate on a particular source of a magnetic field for example drains or postholes. As a result of focusing the dowser will not pick up other fields. The focusing used by the magnetic sense appears to be very similar to that used by the other senses.

Form

The shape of a dowsable field indicates what may be giving rise to the field. Dowsing the form is determining the shape of the hidden object. The shape of an object such as a pipe or the floor plan of a building leaves a chemical stain in the soil which can be followed and the shape of the original object identified. In dowsing the form of an object it is important to follow the same field for example iron, sand or brick and not to move between different fields. A witness for the stain such as brick or iron aids this process.

Grail

(See Cotterell, A. Dictionary of World Mythology). The Grail of mythology is said to be the vessel used at the Last Supper and at the Crucifixion to receive the blood from the spear wound. It is believed to have been brought to Britain by Joseph of Arimathea. The legend of the Grail could be linked to the Celtic cauldron or the pool of water in Druidic Temples both of which were considered to have mystic properties.

Grave Goods

In ancient times and even today objects or clothing of significance to the person being buried are placed in the grave with the body. In the Druidic tradition it is likely that all objects were considered to have a spiritual component. It would be this spiritual component that would go into the next world with the deceased not the physical object. This concept could be common to many cultures where grave goods were placed in tombs.

Green Man

A face emerging from a mass of foliage. Its original meaning has been lost but its complexity indicates that it could be a test of a stone mason's ability. A passing out test. The origins of the Green Man can be traced back to Roman times. He does not have antlers or horns. (Hart, J. The Green Man).

Henge

The term given to a circular earth or stone bank with an internal ditch and opposed entrances. As the authors have to date always found a 6 + 3 temple associated with a visible henge the term henge has been applied to all circular temples. Because these temples can be traced from the Neolithic to the Iron Age this period of history is referred to as the Henge Age.

Human Magnetic Sense   see Magnetic Sense

Hydrogen ion

Acids are compounds that dissociate in water to produce positive hydrogen ions (H+). The hydrogen ion, which is a proton, does not exist on its own but is combined with a molecule of water to produce H3O+. The higher the concentration of hydrogen irons the more acid the solution. The proton or hydrogen ion is paramagnetic.

Image

The magnetic fields of a target can be attracted to a site many metres away. Where the field lines re-enters the ground it generates a magnetic image of the target. The image dowses as if it is the real target. When an image is formed the true target may not be dowsable and disappears.

Ley Lines

There are a large number of ancient sites such as prehistoric tumuli, castles and churches that are connected by straight lines referred to as ley lines. Five sites in line are required to qualify as a ley. The lines can be identified as ancient tracks or roads linking places of importance. The chemical stains left in the ground can be analysed and identified. Other energy lines that may be referred to as ley lines are due to airborne energy lines passing between geological or man-made features of the landscape. A third source of ley lines could be underground geological features or man-made structures such as pipelines.

Long Barrow

Long burial mounds are common in southern Britain with perhaps the best known one being the West Kennet long barrow near Avebury. The archaeologists date their buildings at between 6000 and 5000 years ago. Long barrows can be up to 70m long. The barrows are said to contain burial chambers for the bones left after excarnation. The long barrows investigated by the authors are inside large buildings and are linked by ceremonial routes to Druidic Henge Temples. They therefore formed part of the religious buildings associated with the main temple buildings. Highly decorated chariots were used in the transport of bones between the temple and the building housing the long barrow. As part of a religious building the barrows were terraced with pathways reserved for priests.

Magnetism

The ability to attract other substances. Normally applied to ferromagnetic materials which can attract iron. Phenomena associated with magnetic fields. The magnetic fields arise from the magnetism of the atoms and molecules of a material. There are four main types of magnetic behaviour. The fields responsible for two of them, diamagnetism and paramagnetism, can be sensed by the human magnetic sensory system. (See a Science Dictionary for details of the fields).

Magnetic Axis

The line joining the north and south poles of a magnetic object. Each magnetic field will have an axis.

Magnetic Shadow

The diamagnetic and paramagnetic field lines from a substance will be attracted to and blocked by a piece of the substance (witness) placed at a distance from the source and which in effect cast a shadow.

Magnetic Remanence

Magnetic remanence is where a material retains, at least for a while, a field after it has been magnetised by an applied field. Most paramagnetic materials do not show remanence. Diamagnetic materials show remanence and with the retained diamagnetic field paramagnetic fields will be generated, normally at right angles to the diamagnetic field.

Map Dowsing

Many dowsers employ a method of dowsing using a map. This may be done before or in place of visiting the site. The aim is to identify underground streams and energy lines on the map prior to visiting the site.

Material

Consisting of matter. In the present context all matter is considered to possess diamagnetic fields and most will have paramagnetic fields which are characteristic of the material or matter under consideration.

Mystery

Dowsing is often referred to as a mystery which can be defined as a secret doctrine known only to the initiated. It is also defined as a phenomenon or happening that can not be explained.

Neolithic

The later or New Stone Age covering, in Britain the period 4000 to 2000BC. Agriculture is practiced in the New Stone Age and religion is forming a major component of social activity.

Paramagnetism

When atoms or molecules have unpaired electrons so that magnetic fields from electron spins do not balance out a net field is produced called a paramagnetic field. The net field aligns in the direction of the applied field. Paramagnetism can occur in metals as a result of the magnetic field associated with the conducting electrons spin. If the applied field is a diamagnetic one the paramagnetic axis tends to be at right angles to the diamagnetic field.

Paullinus, Suetonius

A Roman general of high reputation. He was responsible for the attack on Anglesey in 60AD and for the defeat of Queen Boudicca in 60 to 61AD. His troops were almost certainly responsible for attacking and executing Druids in the western midlands and Wales.

Phantom

The term is sometimes used to describe the magnetic field image of a target. The phantom fields give an illusion of the target. The more common term is image. The image of a target can contain considerable detail and be used to analyse the detail of a target.

Polarity
The term 'polarity' is used to indicate opposites such as north and south poles, positive and negative. It is indicated by the rods opening or closing. At the practical level polarity is difficult to determine as responses differ between people and at different times in the same person.

Prayer Stones Rounded water worn stones which have been painted. There is now little visible paint on the stones but when analysed the paints used can be identified. Suitably shaped stones are used as lamps (candles) and for burning incense.

Pre-history
The period before the Romans arrived in the British Isles. It is believed that before the Romans arrived that there was no written record upon which the history of the Henge people could be based. If there were written records they were destroyed by the Romans. The lack of records from Ireland does indicate that the Henge people did not maintain written records. However, the artwork that appears later in Irish Christian manuscripts must have its origin much earlier in history. The Henge people were accomplished artists and could have recorded events and things in painted pictures at least.

Proton
One of the elementary particles which form the nucleus of the hydrogen atom. On its own it is the hydrogen ion and forms $H3O+$. Solar flares eject protons into the solar wind where they produce magnetic fields which may interfere with dowsing.

Proxy-Dowsing
If an observer with rods at the ready watches somebody, the proxy, walk in a line at a distance of say 30m to 50m, when the proxy walks through an energy field the observer's rods will move. The theory is that if a person or a large animal walks through an energy field they create magnetic 'ripples' which the proxy-dowser responds to.

Puddingstone
A conglomerate which geologists say is many millions of years old and was formed in rivers. The research done by the authors indicates that the Henge people made and used a cement that enabled them to make a concrete. Because the formation of the concrete did not involve heat traces of the tools used and ceremonial activities involving the use of bronze and gold implements can be identified in puddingstone. The cement appears to be made from clay, wood ash and lime.

Remanence (retentivity)
When a material is subjected to a magnetic field, if the applied field is removed the material may still maintain a field for a period of time. The classic example is a ferromagnetic field being retained by a permanent magnet. Paramagnetic materials do not show remanence i.e. they do not retain a field. If a copper coin is placed on a concrete floor for a few minutes the copper coin can still be dowsed after it has been removed. The 'phantom' coin disappears after a short while. The presence of a dowsable image of the coin is due to a diamagnetic field being retained by the concrete. The diamagnetic field induces paramagnetic fields in the concrete which are dowsed. As the retained diamagnetic field fades so do the paramagnetic ones.

Science Based Dowsing   See Appendix 1

Scott's Grotto, Ware, Hertfordshire
A series of interconnected chambers extending over 60ft into a chalk hillside. The grotto is said to have been built by John Scott in the 18th Century. The indications are that the grotto is based on a Druidic underground temple.

Search Mode
When looking for the direction of a specific field for example, water or phosphate, the dowser turns through 360°. When facing the target the rods will cross so indicating its direction.

Sensitive
Some people appear to be very sensitive to magnetic fields and can identify magnetic variations in their environment due to contaminants in the ground.

Sensor – sense organ
A part of the body of an animal designed to contain special cells (receptors) which are sensitive to specific stimuli for example light, heat, sound, pressure, smell, taste. The receptors are designed to be very sensitive to the target energy or molecules.

Sensory System
Refers to both the sense organ and the processing of the information by the nervous system and brain.

Sensory System – magnetic field
There is a considerable amount of research on the magnetic sense of animals, fish, and birds. Much of it has been done using ferromagnetic fields and so may be of limited value if the sensory systems studied are similar to the human magnetic sense which is blind to ferromagnetic fields. The human magnetic sense consists of a sensor in the head for diamagnetic fields. The evidence at the moment is that this sensor appears to add diamagnetic fields whatever their source. The sensor is directionally sensitive. The paramagnetic sensor is in the skin and it appears to 'see' different fields as separate and does not add them. There may be some exceptions to this general rule. The paramagnetic sensor is not in the skin of the head. The sensor has a left and right half dividing at the midline of the body. It is directionally sensitive and can identify different fields for example distinguish a copper field from a calcium carbonate field. Both magnetic sensors are extremely sensitive to magnetic fields and match the sensitivity of the other sensors. There does not appear to be a sensor for ferromagnetic fields in humans. Both the magnetic sensors can be saturated by strong fields making it appear to the dowser as if no field is present.

Serial Number – sometimes referred to as 'series'
The number of times the rods will cross on re-entering an object's magnetic field is referred to as the serial number of that object. The idea is that the magnetic sense fatigues after a number of repetitions although this may not be the explanation. Different materials may have different serial numbers which typically range from 1 to 7. The rings on the Mager disc indicate serial number. To identify an object the correct ring for the serial number of the object as well as the correct colour must be held between the finger and thumb.

Shaman
An Asian priest or medicine man. A person who could act as an emissary and contact the spirits or forces that controlled the world. The shamans made mind journeys into the spiritual world using drugs. The shaman saw the world as a spiritual one with everything linked spiritually.

Silica

Silica is the dioxide of silicon, SiO2 best known in its crystalline form as sand and quarts. Silica glass and flint have silica as a main constituent in its amorphous or non crystalline form. Ordinary glass also contains silica.

Silicon (Si)

Is the element

Solar Flare

There are violent explosions in the lower corona of the sun which are referred to as solar flares. They produce ultraviolet, X-ray and solar cosmic radiations. Importantly the flare produces a stream of protons and electrons. If these head for the earth, geomagnetic storms are produced which interfere with terrestrial magnetic fields and may make dowsing impossible.

Solar Wind

An outward flow of charged particles from the sun. Most of the particles are protons and electrons from the sun's corona. The sun's magnetic field controls the particles until they encounter the earth's magnetic field. At high wind speeds above about 450km/sec. dowsers may experience problems.

Spirit Painting

Many ancient societies believed that living animals and in fact all objects possessed a spirit. By placing something of the animal or object into the paints used to paint them the picture would possess the spirit of the object or animal. A picture of a horse or cat would contain something from the horse or cat. The chemical stains from biological materials are easily identified. It is therefore possible to dowse such paintings for what they represent. As the paintings contain something of the 'spirit' of the object they are referred to as spirit paintings.

Target

The term normally used for the object that is being sought or is giving rise to the magnetic fields being detected.

Witness See Box 8.1

Wheelhouse

A large circular house where the roof is supported by more than one circle of posts.

List of Figures

Figure 1.1	Clew Bay, Co Mayo	17
Figure B1.1.1	Movements of 'L' Rods	19
Figure B1.1.2	Dowsing the Form of a Field	19
Figure B1.2.1	Swinside Stone Circle, Cumbria	20
Figure B1.2.2	The Distribution of Ancient Stone Circles	20
Figure B1.3.1	Chemical Stains	23
Figure B1.4.1	The Mager Colour Disc	24
Figure 2.1	Salt marsh and the Killadangan Cromleck Stone Row and Stone Circle	26
Figure 2.2	The Cromleck	26
Figure 2.3	Portal to the Circle	27
Figure 2.4	The Blue Line	27
Figure 2.5	The Beam Splitter	30
Figure 2.6	A Section of the Circle	30
Figure 2.7	Killadangan Magnetic Alignments	31
Figure 2.8	Energy Model of the Killadangan Stone Circle	32
Figure 3.1	Avebury from an Aerial Perspective	36
Figure 3.2	Magnetic Axes of Bricks	37
Figure 3.3	The Rollright Stone Circle	38
Figure 3.4	The Drombeg Stone Circle	39
Figure 3.5	The Nine Ladies Recumbent Stone Circle on Stanton Moor	41
Figure 3.6	The Loan Head Recumbent Stone Circle	42
Figure 4.1	Field Interaction	46
Figure 4.2	Berkhamsted Castle	47
Figure 5.1	A Woodhenge with a Tomb	53
Figure 5.2	A Witness Set	54
Figure 5.3	A Dell	57
Figure 5.4	Children's Graves	58
Figure 5.5	Woodhenge	60
Figure 6.1	St. Albans Cathedral Site	68
Figure 6.2	Magnetic Loop	70
Figure 6.3	Locating the Stains of an Archaeological Feature	72
Figure 7.1	A Yorkshire Stone Circle	75
Figure 7.2	The Magnetic fields of the Yorkshire Stone Circle	76
Figure 7.3	The Prince and Princess	78
Figure 7.4	The Underground Temple	79
Figure 7.5	Schematic Design of the Underground Temple	80
Figure 7.6	The Seven Altars	81
Figure 7.7	Chenies Manor House Temple	82
Figure 8.1a	Ceremonial Footpaths	88
Figure 8.1b	Altar	89
Figure 8.2	The St. Albans Recreation Ground Temple	89
Figure B8.1.1	Circular Sweep Search	86
Figure B8.2.1	The Cross and Thunderbirds Photographs	93
Figure 9.1	Avebury Stone Circles and Village	98
Figure 9.2	Stanton Drew	102
Figure 9.3	Temple Complex	103
Figure 9.4	Stonehenge Bottom	103
Figure 9.5a	Dells	105
Figure 9.5b	Dyke Temple Entrance	105
Figure 9.6	Stanton Drew North gate Area	107
Figure 9.7	Bone Reception	107
Figure 9.8	Execution Lines	109
Figure 10.1	Penang Botanic Gardens	113
Figure 10.2	Berkhamsted Castle - Historical View	113

Figure 10.3	The Castle Ruins	115
Figure 10.4	Bovingdon	119
Figure 10.5	Tunnels	121
Figure 10.6	Dyke Temple	121
Figure 10.7	Dressed Stone	121
Figure 10.8	Prayer Stones	123
Figure 10.9	Druid's House	124
Figure 10.10	Workstations	124
Figure 10.11	Artefacts from Temple Sites	129
Figure 10.12	Stone and Iron Weapons	130
Figure 10.13	The Thornborough Rings	133
Figure B10.1.1	Tree Auras	135
Figure B10.1.2	Star Bursts	135
Figure B10.1.3	Thunderbird Aura	136
Figure B10.1.4	Dolmen Pictogram	136
Figure B10.2.1	A Section of Offa's Dyke	137
Figure B10.2.2	Beech Bottom Dyke, St Albans	137

Figure 11.1	Henge Temple Location	139
Figure 11.2	Temple Pool	139
Figure 11.3	Stone Age Temple	142
Figure 11.4	Tomb Complex	143
Figure 11.5	The Celtic Cross	145
Figure 11.6	The Druidic Honour Guard	146
Figure 11.7	Altars	148

Figure 12.1	Maiden Castle	151
Figure 12.2	Concept Sketch of a Dyke	151
Figure 12.3	Blood Below the West Gate	154
Figure 12.4	East Gate Battle Blood	154
Figure 12.5a	Unloading a HGV	157
Figure 12.5b	Tipping Apparatus	157
Figure 12.6	Sacred Pool	159

Figure 13.1	The Western Dolmen	161
Figure 13.2	The Dowsed Western Dolmen	162
Figure 13.3	The Southern Dolmen	162
Figure 13.4a	The outline of the Eastern Dolmen	163
Figure 13.4b	The Eastern Dolmen Main Features	164
Figure 13.5	Toilets	165
Figure 13.6	Altar Positions	167
Figure 13.7	The Southern Bone Depository	167
Figure 13.8	The Stoney Lane Pool	169
Figure 13.9	The Well and Well house	171
Figure 13.10	The Depth of the Well	172

Figure 14.1	Stoney Lane Tombs	175
Figure 14.2	A Stone Age Tombs	175
Figure 14.3	Temple Doors	176
Figure 14.4	The East Gate and its Chariot Route	179
Figure 14.5	The North Gate and its Chariot Route	180

Figure 15.1	The Tunbridge Wells Temples	184
Figure 15.2	Stone Columns	184
Figure 15.3	A Stone Age Temple	186
Figure 15.4a	The Rocking Stone, Stone Circle at Pontypridd	186
Figure 15.4b	Underground Temple at the Rocking Stone	186
Figure 15.5	The Stonehenge Ceremonial Pool	189
Figure 15.6a	The Pool Plumbing	189
Figure 15.6b	Stonehenge Plumbing	189
Figure 15.7	Stonehenge Execution Lines	191
Figure 15.8	Stonehenge Water supply	191

Figure 16.1	Henge Art	194
Figure 16.2	Dell complex	195
Figure 16.3	Puddingstone	196
Figure 16.4	The Magnetic Patterns of Flowing Water	198
Figure 16.5	The Structure of a Diamond	199
Figure 17.1	Survey Sites at Durrington Walls	203
Figure 17.2	Painted Road Diamonds	204
Figure 17.3	Concept Sketch of the Structure of a Dyke	205
Figure 17.4	Temple Entrance	207
Figure 17.5	The East Gate	208
Figure 18.1	Beaumaris Town and Survey Site	212
Figure 18.2	Beaumaris Beach and Seafront	212
Figure 18.3	Cremlyn Dolmen	217
Figure 18.4	Glastonbury Tor	218
Figure 18.5	Tor Road Terracing	220
Figure 18.6	Tor Temple	223
Figure 19.1	Linteled Bricks	225
Figure 19.2	Trilithon Horseshoe	225
Figure 19.3	Linteled Stone Circle	227
Figure 19.4	The Heel Stone Beam Splitter	227
Figure 19.5	The Model of Present Day Stonehenge	227
Figure 19.6	Blue Beam Generation	228
Figure 19.7	The Polarity of the Stones	230
Figure 19.8	Stonehenge Cursus	232
Figure 19.9	The Little Hay Cursus	232
Figure 19.10	The Arena Symbology	236
Figure 19.11	Entrance to a Hall of a Mountain King	238
Figure 20.1	The Stonehenge Trilithon Energy Model	241
Figure 20.2	The Beam Splitter	241
Figure 20.3	The Second Beam Splitter	242
Figure 20.4	The Stonehenge Blue Beam System	246
Figure 20.5	The Left Split Beam	248
Figure 20.6	A Phantom Person	248
Figure B20.2.1	Polarised Magnetic Fields	244
Figure 21.1	Photomap of the Hell-Fire Caves	252
Figure 21.2	The Entrance to the Caves	252
Figure 21.3	The Banqueting Hall Floor	255
Figure 21.4	Tunnel Entrance	255
Figure 22.1	A Tunnel Entrance in the Dyke	260
Figure 22.2	No Turning Circle for Chariots	260
Figure 22.3a	Painted Stones	264
Figure 22.3b	Stone Pillars	264
Figure 22.4	Dowsing the West Kennet Avenue	264
Figure 22.5	The Wansdyke	266
Figure 22.6	Concept Sketch of Wansdyke	267
Figure 22.7	The Silbury Enigma	270
Figure 22.8	Marburgh Henge	273
Figure 22.9	Marburgh Henge Central Megalith	274
Figure 22.10	Dells in the Scottish Landscape	275
Figure 22.11	Druidic Highway	276
Figure 22.12	The Tomnaverie Stone Circle	276
Figure 22.13	Energy Engineered Recumbent Stone Circle	277
Figure 22.14	A Sacred Stone	278
Figure 23.1	Druidic Roads	283
Figure 23.2	The Merry Maidens, Cornwall	283
Figure 23.3	Pool Area of Kestor Temple	287
Figure 23.4	Kestor	287
Figure 23.5	Tor Energy Engineering	289

Figure 23.6   Dartmoor Stone Circles   291
Figure 23.7   Druidic Engineering   292
Figure 23.8   St. Albans Abbey and Cathedral   292

Figure 24.1   St. Albans Cathedral Druidic Alignments   295
Figure 24.2   The Stoney Lane Well   298
Figure 24.3   The Chapel On Top of a Sealed Well at St Albans Cathedral   300
Figure 24.4   An Underground Chapel at Stonehenge   304
Figure 24.5   Fish and Crucifix   310

Figure 25.1   Bolt Tail   312
Figure 25.2   The Hurlers, Bodmin Moor   313
Figure 25.3   The Long Barrow on Bolt Tail   314
Figure 25.4   The Long Barrow Bone Unloading Points   315
Figure 25.5   Concept Sketch of a Long Barrow Building   315
Figure 25.6   The Thunderbirds of the Tomb   317
Figure 25.7   The General Layout of the North Gate Area of the Temple   317
Figure 25.8   The Crucifix at Stonehenge   319
Figure 25.9   The Processional Way Under the Western Split Beam   321
Figure 25.10   Chamber Wall and Roof Supports   321
Figure 25.11   The Plan of the Bolt Tail Temple   323
Figure 25.12   The Temple Wall   323
Figure 25.13   The Font Area   325
Figure 25.14a   Artists Impression of the Cait-Sith   325
Figure 25.14b   The Cait-Sith   326
Figure B25.2.1   The Snake God   329

Figure 26.1   The Hell-Fire Cave Crucifix   331
Figure 26.2   The Rollright Stones Give up their Secret   334
Figure 26.3   The 'White Horse' at Uffington   335
Figure 26.4   Dragon Hill   335
Figure 26.5   The Manger Valley   338

Figure 27.1   The Entrance to a Tunnel System   345
Figure 27.2   The Proto-Christian Crucifix   346
Figure 27.3   Pottery Shards   354
Figure 27.4   Druidic Road Building   354
Figure 27.5   A Burial Place for the Bones   357
Figure 27.6   The Origin of Henge Artwork   359
Figure 27.7   The Thunderbird Motif   359

Figure B28.1.1a   Beltane - Mythology   369
Figure B28.1.1b   Beltane - Biolocation   369

Figure A1.1   Biolocation Links to the Sciences   376
Figure A2.1   Earth Energies Associated with a Temple Site   381
Figure A3.1   Biolocation Model   383
Figure A3.2   Magnetic Field Lines and Dowsing Responses   384
Figure A3.3   Magnetic Axes   386
Figure A3.4   Dowsing Response to a Diamagnetic Field   386
Figure A3.5   Dowsing Response to a Paramagnetic Field   388
Figure A3.6   Collimated Fields   388
Figure A3.7   Magnetic Field and Target Mass   388
Figure A3.8   Field Shadow from a Witness   391
Figure A3.9   Polarized Fields   391
Figure A3.10   Magnetic Images   394
Figure A3.11   A Diamagnetic Magnet   394

Index

A

abattoir 168

access shaft 12, 61, 65, 67, 80, 81, 82, 83, 84, 90, 96, 117, 119, 187, 288, 296, 299, 300, 302, 303, 304, 305, 312, 313, 316, 318, 319, 320, 342, 357

alignments 28, 34, 350

altar 39, 42, 77, 79, 80, 87, 89, 90, 103, 120, 121, 144, 147, 150, 167, 168, 169, 229, 231, 233, 236, 250, 274, 278, 279, 297, 300, 301, 306, 325, 357, 358

aluminium 29, 31, 33, 35, 71, 87, 170, 224, 225, 240, 243, 245, 248, 379, 386, 389, 390

Anglesey 61, 119, 122, 153, 211, 212, 213, 215, 216, 217, 218, 221, 351, 360, 407

anteroom 58, 69, 77, 94, 100, 103, 127, 144, 167, 186

antlers 145, 146, 166, 192, 268, 297, 299, 301, 309, 330, 351, 353, 406

Arbor Low 41, 49, 50

arenas 232, 235, 236, 301, 359, 362

arrow 128, 141, 142, 144, 145, 280, 353

arsenic 54, 118, 127, 138, 140, 144, 150, 152, 154, 155, 159, 177, 183, 185, 187, 197, 204, 205, 208, 219, 228, 237, 253, 275, 278, 280, 284, 285, 320

art 11, 35, 44, 87, 149, 197, 199, 201, 203, 250, 254, 294, 302, 322, 354, 355, 358, 359, 362, 367, 402, 403

artwork 40, 87, 108, 115, 134, 149, 181, 194, 198, 199, 216, 217, 251, 254, 273, 295, 296, 308, 333, 334, 336, 358, 361, 408

astronomical events 18, 28, 34, 39, 190

astronomy 48, 158, 272, 368, 377

Astronomy 16

aura 91, 92, 94, 111, 114, 181, 211, 217, 249, 250, 341, 358, 361, 389, 401, 402, 405

Aura 16

B

Bandstand 185, 239

barter 355

battle blood 90, 104, 106, 110, 114, 118, 122, 133, 140, 152, 153, 154, 155, 156, 185, 190, 209, 213, 214, 216, 221, 222, 229, 231, 257, 282, 284, 288, 291, 296, 303, 309, 337

beam splitter 28, 30, 32, 33, 35, 226, 227, 229, 240, 241, 242, 243, 247, 248, 249, 320, 401

Beaumaris 212, 213, 214, 215, 217

Beckhampton Avenue 11, 96, 98, 260, 261, 262, 263

Beech Bottom 267

beef fat 94, 95, 100, 144, 152, 170, 192, 193, 204, 257, 263

beer 101, 122, 125, 177, 215, 235, 239, 271, 297, 324, 353, 402

Bell Chamber 12, 256, 258, 272, 285, 303, 305, 306, 309, 311, 320, 330, 351, 401

Beltane Fire 13, 235, 236, 369

Berkhamsted Castle 9, 45, 47, 113, 114, 115, 116, 127, 346, 347

bier 95, 161, 162, 163, 180, 192

bird perches 160, 161, 162, 163, 333

birth table 13, 331, 346, 359, 362

Blairgowrie 275

blind spring 28, 29, 32, 48, 52, 59, 60, 67, 378, 379, 401

blue beam 30, 228, 240, 241, 242, 247, 248, 277, 340, 342

bluebell 126, 128, 145, 174

Blue Energy Lines 13, 243

blue stone 28, 34, 226, 319, 320, 350

Bodmin Moor 12, 311, 312, 313

Bolt Tail 12, 311, 312, 313, 314, 322, 323, 324, 345, 348, 414

bone depositories 61, 68, 71, 78, 79, 83, 84, 94, 95, 99, 100, 103, 111, 132, 141, 143, 144, 160, 166, 175, 187, 192, 236, 257, 286, 296, 302, 345, 356, 361, 401

Bone Reception 107, 411

Boudicca 106, 119, 153, 185, 211, 213, 364, 365, 407

Bowling Green 185

Braemar 275, 276, 277

Brankam Hill 11, 278, 279, 280, 281

Bronze 4, 9, 18, 21, 25, 44, 48, 49, 52, 53, 55, 57, 59, 60, 63, 67, 85, 87, 94, 96, 104, 117, 118, 119, 125, 127, 128, 130, 131, 132, 133, 134, 138, 139, 140, 141, 144, 145, 147, 150, 152, 154, 155, 156, 159, 170, 173, 174, 175, 176, 177, 179, 181, 183, 185, 188, 192, 193, 197, 204, 206, 219, 221, 222, 223, 228, 231, 233, 234, 235, 239, 246, 253, 259, 261, 266, 267, 268, 269, 270, 278, 279, 283, 284, 286, 288, 291, 294, 295, 304, 306, 309, 311, 320, 321, 322, 332, 348, 349, 350, 351, 355, 370

Bronze Age Boats 9, 85

bronze tracks 183, 204, 220, 228, 233, 269, 320, 344

Bury Woods 105, 347

C

cadmium selenium red 254, 309

Caesar 308, 364

Cait-Sith 322, 325, 326, 327, 330, 337, 370, 414

Calverley Grounds Park 10, 118, 183, 184, 185, 186

cannon 104

carbon dioxide 120

Carleon 238

Castlerigg 39, 51

cat 160, 161, 165, 239, 322, 324, 327, 328, 330, 337, 338, 339, 340, 351, 360, 410

Cathedral 12, 67, 68, 69, 120, 122, 200, 292, 293, 294, 295, 297, 298, 299, 300, 312, 318, 320, 364, 411, 414

Celtic Cross 89, 145, 268, 412

Celts 106, 122, 126, 141, 170, 173, 233, 299, 342, 353, 354, 358, 360, 366, 399, 402, 404

cement 129, 155, 165, 196, 197, 201, 203, 205, 207, 208, 220, 250, 253, 262, 272, 282, 284, 294, 295, 337, 343, 344, 347, 349, 353, 360, 375, 408

ceramic 129, 144, 170, 172, 173, 185, 186, 187, 188, 189, 191, 192, 205, 231, 262, 268, 280, 320, 367

ceremonial battles 229, 231, 235, 239

ceremonial footpath 160

chalk mines 193, 251, 343

chapel of rest 180, 296

chariot 66, 95, 99, 100, 103, 115, 116, 117, 118, 131, 134, 149, 150, 152, 154, 155, 156, 160, 163, 166, 178, 180, 185, 192, 193, 194, 195, 202, 204, 207, 214, 219, 220, 230, 232, 233, 234, 237, 238, 246, 247, 253, 254, 257, 258, 259, 261, 262, 265, 266, 268, 269, 270, 275, 277, 278, 282, 283, 285, 286, 288, 309, 313, 314, 315, 316, 317, 322, 345, 351, 358

Chenies Manor House 9, 82, 83, 94, 104, 411

children 56, 58, 60, 61, 112, 114, 120, 121, 122, 125, 139, 146, 153, 156, 165, 168, 176, 177, 185, 216, 238, 268, 280, 291, 316, 337, 360, 364

Children's Graves 8, 56, 58, 411

Chiswell Green 126, 128, 149, 234, 239, 256, 301

Christian 6, 12, 13, 25, 120, 201, 222, 285, 286, 289, 295, 297, 298, 301, 302, 305, 306, 307, 309, 310, 316, 317, 318, 319, 320, 327, 330, 331, 332, 334, 338, 342, 344, 346, 350, 351, 352, 359, 361, 363, 364, 408, 414

Christianity 4, 12, 141, 219, 280, 294, 295, 298, 301, 306, 307, 308, 309, 310, 311, 318, 320, 327, 328, 330, 332, 334, 342, 352, 358, 360, 361, 362, 363, 364, 370

circumcision 168, 324

Civilization 4, 9, 11, 112, 211, 347, 364, 371, 399, 402

clay tiles 125, 131, 273

Clew Bay 16, 17, 21, 28, 341, 411

Clive Beaton 243

communication 33, 65, 85, 116, 224, 243, 249, 271, 378

conjecture 32, 61, 101, 153, 156, 201, 202, 219, 247, 261, 286, 288, 299, 307, 332, 342, 345, 347, 358, 361, 362, 363, 368, 369, 370

copper 54, 55, 85, 118, 125, 127, 131, 138, 140, 150, 152, 154, 177, 183, 185, 187, 197, 204, 205, 222, 253, 261, 278, 286, 328, 355, 389, 408, 409

Cremlyn 215, 216, 217, 413

cromlech 402

Crucifix 12, 13, 301, 303, 310, 319, 331, 346, 402, 414

Cultural traits 4

Culture 9, 13, 123, 259, 281, 282, 336, 337, 378, 403

cups 197

Cursi 11, 139, 228, 231, 233, 234, 236, 237

Cyst Tomb 188

D

Dartmoor 11, 12, 40, 59, 281, 282, 284, 285, 286, 288, 289, 291, 292, 294, 302, 303, 414

deer 94, 118, 125, 126, 128, 145, 155, 166, 192, 194, 195, 196, 197, 204, 215, 221, 222, 228, 229, 233, 234, 235, 237, 246, 253, 257, 258, 261, 265, 268, 280, 286, 296, 314, 339, 343, 344, 351, 353, 369, 378

dells 57, 67, 343

detrimental energy 90

diamagnetic 7, 21, 29, 31, 32, 34, 35, 38, 39, 40, 41, 42, 43, 44, 74, 110, 187, 224, 250, 341, 356, 381, 382, 385, 386, 387, 388, 389, 391, 393, 394, 395, 403, 407, 408, 409

diamonds 2, 11, 119, 193, 194, 198, 199, 200, 201, 203, 204, 220, 221, 231, 232, 235, 246, 247, 251, 254, 263, 265, 266, 273, 275, 276, 277, 278, 279, 282, 285, 286, 309, 311, 316, 317, 318, 320, 322, 337, 344, 353, 358

ditch 34, 117, 151, 158, 202, 205, 228, 229, 230, 231, 233, 235, 375, 404, 406

dog kennel 126, 371

dogs 233

Dolmen 8, 9, 10, 49, 50, 51, 55, 61, 64, 65, 71, 74, 81, 83, 87, 90, 94, 95, 96, 101, 102, 110, 131, 134, 138, 149, 160, 161, 162, 163, 164, 165, 166, 168, 185, 188, 192, 207, 209, 215, 216, 217, 228, 231, 233, 235, 250, 272, 274, 296, 300, 301, 316, 322, 333, 334, 349, 351, 355, 356, 358, 359, 361, 364, 376, 402, 403, 412, 413

Downholme 134

draft animals 66, 94, 155, 166, 192, 234, 327, 339, 378

Dragon Hill 263, 335, 336, 337, 338, 339, 414

drains 3, 45, 112, 125, 333, 374, 384, 405

Drombeg Stone Circle 8, 39, 411

drover's way 52

Druidic Honour Guard 146, 412

Druidic Temples 80, 106, 111, 116, 117, 140, 156, 185, 193, 205, 216, 258, 301, 347, 405

Druidism 4, 91, 100, 111, 126, 140, 141, 181, 201, 209, 211, 281, 282, 295, 302, 306, 307, 308, 310, 316, 318, 345, 347, 360, 361, 362, 365

Druid's house 119, 124, 128

dry stone walling 237, 238, 250, 257, 282, 311, 344

Durrington Walls 11, 13, 133, 202, 203, 205, 206, 207, 208, 209, 215, 221, 226, 230, 347, 351, 413

Dyke 9, 11, 105, 106, 120, 121, 122, 132, 134, 139, 150, 151, 153, 158, 203, 205, 209, 228, 234, 250, 258, 260, 266, 309, 343, 345, 346, 347, 351, 404, 411, 412, 413

Dyke Temple 105, 120, 121, 122, 132, 150, 151, 153, 158, 209, 346, 347, 411, 412

E

earth energies 16, 18, 21, 28, 45, 361

Earth Goddess 18, 174, 175, 194, 222, 238, 254, 256, 257, 258, 297, 342, 343, 348, 356, 358, 360, 366, 378, 404

Egyptian 277

Element Collection 53

Emperor Nero 120

energy 16, 18, 21, 22, 25, 27, 28, 29, 32, 33, 34, 35, 37, 38, 40, 42, 43, 63, 65, 67, 74, 90, 110, 116, 158, 185, 188, 190, 194, 199, 200, 207, 224, 226, 227, 229, 235, 240, 241, 243, 245, 246, 247, 265, 274, 276, 277, 285, 308, 316, 319, 320, 327, 341, 342, 349, 350, 359, 361, 365, 372, 378, 379, 381, 384, 388, 391, 396, 404, 405, 406, 407, 408, 409

energy engineering 21, 25, 28, 34, 35, 38, 40, 43, 65, 67, 74, 158, 188, 190, 207, 224, 226, 227, 246, 276, 316, 319, 327, 341, 342, 349, 350, 361, 365, 372, 404

energy fields 18, 35, 308, 341, 359, 361, 381

Energy Ley 23

Energy Lines 13, 243

energy rings 21, 28, 29, 33, 38, 274

Environmental Studies Centre 52, 53, 55, 237, 238

Eucalyptus 92

Eucharist 297, 300, 305, 306, 316, 318, 324, 325, 352, 370, 402

execution lines 106, 109, 110, 111, 117, 119, 122, 125, 132, 133, 152, 153, 154, 156, 183, 184, 185, 190, 191, 195, 203, 209, 211, 212, 214, 215, 216, 221, 222, 229, 234, 236, 237, 238,

260, 261, 269, 279, 280, 282, 284, 285, 286, 288, 291, 295, 300, 303, 307, 312, 337, 345, 362, 363

F

feathers 87, 95, 104, 131, 142, 144, 165, 174, 176, 177, 178, 188, 234, 235, 257, 280, 282, 285, 286, 291, 315, 328, 348, 349, 355, 370, 402

fish 196, 307, 309, 310, 316, 317, 318, 320, 322, 330, 331, 334, 347, 351, 352, 360, 362, 409

Flowering Cherry 92

font 122, 139, 168, 169, 206, 307, 317, 324, 327, 348, 360

forensic 371

foxglove 144, 163, 164, 168, 176, 177, 187, 188, 280, 357

G

garlic 122, 128, 150, 229, 231, 358

geneticists 362

genocide 9, 153, 211, 218

Genocide 9, 118

Glamis 272, 274, 281

Glastonbury Tor 7, 11, 218, 219, 223, 345, 347, 398, 413

Glebe Stone Circle 8, 40, 51, 65

Gold 8, 55, 57, 182, 267

Grail 13, 307, 405

Green Man 13, 297, 299, 398, 406

gypsum 123, 126, 141, 147, 149, 163, 177, 203, 206, 234, 254, 263, 280, 297, 298, 299, 301, 309, 317, 319, 324, 332, 337, 338, 349

H

Hackforth 134

Heel Stone 11, 158, 226, 227, 228, 229, 230, 240, 241, 243, 245, 246, 248, 305, 317, 318, 321, 413

Hell-Fire Caves 11, 12, 251, 252, 254, 256, 293, 296, 330, 331, 413

hemp 95, 104, 128, 145, 149, 155, 166, 173, 177, 178, 179, 180, 187, 188, 189, 204, 221, 234, 239, 261, 271, 277, 279, 280, 328, 339, 348, 349

Henge Age 11, 13, 47, 68, 130, 133, 134, 140, 158, 185, 192, 197, 201, 209, 229, 251, 257, 265, 266, 273, 278, 280, 281, 282, 291, 295, 296, 306, 308, 311, 313, 322, 326, 328, 336, 342, 344, 347, 349, 350, 353, 358, 359, 360, 367, 368, 370, 378, 392, 406

Henge art 197, 358, 359

heptagon 231

Herne 13, 297, 299, 300, 301, 302, 303, 307, 309, 316, 318, 320, 322, 330, 331, 334, 346, 352, 360, 362, 370

HGV 12, 106, 156, 157, 158, 316, 317, 318, 319, 412

Hill Forts 257, 345, 347

hinges 125, 126, 127, 128, 131, 149, 160, 161, 166, 178, 179, 180, 188, 206, 271, 284, 349, 353, 360

holly 297, 298, 309, 330, 370

hologram 242, 248, 249

Home Farm 301, 302

Hope Cove 257, 309, 311, 312, 313, 322, 353

horses 66, 94, 104, 118, 131, 133, 150, 178, 192, 194, 204, 212, 213, 214, 220, 235, 265, 268, 280, 338, 344, 378

horseshoe cavity 224

human magnetic sense 44, 91, 122, 250, 297, 308, 366, 367, 371, 375, 396, 409

human sacrifice 168, 222, 231, 233, 322, 348, 357, 360, 362, 365, 370

Hurlers 12, 311, 312, 313, 414

I

Images 7, 69, 393, 394, 414

intuitive dowser 285, 303, 374

intuitive dowsing 3, 374, 376, 377, 403

Intuitive Dowsing 374

Iron 3, 4, 10, 13, 25, 44, 45, 47, 48, 52, 54, 55, 59, 60, 61, 63, 64, 65, 67, 71, 82, 87, 92, 96, 97, 99, 100, 104, 110, 111, 115, 116, 117, 125, 126, 130, 131, 132, 139, 144, 145, 150, 152, 155, 158, 166, 169, 173, 174, 179, 181, 192, 197, 204, 205, 208, 220, 221, 231, 246, 247, 257, 259, 261, 262, 264, 266, 268, 269, 270, 275, 284, 291, 294, 296, 297, 306, 308, 309, 310, 311, 316, 318, 319, 320, 321, 333, 334, 347, 348, 349, 350, 351, 352, 353, 360, 362, 370, 404, 406, 412

J

James Randi 45

Jesus 306, 307, 308, 309, 318, 320, 332, 352, 363

Joseph of Arimathea 307, 363, 405

K

Kestor 12, 286, 287, 288, 293, 413

Killadangan 8, 16, 22, 25, 26, 28, 29, 31, 32, 33, 34, 35, 36, 38, 40, 51, 52, 65, 226, 240, 247, 249, 342, 411

Killadangan Stone Circle 22, 32, 411

knives 123, 197, 349

L

l 411

lamps 69, 73, 88, 94, 116, 122, 123, 127, 129, 144, 146, 152, 160, 161, 163, 166, 188, 192, 193, 204, 205, 206, 209, 254, 257, 262, 263, 265, 266, 314, 370, 408

lead 3, 44, 62, 69, 70, 71, 92, 116, 119, 139, 156, 159, 165, 168, 170, 173, 176, 182, 183, 188, 191, 192, 211, 271, 272, 284, 288, 303, 314, 324, 325, 328, 347, 348, 349, 355, 360, 363

leather 49, 87, 90, 95, 104, 108, 118, 126, 128, 141, 143, 144, 146, 149, 155, 174, 177, 178, 179, 204, 214, 215, 221, 234, 237, 239, 247, 253, 261, 275, 278, 280, 331, 347, 348, 349, 353, 378, 403

Ley lines 21, 23, 94, 285, 400

lifts 222, 270, 285, 339

linen 106, 108, 109, 125, 130, 131, 141, 142, 144, 145, 146, 147, 152, 166, 168, 174, 175, 176, 177, 178, 182, 185, 187, 188, 214, 235, 237, 251, 257, 268, 277, 279, 280, 282, 286, 291, 299, 301, 305, 313, 328, 333, 339, 348, 349, 351, 364, 370, 405

linseed oil 127, 128, 141, 145, 280, 286, 328, 405

lintel stone 224, 230, 342, 349

Little Hay 232, 234, 235, 239, 256, 301, 330, 331, 413

Loan Head Recumbent Stone Circle 41, 42, 411

Lodge Park 9, 117

Long Barrows 313, 322, 344, 345, 348, 355

lubrication 94, 106, 115, 116, 149, 193

M

magnetic axis 21, 29, 31, 34, 37, 43, 224, 388

magnetic beacon 22

magnetic bearing 34, 177, 280

magnetic field 22, 29, 38, 40, 42, 70, 91, 181, 197, 198, 226, 230, 231, 241, 243, 244, 246, 247, 249, 250, 278, 349, 356, 367, 370, 375, 377, 381, 386, 387, 388, 390, 391, 392, 395, 401, 402, 403, 404, 405, 407, 408, 409, 410

magnetic harmony 21, 40, 51, 99

Magnetic Loop 70, 411

magnets 21, 341, 385, 386, 388, 395, 403, 405

Maiden Castle 10, 150, 151, 152, 153, 155, 156, 158, 221, 223, 233, 345, 346, 347, 351, 412

Malachite 145, 163, 174

Marburgh Henge 11, 272, 273, 274, 413

mass grave 90, 106, 114, 132, 133, 152, 154, 155, 185, 191, 195, 209, 214, 215, 284, 371

Maud Cunnington 59

Maypole 11, 239, 309

Merry Maidens 11, 283, 284, 285, 413

Michael Line 11, 285, 286

mine galleries 193

mistletoe 91

moat 115, 117

mortar 87, 95, 126, 127, 128, 131, 134, 138, 141, 147, 155, 166, 174, 177, 203, 207, 220, 231, 253, 262, 272, 294, 299, 344, 349, 353, 375

Mott 113, 114, 115, 116

murrisk friary 114
mystic 16, 18, 21, 32, 35, 37, 44, 87, 91, 190, 307, 308, 316, 405

N

Neolithic 4, 25, 44, 48, 49, 52, 53, 60, 63, 67, 96, 97, 112, 117, 126, 132, 133, 140, 144, 145, 202, 206, 208, 209, 228, 231, 234, 280, 289, 311, 326, 332, 343, 347, 348, 350, 351, 353, 355, 361, 364, 370, 378, 404, 406, 407
nettle 104, 126, 128, 149, 174, 178, 179, 222, 234, 239, 261, 271, 277, 279, 280, 316, 324, 328, 339, 348, 349, 402
New Grange 194, 358
Nine Ladies Stone Circle 41, 50
Nine Stones Circle 40, 303
North America 101, 140, 181, 366

O

oak 83, 85, 91, 92, 93, 94, 95, 111, 114, 123, 126, 128, 141, 144, 160, 170, 177, 178, 179, 181, 187, 192, 196, 197, 211, 217, 235, 250, 256, 261, 272, 281, 285, 291, 297, 299, 304, 309, 313, 319, 320, 330, 341, 346, 351, 358, 359, 361
oak bridges 272
oats 125, 126, 298, 324
obelisk 139, 183, 188, 189, 190, 348, 361
Offa's Dyke 266, 412
olive oil 126, 177, 205, 263
oxen 94, 97, 118, 151, 152, 158, 166, 170, 171, 172, 173, 190, 192, 208, 220, 222, 256, 265, 280, 294, 296, 304, 343

P

paramagnetic 7, 21, 29, 31, 32, 33, 34, 35, 39, 40, 41, 42, 44, 92, 224, 250, 341, 356, 370, 381, 382, 385, 386, 387, 388, 389, 390, 391, 392, 393, 394, 395, 404, 406, 407, 408, 409
Paullinus 61, 407
Penang 9, 63, 64, 90, 112, 113, 411
phosphate 35, 50, 53, 55, 56, 59, 61, 70, 71, 90, 95, 101, 104, 106, 110, 112, 120, 125, 127, 160, 166, 188, 196, 215, 216, 249, 254, 356, 375, 376, 379, 404, 409
photon 240
pictures 85, 123, 127, 144, 149, 150, 156, 181, 250, 294, 297, 299, 301, 330, 342, 359, 360, 361, 367, 408
pillow 144, 188
plumbing 125, 159, 170, 173, 188, 192, 206, 271, 284, 302, 303, 325, 343, 350, 360
Polarised Magnetic Fields 413
pond water 207
Pontypridd 6, 10, 117, 183, 186, 187, 219, 237,

257, 372, 412
pool 55, 72, 75, 77, 80, 81, 82, 87, 106, 117, 122, 131, 132, 138, 139, 140, 143, 147, 150, 156, 158, 159, 168, 169, 170, 173, 174, 175, 178, 179, 181, 183, 185, 186, 187, 188, 189, 190, 191, 192, 197, 206, 207, 217, 222, 223, 254, 268, 269, 271, 273, 277, 280, 281, 284, 286, 291, 296, 300, 302, 303, 307, 313, 316, 318, 319, 324, 342, 344, 347, 348, 349, 355, 360, 361, 379, 405
portal stones 25, 117, 186, 188, 280, 350
postholes 4, 44, 45, 48, 49, 50, 51, 52, 53, 55, 56, 57, 59, 60, 61, 63, 64, 65, 66, 67, 72, 74, 75, 78, 81, 82, 84, 88, 96, 99, 100, 106, 108, 110, 111, 116, 120, 126, 132, 138, 140, 152, 157, 161, 162, 166, 170, 174, 183, 188, 193, 202, 206, 208, 214, 221, 222, 232, 233, 238, 250, 257, 260, 262, 263, 265, 268, 269, 274, 277, 280, 284, 285, 286, 291, 313, 314, 315, 320, 321, 323, 332, 337, 339, 351, 368, 405
potassium 31, 32, 45, 69, 109, 122, 146, 152, 156, 183, 187, 196, 206, 214, 234, 235, 237, 260, 262, 263, 264, 285, 291, 299, 315, 337
prayer stones 123, 150, 194, 358
Processional Way 12, 66, 320, 321, 414
projectiles 128
Proto-Christianity 308, 310, 318, 320, 327, 330, 334, 342, 352, 358, 360, 362, 363
protons 42, 243, 392, 408, 410
Puddingstone 10, 155, 196, 197, 408, 413

Q

Quantum Entanglement 11, 13, 240
quiver 141, 144, 174, 178, 280

R

Raven 355
Recumbent Stone Circle 39, 40, 41, 42, 411, 413
red ochre 55, 71, 73, 81, 83, 90, 94, 95, 99, 104, 109, 110, 111, 116, 129, 131, 141, 146, 149, 160, 161, 162, 170, 178, 179, 180, 181, 187, 192, 196, 203, 206, 207, 209, 220, 235, 254, 263, 273, 279, 280, 284, 296, 297, 301, 314, 316, 319, 333, 345, 355, 356, 357, 376, 403
Reindeer 146, 234
religion 4, 63, 100, 112, 122, 134, 141, 181, 223, 296, 297, 299, 302, 306, 307, 308, 309, 310, 320, 322, 327, 342, 352, 353, 361, 362, 363, 364, 365, 368, 369, 370, 407
Rempstone Stone Circle 51
resin 91, 104, 127, 145, 173, 174, 187, 188, 189, 328
River Ure 133, 345
River Ver 200
roads 4, 12, 22, 23, 44, 47, 48, 80, 94, 95, 100, 116, 118, 128, 133, 134, 139, 155, 183, 187, 188,

197, 203, 216, 219, 220, 221, 223, 231, 236, 237, 238, 246, 247, 250, 261, 265, 267, 271, 273, 275, 277, 282, 284, 285, 291, 293, 299, 302, 308, 309, 310, 311, 316, 317, 321, 327, 337, 339, 343, 344, 348, 349, 351, 352, 353, 355, 358, 360, 362, 375, 379, 403, 406

rocking stone 186, 273

Rollright Stone Circle 8, 9, 12, 38, 110, 332, 333, 411

Roman armour 353, 365

Roman attack 153, 217, 218, 300

Roman attitude to Druids 365

Roman casualties 109, 217

Roman invasion 229, 297, 350, 355

Roman roads 348, 375

Roman soldier 129, 132, 286

Roman swords 129

Roman weapons 129, 365

Round Barrow 116, 346, 347

Roundhouse 8, 45, 87

Round Pound 12, 286, 288, 291

Royal Box 100, 103

Ryton Organic Gardens 56

S

sable 142, 174, 177

sarsens 224, 316, 320

Saxon Church 9, 67, 68, 69

Scorhill Stone Circle 12, 289

Scott's Grotto 251, 409

Seahenge 75, 351

secular 119, 120, 130, 133, 134, 152, 153, 156, 165, 188, 192, 211, 213, 233, 308, 343, 355, 360, 362, 363, 364, 365, 369

semi-precious stones 144, 145, 146

sensitive 21, 34, 43, 53, 90, 108, 140, 141, 249, 297, 381, 384, 386, 391, 395, 409

Shaman 21, 361, 409

Shamrock 59

Silbury Hill 11, 259, 261, 263, 269, 270, 271, 272, 293, 338, 339

silica 29, 31, 43, 54, 69, 87, 144, 225, 341, 410

silver 54, 55, 59, 146, 156, 159, 168, 170, 173, 176, 182, 185, 187, 188, 211, 217, 254, 328, 360

Snake 414

solar activity 42, 375

solstice 28, 63, 79, 211

spear 45, 128, 131, 141, 142, 144, 145, 174, 177, 185, 187, 280, 291, 303, 306, 309, 318, 328, 330, 332, 405

spirals 2, 11, 194, 195, 198, 199, 200, 201, 316, 358

spirits 18, 22, 33, 42, 43, 56, 78, 108, 111, 116, 119, 122, 174, 196, 224, 247, 271, 341, 361, 365, 368, 378, 409

staffs 50, 61, 65, 66, 73, 95, 108, 110, 116, 131, 144, 146, 165, 183, 204, 221, 246, 247, 316,

351

stain 4, 21, 22, 23, 53, 55, 56, 63, 67, 71, 72, 75, 79, 80, 81, 83, 85, 87, 90, 91, 94, 104, 116, 117, 127, 131, 141, 149, 152, 165, 169, 173, 174, 177, 178, 179, 183, 188, 206, 207, 217, 222, 231, 234, 261, 262, 268, 273, 297, 303, 314, 324, 333, 355, 375, 405

St Albans Cathedral 414

Stanton Drew 9, 13, 48, 49, 50, 61, 100, 101, 102, 104, 107, 108, 109, 110, 116, 181, 350, 351, 411

starburst 149, 358, 359

Stone Age Temple 10, 12, 142, 145, 146, 147, 180, 185, 186, 192, 267, 322, 339, 412

Stone Age tomb 140, 144, 174, 176, 177, 185, 348

stone columns 116, 117, 127, 131, 132, 134, 140, 146, 147, 174, 183, 184, 185, 188, 343, 349, 359

Stonehenge Bottom 100, 103, 411

Stone Pillars 264, 413

Stoney Lane 10, 11, 12, 13, 119, 138, 139, 140, 168, 169, 170, 174, 175, 178, 180, 181, 182, 190, 234, 256, 295, 296, 297, 298, 301, 350, 356, 358, 412, 414

storage pit 45

Swallow Hole 251

Swinside Stone Circle 411

T

tannin 49, 91, 92, 112, 141, 144, 174, 177, 178, 215

telepathy 22

Temple Design 9, 72

Temple Guard 10, 106, 145, 233

temple pool 168, 173, 207, 307, 324, 355, 379

terraced 95, 133, 192, 223, 267, 339, 343, 406

terraces 120, 151, 183, 185, 205, 216, 219, 221, 222, 223, 344, 351

terracotta 77, 144, 147, 161, 163, 170, 193, 254, 279, 324, 328

the Sanctuary 96, 263, 265, 268

Thornborough Rings 132, 133, 412

Thornhill Strand 22, 29

Thunderbirds 93, 143, 181, 192, 193, 216, 235, 236, 238, 273, 301, 317, 322, 323, 328, 334, 341, 358, 359, 361, 411, 414

tin 54, 56, 118, 125, 127, 131, 144, 152, 154, 155, 177, 183, 185, 197, 204, 208, 219, 222, 228, 234, 237, 253, 261, 275, 278, 280, 284, 286, 307, 320, 348

tipping bench 208

toadstool 122, 125, 126, 144, 163, 164, 176, 187, 188, 215, 229, 231, 235, 280, 357

toilets 47, 112, 124, 125, 126, 160, 165, 166, 209, 362, 363, 371

tomb 53, 54, 55, 56, 57, 58, 59, 60, 64, 65, 67, 68, 71, 73, 74, 75, 77, 78, 79, 82, 83, 91, 94, 95,

96, 100, 103, 104, 108, 116, 117, 125, 126,
127, 130, 132, 134, 138, 140, 141, 142, 143,
144, 145, 146, 147, 149, 152, 174, 175, 176,
177, 178, 181, 183, 185, 187, 188, 192, 193,
201, 206, 208, 222, 231, 234, 236, 257, 267,
268, 269, 271, 274, 276, 277, 279, 280, 283,
284, 286, 288, 289, 291, 312, 313, 316, 317,
322, 324, 328, 339, 342, 348, 350, 351, 352,
356, 357, 359, 361, 367, 401

tomb complex 75, 83, 91, 94, 96, 100, 108, 116,
117, 127, 141, 143, 145, 149, 185, 187, 188,
193, 222, 231, 236, 257, 274, 284, 288, 348,
350, 351, 356, 357, 361

Tomnaverie Stone Circle 276, 413

Tree Auras 9, 91, 412

trilithon 11, 183, 207, 208, 209, 224, 225, 227, 228,
230, 241, 246, 339

tropics 91, 134

Tunbridge Wells 9, 10, 36, 118, 134, 183, 184, 185,
412

U

Uffington 12, 335, 336, 337, 338, 347, 414

Ulshaw Bridge Earthworks 133

underground temple 58, 59, 61, 64, 65, 66, 67, 69,
73, 74, 78, 80, 81, 90, 111, 143, 175, 208,
251, 253, 256, 257, 258, 268, 271, 272, 284,
285, 286, 288, 294, 296, 304, 312, 316, 318,
330, 356, 409

urine 49, 114, 125, 126, 127, 128, 138, 141, 144,
155, 174, 177, 178, 196, 197, 203, 207, 214,
221, 234, 253, 280, 294, 299, 327, 331, 348,
353

W

Waden Hill 272

Wansdyke 11, 261, 265, 266, 267, 413

warriors 118, 122, 132, 153, 156, 211, 213, 214,
215, 216, 217, 231, 233, 235, 284, 327, 354,
370

wattle 45, 83, 94, 104, 110, 123, 125, 126, 127,
138, 147, 156, 158, 160, 161, 163, 166, 171,
174, 183, 221, 265, 278, 279, 285, 286, 291,
293, 294, 298, 299, 312, 323, 337

weaving loom 126

well house 170, 294

Welwyn 197, 200

West Kennet Avenue 11, 96, 97, 98, 99, 259, 263,
264, 265, 268, 269, 272, 413

Westport 16, 21, 22, 25, 28, 34, 35, 226

Wheathampstead 192

wheel 23, 24, 28, 33, 35, 45, 48, 74, 76, 94, 99,
100, 115, 116, 117, 127, 149, 150, 152, 155,
158, 163, 166, 168, 170, 180, 208, 219, 220,
221, 226, 228, 232, 233, 240, 253, 257, 265,
277, 278, 280, 286, 302, 313, 314, 316, 320,
321, 337, 378, 402

Wheelhouse 410

White Horse 12, 335, 336, 337, 338, 347, 414

willow 141, 144, 280

wine 177, 214, 229, 299, 300, 301, 305, 306, 316,
324, 327, 402

witnesses 4, 29, 45, 53, 54, 55, 57, 65, 69, 70, 71,
73, 85, 87, 90, 91, 94, 95, 104, 106, 109, 110,
111, 117, 122, 125, 126, 134, 141, 144, 147,
149, 156, 159, 160, 162, 163, 168, 170, 178,
180, 185, 196, 197, 198, 199, 200, 203, 205,
213, 221, 224, 229, 231, 235, 254, 256, 279,
284, 294, 297, 298, 301, 306, 313, 318, 330,
331, 337, 357, 367, 375, 376, 377, 378, 379,
389, 391, 394

wolf 142, 144, 174, 177, 185, 286, 291

Woodhenge 8, 9, 11, 48, 53, 55, 59, 60, 64, 65, 66,
71, 73, 77, 82, 91, 94, 96, 99, 127, 156, 202,
203, 226, 229, 230, 244, 245, 246, 247, 250,
257, 342, 348, 349, 411

wool garments 144, 178, 185, 308, 348

Workstations 124, 412

Y

yew 123, 128, 141, 144, 174, 187, 280, 298, 309

Yorkshire stone circle 85, 349

Z

zigzag 2, 116, 198, 199, 200, 201, 358, 359A

Stone Row Linear Accelerator - A Stairway to Heaven

7th Century Chrisitan Art - Lindisfarne Gospels